Alphabetical Table of the Elements

Element	Symbol	Atomic number	Atomic* weight	Element	Symbol	Atomic number	Atomic* weight	Element	Symbol	Atomic number	Atomic* weight
Actinium	Ac	89	(227)	Holmium	Ho	67	164.9303	Rhenium	Re	75	186.207
Aluminum	Al	13	26.9815	Hydrogen	H	1	1.0079	Rhodium	Rh	45	102.9055
Americium	Am	95	(243)	Indium	In	49	114.82	Rubidium	Rb	37	85.4678
Antimony	Sb	51	121.75	Iodine	I	53	126.9045	Ruthenium	Ru	44	101.07
Argon	Ar	18	39.948	Iridium	Ir	77	192.22	Samarium	Sm	62	150.36
Arsenic	As	33	74.9216	Iron	Fe	26	55.847	Scandium	Sc	21	44.9559
Astatine	At	85	(210)	Krypton	Kr	36	83.80	Selenium	Se	34	78.96
Barium	Ba	56	137.327	Lanthanum	La	57	138.9055	Silicon	Si	14	28.0855
Berkelium	Bk	97	(247)	Lawrencium	Lr	103	(260)	Silver	Ag	47	107.8682
Beryllium	Be	4	9.0122	Lead	Pb	82	207.2	Sodium	Na	11	22.9898
Bismuth	Bi	83	208.9804	Lithium	Li	3	6.941	Strontium	Sr	38	87.62
Boron	B	5	10.811	Lutetium	Lu	71	174.967	Sulfur	S	16	32.066
Bromine	Br	35	79.904	Magnesium	Mg	12	24.3050	Tantalum	Ta	73	180.9479
Cadmium	Cd	48	112.411	Manganese	Mn	25	54.9380	Technetium	Tc	43	(98)
Calcium	Ca	20	40.078	Mendelevium	Md	101	(258)	Tellurium	Te	52	127.60
Californium	Cf	98	(251)	Mercury	Hg	80	200.59	Terbium	Tb	65	158.9253
Carbon	C	6	12.011	Molybdenum	Mo	42	95.94	Thallium	Tl	81	204.3833
Cerium	Ce	58	140.115	Neodymium	Nd	60	144.24	Thorium	Th	90	232.0381
Cesium	Cs	55	132.9054	Neon	Ne	10	20.1797	Thulium	Tm	69	168.9342
Chlorine	Cl	17	35.4527	Neptunium	Np	93	(237)	Tin	Sn	50	118.710
Chromium	Cr	24	51.9961	Nickel	Ni	28	58.69	Titanium	Ti	22	47.88
Cobalt	Co	27	58.9332	Niobium	Nb	41	92.9064	Tungsten	W	74	183.85
Copper	Cu	29	63.546	Nitrogen	N	7	14.0067	Unnilennium	Une	109	(266)
Curium	Cm	96	(247)	Nobelium	No	102	(259)	Unnilhexium	Unh	106	(263)
Dysprosium	Dy	66	162.50	Osmium	Os	76	190.2	Unniloctium	Uno	108	(265)
Einsteinium	Es	99	(252)	Oxygen	O	8	15.9994	Unnilpentium	Unp	105	(262)
Erbium	Er	68	167.26	Palladium	Pd	46	106.42	Unnilquadium	Unq	104	(261)
Europium	Eu	63	151.965	Phosphorus	P	15	30.9738	Unnilseptium	Uns	107	(262)
Fermium	Fm	100	(257)	Platinum	Pt	78	195.08	Uranium	U	92	238.0289
Fluorine	F	9	18.9984	Plutonium	Pu	94	(244)	Vanadium	V	23	50.9415
Francium	Fr	87	(223)	Polonium	Po	84	(209)	Xenon	Xe	54	131.29
Gadolinium	Gd	64	157.25	Potassium	K	19	39.0983	Ytterbium	Yb	70	173.04
Gallium	Ga	31	69.723	Praseodymium	Pr	59	140.9076	Yttrium	Y	39	88.91
Germanium	Ge	32	72.61	Promethium	Pm	61	(145)	Zinc	Zn	30	65.39
Gold	Au	79	196.9665	Protactinium	Pa	91-	231.0359	Zirconium	Zr	40	91.224
Hafnium	Hf	72	178.49	Radium	Ra	88	(226)				
Helium	He	2	4.0026	Radon	Rn	86	(222)				

* Values in parentheses are atomic masses. They are not atomic weights because they represent the masses of the longest-lived isotopes of the elements.

Introduction to Chemistry

Introduction
to
Chemistry

E. RUSSELL HARDWICK
University of California, Los Angeles

JOAN BOUILLON
Metropolitan State College of Denver

SAUNDERS GOLDEN SUNBURST SERIES
SAUNDERS COLLEGE PUBLISHING
HARCOURT BRACE JOVANOVICH
PUBLISHERS

Fort Worth   Philadelphia   San Diego   New York   Orlando   Austin
San Antonio   Toronto   Montreal   London   Sydney   Tokyo

Requests for permission to make copies of any part of the work should be mailed to Permissions
Department, Harcourt Brace Jovanovich, Publishers, 8th Floor, Orlando, Florida 32887.

Text Typeface: Times Roman
Compositor: York Graphic Services
Acquisitions Editor: John J. Vondeling
Developmental Editor: Sandra Kiselica
Managing Editor: Carol Field
Project Editor: Margaret Mary Anderson
Copy Editor: Becca Gruliow
Manager of Art and Design: Carol Bleistine
Art Director: Anne Muldrow
Art Assistant: Caroline McGowan
Text Designer: Becky Lemna
Cover Designer: Lawrence R. Didona
Text Artwork: J/B Woolsey and Associates
Layout Artist: B. J. Crim
Director of EDP: Tim Frelick
Production Manager: Charlene Squibb
Marketing Manager: Marjorie Waldron

Cover Credit: Yoav Levy/Phototake, Inc.

Printed in the United States of America

INTRODUCTION TO CHEMISTRY

0-03-035354-8

Library of Congress Catalog Card Number: 92-050576

2345 69 987654321

Preface

This textbook has been designed for an introductory course in chemistry. It was written for students who expect to continue in science as well as for those who may be taking chemistry as their only college level science course. We expect that students in the course will have little or no background in science, particularly in chemistry, but that they will have arithmetic skills and will be able to perform simple algebraic operations.

A major objective of our text is to give students an understanding of many of the fundamental concepts of chemistry as well as the observations and reasoning that led to those concepts. We intend that students using this textbook will learn how to use many of the tools of chemistry, in particular that they will learn how to

1. approach and solve several important kinds of chemistry problems;
2. use chemistry vocabulary;
3. name and write the chemical formulas for simple compounds;
4. complete and balance chemical equations.

Approach

To learn science, chemistry included, is to learn a new vocabulary as well as a new approach to thinking about the world. This is a daunting task for anyone just starting, but it is far more difficult for those students without adequate reading skills or for whom English is a second language. Our text has been written in such a way as to help such individuals and to make reading it easier for everyone. We have made a special effort to keep our language as simple as possible. We do not use a complex word when a simple one will do.

Common items are routinely used to explain chemical concepts. For example, nuts and bolts help to describe how elements combine to form compounds, the volume of a basketball is used as a comparison to the volume of a gas at STP, and the concept of limiting reagents is explained in terms of preparing sandwiches with given numbers of slices of bread, cheese, and ham.

Even though our objective is to keep the language as simple and friendly as possible, we have taken extreme care to keep the science exact and accurate, and to be precise in our definitions and explanations of concepts. We want to emphasize that we have not watered down the chemistry; no liberties are taken with rigor in an attempt to simplify.

Organization

The introduction sets the stage for understanding the scientific method. Chapter 1 begins with the concept of experimental measurement, precision, and the use of significant figures. It introduces the metric system, defines important terms and explains approaches to problem-solving.

We begin our study of chemistry as such in Chapter 2. After establishing a basic understanding of matter, energy, heat, and temperature, we begin our exploration of elements and compounds and their properties. As a springboard to our encounter with the atomic theory in Chapter 3, we study the simple but crucial law of definite proportions.

Beginning with the earliest views of the nature of matter, Chapter 3 leads us to follow Dalton in concluding that matter must be made of atoms. Given this, we look inside atoms and meet electrons, protons, and neutrons. Chapter 4 takes us deeper inside the atom to explore the behavior of electrons, discover quantum numbers, and learn to draw Lewis diagrams.

Chapter 5, the periodic table discussion, applies the knowledge from the preceding two chapters to explore the relations between atomic structure and elemental properties. This prepares the way for Chapter 6, which introduces chemical bonding and Lewis diagrams of molecules and polyatomic ions.

Following Chapter 7, which deals with naming compounds, the next three chapters impart a firm foundation to stoichiometry. Chapter 8 introduces atomic mass, atomic weight, and the mole concept. Chapters 9 and 10 deal with chemical calculations, carrying the student through limiting reagent problems.

Introduced by fundamental discussions of kinetic theory, Chapter 11 explains the gas laws in both their basic and combined forms and reinforces those concepts with numerous problems of various kinds. The gas law chapter is followed by a detailed chapter on the condensed phases. After a thorough explanation of the forces that exist between molecules, the chapter closes with a comprehensive overview of the properties of crystalline solids.

Chapters 13 and 14 introduce solution chemistry and acid-base reactions. Solubility and the mechanism of ionic dissociation is examined, giving the groundwork for understanding why and how ions react in solution and predicting whether a reaction will take place. Chapter 13 covers molarity, percent concentration, molality, and colligative properties. Chapter 14 treats acid-base definitions and reactions, and, in the process, covers pH and titration calculations.

Chapter 15 introduces kinetics, reversible reactions, and chemical equilibrium. Le Chatelier's principle, equilibrium constants, and equilibrium calculations are covered, finishing with buffers and solubility product. The sections of Chapter 15 are written to be as independent as possible so that an instructor can make an abbreviated offering by choosing desired topics and omitting others.

The final chapter reviews redox reactions, a topic that was first introduced in Chapter 6. Much of the chapter is devoted to balancing redox reactions. That discussion is written so that it can be lifted from the chapter and used much earlier if so desired. Most of the remainder of the chapter describes important redox systems used in laboratories, industry, or biological organisms. The chapter concludes with explanations of redox reactions in batteries and electrolytic cells.

Flexibility

At the beginning level, there traditionally have been two general sequences in which topics are presented. Many instructors feel that an understanding of atomic structure and bonding must be attained before stoichiometry is introduced. Because many introductory courses include no laboratory, we have opted to present atomic structure before stoichiometry; however these sets of chapters could actually have been interchanged and renumbered without damage. For those courses that do have a laboratory and for those instructors (including one of the authors) that prefer to cover stoichiometry before atomic theory and structure, we have been careful to write the stoichiometry chapters, Chapters 8, 9, and 10, so that they can be presented following Chapter 3. The chapter on gases, and parts of the chapters covering solutions and acids and bases, Chapters 11, 13 and 14, may be presented after chapter 10, leaving the discussions on condensed states for later.

We do not expect that all the material in the book will be covered in a one-semester course; in fact, there is probably enough for a relaxed two-semester offering. We have written in such a way, however, that certain sections can be skipped. Many instructors will not have time to cover the last two chapters on equilibrium and redox reactions, but there are certain sections in those chapters, balancing equations for oxidation-reduction reactions, for example, that can be taught independently during the course.

Features

Problem solving has traditionally depended heavily on setting up the solution by using unit analysis. But unit analysis is no substitute for understanding. To prevent students from letting it do the work, we employ a stepwise strategy that includes a "solution map" so that students know where they are going and when they have gotten there. The objective is to learn chemistry, not merely a technique for problem solving. We follow this stepwise sequence in many of our worked-out examples, then, in others, allow the student to supply some of the direction.

Examples and Exercises

Believing that practice promotes understanding, we have included a very large number of examples and exercises. The exercises, one or two of which immediately follow each example, have worked-out solutions at the ends of the chapters. We expect students to follow each example, working along with the text, then to work the exercise, consulting the solution only after finishing or when stuck.

Questions and Problems

Each chapter ends with a group of study questions and problems divided by section. The questions test students' knowledge of the material. For reinforcement, many of the problems are identical to the examples and exercises. A few chapters end with a section of "More Difficult Problems" designed to challenge students. The answers to all odd-numbered problems are found in the back of the book.

Outlines, Objectives, and Summaries

Each chapter begins with an outline of what is covered in that chapter. Then, to guide students through the discussion, each section begins with a list of learning objectives. To help the students further, each chapter ends with a summary of important concepts and a list of key terms referenced by page number.

Margin Remarks

We have included frequent margin notes to highlight important points and interesting facts or to help locate related topics.

Boxed Essays

To answer the question asked by so many students, "What does this have to do with the real world?", we have written short essays relating the principles of chemistry to interesting real life applications. "Science in Action" boxes such as "Home Heating with Hydrates" appear throughout the book.

"Profiles in Science" boxes, highlighted in each chapter, help to humanize the science. These vignettes describe the lives and interests of personalities in chemistry and physics.

Use of Color

More than 200 full-color photographs are used to show chemistry as it appears in the laboratory and the world around us. Color also makes the diagrams more meaningful as well as attractive. Color is used consistently throughout. For example, gray is used for carbon atoms and red for oxygen.

Appendices

The appendices include a review of mathematics, a review of graphing techniques, and a table of water vapor pressures at temperatures from 0°C to 100°C, and the solutions to odd-numbered problems. A glossary of important terms can also be found at the back of the book.

Extended Edition

For those instructors who would like to teach topics in nuclear, organic, and biochemistry, Introduction to Chemistry is also available in an extended version containing these chapters.

Accuracy

To help insure the accuracy of this book, Steven Strauss (Colorado State University) and Phil Silberman (Scottsdale Community College) read all galleys and page proofs and provided us with corrections. In addition, Bill Vining (Hartwick College) solved all of the end-of-chapter problems so that we could cross-check our answers.

Supporting Materials

This textbook is accompanied by a set of supporting materials for the student and the instructor.

A **Study Guide** by Joseph Ledbetter (Contra Costa College) provides a chapter outline and learning objectives, solutions to all in-text exercises and self-study tests with answers.

A **Student Solutions Manual** by Joan Bouillon (Metropolitan State College of Denver) contains complete solutions to odd-numbered end-of-chapter problems, as well as hints for problem-solving.

A laboratory manual, *Introduction to Chemical Principles: A Laboratory Approach,* 4th edition, by Susan Weiner (West Valley College) and Edward Peters provides 33 experiments, including the collection and analysis of experimental data. It is supplemented with an instructor's manual.

An **Instructor's Manual** by Russ Hardwick and Joan Bouillon includes teaching objectives and solutions to all the even-numbered end-of-chapter problems.

A **Test Bank** by Robin Horner (Fayetteville Technical Institute) is an extensive file of multiple-choice questions for each chapter. It is available in both a printed as well as computerized version (Apple and IBM PC). Included with the computerized text bank is ExamRecord, which allows the instructor to record, curve, graph and printout grades.

A set of **100 overhead transparency acetates** of four-color illustrations and photographs from the textbook is available for classroom and laboratory use.

Shakhashiri Chemical Demonstration Videotapes are also available. This set of 50, 3–5 minute classroom experiments bring the imagination and vitality of Bassam Shakhashiri to the classroom. Accompanying the videotapes is an **Instructor's Manual** containing a description of each demonstration as well as follow-up discussion questions.

Acknowledgments

Many people have helped contribute to the final product. While we cannot single each of them out to say thanks for the help, understanding, or encouragement, we have appreciated one and all. There are a number of individuals, however, to whom we must give special thanks.

We are endlessly grateful to Pat Hardwick who acted as our primary editor without much thanks or any remuneration. She read every page, rewriting and making essential changes to presentation, grammar, and content. This is as much her book as ours.

Lincoln Bouillon and Jeeves cannot be thanked enough for standing by us with their handholding and sympathizing and for putting up with several years of neglect.

The staff of Saunders College Publishing have applied their impressive talents and endless hard work to create the final product. We thank publisher John Vondeling, Sandi Kiselica, Tim Frelick, Margaret Mary Anderson, and Anne Muldrow. They have been a pleasure to know and with whom to work.

Many of the photographs were taken by Charles D. Winters of SUNY at Oneonta. In spite of numerous retakes he has remained enthusiastic and charming. The line art was provided by J/B Woolsey and Associates, who created drawings from our descriptions and stick figures. Their artistic efforts will in no small way make the chemistry more understandable.

Reviewers

Without the important and helpful comments of our many dedicated reviewers, this textbook would be of far less value. While we are not able to thank them individually, we do gratefully acknowledge their efforts and contributions.
Jack Ballinger, St. Louis Community College, Florissant Valley
Hal Bender, Clackamas Community College

Larry Bray, Miami-Dade Community College, South Campus
Robert Byrne, Illinois Valley Community College
Henry Derr, Laramie Community College
Jerry Driscoll, University of Utah
Robert Farina, Western Kentucky University
Arthur Friedel, Indiana University/Purdue University at Fort Wayne
Kirk Hunter, Texas State Technical Institute
Olaf Larson, Ferris State University
E. Jerome Maas, Oakton Community College
Robin Monroe, Southeast Community College-Lincoln
Raymond O'Donnell, Oswego State University
Terri Prichard, Merced College
Barbara Rainard, Community College of Allegheny County
Fred Redmore, Highland Community College
Ruth Sherman, Los Angeles Community College
Phil Silberman, Scottsdale Community College
Steven Strauss, Colorado State University
Ruth Ann Summers, Tacoma Community College
William Wasserman, Seattle Central Community College

Contents Overview

Contents

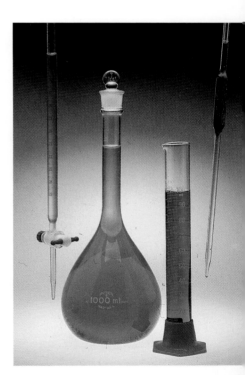

CHAPTER 3   The Atomic Theory   83

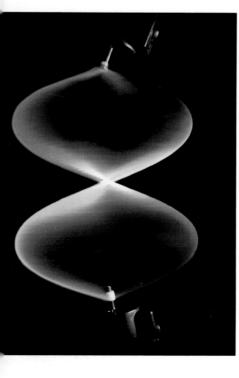

CHAPTER 4   Electron Structure of the Atom   105

CHAPTER 5   Groups of Elements and the Periodic Table   133

CHAPTER 6   How Elements Form Compounds   157

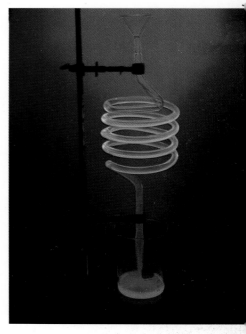

CHAPTER 15   Rates of Chemical Reactions
and Chemical Equilibrium   437

CHAPTER 16   Oxidation-Reduction Reactions   473

Introduction to Chemistry

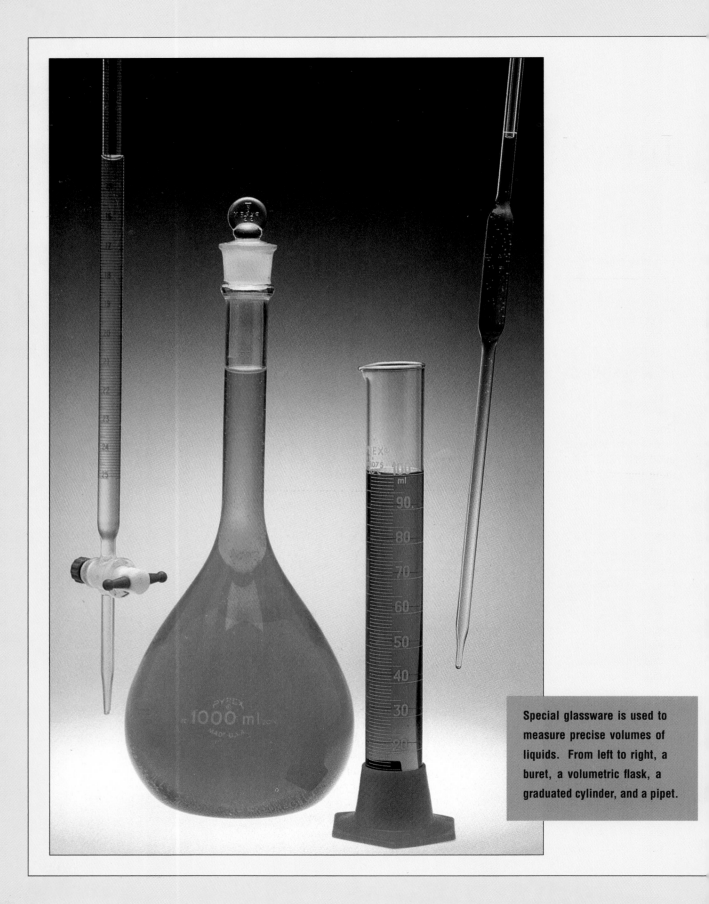

Special glassware is used to measure precise volumes of liquids. From left to right, a buret, a volumetric flask, a graduated cylinder, and a pipet.

Introduction

I.1 Why Study Chemistry / all around us

"Why study chemistry?" beginning students sometimes ask. "This is all mysterious stuff that has nothing to do with me; why should I waste my time on it?" Good question. This short section gives some answers.

Close your eyes, reach out, and touch something. Okay. The chances are good that what you touched was a substance that doesn't exist in nature, something that wasn't even heard of until a few decades ago. Perhaps you touched a piece of clothing. With the exception of cotton, wool, and a few exotic fibers like silk or linen, most fabrics today contain nylon, polyester, or some other synthetic fiber created chemically and dyed chemically. Did you touch a piece of furniture? Much upholstered furniture is covered by leather-like plastics made by chemical processes; furniture is assembled with glues made chemically. Was it a metallic object? Metals are refined from ores and later finished by processes that are chemical reactions. Was it this book? It contains paper, a substance made from wood by chemical processes. The colors in the pictures are synthetic dyes. The best, the most complex, and entirely the most wonderful example of chemistry is the human body, a machine created and operated by thousands of continually running chemical reactions.

Studying chemistry will give you a window on the processes and substances without which civilization, even life itself, could not exist. Chemistry is inside us and all around us. We could not do without it.

Science, and chemistry in particular, has already produced an unbelievable wealth of knowledge, so much that no one person could master even a substantial fraction of it. Everyday in chemistry alone, far more new information is reported in print than can be contained in a shelf of books the size of this one. Because of this, learning chemistry at first may seem to you an impossible task, but be of good heart. The fundamentals, the part that will introduce you to an understanding of much of the physical world around you, can be mastered fairly easily and, in fact, enjoyed. We hope it will be exciting for you to see and understand the explanations for many everyday processes you now take for granted. Work hard on your chemistry course; it will pay you back manyfold.

Star trails around the South celestial pole. Stars and other heavenly bodies obey the laws of mechanics in their movements.

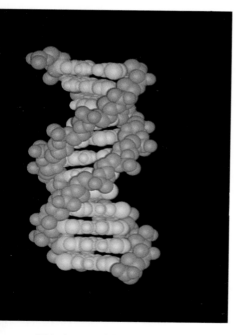

This is a model of a very small part of a DNA (deoxyribonucleic acid) molecule, the complex structure that stores genetic information.

I.2 Science and the Scientific Method

Before discussing the science of chemistry, we should know what science itself is. **Science** is the process by which humans seek in an organized way to understand and explain the natural world. In their search, scientists use an approach sometimes called the scientific method.

Science is a venture into the unknown. Guided only by what is already known, a scientist attempts to push beyond the frontiers of knowledge to discover new explanations for the way the world operates. Constantly questioning, wondering, asking, experimenting, a scientist works, not knowing what lies beyond the next experiment. Science is adventure. Imagine the thrill of discovering some fundamental secret of nature, being the first person ever to know it.

A large proportion of all human effort in our modern world is devoted to science at one level or another. Science is divided into several somewhat overlapping fields, of which physics, chemistry, and biochemistry are examples. Let's briefly look at the subject matter of these three related fields.

Physics is *a science devoted to the study of matter and energy; it seeks to develop laws and theories that apply to samples of any kind of matter or energy.* For example, physics gives us the laws of mechanics. These govern the movement of bodies in relation to one another—balls striking one another on a billiard table perhaps, or planets moving in gravitational fields. The laws are valid for any kind of body in any situation, including the moon and the earth. They apply to all kinds of matter regardless of differences.

A dictionary defines **chemistry** as *"a science that deals with the composition, structure, and properties of substances, of the transformations that they undergo and of the energies of those transformations."* Chemistry is less general than physics. It is the goal of chemistry to explain the properties of *specific* kinds of matter and to unravel the complexities of the ways in which various kinds of matter interact. Chemistry is concerned with the differences between kinds of matter and the reasons for those differences, as much as with generalizations that hold for any kind of matter.

Biochemistry is *the study of the chemical composition and the chemical reactions of living organisms.* Biochemistry seeks eventually to describe in intimate detail how the components of living organisms behave, and more important, to explain why they behave as they do. To do this completely, the biochemist must apply sophisticated knowledge derived from chemistry and physics to the analysis of incredibly complex systems. How, for example, does the brain work?

I.3 Science and Technology

Science and **technology** are often taken to mean the same thing, although they are quite different. Science, by means of research, deals in ideas, concepts, understanding, and explanation. Technology, closely associated with engineering, puts scientific discovery to work. Technology is *applied science.*

The line between science and technology is not clearly drawn. Pure research is done entirely for the purpose of increasing knowledge. Pure research is vital because the new knowledge provides a basis for more applied investigation and eventual technological use. Some research, however, is performed with a definite objective in mind, perhaps the creation of a new polymer or a better method

An alchemist in his workshop. Alchemy was more superstition than science.

of making fuel. Such research adds to knowledge while also accomplishing a technological goal. In technology as such, there is little pure research.

I.4 **The Beginnings of Chemistry**

The deliberate use of chemistry dates back to about the fourth century BC, probably in China, with the beginning of the practice of **alchemy**, a blend of mysticism and experimentation with chemicals. Spreading throughout the world, alchemy was practiced well into the beginning of the Renaissance era. The objective of alchemists was to develop an "elixir" that would prolong life; from the beginning, elixirs contained gold in one form or another. When alchemy was introduced into Europe many centuries later, gold was scarce, and so one of the main objectives of alchemy was to change cheaper metals into gold.

Alchemy little resembled modern science. Most alchemists merely mixed various materials, usually in furnaces, to see what might result. Ritual and incantation played a large role, and the work was sporadic and disorganized. Nonetheless, many useful experimental techniques in chemistry evolved from these early efforts. In some parts of the world, medicines discovered in the days of alchemy are still in use.

While alchemy was flowering in the East and in Europe, the Egyptians were making practical studies of some of the chemistry of copper, gold, silver, and iron, developing what is today called metallurgy. In addition, these early Egyptian researchers discovered secrets of embalming and of dyes and paints.

In Greece, about 400 BC, philosophers suggested detailed and realistic explanations of the natural world. Democritus and other philosophers speculated that all matter was made of indivisible particles that he named *atomos,* from which came the term we use today. However, because Aristotle championed

銅鉱と吹分る図

A Japanese woodcut showing a worker extracting copper from its ore, using a method more than two thousand years old.

another idea, namely that matter consisted of the four "elements" fire, air, earth, and water, atomic theory was little heard from for more than 2,000 years. What the Greeks did was not science, however, because it did not seek verification through observation and experiment.

Serious and fruitful inquiry into the physical nature of our world using the scientific method began with simple, carefully performed, reproducible measurements. With ideas based on repeated observations, Copernicus startled the world in the sixteenth century. His theory that the earth revolved around the sun stimulated research in astronomy and other sciences. Later, Galileo was said to have dropped weights from the Leaning Tower of Pisa to show that the rate of descent did not depend on the size of the weight. In the seventeenth century, Isaac Newton made brilliant fundamental studies of motion and gravitation.

Chemistry as we now know it began with Robert Boyle (see Section 11.4) and his careful measurements of the behavior of gases. The discovery of oxygen by Priestley in 1774 and the analytical work of Lavoisier, one of the first to use a balance, established many of the methods and approaches chemistry uses today. On this firm experimental base, John Dalton proposed his atomic theory of matter in the first decade of the nineteenth century.

beginning

As the base of scientific knowledge grew, more complex studies became possible. Nowadays the physicist probes the interior of the atomic nucleus, and the organic chemist confidently designs totally new kinds of matter. Science now seems actually to be on the threshold of unlocking the secrets of life itself.

I.5 The Scientific Method

Chemistry, along with other sciences, employs what is often called the **scientific method.** Because it is at the very heart of science, we can start our study of chemistry by learning how to use the scientific method.

Let's make up a story.

You introduce yourself to the scientific method on the day that you go out to ride your bike to class and find that it has a flat tire. Still half asleep, you attach your little tire pump, and as you begin pumping away you start to notice the way the pump behaves. Your curiosity wakes you up (curiosity is one of the most important characteristics of a scientist), and you begin to pay careful attention to what is going on. At the beginning of a stroke on the pump, it is easy to press the plunger down, but as you near the bottom this takes much more force. Funny. Why is that? You have made an observation; now you theorize a bit and devise a little experiment. Maybe the effect is caused by the elasticity in the tire. You pinch the air tube of the pump so that no air can get out, and you push the plunger again. The effect is now even more pronounced and obviously has nothing to do with the tire. So much for that theory. You finish filling the tire and go to class, but you think about this all day (persistence is also a valuable quality for scientists).

Late that afternoon, you perform a careful experiment. You fasten the pump upright and rig a little platform on the top of the plunger. Now you pinch the tube again to trap air in the pump and then put a brick on the platform. The plunger moves down somewhat (Figure I.1). Then you put another brick on the platform and yet another. With the increased downward force caused by the bricks, the plunger moves downward again and again (Figure I.2).

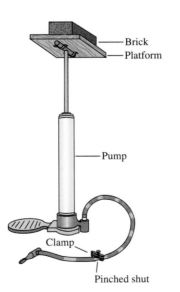
Brick
Platform
Pump
Clamp
Pinched shut

Figure I.1 Placing a brick on the platform causes the plunger to move downward.

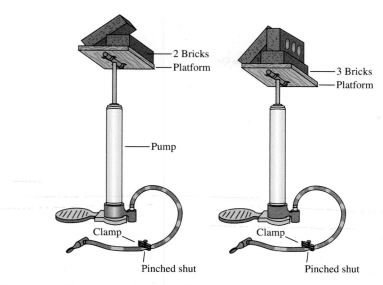

Figure I.2 When more bricks are added, the plunger moves farther down.

As a result of many observations, here is what you find: As you increase the number of bricks on the platform, thus increasing the force on the plunger, the position of the plunger lies farther and farther down, compressing the trapped air into a smaller and smaller volume. By repeating the experiment and remembering what you have observed about tires, pumps, and similar equipment, you are able to write a general statement of what you have found.   Claim

The volume of air trapped in a bicycle pump gets smaller as the force on the plunger gets larger.

You have just stated a natural law. True, it is limited to air in bicycle pumps; it stops short of saying exactly how much force is needed to produce various changes in volume. However, as far as you can tell, the law is valid within its limitations. The validity is soon reinforced by a friend who makes similar experiments. She uses a different kind of pump but gets the same results. Your law has been tested and verified.

What's next? Well, you might continue your investigation in two ways. First, you will want to accumulate more facts about the phenomena you have observed. Exactly how does the volume of the air change in relation to the number of bricks on the platform? What happens if you let some of the air out? On a hot day, you notice that more bricks are needed to get the plunger to go down a particular distance; what about that? In addition to gathering physical facts from your experiment, you will want an explanation for them. Why does the plunger behave that way?

You will eventually want to explain what you have seen by making some predictions about things you can't see: So you begin to devise a theory. Perhaps a tiny little man in the pump pushes back on the plunger as you squeeze him. (The history of science includes such imaginary little men and similar creatures.) The little man seems like a fruitless idea, however, so you look further. Maybe there is a spring inside the pump. You open the pump (an experiment) and look inside. No spring. Back to the drawing board.

What other kinds of forces are there? Gravity, perhaps. True, the bricks are attracted by gravity, but you can't see how something pushing *up* on the plunger works by gravity. One evening on television you see pictures of a hurricane and notice that moving air can exert tremendous force. Can something be moving and colliding with the plunger even though there is no wind inside the pump? Coming up with a theory is going to require a lot of thought and effort.

We can't pursue our story further now without going into much more detail; however, we will spend considerable time later on just this subject. The purpose of this story was to demonstrate the scientific method. First you made observations of a physical phenomenon. Then you organized and generalized your findings into a statement describing what you had seen: You proposed a **natural law.** You, and others, used experiments to verify the law. You then tried to design a **theory,** a reasonable explanation of the natural law in terms of things that you were not able to observe. You made further observations and experiments to test the theories.

An essential feature of the scientific method is reproducibility. If no one else can observe what you claim to have observed, few scientists are likely to take your experiments seriously. Reproducibility is an important difference between the work of scientists and that of fakes and charlatans.

Unless other scientists know exactly what one of their colleagues has done, it will be impossible for them to reproduce the findings and verify them. For this reason, communication is another essential part of the workings of science. Scientists communicate with one another using scientific publications. The list, *merely a list,* of the names of scientific journals published regularly throughout the world requires several pages of fine print; a brief description of each paper requires a stack of journals (Figure I.3)! Additionally, scientists attend meetings with others doing similar studies. At meetings, researchers describe their work by reading papers they have written; these are then discussed and questioned. Scientific meetings are an essential part of the communication system of science.

This, then, is the way science works: observations, laws, theories, experiments to test the theories, modifications of the theories, more experiments, and on and on. As you continue through the chapters that follow, you will see the process repeated again and again.

I.6 How to Study Chemistry

Finally, a word about studying. You have undoubtedly been advised many times to set aside a scheduled time for study and to discipline yourself to work during that time. Although that is good advice, it isn't enough. If you use the allotted time merely to pass your eyes over the pages, thinking meanwhile of what you will be doing when your study period is over, you will have wasted your time and effort. Set yourself a goal based on understanding, not on the clock. You can learn more and perhaps even save some time. Choose a part of the chapter that you want to learn. Take a piece that will just about fit the time you have, even if it is only a paragraph or one difficult equation. Don't take too much. Promise yourself that you will not stop until you have learned what you set out to learn. Then test yourself as you go. You have not accomplished your objective unless you pass your test. At first, take the section paragraph by paragraph (or, if need be, sentence by sentence). Stop after each paragraph, look away, and say out loud what the paragraph meant to you. (Whisper if you are afraid to be

hypothesis

A natural law can be verified independently by persons other than those who proposed it.
A theory is a concept or model used to explain a natural law.

Figure I.3 For each paper published in chemistry and closely related fields, a brief abstract is published, usually only four or five lines of print. For one year, 1990, those abstracts filled this stack of books.

caught talking to yourself). Then read the paragraph again and check. Did you have it? If not, stay on that paragraph until you do understand it and can pass your test. When you have covered what you assigned yourself, test yourself again. Before you decide the session is over, *write* a brief summary of what you have covered. Do it without looking at the book. If you can't, you did not learn it. Start new work only when you feel you have mastered what you set out to do. Try the constant testing method; you will find you are learning more and having to work less.

Not all sections of the text are equally difficult or equally long. Sometimes you will be able to sail through a page or two without having to read and reread. At other times, you will need to spend several minutes on just one sentence or one equation.

Make an effort not to skip equations. An equation is just a compact way of making a statement, and one equation can contain as much information as an entire paragraph or page. Read equations carefully, term by term, and understand what they are saying before you go on.

It is absolutely necessary to keep up with your class, because chemistry is cumulative. Each new thing learned rests on the knowledge of something previously covered. It is important to study assigned sections, usually several times, *before* the class meets. Then you can relax and follow the reasoning of your instructor without needing to write down every word. Falling behind, even by a day or two, can be fatal.

Much of chemistry requires learning words and symbols that are unfamiliar to you. It is essential that you learn and understand the new language because what follows is explained in these new terms. Pay particular attention to the words signalled by heavy print **like this** and note their definitions carefully. At the end of each chapter is a list of the significant new words defined in the chapter, the "key terms." There is a glossary with a complete set of those terms and their definitions in the Appendix.

Chemistry, like other fields of science, often depends on the mathematical manipulation of numerical data from measurements. For this reason, careful problem solving is essential to your success in chemistry. Chapter 1 discusses in detail an organized approach to problem solving. To check your solutions to problems, you will want to see answers. Answers to exercises appear at the end of each chapter.

You'll find that chemistry can be fun if you just relax and enjoy it. If you take it step by step and don't leave any gaping holes in your knowledge, you will find it easy. Good luck, and have fun (really!).

In chemistry we study chemical reactions. In this reaction, nickel metal (the gray bar) reacts readily with a solution containing copper. The products of the reaction are copper metal (orange) and a green solution containing nickel.

Measurement and Problem Solving in Chemistry

As science advanced from its early days, when dropping weights from a tower or floating in the bathtub could produce significant new insights, the task of observing physical phenomena grew more complex and demanding. It became increasingly important to record numerical data. Before an observation could be of much use, it was necessary to record not only what was seen but also how much was seen. Precise and well-defined systems of measurement began to develop.

Now we live in a world in which a large part of science is quantitative. The metric system is used by scientists worldwide to communicate the numerical results of their measurements, from the speed of escaping nuclear particles to the intensities of cosmic rays. Chapter 1 introduces us to the concept of measurement and the metric system.

In Chapter 1 we also tackle problem solving: what to do, what not to do, and how to use unit analysis, a helpful method of setting up problems. This chapter also covers the subjects of temperature and density. Mastering the information contained here will give us the necessary tools to begin the serious study of chemistry.

1.1 Measurement and Uncertainty

After studying this section, you should be able to:

- Identify the two components of a value resulting from a measurement.
- Understand that all measurements have some uncertainty.

Chemistry is a physical science; it is concerned with the quantitative measurement of the properties of matter. To measure something is simply to find the numerical value of one of its properties, perhaps its length or its temperature. The results of quantitative measurements are called **values.** A value consists of a numerical quantity and the set of units in which the measurement was made. **Units** are the standard quantities that have been defined for making measurements. If the speed of a car is measured, the units of the result could be miles/hour (that is, miles per hour) or any of several other sets. It is meaningless to say that the distance to a friend's house is 60. How far is 60? It is too far to walk if it is 60 miles, but if it is 60 inches, you can almost reach out and touch it. When giving the result of a measurement, it is always necessary to state the units as well as the numerical quantity. The importance of conscientious and careful attention to units cannot be overemphasized. It is necessary for communication and is a tremendous aid to problem solving. In Section 1.5, we will explore in detail the technique of using units as a guide to working problems.

The value resulting from a measurement must always contain a numerical quantity and its units.

That science is exact, that scientific experiments are so precise as to be free of error, is a popular misbelief. Even the simplest results contain an element of uncertainty. Even if an experiment is designed to obtain a simple yes or no answer, the chance for error still exists. Some inescapable uncertainty is always present in measurements themselves, regardless of what the true values are.

Suppose, for example, that you are asked to measure the length of a book (Figure 1.1). You are given a ruler on which there are marks one inch apart. You place your ruler next to the book and say that the length is 9.7 inches. The 9 you know for sure, but the .7 is only an estimate. Making the same measurement again, you might estimate the length to be 9.6 or 9.8 inches, while another person might make other similar estimates. Everyone would agree on the 9, but there would be some doubt about the number of tenths of an inch. Now suppose that the scale is marked in inches and tenths of inches. Then you might report that the book is 9.73 inches or 9.74 inches, estimating the last hundredth of an inch. If the ruler were marked with hundredths of an inch, you probably could not do any better, because the roundness of the edge of the book and the way in which you positioned the ruler would cause more than a hundredth of an inch of uncertainty. The improvement in the ruler would not necessarily improve your measurement. No matter how finely a scale is marked, there will always be some uncertainty in a measurement. There is no such thing as an exact measurement. Only when you are counting small numbers of individual units such as coins, people, or automobiles can you get an exact answer.

In most cases, then, we cannot know an exact value. Because we do not know the exact value, we cannot know exactly how much the measurement differs from it. The best we can do is to measure as carefully as possible and attempt to eliminate errors by using many different methods of measurement and improving our instruments and techniques.

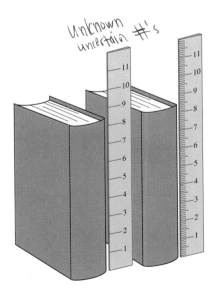

Figure 1.1 A measurement using a ruler with a large uncertainty (*left*) and one with less uncertainty (*right*).

1.2 **Significant Figures**

After studying this section, you should be able to:

- Determine how many significant figures are in a value resulting from a measurement.
- Interpret the number of significant figures in a reported measurement.
- Round off a given number to a specified number of significant figures.
- Express the result of a multiplication or division to the correct number of significant figures.
- Express the result of an addition or subtraction to the correct number of significant figures.

When you make measurements, you know that the measurements have some degree of uncertainty. You should indicate how much uncertainty there is because others who may rely on the measurements need to know how precise the values are.

A frequently used method of indicating the precision of measurements is the system of **significant figures.** In reporting the result of a single measurement, it is not too difficult to decide how many significant figures to use; give the number of digits that are definitely known along with one digit that is uncertain or estimated. For example, let's return to your measurement of the book. When you reported the result as 9.7 inches, the last digit, 7, was an estimate; it was uncertain. With the better ruler, you reported 9.73 inches, and the uncertain digit was the 3. In the first case, you reported two significant figures; in the second case, three significant figures. In the second case, you might have found the measurement to lie *on* the .7 line (9.70), in which case your third significant figure would have been a zero. When you make a measurement, write all the known digits and one estimated digit, even if the last one is a zero.

Here are some rules for writing and interpreting significant figures.

1. *All nonzero digits are significant.* There are two significant figures in 15 and five in 23.784.

2. *Zeros between nonzero digits are significant.* The zeros in 3.102 and 2801 are significant; therefore, there are four significant figures in each of these numbers.

3. *Zeros beyond the decimal point are significant if preceded by a nonzero digit.* In 37.0 and 6.00, the zeros are significant. In 0.0030, only the last two digits, 3 and 0, are significant.

4. *Zeros preceding the first nonzero digit are not significant.* If there is *only* a zero to the left of a decimal point, as in 0.002 for instance, that zero is there merely to emphasize the decimal point and therefore is not significant. Beyond the decimal point, such zeros are there as "place holders" and serve only to fix the position of the decimal point. Therefore, 0.002 has one significant figure, the 2.

5. *If there are only zeros to the right of a digit, they are called place holders, unless preceded or followed by a decimal.* For instance, 2,000 has only one significant figure, the 2. However, in the number 2.000 there is no need for place holders, but the zeros have been shown nonetheless. Therefore the zeros in 2.000 are significant. The number 3,000. has a decimal point at

Scientific notation is explained in the next section.

A counted quantity has an infinite number of significant figures.

One result of several sequential operations on a calculator is usually many nonsignificant digits.

the end purposely to make the zeros significant. To indicate that one or two of the zeros in 3,000 is significant requires that the number be written in scientific notation.

6. *An integer, an exact number, that results from a counting of individual objects, such as 30 students, or that arises from a definition (example: 1 ft = 12 in.), may be considered for purposes of calculation to have an infinite number of significant figures, that is, it need not be considered when rounding.* The subscripts in chemical formulas, such as the 2 in H_2O, are taken to be exact because they represent ratios between counted numbers of individual atoms.

When you make calculations, especially with calculators and computers, the number of digits in your answer is not likely to have the appropriate number of significant figures for your measurements. Although occasionally a calculator will drop significant figures that must later be added (example: The answer a calculator gives to 0.250×4.00 is 1), usually the number of digits must be decreased by a process called **rounding.** Because rounding numbers before or during a calculation can lead to errors, you should *round only after the calculation is finished.* The rules for rounding follow.

1. *If the first nonsignificant digit is greater than five, round the last significant figure up.* For example, if 2.87 has only two significant figures, round to 2.9.
2. *If the first nonsignificant digit is less than five, the last significant figure remains unchanged.* For example, if 2.84 has only two significant figures, round to 2.8.
3. *If the first nonsignificant digit is exactly five (or five followed by zeros), the last significant figure remains unchanged when it is even or increased by one when it is odd.*

Study these examples in which each number has three significant figures.

Number	Rounded Correctly
43.82	43.8
43.85	43.8
43.45	43.4
43.78	43.8
43.75	43.8
43.95	44.0

After a calculation has been made, you must determine how many significant figures should be in your answer. Here are two rules to help determine how far to round after a calculation.

1. *In multiplication and division, the measured value having the <u>fewest significant figures</u> determines the number of <u>significant figures in the answer</u>.*

EXAMPLE 1.1

How many significant figures should be in the answer to 2.0×0.003?

Solution: The first number has two significant figures, but 0.003 has only one; therefore the answer is 0.006, which has only one significant figure.

EXAMPLE 1.2

Determine the answers to the correct number of significant figures for each of the following calculations.

(a) $\dfrac{9.00}{3.0}$

(b) 1.21785×2.0

(c) $1.21785 \times 2,000$

Solution: (a) 9.00 has three significant figures; 3.0 has two significant figures. The answer, 3.0, has two significant figures. (b) 1.21785 has six significant figures; 2.0 has two significant figures. The answer, 2.4, has two significant figures. (c) 1.21785 has six significant figures; 2,000 has one significant figure. The answer, 2,000, has one significant figure.

Exercise 1.1

Do the following calculations and express the answers using the correct number of significant figures.

(a) 4.2×0.311

(b) $\dfrac{74}{8}$

(c) $\dfrac{4.42 \times 0.8843 \times 19}{36257}$

(d) $0.81 \times 788.3 \times 0.000344$

EXAMPLE 1.3

How many pounds of flour are contained in four bags of flour, each containing 2.113 pounds?

Solution:

$$4 \text{ bags} \times \frac{2.113 \text{ pounds}}{\text{bag}} = 8.452 \text{ pounds}$$

Do not round 8.452 to 8. The bags of flour are counted; therefore the 4 is exact and need not be considered in determining the number of significant figures in the answer.

---

2. *In subtraction and addition, the number of digits beyond the decimal in the answer must be no greater than the fewest number of digits beyond the decimal in any of the numbers to be added or subtracted. If one or more of the numbers to be added or subtracted has no digits after the decimal, the answer must not have any significant figures to the right of the column containing an uncertain digit.*

EXAMPLE 1.4

Add

$$
\begin{array}{r}
3.4560 \\
221.75 \\
9.421 \\
\underline{9.00} \\
243.6270 = 243.63
\end{array}
$$

Solution: Because the numbers 221.75 and 9.00 have the fewest number of digits beyond the decimal (two), these numbers determine the number of significant figures in the answer. The answer, 243.6270, therefore can have only two digits beyond the decimal. Round it to 243.63.

EXAMPLE 1.5

Add

$$
\begin{array}{r}
23{,}400 \\
4{,}613.55 \\
9{,}828
\end{array}
$$

Solution: The sum is 37,841.55, but it must be rounded to 37,800. The number 23,400 has three significant figures; the two zeros are placeholders. Because the third column of 23,400 contains an uncertain digit, the answer, 37,841.55 must be rounded off to the third column. The correct answer is 37,800.

Exercise 1.2

Perform the following operations.

(a) $33.482 + 0.88 + 2.9993 + 410.33 =$
(b) $553.1 + 0.22 + 25.334 =$
(c) $51.334 - 7.8832 =$
(d) $344.28544 - 91 =$
(e) $3{,}432 + 240 + 21.7 =$

EXAMPLE 1.6

Make the following calculation and express the answer in the correct number of significant figures.

$$
\frac{22.0}{6.008} \times (2.117 + 34.74)
$$

Strategy and Solution:

Step 1. Perform the operation in the parentheses first: Add 2.117 and 34.74. Because 34.74 has the fewest number of decimal places (two), the answer,

36.857, must have only two decimal places. Don't round off, however, until the entire calculation is complete.

Step 2. Perform the multiplication and division.

$$\frac{22.0}{6.008} \times 36.857 = 134.962$$

Step 3. Because the fewest number of significant figures in the operations is three (22.0), the answer must be rounded off to three significant figures, 135.

Exercise 1.3

Perform the calculation and round the result to the correct number of significant figures.

$$(2.3 + 4.00) \times (3.6 + 9)$$

1.3 Scientific Notation

After studying this section, you should be able to:

- Convert numbers to scientific notation.
- Convert numbers written in scientific notation to decimal form.

In chemistry we meet some exceptionally large and exceptionally small numbers; 602,000,000,000,000,000,000,000 and 0.00000000000000000000000000911 are examples. Because it would be inconvenient to write such numbers in full, a system called **scientific notation** was developed. A number written in scientific notation is a number equal to or greater than 1 and less than 10, multiplied by 10 raised to an appropriate power. The power to which the 10 is raised is shown by an exponent, a small number printed near the top and to the right of the 10. Example: 10^7 is 10 raised to the seventh power.

Raising a number to some power means multiplying it by itself that number of times. For example, to raise 10 to the third power, 10^3, we multiply 10 by itself three times:

$$10^3 = 10 \times 10 \times 10 = 1,000$$

We could have accomplished the same thing by taking 1.000 and moving the decimal point to the right three places to give 1,000.

$$10^3 = 1\ 0\ 0\ 0 = 1,000$$

Multiplying any number by 10 raised to a positive exponent is the same as moving the decimal point to the *right* as many places as the exponent.

Ten raised to a negative power is the same as $\frac{1}{10}$ multiplied by itself some number of times. Then 10^{-5}, for example, is

$$\frac{1}{10} \times \frac{1}{10} \times \frac{1}{10} \times \frac{1}{10} \times \frac{1}{10}$$

Multiplying a number by 10 raised to a negative exponent is the same as moving the decimal point to the *left* as many places as the exponent.

$$10^{-5} = \underset{\longleftarrow}{0.0\ 0\ 0\ 0\ 1.} = 0.00001$$

Let's look at an example. Suppose we wish to express the number 0.0003141 in scientific notation. By moving the decimal point four places to the right, we convert this to a number between 1 and 10, that is, 3.141. Because the decimal has been moved to the right four places, we multiply by 10 raised to the -4, or 10^{-4}.

$$0.0003141 = 3.141 \times 10^{-4}$$

Not only does scientific notation make it easier to read a number, it also makes arithmetic with decimal numbers vastly easier. Here is the step-by-step way to convert numbers to scientific notation.

1. *Move the decimal point in the original number to the right or left to convert the number to a number between 1 and 10. If the decimal point does not appear at the right of a whole number, imagine it to be there.*

2. *If you moved the decimal to the* left, *multiply the new number by 10 raised to a* positive *power, the exponent to be equal to the number of places you moved the decimal. If the decimal was moved to the* right, *multiply the new number by 10 raised to a* negative *power, the exponent to be equal to the number of spaces you moved the decimal.*

Here are some examples to follow.

EXAMPLE 1.7

Write 10,000 in scientific notation.

Strategy and Solution:

Step 1. Move the decimal point in 10,000 to the left until the 10,000 is converted to 1.

$$1\overset{\curvearrowleft}{.}0\,0\,0\,0$$

Step 2. Because the decimal was moved four places to the left, multiply the new number by 10 raised to the fourth power.

$$1 \times 10^4$$

EXAMPLE 1.8

Write 0.020 in scientific notation.

Strategy and Solution:

Step 1. Move the decimal point in 0.020 to the right until the 0.020 is converted to a number between 1 and 10, that is, to 2.0.

$$0 \cdot 0 \;\; 2 \;\; 0$$

Step 2. Because the decimal was moved two places to the right, multiply the new number by 10 raised to the negative two power.

$$2.0 \times 10^{-2}$$

EXAMPLE 1.9

Write 354.68 in scientific notation.

Strategy and Solution:

Step 1. Move the decimal point in 354.68 to the left to convert to a number between 1 and 10, that is, to 3.5468.

$$3 \;\; 5 \;\; 4 \cdot 6 \;\; 8$$

Step 2. Because the decimal was moved two places to the left, multiply the new number by 10 raised to the second power.

$$3.5468 \times 10^2$$

Go back to the numbers at the beginning of this section and express them in scientific notation. You will be seeing them again.

To convert from scientific notation to decimals is easy. Look at the exponent on the 10. The value of the exponent tells you how many places to move the decimal point. If the exponent is positive, move the decimal point to the *right*. For example, if your number in scientific notation is 5.22×10^5, move the decimal point to the right five places.

$$5.22 \times 10^5 = 5 \cdot 2 \;\; 2 \;\; 0 \;\; 0 \;\; 0 = 522{,}000$$

Notice that to do that, you had to add zeros where there were originally none. These trailing zeros, of course, are not significant; they are merely place holders.

If the exponent is negative, you move the decimal point to the *left*. If your number had been 5.22×10^{-3}, you would have moved the decimal left three places to produce 0.00522.

$$5.22 \times 10^{-3} = 0 \;\; 0 \;\; 0 \;\; 5 \cdot 2 \;\; 2 = 0.00522$$

Again, the added zeros are not significant. The number of significant figures must not be allowed to change as a result of converting to or from scientific notation.

EXAMPLE 1.10

Write 2.06×10^{-2} in decimal form.

Solution: The exponent indicates that the decimal must be moved to the left two places.

$$2.06 \times 10^{-2} = 0 \;\; 0 \;\; 2 \cdot 0 \;\; 6 = 0.0206$$

Exercise 1.4

Convert the following numbers to decimal notation.

(a) 7.04×10^{-4}

(b) 5.20×10^{-2}

(c) 8.03×10^{5}

(d) 1.003×10^{2}

Scientific notation is useful when reporting significant figures. Recall that 2,000 has only one significant figure, the 2. The number 2,000. has a decimal point at the end purposely to make the zeros significant. How can you indicate that two of the zeros in 2,000 are significant? Use scientific notation. The number 2,000 with two significant zeros is written as 2.00×10^{3}. Whenever you write a number in scientific notation, the number of significant figures will equal the number of digits in the number multiplied by 10. Because there are three digits in 2.00, there are three significant figures in 2.00×10^{3}.

1.4 Solving Problems in Chemistry

After studying this section, you should be able to:

- Use an organized thought process to approach chemistry problems.
- Analyze your answers to problems in terms of reasonableness, precision, and the units used.

General Principles

Nearly everyone who has been through first-year college chemistry feels that solving numerical problems is the toughest part. By attacking problems in an organized way, however, you can make the task of problem solving considerably easier. This section offers a general approach to problem solving, showing what to do and what not to do. Once you understand the basic principles, you will be ready for the next section, which describes a detailed method, called *unit analysis*, of finding and setting up the relations that will lead to desired answers.

Chemistry problems require that measured values be put together mathematically to produce a wanted result. To achieve this, several steps are necessary:

On exams, students frequently lose substantial credit because they misread questions.

Organizing your thoughts in a solution map leaves nothing to do but the mechanical processes of setup and arithmetic.

1. *Read the problem carefully. Be sure you understand it.*
2. *Write as conversion factors the information given in the problem.*
3. *Make a mental or written plan for working the problem (a "solution map").*
4. *Make sensible approximations when you can.*
5. *Set up and make your calculation, being sure to include the units.*
6. *Check your answer—both the numerical quantity and the units.*

How to Begin

1. *Read the problem and be sure you understand it.* Start by reading slowly. Pause to think about what you read; take care to see and understand the

meaning of every word and symbol. Be sure that you know what is being asked for and what the units of the answer should be.

2. *Write mathematical versions of the values given.* It is often difficult to decide how a problem should be set up until the data statements are organized in mathematical form, complete with units. Make a list of the relations, similar to the one below. This step is worth the time it takes, just to avoid confusion.

Suppose you are told that there are seven oranges in a box, that a car can run 21 miles on a gallon of gasoline, and that three eggs, two golf balls, a potato, and a bottle of beer make 4 pints of soup. Here is the first of those relations:

$$\frac{7 \text{ oranges}}{1 \text{ box}}$$

Often, when exactly one of something appears in an equation, the numeral 1 is not shown. We would write the first relation as

$$\frac{7 \text{ oranges}}{\text{box}}$$

This statement is read as "seven oranges per box," the horizontal line meaning *per* and indicating division. (For example, the number of oranges per box could have been determined by counting all the oranges from several boxes and dividing by the number of boxes. You know the old joke: To count the number of sheep in a field, count the number of legs and divide by four.)

The rest of the relations are

$$\frac{21 \text{ mi}}{\text{gal gasoline}}$$

$$\frac{3 \text{ eggs}}{4 \text{ pt soup}}$$

$$\frac{2 \text{ golf balls}}{4 \text{ pt soup}}$$

$$\frac{\text{potato}}{4 \text{ pt soup}}$$

$$\frac{\text{bottle beer}}{4 \text{ pt soup}}$$

The process of division, expressed in words as per *or* for every, *can be shown mathematically in different ways:* $1/2$, $1 \div 2$, $\frac{1}{2}$.

The same mathematical relations may be written in the inverse form if necessary. If there are 7 oranges in every 1 box, it is surely true that there is a box for every 7 oranges, or

$$\frac{\text{box}}{7 \text{ oranges}}$$

flip at will

Which form to use in your mathematical statements depends on the nature of the problem and the way in which you plan to work it. Once the statements are written in mathematical form, they may later be inverted if you wish. In any case, write down all the relations you find in the problem; it is best to have them where you can easily refer to them.

3. *Make a mental or written plan for working the problem.* Begin by deciding which chemical principles or concepts apply and the order in which they must be used. Plan the steps you will use in getting from the initial data to the final answer. Make what we will call a solution map. Study Examples 1.12 and 1.13 to see how to make solution maps. (Using the units as a guide, as explained in Section 1.5, will help you in this step.)

   Next, check to see that you have all the data necessary to work the problem. Sometimes you are expected to supply some of the information yourself. Do not begin to use formulas or make calculations until your solution map or "roadmap" to the answer is complete. (When instructors grade papers, they often give substantial credit for a step-by-step plan for working a problem even if no answer is given.)

Scientists often talk in "orders of magnitude," which is an abbreviation for saying "about how big," a rough approximation to the nearest factor of ten.

We can approximate that adding a cup of water to the Pacific Ocean does not significantly change the total volume of the ocean.

4. *Make sensible approximations when you can.* You can often spare yourself much work and confusion by leaving out part of a calculation or by making an easy but inexact calculation when the exact one would be difficult and unnecessary. If properly done, such approximations will not affect the answer. Shortcuts like this are especially appropriate when small numbers are to be added to or subtracted from large numbers of low precision.

Now you are ready to work out a mental or scratch paper estimate of what the numerical part of the answer will be. This exercise helps you check your solution map and will also be valuable when you finish the problem. Write down your estimated answer.

EXAMPLE 1.11

How many pounds of cake mix can be made from 33 pounds of flour, 17 pounds of sugar, 4 pounds of butter, and 0.01 pounds of salt?

Solution: The flour, sugar, and butter together weigh 54 pounds. Because it is possible to have an error of a pound or so in each of these quantities, the 54-pound total could have a total error of as much as 3 or 4 pounds. Compared to this, the 0.01 pound of salt is negligible, that is,

$$54 \text{ lb} + 0.01 \text{ lb} = 54.01 \text{ lb}$$

But this answer must be rounded to 54 pounds, and the 0.01 is lost. As an approximation, merely leave the salt out at the start. Although this is a particularly simple example, later we will encounter problems in which making an approximation at the right time can save an enormous amount of work.

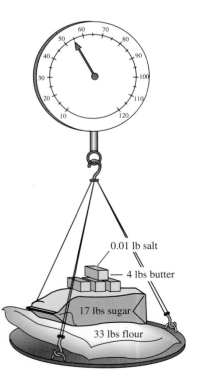

0.01 lb salt

4 lbs butter

17 lbs sugar

33 lbs flour

5. *Set up and make the calculation.* After deciding what steps are necessary to get from data to answer, write down the data in the mathematical form that will lead to the answer. Put the data into your solution map. Do the arithmetic.

6. *Check your answer.* A most important step. Are the units correct? Probably so if you have used them in your setup. Is the number of significant figures appropriate? If not, round off. Finally, is the answer reasonable? Students working first-year chemistry problems sometimes devise a correct setup but elsewhere make a mistake that leads to a ridiculous answer. Here is an example of how you can go wrong by not checking your answer.

Question: How many gallons of coffee can be made with 2.35 pounds of ground coffee and as much water as needed, if each pound of ground coffee will make 3.41 gallons? Answer: 801.35, right? Wrong.

The answer is wrong on all three counts. First, it should carry its units, gallons. Second, even assuming that "each pound" means exactly one pound, there still should be only three significant figures in the answer. Finally, it should be clear that the answer ought to have been nearer 8 gallons than 800. There is no way to get more than a few gallons of coffee out of less than 4 pounds of ground coffee. The correct answer is 8.01 gallons.

Now is the time to look back at your estimated answer. How does the calculated answer compare? If it is close, then everything is probably all right. If it is way off, then what went wrong? Don't go on until you are satisfied with your answer.

What to Avoid

It is essential that you try to understand a problem before you begin to solve it; any other approach only hinders learning. Be sure you know the fundamental principles that the problem is designed to teach, then use problem-solving techniques only to help devise the setup and to check your answer. Students often attempt to understand chemistry by memorizing countless equations and planning to use them without really knowing what they mean. Little good comes from memorizing equations unless the concepts that underlie them are understood. Remember that your goal is to achieve a better understanding of chemistry. The equations are merely tools to allow you to employ that understanding.

If you understand something, it will always be yours. If you merely memorize a setup for a problem, you'll eventually forget it.

1.5 Using Unit Analysis to Solve Problems

After studying this section, you should be able to:

- Use unit analysis to set up mathematical solutions to problems.

In our discussion of problem solving, we touched on what is probably the most critical and at the same time the hardest part: putting the fragments of the problem into a mathematical setup that will lead to a solution. This section describes a method that can guide you in setting up many kinds of problems. The method, called **unit analysis**, is based on the fact that arithmetic and algebra work on units in exactly the same way that they work on numerical values.

$$3 \times 3 = 3^2$$

$$\text{feet} \times \text{feet} = \text{feet}^2$$

Conversions from one set of units to another can be accomplished using **conversion factors**. A conversion factor is a ratio of two related terms. There is, for example, a conversion factor that converts feet to inches.

$$\text{feet} \times \text{conversion factor} = \text{inches}$$

This conversion factor must cancel or eliminate *feet* and must introduce *inches*. To accomplish this, it must have *feet* in the denominator (the bottom part of the fraction) and *inches* in the numerator (the top part).

Conversion factors are also sometimes called unit factors, for a reason that we will soon see.

$$\text{feet} \times \frac{\text{inches}}{\text{feet}} = \text{inches}$$

A conversion factor can be written in either of two ways, each of which is the reciprocal of the other. Although the relation is the same regardless of which way the factor is written, it is convenient to choose in advance the form you plan to use. For example, there are 12 inches in a foot. This relation is written in shorthand form as

$$\frac{12 \text{ in.}}{\text{ft}}$$

or turned upside down:

$$\frac{\text{ft}}{12 \text{ in.}}$$

Notice that periods are not used after abbreviations for units except in the case of the abbreviation for inches, *which is written as* in. *to distinguish it from the preposition* (in).

These two conversion factors contain the same information, but one converts feet to inches and the other converts inches to feet.

$$2.5 \text{ ft} \times \frac{12 \text{ in.}}{\text{ft}} = 30. \text{ in.}$$

$$30. \text{ in.} \times \frac{\text{ft}}{12 \text{ in.}} = 2.5 \text{ ft}$$

Remember that $30 \times \frac{1}{12}$ *equals* $30 \div 12$.

Because these particular conversions are exact relationships, they will not influence the number of significant figures in the answer. That is, they can be ignored when you are determining how many significant figures to keep. As we will see somewhat later, exact numbers arise out of chemical equations as well as from definitions.

It is possible that we might have been given the conversion 1 in. = 0.0833 ft, for which we would write the factors as

$$\frac{1 \text{ in.}}{0.0833 \text{ ft}}$$

or

$$\frac{0.0833 \text{ ft}}{1 \text{ in.}}$$

In this case, the 0.0833 was obtained by dividing 1 by 12 and rounding to three digits. It is not exact. If you make a calculation using the conversion in this form, you must take notice of its three significant figures when you round your answer.

To see why conversion factors are called unit factors, look at the conversion factor

You may understand this more easily put this way. You are told that 1 ft = 12 in. Then,

$$\frac{1 \text{ ft}}{12 \text{ in.}} = \frac{12 \text{ in.}}{12 \text{ in.}}$$

therefore

$$\frac{1 \text{ ft}}{12 \text{ in.}} = 1$$

$$\frac{1 \text{ in.}}{0.0833 \text{ ft}}$$

Because the length of 1 in. is stated to be the same as the length of 0.0833 ft, we can substitute 0.0833 ft for the 1 in. of that equation and obtain a result of 1.

$$\frac{0.0833 \text{ ft}}{0.0833 \text{ ft}} = 1$$

True conversion factors all have the numerical value of 1. Because this is so, we can multiply or divide anything by a conversion factor without changing its actual amount, although we will indeed change its units and its numerical value. Watch the change of units happen as we work the problems that follow.

Almost any quantitative relation can be written as a conversion factor.

EXAMPLE 1.12

A shelf is 4.42 ft high. How high is it in inches?

Solution Map: To make your solution map, first write the units of the initial information, *feet*. Find the units in which the answer is to be given, *inches*. Then, step by step, work out and write in the conversions needed to reach the answer. In this case, only one conversion is needed.

$$\boxed{\text{feet}} \longrightarrow \boxed{\text{inches}}$$

Strategy and Solution: Of the two possible relations between foot and inch, which form should we use? One gives an answer in the correct units and one leads to nonsense. Let's look at both.

$$4.42 \text{ ft} \times \frac{1 \text{ ft}}{12 \text{ in.}} = 0.368 \frac{\text{ft}^2}{\text{in.}}$$

$$4.42 \text{ ft} \times \frac{12 \text{ in.}}{\text{ft}} = 53.0 \text{ in.}$$

The first answer can't be correct because it has the wrong units, $\text{ft}^2/\text{in.}$ Our check of the units in the first calculation tells us to go back and set up the problem correctly. We accept our second answer, 53.0 in., because it is reasonable and has the correct units, inches.

Obtaining a solution to a problem often requires more than a single conversion factor. Here is a problem that uses a conversion in the metric system, with which we will soon become more familiar.

EXAMPLE 1.13

A room is 15.3 ft wide. How many centimeters (cm) wide is it? One inch equals approximately 2.54 cm.

Solution Map: To make the solution map, two conversions are needed.

$$\boxed{\text{feet}} \longrightarrow \boxed{\text{inches}} \longrightarrow \boxed{\text{centimeters}}$$

Strategy and Solution:

Step 1. Write conversion factors from the information you have been given in the problem.

$$\frac{\text{ft}}{12 \text{ in.}}$$

$$\frac{\text{in.}}{2.54 \text{ cm}}$$

Step 2.   Write the initial information with its units, then multiply by the proper conversion factors to obtain the answer in the desired units. Although we already know how to convert feet to inches, we must also use the conversion from inches to centimeters.

This setup is merely a string of conversion factors.

$$15.3 \text{ ft} \times \frac{12 \text{ in.}}{\text{ft}} \times \frac{2.54 \text{ cm}}{\text{in.}} = 4.66 \times 10^2 \text{ cm}$$

Step 3.   We check and accept the answer. If either conversion factor had been used incorrectly, the units would not have been cm, and the answer would have been incorrect.

Exercise 1.5

Four teaspoons of cocoa are needed for one cup of hot chocolate. How many tablespoons do you need for five cups if one tablespoon equals three teaspoons?

Because it arises from a definition, 5,280 can be considered, for the purposes of calculation, to have an infinite number of significant figures.

Exercise 1.6

A mile is 5,280 ft. How many yards are in 0.242 mi?

Although it is essential that you become familiar with the units of measurement used in chemistry, conversion factors work whether you know the meaning of the units or not.

EXAMPLE 1.14

Here we will solve a problem using imaginary units from the planet Xanthu. No one on earth knows the meaning of these units, but anyone can do the problem using unit analysis. There are 3.21 morks to every neb. How many morks are there in 7.2 nebs?

Solution Map: $\boxed{\text{nebs}} \longrightarrow \boxed{\text{morks}}$

Strategy and Solution:

Step 1.   Write conversion factors from the information given in the problem.

$$\frac{3.21 \text{ morks}}{\text{neb}}$$

Step 2.   Write the initial information with its units, then multiply by the proper conversion factors to obtain the answer in the desired units.

$$7.2 \text{ nebs} \times \frac{3.21 \text{ morks}}{\text{neb}} = 23 \text{ morks}$$

Step 3. Check your answer. Is it reasonable? Yes, if there are 3.21 morks for every neb, then 7.2 nebs must equal more than 7×3 or 21 morks.

Conversion factors can be used more than once in a single calculation; that is, it may be necessary to use the square or cube of a factor.

EXAMPLE 1.15

How many cubic inches are there in a box that measures 2.5 ft × 1.5 ft × 3.7 ft?

Solution Map: $\boxed{\text{feet} \times \text{feet} \times \text{feet}} \longrightarrow \boxed{\text{inches}^3}$

Solution: To convert the volume of the box to cubic inches, we could multiply each of the sides by the conversion factor from feet to inches.

$$2.5 \text{ ft} \times \frac{12 \text{ in.}}{\text{ft}} \times 1.5 \text{ ft} \times \frac{12 \text{ in.}}{\text{ft}} \times 3.7 \text{ ft} \times \frac{12 \text{ in.}}{\text{ft}} = 2.4 \times 10^4 \text{ in.}^3$$

This is the same as

$$2.5 \text{ ft} \times 1.5 \text{ ft} \times 3.7 \text{ ft} \times \left(\frac{12 \text{ in.}}{\text{ft}}\right)^3 = 2.4 \times 10^4 \text{ in.}^3$$

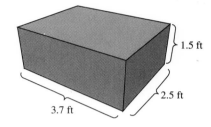

1.5 ft

2.5 ft

3.7 ft

To convert a value in which some of the units appear raised to a power, you must raise the *entire* conversion factor to a power. And, of course, raising the factor to a power means raising *both* the numerical part and the units. Here is a somewhat more difficult problem.

EXAMPLE 1.16

A plot of ground contains 12,555 square rods. If a rod contains 16.5 ft (not an exact number) and an acre contains 43,560 square ft, how many acres are there in the plot?

Solution Map: $\boxed{\text{rods}^2} \longrightarrow \boxed{\text{feet}^2} \longrightarrow \boxed{\text{acres}}$

Strategy and Solution:

Step 1. Write conversion factors from the information given in the problem.

$$\frac{\text{rod}}{16.5 \text{ ft}} \qquad \frac{\text{acre}}{43,560 \text{ ft}^2}$$

Step 2. Write the initial information with its units, then multiply by the proper conversion factors to obtain the answer in the desired units.

$$12,555 \text{ rods}^2 \times \left(\frac{16.5 \text{ ft}}{\text{rod}}\right)^2 \times \frac{\text{acre}}{43,560 \text{ ft}^2} = 78.4688 \text{ acres}$$

Step 3. Check the answer. Notice how the number of digits increased as we did the calculation. We round to 78.5 acres. Then, having checked our answer to see if it is reasonable and in the correct units, we accept it.

Exercise 1.7

A rectangular tank measures 1.32 yd by 3.55 yd. If the tank is filled with water to a depth of 0.23 yd, how many gallons of water are in it? A gallon contains 231 cubic in.

Exercise 1.8

You are going to paper a wall that is 10.5 ft by 8.0 ft. If one roll of wallpaper covers 4.0 square yd, how many rolls do you need?

The next problem requires that you make several conversions and that one of the conversion factors be raised to the third power if the units are to cancel properly. Because the problem is somewhat more complex than the preceding ones, we will approach it with the full problem-solving technique discussed in the preceding section.

EXAMPLE 1.17

Given that there are exactly 36 in. in 1 yd, that 1 in.^{3} of water has a mass of 0.554 oz, and that a pound contains exactly 16 oz, what is the mass in pounds of a cubic yard, yd^3, of water?

The words "one" or "a" are often used in problems. Like any number that is written as a word, they are taken to mean exactly that amount. With such defined quantities, you can use as many significant figures as you wish.

Strategy and Solution:

Step 1. Read the question and be sure you understand it. The units of the answer will be pounds.

Step 2. Write conversion factors for the information given in the problem.

$$\frac{36 \text{ in.}}{\text{yd}} \qquad \frac{0.554 \text{ oz water}}{\text{in.}^3} \qquad \frac{\text{lb}}{16 \text{ oz}}$$

Step 3. We set up our solution map. We are to calculate the number of pounds in a cubic yard of water. (Units are sometimes stated in plural form. You need not worry about that. Whether we say pound or pounds makes no difference in the calculation.) We look at the starting place, in this case it is a cubic yard of water. Now we check the conversion factors. There is one that converts from oz to in.^{3}, one from lb to oz, and one from in. to yd. Since our starting point is in yards, we use 36 in./yd first. Then convert the resulting in.^{3} to oz and finally the oz to lb. Follow the same procedure every time you make a solution map. Begin with the initial data, look at the conversion factors, and line them up in the proper order. Here is what the finished map looks like.

Solution Map: $\boxed{\text{yd}^3} \longrightarrow \boxed{\text{in.}^3} \longrightarrow \boxed{\text{oz}} \longrightarrow \boxed{\text{lb}}$

Step 4. Although we may not have seen some of these units before, we have
read and understood the question. There do not seem to be any approximations
we can make, so we will guess at an answer. A cubic yard might be a box about
as big as a card table. Imagine such a box filled with water; lift it. You can't.
It must weigh hundreds of pounds. That's our guess: hundreds of pounds.

Try to visualize the physical arrangement whenever possible. It will help when you check your answer.

Step 5. Write down the setup and calculate the answer.

Because it would be inconvenient to include the conversion factor, inches to
yards, three times in the calculation, we will raise it to the third power when
we put it in our equation.

$$\left(\frac{36 \text{ in.}}{\text{yd}}\right)^3$$

$$1 \text{ yd}^3 \times \left(\frac{36 \text{ in.}}{\text{yd}}\right)^3 \times \frac{0.554 \text{ oz}}{\text{in.}^3} \times \frac{\text{lb}}{16 \text{ oz}} = 1{,}615.464 \text{ lb}$$

Step 6. Now check our work. The units are correct. The answer is in the same
ballpark as our guess, many hundreds of pounds. How many significant figures
should the answer have? Except for the 0.554 oz per in.^{3}, all the values were
defined (and have as many significant figures as we need); therefore, 0.554 is
our limiting value. After rounding to 1.62×10^3 lb of water, we accept the
answer.

Solved problems are easier to read when the units have not been cancelled. Beyond this point, we have not cancelled units, but you should continue to do so whenever you check a calculation.

Exercise 1.9

If a cubic inch of metal weighs 4.55 oz, how much will a cubic foot of the
metal weigh in pounds? (See Example 1.17 for conversions.)

Exercise 1.10

A bag contains 0.050 cubic yards of topsoil. If it is spread 0.25 in. thick,
how many square ft will a bag of topsoil cover?

Most chemical problems involve converting values having one kind of units
to values having another kind. With some exceptions, such as temperature
conversions, such calculations can be set up simply by doing the arithmetic on
the units themselves and letting the numbers follow along. Unit analysis is the
method of approaching problems this way. But unit analysis will not substitute
for understanding. Although unit analysis is a powerful tool to help us set up
problems, it falls short of being magic! It cannot think for us. We can use
the method as a valuable guide—especially to check answers—but we must not
yield to the temptation of letting it do the work for us. Our objective is to learn
chemistry, not merely a technique for automatic problem solving.

In the chapters that follow, we will encounter numerous problems, many
of which will be worked-out examples. Occasionally, the examples will use
the entire problem-solving sequence explained in Section 1.4. As we become

increasingly familiar with the technique, only parts of the sequence will appear. When you do your own problems, however, you should continue to use the entire method until you are thoroughly at home with problem solving and have developed your own approach.

1.6 Metric System of Units and SI

After studying this section, you should be able to:

- Identify five SI units of measurement.
- State, for any metric prefix, its corresponding numerical value.

The metric system is used exclusively in most countries, the outstanding exception being the United States. All scientists use the metric system.

The **metric system** of units originated in France and was officially adopted there in 1799. It was generally based on two units, the meter and the kilogram.

In 1960, the General Conference on Weights and Measures, an international organization that defines the units of the metric system, set up the system called the Système Internationale d'Unites, **SI,** in which certain of the metric units were to be used in preference to others, whenever convenient. SI is now used generally by most of the countries of the world and by scientists. SI defines seven independent basic units, five of which will become familiar to us: the meter (m), the kilogram (kg), the second (s), the kelvin (K), and the mole (mol). The SI units form the core of the larger metric system, in which the other units of measurement are derived primarily from the basic SI units.

Five SI units are meter, kilogram, second, Kelvin, and mole.

Unlike the English system, which is based on a series of units having no organized relation to one another (foot, pound, quart, and so on), the metric system relates the units of length, volume, and weight in an easily understood way. Moreover, the different sizes of units for each kind of measurement (length, for instance) are related to each other by simple powers of ten. Its structure and organization makes the metric system much easier to learn than the English system.

Examples of the use of prefixes in the metric system:

1 kilogram = 1,000 grams

1 centimeter = 0.01 meters

1 milliliter = 0.001 liters

To express larger or smaller quantities of any of the basic units, the metric system uses a series of prefixes that are added to the name of the unit. Because each prefix represents multiplication or division by a factor of ten, the metric system is entirely decimal in nature. Table 1.2 shows the names, symbols, and numerical values of the most commonly used metric prefixes.

In the strict use of SI, the basic units themselves, or units derived directly from those basic units, are to be used. The prefixes milli (0.001) and kilo (1000) are to be used whenever possible. In chemistry, however, some of the basic SI units are inconveniently large. In the laboratory, for example, there is wide use of the gram rather than the kilogram, because the gram is a smaller and more

Table 1.1	Some SI Units of Measurement	
Quantity	Name	Abbreviation
Length	meter	m
Mass	kilogram	kg
Temperature	Kelvin	K
Time	second	s
Amount of substance	mole	mol

Table 1.2   Some Metric Prefixes

Prefix	Symbol	Numerical Value
giga	G	1,000,000,000 or 10^9
mega	M	1,000,000 or 10^6
kilo	k	1,000 or 10^3
deca	da	10 or 10^1
deci	d	0.1 or 10^{-1}
centi	c	0.01 or 10^{-2}
milli	m	0.001 or 10^{-3}
micro	μ	0.000001 or 10^{-6}
nano	n	0.000000001 or 10^{-9}
pico	p	0.000000000001 or 10^{-12}

easily measured unit. Also, in chemistry the centimeter is generally used rather than the meter.

In the chapters that follow, we will have little occasion to make conversions between the English and metric systems; however, problems involving such conversions do appear in this chapter. The problems are included to familiarize you with the sizes of metric units and to provide practice in doing unit conversions.

We will use most of the SI units; however, chemists often use centimeters instead of meters and grams instead of kilograms.

1.7 Measurement of Length

After studying this section, you should be able to:

- Convert length units to other length units within the metric system.

The standard unit of length in the metric system is the **meter.** Until recently, the meter was defined by two marks placed on a platinum bar stored in Sèvres, France. The distance between the two marks was at first intended to be equal to 1/10,000,000 of the distance from either pole of the earth to the equator. Recently, however, older definitions for many of the metric units have been replaced with more sophisticated and precise ones. For example, the new definition of the meter is the distance light travels through a vacuum in 1/299,792,458 of a second. Some useful metric units of length are listed in Table 1.3.

Table 1.3   Units of Length

Unit	Abbreviation	Equivalent in Meters
Kilometer	km	1,000 m
Meter	m	1 m
Decimeter	d	0.1 m
Centimeter	cm	0.01 m
Millimeter	mm	0.001 m
Micrometer (micron)	μm	0.000001 m or 10^{-6} m
Nanometer	nm	0.000000001 m or 10^{-9} m
Angstrom	Å	0.0000000001 m or 10^{-10} m
Picometer	pm	0.000000000001 m or 10^{-12} m

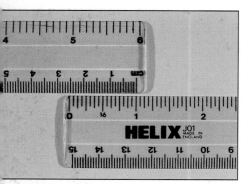

Figure 1.2 These two scales show the relationship of centimeters and millimeters to inches. One inch equals 2.54 centimeters.

A meter is approximately 39.37 in., which is, of course, a little more than a yard in the English system. One thousand meters is a **kilometer,** about 0.62 miles. A convenient unit of length for working in the laboratory is the **centimeter.** There are 100 centimeters in a meter. One inch is 2.54 cm (Figure 1.2).

Other units frequently used in chemistry are the **millimeter,** mm, and the **angstrom,** Å. One thousand millimeters equal a meter. There are 10^{10} Å in a meter or 10^8 Å in a centimeter. A small atom is about 1 Å in diameter.

Conversion factors among the metric length units are

$$\frac{1000 \text{ m}}{\text{km}} \qquad \frac{100 \text{ cm}}{\text{m}} \qquad \frac{10^{10}\,\text{Å}}{\text{m}}$$

EXAMPLE 1.18

A typical track and field event is the 10 km run. How long is this in centimeters?

Solution Map: $\boxed{\text{km}} \longrightarrow \boxed{\text{m}} \longrightarrow \boxed{\text{cm}}$

Solution:

$$10 \text{ km} \times \frac{1000 \text{ m}}{\text{km}} \times \frac{100 \text{ cm}}{\text{m}} = 1 \times 10^6 \text{ cm}$$

> **Exercise 1.11**
>
> A small atom is about one Å in diameter. How many centimeters is this?

EXAMPLE 1.19

We are traveling in Europe and decide to buy some clothes. One of us has a neck measurement of $15\frac{1}{2}$ in., but the shirt collar sizes are given in centimeters. What collar size shall we ask for?

Solution Map: $\boxed{\text{in.}} \longrightarrow \boxed{\text{cm}}$

Solution:

$$15.5 \text{ in.} \times \frac{2.54 \text{ cm}}{\text{in.}} = 39.4 \text{ cm}$$

We'll try a 39-cm collar and see if it will fit.

> **Exercise 1.12**
>
> You are buying a cut glass vase that measures 290 mm high. Will it fit on your bookshelf at home, where the shelves are spaced one foot apart?

Exercise 1.13

The earth is approximately 25,000 mi in circumference. How many meters is this? Kilometers?

1.8 **Units of Volume**

After studying this section, you should be able to:

● Convert from volume units to other volume units within the metric system.

Although the standard unit of volume in SI is the cubic meter, m^3, a more convenient unit for use in the laboratory is the **cubic centimeter**, cm^3. A cubic meter contains one million cm^3.

Another more widely used unit of volume is the **liter**, L, which is 1,000 cm^3. The liter is roughly equal to the English quart; 1 qt = 0.946 L.

A liter contains 1000 **milliliters**, and because the milliliter is the same as a cubic centimeter, the two are used interchangeably.

Two useful metric-to-metric conversion factors for volume are

$$\frac{1,000 \text{ mL}}{L} \qquad \frac{1,000 \text{ cm}^3}{L}$$

This shows a quart jar and a liter flask. The quart is slightly smaller. One quart is 0.946 liters.

EXAMPLE 1.20

How many L are there in 1.000 m^3?

Solution Map:

$$\boxed{\text{cubic meters}} \longrightarrow \boxed{\text{cubic centimeters}} \longrightarrow \boxed{\text{liters}}$$
$$\boxed{m^3} \longrightarrow \boxed{cm^3} \longrightarrow \boxed{L}$$

Solution:

$$1.000 \text{ m}^3 \times \left(\frac{100 \text{ cm}}{m}\right)^3 \times \frac{L}{1000 \text{ cm}^3} = 1.000 \times 10^3 \text{ L}$$

Notice that it was necessary to raise the factor, 100 cm/m, to the third power and that it was taken as being a defined term, with as many significant figures as needed.

EXAMPLE 1.21

You are camping your way around Europe. You are carrying a 2.0-liter water bottle. At a sidewalk cafe, the price of lemonade is one German mark per 5.0 deciliter. How many marks will it cost to fill your jug with lemonade?

Solution Map: $\boxed{L} \longrightarrow \boxed{dL} \longrightarrow \boxed{\text{marks}}$

Solution:

$$2.0 \text{ L} \times \frac{10 \text{ dL}}{L} \times \frac{\text{mark}}{5.0 \text{ dL}} = 4.0 \text{ marks}$$

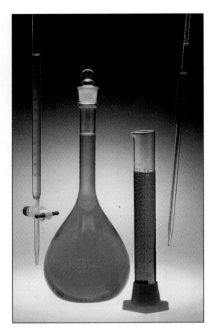

Figure 1.3 Examples of laboratory glassware: (from left to right) a buret, a volumetric flask, a graduated cylinder, and a pipet.

EXAMPLE 1.22

Your friend has a tin box that holds cookies. The box is 3.3 in. × 2.5 in. × 4.0 in. deep. How many deciliters of lemonade will it hold?

Solution Map: $\boxed{\text{in.}^3} \longrightarrow \boxed{\text{cm}^3} \longrightarrow \boxed{\text{L}} \longrightarrow \boxed{\text{dL}}$

Solution:

$$3.3 \text{ in.} \times 2.5 \text{ in.} \times 4.0 \text{ in.} \times \left(\frac{2.54 \text{ cm}}{\text{in.}}\right)^3 \times \frac{\text{L}}{1000 \text{ cm}^3} \times \frac{10 \text{ dL}}{\text{L}} = 5.40773 \text{ dL}$$

Round to 5.4 dL.

Exercise 1.14

Because you were unable to pay for the lemonade, the proprietor has put you to work. You are required to fill a large tank with water using your 2.0-L jug. The tank holds 0.75 m³. How many trips must you make with the jug?

Exercise 1.15

In Italy a soft drink costs about 1,400 lire for 5.0 dL. How many lire would you pay for a 12.0-oz can of the drink? (There are 32 oz in 1 qt.)

In the laboratory, the volumes of liquids are measured using special glassware (Figure 1.3). Graduated cylinders are used for rough measurements. Pipets and volumetric flasks designed to minimize error are used for more precise work. Volumetric glassware is made in various sizes, pipets are ordinarily used for amounts less than 25 mL and volumetric flasks for volumes from 50 mL to 1 or 2 L. A buret is a long, graduated tube with a valve at the bottom.

Figure 1.4 When you blow on a Ping-Pong ball and a golf ball, the Ping-Pong ball moves faster because it has less mass than the golf ball.

1.9 Mass and Weight

After studying this section, you should be able to:

- Distinguish between mass and weight.
- Convert mass units to other mass units within the metric system.

Mass is that property of matter that causes it to resist changes in motion; it is a measure of the amount of matter in a sample. To get an intuitive feel for mass, imagine a smooth tabletop with a Ping-Pong ball and a golf ball on it. You blow gently on each ball (Figure 1.4). Which ball moves faster? The Ping-Pong ball, of course. Why? Because it has much less mass than the golf ball and is therefore much easier to get into motion.

Table 1.4	Units of Mass	
Unit	Equivalent in Grams	Abbreviation
Kilogram	1000 g	kg
Gram	1 g	g
Decigram	0.1 g	dg
Centigram	0.01 g	cg
Milligram	0.001 g	mg
Microgram	0.000001 g	μg

The weight of Astronaut Edwin E. Aldrin, Jr. on the moon is less than it is on earth. His mass, however, stays the same.

Because of its mass, a sample of matter in motion tends to remain in motion, and a sample of matter at rest tends to stay at rest. The greater the mass of the sample, the more difficult it is to bring about a change in its motion. Ocean liners, moving only at a fast walk, have crunched into docks and adjoining buildings, slowly but irresistibly destroying them before finally coming to a stop. The huge mass of a large ship has an equally huge tendency to remain in motion.

A commonly used metric unit of mass is the **gram,** g. A gram equals the mass of 1 cm³ of water at a temperature of 4°C. Although the gram is commonly used in the laboratory, the SI unit of mass is the **kilogram,** kg, which equals 1,000 grams. The kilogram is defined as the mass of a block of metal stored in Sèvres, France. A useful unit for very small samples is the **milligram,** mg, which is 1/1000 of a gram. Here are some useful mass conversions.

$$\frac{1000 \text{ g}}{\text{kg}} \qquad \frac{1000 \text{ mg}}{\text{g}}$$

Imagine holding a heavy object in your hand. It is pressing down on your palm; gravity is causing it to exert a force downward. What we call the **weight** of an object is the force exerted on its mass by gravity. Although the mass of an object is independent of its location, its weight depends both on the amount of mass and on the strength of gravitational attraction where the object is located. For example, if the mass of a man is 80 kg on the earth, it would also be 80 kg on the moon. But because the gravitational attraction on the moon is about one sixth that of earth, his weight on the moon is far less than his weight on earth.

Balances provide a way to measure mass. A balance measures mass by comparing the weight of one object with the weight of another object of known mass. Because both objects are subject to the same gravitational field, if they have equal weights, they will have equal masses. Although what is actually being determined by a balance is mass, using the balance is often called *weighing,* and mass so determined is often called *weight.* Actually, one uses the balance to *mass* a sample, a term less often used but more correct.

For large samples or for weighings not requiring high precision, a double-beam pan balance is often employed. The object to be weighed is placed on the pan, and the moveable weights are positioned by hand until the instrument balances. The amounts of the weights can then be read. Triple-beam balances can be precise to about 0.01 g.

Weight and mass are often misunderstood. Mass *is a property of matter.* Weight, *strictly speaking, is a force.*

A triple-beam balance can be read to a precision of about 0.01 g.

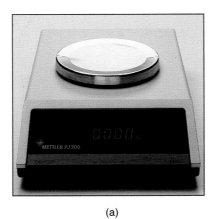

(a)

(b)

Figure 1.5 (a) An electronic top-loading balance has a digital readout to 0.001 g. (b) An electronic analytical balance has a digital readout to 0.0001 g.

For small samples or for measurements of greater precision, several kinds of balances are available (Figure 1.5), but the single-pan analytical balance is most often used. For measurements with this instrument, the operator places the sample on the pan, internal weights are positioned to balance it, and the final value is indicated by a digital readout. Balances like this are generally used for measurements in which the precision must be in the milligram range or slightly better. The machines are delicate and expensive. They do not have the capacity to handle large amounts.

The English-to-metric conversion that we will find most useful is

$$\frac{453.6 \text{ g}}{\text{lb}}$$

EXAMPLE 1.23

After riding a bicycle over a hundred miles of French roads, you step on a scale to weigh yourself. The French scale is marked in kilograms. The number you read is 47.7, but at home, you weighed 112 lb. Could you possibly have lost that much?

Solution Map: $\boxed{\text{kilograms}} \longrightarrow \boxed{\text{grams}} \longrightarrow \boxed{\text{pounds}}$

Solution: Your weight in pounds is

$$47.7 \text{ kg} \times \frac{1000 \text{ g}}{\text{kg}} \times \frac{\text{lb}}{453.6 \text{ g}} = 105 \text{ lb}$$

You have lost only a few pounds.

Science in Action

In prehistoric days, primitive commerce probably relied mostly on counting objects, but even counting apparently developed quite slowly. Further steps in trading goods came later and began to involve crude measurements. The earliest method compared sizes of objects to parts of the body, often a foot, from which our present unit evolved. Primitive weighing needed only the use of the hands to hold an object on each hand and estimate the relative weights.

Even so, the idea of weighing machines, balances, and weights had originated long before. As early as 2,000 BC, Egyptians used several kinds of balances in trade and construction. Egyptian weights were ceramic disks shaped somewhat like cupcakes.

To give the weight of an object is to communicate information about its mass. But to have any general use, that information must be given in terms of a unit that has widespread acceptance. The idea of standardizing weights apparently first emerged in Babylonia about 1,000 BC. Centuries later, the shekel was an accepted weight, although different varieties of shekels had different uses. Later still, the Romans introduced and carefully monitored a single weight unit, the libra, which was used for all purposes. The Romans used weights as small as 0.2 gram, indicating that there existed balances capable of weighing to at least that precision. Some evidence exists that the Arab societies of only a few centuries later were able to standardize their coins to weights of less than a milligram.

Only during the last two centuries, however, has routine weighing reached high precision. The importance of careful weighing in science became evident after the experiments of Lavoisier in the 1770's. Lavoisier was also closely associated with the beginnings of the metric system, which became the legal system of weights and measures in France some 40 years after his death.

By the beginning of the 1900's, analytical balances, capable of reproducible weighing to milligram precision or better, were commercially available. These beautifully crafted instruments were housed in glass cases to protect them from dirt and corrosion.

Today, one can buy at a reasonable price a single-pan automatic analytical balance of milligram precision. Using such an instrument requires only that the sample be placed on the pan. The weights, which are contained inside the balance case, work automatically, and the result is reported on an electronic screen.

Using specially built instruments, weighings have now been made to thousandths of a milligram. Even so, such precision is not nearly enough to weigh atoms. A thousandth of a milligram of water, for example, contains about a quintillion atoms.

Weighing and Balances

Until about the mid-1900's, analytical balances like this one were used widely to measure masses to a precision of about one milligram.

Exercise 1.16

At home you usually eat about half a pound of meat. When you go to the local French market, the meat is sold in units of hundreds of grams. About how much should you buy?

EXAMPLE 1.24

A tablet contains 500 milligrams of aspirin. How many tablets can be made from 4.0 kilograms of aspirin?

Solution Map: | kilograms | $\longrightarrow$ | grams | $\longrightarrow$ | milligrams | $\longrightarrow$ | tablets |

Solution:

$$4.0 \text{ kg} \times \frac{1000 \text{ g}}{\text{kg}} \times \frac{1000 \text{ mg}}{\text{g}} \times \frac{\text{tablet}}{500 \text{ mg}} = 8 \times 10^3 \text{ tablets}$$

Exercise 1.17

The corrosive compound sulfuric acid is sometimes sold in large, carefully protected jugs called carboys. A particular carboy holds 76 kilograms. How many grams is this?

1.10 Measurement of Temperature

After studying this section, you should be able to:

- Convert temperatures from one scale to another.

In the English system, the standard for measuring temperature is the **Fahrenheit** temperature scale. In our study of chemistry, however, we will use both the **Celsius** (formerly called Centigrade) scale and the **Kelvin**, or **absolute**, scale. The unit of both the Celsius and Fahrenheit scales is the **degree**, but a Celsius degree is not the same size as a Fahrenheit degree. The unit of the Kelvin scale is the same size as a degree on the Celsius scale and is called a **kelvin**. The abbreviations for these units are

$$°C = \text{degrees Celsius}$$

$$K = \text{kelvins}$$

$$°F = \text{degrees Fahrenheit}$$

On the Celsius scale, the freezing temperature of water under carefully defined conditions is assigned a value of 0°C, and the temperature of boiling water is similarly defined as 100°C. The interval between the freezing and boiling points of water, then, contains 100 degrees. A temperature that is 20/100 of the total difference between freezing and boiling, for instance, is 20°C. On the Celsius scale, temperatures below freezing are designated in the same way as those above freezing, but they are given negative signs (Figure 1.6).

20°C is approximately room temperature. Water that is so hot you can hardly put your hand in it is about 50°C.

The Fahrenheit scale is also arbitrarily defined, but with different values for the freezing and boiling points of water. Water freezes at 32°F and boils at 212°F. Thus while there are 100 Celsius degrees between the freezing and boiling points of water, there are $212 - 32$ or 180 Fahrenheit degrees between the same two points. Clearly, Fahrenheit degrees are smaller than Celsius degrees. More exactly, 100 Celsius degrees are the same as 180 Fahrenheit degrees; therefore a Celsius degree is $\frac{180}{100}$ or $\frac{9}{5}$ as big as a Fahrenheit degree. Using this information, we can derive mathematical formulas to convert from the Fahrenheit scale to the Celsius scale and vice versa.

At $-40°C$, the Fahrenheit and Celsius temperatures have the same value: $-40°C = -40°F$.

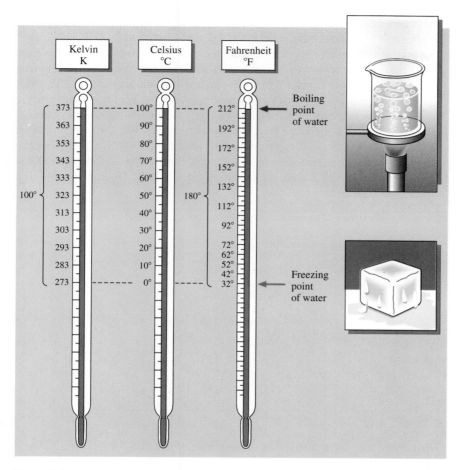

Figure 1.6 The Kelvin, Celsius, and Fahrenheit temperature scales.

$$\text{Fahrenheit temperature} = 32 + \left(\frac{9}{5} \times \text{Celsius temperature}\right)$$

$$\text{Celsius temperature} = \frac{5}{9} \times (\text{Fahrenheit temperature} - 32)$$

In the first equation, converting from Celsius temperature to Fahrenheit temperature, start at the freezing point of water on the Fahrenheit scale, 32°F. Next, convert the number of Celsius degrees above freezing (actually the Celsius temperature, since that scale puts freezing at zero) to Fahrenheit-sized degrees ($\frac{9}{5} \times$ Celsius temperature). Now add thirty-two. This gives the Fahrenheit temperature.

In the second case, converting Fahrenheit temperature to Celsius temperature, reverse the process. Find how many degrees the Fahrenheit temperature is above freezing (Fahrenheit temperature -32). Now multiply that by $\frac{5}{9}$ to give the Celsius temperature. How do you know to use $\frac{5}{9}$ here instead of $\frac{9}{5}$? Since the

Celsius degrees are larger, there will be fewer of them; therefore multiply the number of Fahrenheit degrees by the fraction smaller than 1, $\frac{5}{9}$.

Because Fahrenheit temperatures are rarely used in science, Celsius-to-Fahrenheit conversions are not very useful to us, although practice in converting from one scale to the other is helpful in understanding how the scales work. The most common temperature conversion you will encounter in beginning chemistry is the change from Celsius to Kelvin and vice versa. The conversion is quite easy, because kelvins are the same size as Celsius degrees. The difference between the two scales is that the zero of the Kelvin scale lies at -273 °C. The freezing temperature of water therefore is 273 K, and the boiling temperature 373 K. To go from Celsius to Kelvin temperature, just add 273. To go from Kelvin to Celsius, subtract 273.

The process of adding or subtracting 273 changes a temperature from Celsius to Kelvin or vice versa. For this reason, the units do not cancel in conversions of this kind.

$$\text{Kelvin temperature} = \text{Celsius temperature} + 273$$

$$\text{Celsius temperature} = \text{Kelvin temperature} - 273$$

EXAMPLE 1.25

The normal temperature of the human body is 98.6°F or 37°C. What is the Kelvin equivalent of 37°C?

Solution:

$$37°C + 273 = 310 \text{ K}$$

EXAMPLE 1.26

What is the Celsius equivalent of a temperature of 258 K?

Solution:

$$258 \text{ K} - 273 = -15°C$$

Exercise 1.18

What are the Kelvin equivalents of 32°C and 178°C?

Exercise 1.19

What are the Celsius equivalents of 212 K and 298 K?

1.11 Density

After studying this section, you should be able to:

- Calculate the density of a substance given its mass and volume.
- Use density to convert from volume to mass and vice versa.

Profiles in Science

William Thomson, Lord Kelvin of Largs (1824–1907)

Born in Belfast, Ireland, William Thomson was astonishingly brilliant as a child. At age 10, he entered Glasgow University. Following his undergraduate work at Glasgow, he went to Cambridge University for advanced study, after which he returned to Glasgow as Professor of Natural Philosophy, a position he held for 53 years. For his contributions to science, Thomson was raised to the peerage, becoming Lord Kelvin of Largs in 1892.

At Glasgow, Kelvin organized and ran one of Britain's first adequately equipped physics laboratories, the place in which his fundamental research was carried out. Kelvin pursued two general fields of inquiry in physics, the theory of electromagnetism and the interaction of energy and atoms (thermodynamics). When only 28 years old, he proposed the concept of absolute zero. Although he is best known to chemists for the temperature scale named in his honor, he also introduced the law of conservation of energy, on which much of the study in physics was based in the late nineteenth and early twentieth centuries.

The ratio of the mass of a sample to its volume is called the **density**, *d*, of that sample. To obtain the density of a substance, divide the mass of the substance by its volume.

$$\text{density} = \frac{\text{mass}}{\text{volume}}$$

EXAMPLE 1.27

A can of motor oil contains 755 cm³ of oil. The mass of the oil is 733 grams. What is the density of the oil?

Solution: Divide the mass by the volume.

$$\text{density of oil} = \frac{733 \text{ g}}{755 \text{ cm}^3} = 0.971 \text{ g/cm}^3$$

Exercise 1.20

What is the density of an unknown substance, a sample of which has a mass of 32.4 g and a volume of 21.0 mL?

If equal-volume samples are taken from substances of different densities, the samples will have different masses. Density is one of the characteristic properties of matter and is one of the means of identifying substances. For example, a cube of gold measuring 10 cm on a side will weigh about 20 times as much as a cube of ice of exactly the same size (Figure 1.7), that is, the density of gold is about 20 times as great as that of ice.

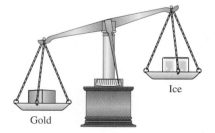

Figure 1.7 These two cubes are the same size, but because the gold is more dense than the ice, the cube of gold is much heavier than the ice cube.

Ping Pong balls

Golf balls

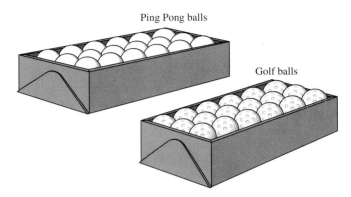

Figure 1.8   Even though the number of balls in each box is the same, the box of golf balls is heavier than the one with Ping-Pong balls because a golf ball is heavier than a Ping-Pong ball.

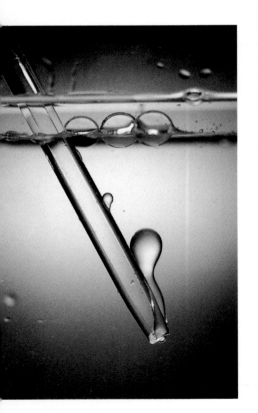

Figure 1.9   Because oil is less dense than water, oil floats on water.

The density of a substance depends on both the mass of the individual particles of the substance and the closeness with which those particles are packed together. Let's consider both effects. Suppose you have some shoe boxes, all of the same volume. Into one box you place some golf balls, each carefully packed in styrofoam (which, we will assume for our example, has no mass). One box will hold 15 golf balls. Into the other box you pack golf balls as tightly as you can, managing to put 45 golf balls into it. The second box will weigh three times as much as the first. Because the boxes have equal volumes, the ratio of mass to volume for the second box (its density) will be three times greater than that of the first. The fact that the balls are packed more closely into the second box causes the difference. Suppose you now take a third box and place 45 Ping-Pong balls in it. The third box will weigh far less than the second, even though it has the same number of balls. The density of the third box is less than that of the second because each ball in the third box has less mass than each ball in the second box (Figure 1.8).

The density of gold is greater than that of ice because the individual particles of gold are packed about twice as tightly as are those of water and because each particle of gold has about ten times the mass of a particle of water.

The most commonly used density units for liquids and solids are grams per cubic centimeter, g/cm^3. (The SI units of kilograms and cubic meters are inconvenient.) Because milliliters and cubic centimeters are the same, sometimes you will see density written in units of g/mL.

The density of liquid mercury at 20°C is 13.6 g/cm^3, a value indicating that mercury is more than 13 times as dense as water. If we were to place some mercury in a beaker and put some water in with it, the water would float on the surface of the mercury, just as oil, which is less dense than water, floats on water (Figure 1.9).

Density changes with temperature. Therefore it is important to specify the temperature at which the measurements were taken when quoting values of density. In writing density values, a subscript is used to identify the substance, and a superscript gives the Celsius temperature at which the measurement was made. The density of water at 4°C for example is written as

$$d_{H_2O}^{4°C} = 1.000 \ \frac{g}{cm^3}$$

At other temperatures, water has different densities. For example,

$$d_{H_2O}^{25°C} = 0.999 \ \frac{g}{cm^3} \qquad \textit{notation}$$

$$d_{H_2O}^{95°C} = 0.962 \ \frac{g}{cm^3}$$

Exercise 1.21

You are studying the way in which the density of water changes with temperature. You find that a sample of water weighing 32.7 grams has a volume of 33.1 mL at 50°C. What is the density of water at 50°C? Is this more or less than the density of water at 4°C?

Table 1.5 lists values for the densities of some substances. Notice that silicon dioxide, one of the more plentiful components of the earth's crust, has a density not quite three times that of water, whereas iron is about eight times as dense. The only density to memorize from this table is that of mercury, which we will use from time to time. Other density values can be looked up when necessary.

It is useful also to measure the densities of gases. Gases are not nearly so dense as liquids and solids. One or two grams of a gas will ordinarily occupy a liter or so of volume at room temperature and pressure; thus it is convenient to use the units g/L. Because the densities of gases depend on pressure as well as temperature (a property that we will examine in detail somewhat later), it is necessary when reporting gas densities to specify both the temperature and the pressure. Table 1.6 gives some sample gas densities, all taken at 0°C and sea-level atmospheric pressure.

Hot air balloons float on air, because the density of hot air is less than the density of cool air.

Table 1.5	Densities of Some Liquids and Solids at 20°C
Substance	Density (g/cm³)
Gold	19.32
Mercury	13.55
Lead	11.34
Iron	7.87
Carbon, diamond	3.51
Aluminum	2.70
Silicon dioxide, sand	2.65
Carbon, graphite	2.25
Water	0.998

The average density of the earth is about 5.5 g/cm³.

Table 1.6	Densities of Some Gases at 0°C and Sea Level Pressure
Gas	Density (g/L)
Freon	6.11
Chlorine	3.17
Carbon dioxide	1.963
Oxygen	1.43
Air	1.29
Methane	0.714
Helium	0.18
Hydrogen	0.090

The density of a substance does not depend on the size of the sample. Metallic iron has a density of 7.87 g/cm³. A sample of iron as large as a peach will weigh about a kilogram, and its density will be 7.87 g/cm³. An ingot of iron at a foundry will weigh several metric tons, but its density will nevertheless be 7.87 g/cm³. Mass and density are often confused. Meaningless statements such as "lead is heavier than wood" (meaning, of course, that lead is more dense than wood) are common. Be careful.

Density calculations often appear in chemistry problem solving. Although many laboratory techniques depend on measurement of volume, many calculations involve mass. It is therefore necessary to convert from mass to volume and the reverse. Density, used as a conversion factor, is the key to making such calculations.

EXAMPLE 1.28

You wish to buy enough honey to fill your honey jar, which holds 460 mL. Honey costs $2.25 per kg and has a density of 1.44 g/cm³. How much will it cost to fill your jar?

Solution Map: $\boxed{\text{mL}} \longrightarrow \boxed{\text{cm}^3} \longrightarrow \boxed{\text{g}} \longrightarrow \boxed{\text{kg}} \longrightarrow \boxed{\$}$

Solution:

$$460 \text{ mL} \times \frac{\text{cm}^3}{\text{mL}} \times \frac{1.44 \text{ g}}{\text{cm}^3} \times \frac{\text{kg}}{1000 \text{ g}} \times \frac{2.25 \text{ dollars}}{\text{kg}} = 1.5 \text{ dollars}$$

> **Exercise 1.22**
>
> A rectangular piece of copper metal measures 3.5 cm by 7.8 cm and weighs 12.74 grams. The density of copper is 8.92 g/cm³. What is the thickness of the copper in centimeters?

> **Exercise 1.23**
>
> A box contains 3.31 kg of salt, the density of which is 2.16 g/cm³. What is the volume of the salt?

Summary

Quantitative measurement is essential to science, but measurements are meaningless unless results can be reliably communicated. Because there is no such thing as an exact measurement, reports of results must contain some indication of the precision of the measurements themselves. The precision of a set of data is often expressed numerically through the use of significant figures.

Part of the language of scientific communication is the system of units by which numerical values are related to one another. Science uses the metric system of units, which contains the subset called SI. Chemists often use some of the SI units along with others from the metric system. Reports of measured values must always be accompanied by the units in which the measurements were made.

In chemistry, as in other sciences, it is necessary to make mathematical calculations using values from experimental measurements. Such calculations can be approached in an organized way by using a series of conversion factors or "unit factors." It is possible to use the units of the factors as an aid in setting up and checking the calculations.

A good method of approaching a problem is to follow these steps:

1. *Read and understand the problem.*
2. *Write conversion factors for the statements given.*
3. *Make a plan, or solution map, for working the problem.*
4. *Make sensible approximations when appropriate.*
5. *Set up the calculation and do the arithmetic.*
6. *Check your answer thoroughly.*

In science, temperature is usually reported in units of degrees Celsius or in kelvins. Celsius temperatures can be converted to Kelvin temperatures by adding 273 to the Celsius value; the reverse is accomplished by subtracting 273 from the Kelvin temperature.

Density, a property of matter, is the ratio of the mass of a sample to its volume (example: grams/liter). Calculations in chemistry often involve density.

Key Terms

values	*(p. 10)*	cubic centimeter	*(p. 31)*
units	*(p. 10)*	liter	*(p. 31)*
significant figures	*(p. 11)*	milliliter	*(p. 31)*
rounding	*(p. 12)*	mass	*(p. 32)*
scientific notation	*(p. 15)*	gram	*(p. 33)*
unit analysis	*(p. 21)*	kilogram	*(p. 33)*
conversion factor	*(p. 21)*	milligram	*(p. 33)*
metric system	*(p. 28)*	weight	*(p. 33)*
SI	*(p. 28)*	Fahrenheit scale	*(p. 36)*
meter	*(p. 29)*	Celsius scale	*(p. 36)*
kilometer	*(p. 30)*	Kelvin scale	*(p. 36)*
centimeter	*(p. 30)*	degree	*(p. 36)*
millimeter	*(p. 30)*	kelvin	*(p. 36)*
angstrom	*(p. 30)*	density	*(p. 39)*

Questions and Problems

Section 1.1 Measurement and Uncertainty

Questions

1. Is it ever possible to make a measurement without error? Is it ever possible to make a measurement without error and know that there is no error? Explain.

Section 1.2 Significant Figures

Questions

2. Why is it important to report the result of a measurement to the correct number of significant figures?

3. All digits in a numerical value are considered to be significant, subject to several rules pertaining to zeros. Explain why zeros may or may not be significant.

4. List the rules for rounding.

Problems

5. How many significant figures are there in each of the following numbers: 225.0; 1000; 0.0003210; 0.000312; 1,000,000.; 0.121000; 2.00001?

6. How many significant figures are there in each of the following numbers: 41.7; 22.150; 0.0030; 3000; 1,000,000.1; 300.0?

7. Perform the indicated arithmetic operations, and round the results to the appropriate number of significant figures.
   (a) 77.981×2.33
   (b) 4×0.0665
   (c) $(4.1 - 0.0093) \times (0.21 + 0.19)$
   (d) $\dfrac{88.7}{32}$
   (e) $\dfrac{22.414}{0.082 \times 273}$

8. Perform the indicated arithmetic operations, and round the results to the appropriate number of significant figures.
   (a) $(4001 + 0.043 + 3.3) \times (0.0001)$
   (b) $11.31 + 453.2 + 0.22 + 0.0003 + 7.7$
   (c) $\dfrac{441.332}{0.224}$
   (d) $\dfrac{24}{6.002} \times (0.000056 + 35.742)$
   (e) $\left(\dfrac{319.61}{35.67}\right) + 123$

Section 1.3 Scientific Notation

Questions

9. Why was the system of scientific notation developed?

10. In terms of significant figures, how can scientific notation help report the correct number?

Problems

11. Put the following numbers into scientific notation: 0.225; 2.5; 44163; 2,000,019; 0.00000321.

12. Put the following numbers into scientific notation: 0.44; 415.32; 2,115; 502,331; 0.00000000991.

13. Convert the following fractions into scientific notation, rounding to the appropriate number of significant figures.

   $$\frac{12}{335} \qquad \frac{441}{2.3} \qquad \frac{11}{33 \times 215} \qquad \frac{515 + 1{,}032}{0.0011}$$

14. Convert the following fractions into scientific notation, rounding to the appropriate number of significant figures.

   $$\frac{9}{33} \qquad \frac{29}{0.11} \qquad \frac{52}{188 \times 117} \qquad \frac{3122 \times 665}{0.00055}$$

15. Perform the following calculations. Put the answers into scientific notation, rounding to the appropriate number of significant figures.
   (a) $(1.15 \times 10^{-3}) \times (2.4 \times 10^{5})$
   (b) $(3.5 \times 10^{-7}) \times (2.55 \times 10^{-3})$
   (c) $\dfrac{2.15 \times 10^{3}}{9.93 \times 10^{5}}$

(d) $\dfrac{1.11 \times 10^{-4}}{5.32 \times 10^{7}}$

(e) $\dfrac{2.24 \times 7.63}{5.2 \times (8.21 \times 10^{-1}) \times 273}$

16. Perform the following calculations. Put the answers into scientific notation, rounding to the appropriate number of significant figures.
   (a) $(3.32 \times 10^{2}) \times (1.1 \times 10^{-1})$
   (b) $(9.0 \times 10^{-28}) \times (6.0 \times 10^{23})$
   (c) $\dfrac{22.797}{6.0 \times 10^{23}}$
   (d) $\dfrac{4.14 \times 10^{-3}}{5.2 \times 10^{5}}$
   (e) $\dfrac{2.95 \times 10^{-7}}{1.64 \times 10^{-23}}$

Section 1.4 Solving Problems in Chemistry

Question

17. List the six steps used in the approach to problem solving.

Problems

18. How many pounds of chocolate chip cookie dough will you obtain from 5.0 lb of flour, 3.0 lb of white sugar, 3.0 lb of brown sugar, 8.0 lb of butter, 10.0 lb of chocolate chips, 0.03 lb of baking powder, and 0.02 lb of salt?

19. Assuming that the volumes are additive, how many gallons of solvent will result if you mix together the following liquids: 2.0 qt of heptane, 6.5 gal of hexane, 3.5 qt of pentane, and 1 tablespoon of methanol?

20. You are wrapping gifts for five children. If a package requires 8.0 ft of ribbon, a bow requires 2.0 ft, and the card is attached with 2.0 in. of ribbon, how many rolls of ribbon do you need for five packages? Each roll contains 54 yd.

Section 1.5 Using Unit Analysis to Solve Problems

Question

21. What are the mathematical relations (all the possible conversion factors) given in Problem 34 below?

Problems

22. If one box of grass seed covers 250 square ft, how many boxes will you need to plant one acre? One acre contains 43,560 square ft.

23. A recipe calls for $\frac{3}{4}$ cup of vinegar, but the cook has only a single measuring spoon, which holds 2 teaspoons. How many of these spoonfuls of vinegar will be required

if a cup holds 16 tablespoons, and a tablespoon holds 3 teaspoons?

24. A German shepherd eats five cups of dry food a day. A pound of food contains 10.5 cups. How much does it cost to feed the dog each day if a 50.0-lb bag costs $22.50?

25. An American college student is in Paris and has 550 German marks that she now wants to change into French francs. A franc equals 0.185 dollars, and 0.49 dollars equals 1.41 marks. Does she have enough money to stay two days in a hotel that costs 145 francs per day?

26. If a stopped car takes up 13.5 ft on a freeway, how many cars will there be in a 12.6-mi traffic jam on a four-lane freeway?

27. A mason can carry 12 bricks at one time. If four bricks cover a wall area of 0.22 square yd, how many trips up the ladder must the mason make to cover a panel that is 1.7 yd by 3.25 yd?

28. Here is another nonsense problem. Besides illustrating that unit analysis will work even on units with which you are not familiar, it provides practice in inverting conversion factors and raising them to powers.

    You have a fream, which is 21 nuds wide × 75 nuds long. A greep has 41 s, there are 17 glors to every nud, every square glor contains 22 snaff, but a snaff can be accomplished in 0.16 s. How many greeps are used in this fream?

    You can use unit analysis to do this apparently complex problem. Start by writing down the conversion factors. Here are the first two of them:

    $$\frac{17 \text{ glor}}{\text{nud}} \qquad \frac{22 \text{ snaf}}{\text{glor}^2}$$

    (See how the words and phrases "to every," "can be accomplished in," "has," and "contains" all become simple conversion factors.) Now take it from here.

29. Painted lines on an interstate highway are 2.5 ft long and 4.0 in. wide. One qt of paint covers 43 ft². How many lines can be painted with 15 gal of paint?

30. A contractor is constructing a concrete sidewalk that is to be exactly 4 ft wide, 30 yd long, and 4.0 in. thick. Each cubic yard of concrete contains 6.6 sacks of cement. How many sacks of cement will be required for the sidewalk?

31. How many quarts of paint will you have to buy to paint all four walls and the ceiling of a room that is 24 ft 9.25 in. long × 19 ft 4.75 in. wide and has a ceiling height of 8.00 ft? A quart of paint covers 43 ft² of surface. Paint is sold only by the quart.

32. A rectangular park measures 0.22 mi by 0.074 mi. How many acres are there in the park if an acre contains 43,560 square ft?

33. A brewery fills 8.2 million 12-oz cans of beer per can line per day. If there are ten can lines, how many gallons of beer will be canned in one day?

34. Approximately 26 million cupcakes are produced each year. One cupcake has 0.70 oz of cake and 1.30 oz of cream filling. How many pounds of sugar are used per year for cupcakes? Each pound of cake contains 4.0 oz of sugar, and each pound of filling contains 12.0 oz of sugar.

Section 1.6 Metric System of Units and SI

Questions

35. What is the relationship of the International System, SI, to the metric system?

36. Name five of the basic SI units. Which of these has an unusual characteristic that makes it different in kind from the others?

37. Without looking them up, name the metric prefixes that indicate multiplication and division by powers of ten, as listed in Table 1.2.

Section 1.7 Measurement of Length

Problems

38. Fill in the table below by expressing the values given in the units shown at the heads of the columns. The second number in the second line is given as an example. Fill in the empty spaces.

m	cm	mm	km	in.
2.25				
	153.7	1537		
		339		
				17
			33.876	

39. For each of the following, convert the given value into the units requested. Show your calculations for each in unit analysis form.
    (a) 2.3 km = mi
    (b) 4 ft 2.2 in. = cm
    (c) 223 Å = mm
    (d) 1.433 cm = Å
    (e) 4.11 m = in
    (f) 55.66 cm = ft
    (g) 44 mm = in.
    (h) 3.2 yd = m
    (i) 9.9 ft = m
    (j) 32.113 mi = km
    (k) 32 m = ft
    (l) 22 cm = in.

40. Your waist is 32.5 in., and you want to buy a pair of pants in Berlin. What size in centimeters should you try on?

41. The distance between Los Angeles and Denver is 1182 mi. How many kilometers is it?

42. The distance between the city of Munich and the village of Holzkirchen is 12.5 km. How many miles is it?

43. An airline is offering a roundtrip ticket to Hawaii from Los Angeles for $100, if you buy another roundtrip ticket for a distance greater than 550 km each way. Can you get the Hawaiian ticket if you fly from Los Angeles to Fresno, California, and back? The distance between those cities is 382 mi.

44. A chemist is 5 ft 6.5 in. tall. What is her height in cm? Write your own height in feet and inches, then convert it to centimeters.

45. Visible light, as well as ultraviolet, infrared, X-ray, and other radiation, is characterized by what is called wavelength. The wavelength of certain infrared light is 30 μm. How many meters is this? Millimeters?

46. Visible light has wavelengths between 3.75×10^{-7} m and 8.0×10^{-7} m. Can we see light with a wavelength of $5.5 \times 10^{-2} \mu$m?

47. X-rays have wavelengths between 180 Å and 500 Å. What are these wavelengths in centimeters and in millimeters?

48. A molecule is approximately 1.00 Å in length. How many molecules will fit end-to-end in a row 1.00 m long?

Section 1.8 Units of Volume

Question

49. Explain the relation between the length unit, meter, and the volume unit, liter. Does such a defined relation exist in the units of the English system?

Problems

50. Fill in the table below by expressing the values given in the units shown at the heads of the other columns.

Cubic Meters	Cubic Centimeters	Liters	Quarts
	435		
0.75331			
		1.005	
			2.99

51. For each of the following, convert the given value into the units requested. Show your calculations for each in unit analysis form.
    (a) 22.4 L = qt   (d) 2 dL = fl oz
    (b) 8 fl oz = mL   (e) 448 mL = qt
    (c) 432 mL = pt   (f) 9.9 gal = m^3

52. How many milliliters are in 4.76×10^{-2} L?

53. How many cubic centimeters are in 5.89×10^2 L?

54. How many liters equals 432 mL of solution?

55. A gas has a volume of 250 mL. What is its volume in liters?

56. A recipe requires 2.5 cups of milk. In Europe, milk is usually sold in 750-mL bottles. Will one bottle be enough milk for the recipe?

57. Will a hollow cube with sides equal to 0.500 m hold 8.5 L of solution?

58. A student health center will vaccinate 1,555 students. Each vaccine contains 10.0 mL of serum. Will 1.75 L of serum be enough to vaccinate all the students?

59. A sip of wine is about 2.0 mL. If a 1965 bottle of Laffite Rothschild is $335 and contains 750 mL, how much does just one sip cost?

60. In Germany you can buy a liter of lemonade for 2.00 German marks. In the United States you can buy 1.5 cups of a soft drink for $0.40. Which drink costs more per quart if one dollar is 1.90 German marks?

Section 1.9 Mass and Weight

Questions

61. Explain how mass differs from weight.

62. If mass and weight are different, how can it be that we measure mass by weighing?

63. Design a simple experiment by which you could measure mass in an environment free of gravity–in a manned satellite, for example.

Problems

64. Fill in the table below, by expressing the values given in the units shown at the heads of the columns.

Grams	Milligrams	Kilograms	Pounds
	4.32×10^6		
0.0059			
		7.52	
			2.643

65. For each of the following, convert the given value into the units requested. Show your calculations for each in unit analysis form.
    (a) 155 lb = kg   (d) 0.065 lb = mg
    (b) 44 kg = lb   (e) 14 g = oz
    (c) 3.10 lb = g   (f) 12 oz = g

66. How many mg of copper are in a copper sample weighing 2.75×10^{-3} kg?

67. You need 5.75×10^{-2} kg of the compound heptane. How many milligrams of heptane do you need?

68. For a particular chemical equation you need to calculate the mass of a gas in kilograms. What is 4.52×10^3 mg of gas in kilograms?

69. A chemist wishes to know how many grams of the compound sodium carbonate are contained in a 55-lb drum. Make the calculation.

70. You need 300.0 grams of lead chloride for a chemical reaction. Will a 12.0-oz jar of lead chloride be sufficient?

71. You have a French friend who weighs 65 kg. What is her weight in pounds? What is your weight in kilograms?

72. A mechanic uses a lift to raise automobiles so that he can work on them. The lift has a capacity of no more than 3,000 lb. A man drives in a German limousine that needs some work. His car might weigh more than the lift will raise. The owner's manual says that the car weighs 1,568 kg. One kilogram contains 2.205 lb. Will the lift work?

73. A pharmacist needs to fill five prescriptions of a drug. Each prescription is for 40 capsules, each containing 50.0 mg of drug. He has 2.5 grams of the drug. Will there be enough?

74. Rice in Japan costs 12 yen per kilogram. In the United States, Japanese-style rice costs about 32 cents per pound. If 145 yen equal one dollar, where is the rice more expensive?

75. In France, gold is sold by the gram. How much does a gram of gold cost in French francs if there are 4.25 francs in a dollar and if gold costs $400 per ounce? (There are 16.0 oz per lb.)

Section 1.10 Measurement of Temperature

Questions

76. Describe the differences in the sizes of the units in the Celsius, Kelvin, and Fahrenheit temperature scales.

77. Why are there 180 Fahrenheit degrees for every 100 Celsius degrees?

Problems

78. Calculate the following temperature conversions:
    (a) 331°C = K   (d) 11.4°C = K
    (b) 285 K = °C   (e) 520 K = °C
    (c) −221°C = K   (f) 33 K = °C

79. A gas at 328 K has what Celsius temperature?

80. Would an object at 280 K feel cold?

Section 1.11 Density

Questions

81. Explain the relation between the volume unit, cubic meter (m^3), and the mass unit, kilogram, in the metric system. Why must temperature be involved in this relation?

82. What are the two phenomena on which density depends?

Problems

83. Complete the table by performing the following density conversions.

g/cm^3	kg/m^3	lb/gal	g/L
1.000			
	0.331		
		6.002	
			2.55

84. A block of metal measures 3.3 cm × 0.51 cm × 19 cm. Its mass is 0.198662 kg. What is its density?

85. The density of pure sulfur is 2.07 gm/cm^3. What is the volume of a sample of sulfur with a mass of 1.44 kg?

86. Having finished his lemonade (Example 1.22), your friend wants to fill his cookie box (dimensions are 8.4 cm × 6.4 cm × 10.2 cm) with honey, the density of which is 1.44 g/cm^3 and which costs 5.00 German marks per kg. How much must he pay for his honey?

87. An alloy for making tire weights is prepared by melting and mixing 500 grams of lead (density 11.34 g/cm^3), with 45 grams of antimony (density 6.69 g/cm^3), and 39 grams of tin (density 7.31 g/cm^3). Assuming that there is no loss in the volume of the metals, calculate the volume of the finished alloy and its density.

88. An empty graduated cylinder has a mass of 44.113 g. Into it is put 19.6 cm^3 of a liquid. The mass of the cylinder with the liquid in it is 91.552 g. Give the density of the liquid, using the appropriate number of significant figures and the correct units.

89. The graduated cylinder of Problem 88 is cleaned, and into it is put 16.3 cm^3 of mercury. What is the mass of the cylinder with the mercury?

90. Does the saying, "A pint's a pound the world around," refer to a pint of any particular substance? Must it do so? How many pounds would a pint of mercury weigh?

More Difficult Problems

91. On the planet Xanthu a neb contains 3.21 morks, a bip equals a square neb, 4.576 curps have 8.43 bips, and 2.2 curps make a square blump. A space explorer discovers that a blump equals 0.42 mi on earth. How many inches are there in 5.3 morks?

92. A pet shop owner is planning to buy three tanks of goldfish, which he will keep for 31 days before selling. Each tank contains 144 fish. Each fish will eat exactly an ounce of fish food mixture every 83 days (that is, 1 oz of food per fish per 83 days or 1 oz food/83 day fish). To each ounce of fish food is added 0.000355 oz of an antibiotic called Xot to keep the fish healthy. How much antibiotic will the owner need?

    Solution Map:

    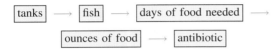

93. The state of Wyoming is 400 mi × 500 mi in size. If an average snowflake is 0.0000223 in. thick and 0.250 in.^{2} in area, how deep would the snow be (in feet) if 6.0 × 10^{23} snowflakes covered all of Wyoming? Would the snow be deep enough to ski on?

94. If each man will drink 7 oz of punch and each woman 5 oz, how many couples may be served by a punch made with the juice of three dozen oranges mixed with 1.10 gal of ginger ale, 0.5 lb of sugar, and one teaspoon of mint? A dozen oranges will make 1.7 pt of juice. A pound of sugar adds 0.33 pt of volume. One cup = 48 teaspoons, 1.00 pt = 16.0 oz, 1.00 gal = 128 oz, and 1.00 cup = 8.00 oz. Assume that 2% of the punch will be lost in mixing and pouring.

95. In a chemical reaction, 3.45 L of one solution reacts exactly with 8.56 L of another solution. If you have 150. mL of the first solution, how many milliliters of the second solution do you need for an exact reaction?

96. You need twice the weight of hydrochloric acid as of sodium hydroxide for an experiment. You have 4.75 × 10^{-2} grams of acid. How many milligrams of sodium hydroxide do you need?

97. You are going to run a chemical reaction in which you need 16 grams of oxygen gas for every 7.0 grams of nitrogen gas that will be used. If you have 0.554 kilograms of oxygen, how many milligrams of nitrogen do you need?

98. If 36 grams of water yield 4.0 grams of hydrogen gas after reaction, how many kilograms of hydrogen will you obtain from 4.35 × 10^4 milligrams of water?

99. A tall glass vessel with a rectangular cross section has inside dimensions of 15.2 cm by 22.9 cm and it is about a meter tall. Water is put into the vessel to a height of 55.0 cm. A jagged piece of rock weighing 5.21 kg is placed in the vessel; it sinks to the bottom, and the water level rises to 58.3 cm. What is the density of the rock?

100. A zeppelin, which rises by floating on air, contains 2.4 × 10^4 kg of helium gas in its gas bags. Why would it not be better to carry that same weight in lead and save space? How many liters of volume would be occupied by a piece of lead weighing 2.4 × 10^4 kg? How many liters are occupied by the helium? How many cubic feet? (See Tables 1.5 and 1.6 in Section 1.11 and the conversions earlier in the chapter.)

Solutions to Exercises

1.1 (a) 1.3, (b) 9, (c) 0.0020, (d) 0.22

1.2 (a) 447.69, (b) 578.7, (c) 43.451, (d) 253, (e) 3,690

1.3 79

1.4 (a) 0.000704, (b) 0.0520   (Notice that the final zero, which is significant, is kept.), (c) 803,000, (d) 100.3

1.5 $5 \text{ cups} \times \dfrac{4 \text{ teaspoons}}{\text{cup}} \times \dfrac{\text{tablespoon}}{3 \text{ teaspoons}} = 7 \text{ tablespoons}$

1.6 $0.242 \text{ mi} \times \dfrac{5,280 \text{ ft}}{\text{mi}} \times \dfrac{\text{yd}}{3 \text{ ft}} = 426 \text{ yd}$

1.7 $1.32 \text{ yd} \times 3.55 \text{ yd} \times 0.23 \text{ yd} \times \left(\dfrac{36 \text{ in.}}{\text{yd}}\right)^3 \times \dfrac{\text{gal}}{231 \text{ in.}^3} = 220 \text{ gal}$

1.8 $10.5 \text{ ft} \times 8.0 \text{ ft} \times \left(\dfrac{\text{yd}}{3 \text{ ft}}\right)^2 \times \dfrac{1 \text{ roll}}{4.0 \text{ yd}^2} = 2.3 \text{ rolls}$

1.9 $1 \text{ ft}^3 \times \left(\dfrac{12 \text{ in.}}{\text{ft}}\right)^3 \times \dfrac{4.55 \text{ oz}}{\text{in.}^3} \times \dfrac{\text{lb}}{16 \text{ oz}} = 491 \text{ lb}$

1.10 $1 \text{ bag} \times \dfrac{0.050 \text{ yd}^3}{\text{bag}} \times \left(\dfrac{3 \text{ ft}}{\text{yd}}\right)^3 \times \dfrac{12 \text{ in.}}{\text{ft}} \times \dfrac{1}{0.25 \text{ in.}} = 65 \text{ ft}^2$

1.11 $1 \text{ Å} \times \dfrac{\text{m}}{10^{10} \text{ Å}} \times \dfrac{100 \text{ cm}}{\text{m}} = 10^{-8} \text{ cm}$

1.12 $290 \text{ mm} \times \dfrac{\text{cm}}{10 \text{ mm}} \times \dfrac{\text{in.}}{2.54 \text{ cm}} \times \dfrac{\text{ft}}{12 \text{ in.}} = 0.95 \text{ ft}$

1.13 $25{,}000 \text{ mi} \times \dfrac{5{,}280 \text{ ft}}{\text{mi}} \times \dfrac{12 \text{ in.}}{\text{ft}} \times \dfrac{2.54 \text{ cm}}{\text{in.}} \times \dfrac{\text{m}}{100 \text{ cm}} = 4.0 \times 10^7 \text{ m}$

$4.0 \times 10^7 \text{ m} \times \dfrac{\text{km}}{1000 \text{ m}} = 4.0 \times 10^4 \text{ km}$

1.14 $0.75 \text{ m}^3 \times \left(\dfrac{100 \text{ cm}}{\text{m}}\right)^3 \times \dfrac{\text{L}}{1000 \text{ cm}^3} \times \dfrac{\text{trip}}{2.0 \text{ L}} = 3.8 \times 10^2 \text{ trips}$

1.15 $12.0 \text{ oz} \times \dfrac{\text{qt}}{32 \text{ oz}} \times \dfrac{0.946 \text{ L}}{\text{qt}} \times \dfrac{10 \text{ dL}}{\text{L}} \times \dfrac{1400 \text{ lire}}{5.0 \text{ dL}} = 9.9 \times 10^2 \text{ lire}$

1.16 $0.500 \text{ lb} \times \dfrac{453.6 \text{ g}}{\text{lb}} = 227 \text{ g}$ You should buy 200 g.

1.17 $76 \text{ kg} \times \dfrac{1000 \text{ g}}{\text{kg}} = 7.6 \times 10^4 \text{ g}$

1.18 $32°C + 273 = 305 \text{ K}; \ 178°C + 273 = 451 \text{ K}$

1.19 $212 \text{ K} - 273 = -61°C$ $298 \text{ K} - 273 = 25°C$

1.20 $\dfrac{32.4 \text{ g}}{21.0 \text{ mL}} = 1.54 \dfrac{\text{g}}{\text{mL}}$

1.21 $\dfrac{32.7 \text{ g}}{33.1 \text{ mL}} = 0.988 \dfrac{\text{g}}{\text{mL}};$ less

1.22 $12.74 \text{ g} \times \dfrac{\text{cm}^3}{8.92 \text{ g}} = 1.43 \text{ cm}^3$ $\dfrac{1.43 \text{ cm}^3}{3.5 \text{ cm} \times 7.8 \text{ cm}} = 0.052 \text{ cm}$

1.23 $3.31 \text{ kg} \times \dfrac{1000 \text{ g}}{\text{kg}} \times \dfrac{\text{cm}^3}{2.16 \text{ g}} = 1.53 \times 10^3 \text{ cm}^3$

The enormous energies required to lift the shuttle into space are provided by chemical reactions.

Matter and Energy

L ook around you. Everything you see and feel is made of matter. Chemistry is the study of matter. Chemistry is the study of matter and of the interactions—called chemical reactions—between samples of matter. Because chemical reactions usually involve energy changes, we must understand both matter and energy.

We are about to study two kinds of energy, stored energy (potential energy) and energy of motion (kinetic energy). We learn how energy can be changed from one form to another.

In this chapter we meet for the first time the natural law that says that neither matter nor energy can be created or destroyed, although each can be converted to the other.

We also learn a system by which matter can be classified, as well as the definitions of two important terms, element and compound. We will study some of the ways in which samples of matter can be identified and find out how to purify a sample of matter.

2.1 Matter

After studying this section, you should be able to:

- Distinguish between the states of matter, explaining how they differ from one another.
- Give examples of the different states of matter.

Matter is anything that occupies space and has mass. Mass, as we learned in Section 1.9, is that measurable quantity that causes matter to resist changes in motion. The amount of mass in a sample is often determined by weighing.

Although it appears to be continuous, matter is made of tiny particles called **atoms.** Atoms are so small that they can't be seen individually, even with microscopes. Chemistry investigates atoms and the ways in which they give rise to various characteristics of matter.

51

Table 2.1	Common Solids, Liquids, and Gases	
Solids	Liquids	Gases
Ice	Molasses	Methane (natural gas)
Sugar	Mercury	Air
Iron	Gasoline	Carbon dioxide
Aluminum	Alcohol	Steam
Carbon	Water	Oxygen

What are ordinarily called microscopes use light. Scanning tunnelling microscopes (described on p. 88), which do not use light, can make pictures that distinguish individual atoms. In reality, however, the atoms are not "seen."

Matter can exist in any of the three familiar states: solid, liquid, or gas. A **solid** has a definite volume and a definite shape. A **liquid** has a definite volume and takes the shape of the part of the container it occupies. On the other hand, a **gas** has neither definite shape nor volume; it will fill its entire container, pressing everywhere against the walls. Gases can be compressed into small volumes or expanded into large ones; liquids and solids, however, are practically incompressible. Table 2.1 lists some solids, liquids, and gases.

It is easy to think about solids and liquids as having mass and occupying space, but seeing gases that way is more difficult. Consider, however, the fact that moving air—wind—can pick up an entire house. What keeps the liquid shown in Figure 2.1 from collapsing into the bubbles under its surface? It is the gas in the bubbles occupying its own space. If you drop a 5-lb piece of dry ice (solid carbon dioxide) on your toe, you will surely notice that it has mass. And, when that dry ice vaporizes into carbon dioxide gas, the gas will retain all the mass it had as a solid.

Ice, a solid, has a definite shape. Liquid water takes the shape of its container.

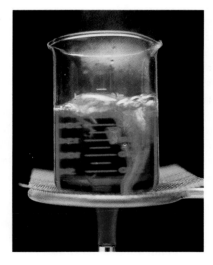

Figure 2.1   The gas in the bubbles prevents the liquid from collapsing into the bubbles.

A crystalline solid.

When heated, glass becomes softer and softer until it can be reshaped.

The atoms in solids are held close together in fixed locations by strong forces, causing the solids to have definite shapes. The individual particles in liquids are also close together, but the forces are weaker, so that the particles can move in relation to one another: That is why liquids can flow. The particles of gases are far apart and have no fixed positions relative to one another, as is illustrated by the ability of gas to expand to fill the space available.

A solid can be either crystalline or glassy. A **crystal** is a solid in which the atoms are arranged in an orderly way, according to a pattern. A **glass** is a solid in which they are not. Crystals and glasses differ in the way they melt. The melting of a crystal into a liquid occurs at a definite, characteristic temperature called the *melting point,* or *melting temperature*. When the crystal melts, there occurs a sharp transition in which the organized pattern of its atoms is lost. The change from a glass to a liquid, however, is gradual; there is no pattern to be lost. As glass is heated, it becomes softer and softer, until it flows. Because it has no definite melting point, a glassy solid may be simply thought of as a stiff liquid. Glass (even window glass) will change its shape (flow) over a long time. Crystals do not do this.

Most kinds of matter can exist in any of the three states, although we are usually familiar with only one, or possibly two, forms of most matter we encounter every day. The state of a particular kind of matter depends on its pressure and temperature. We know, for instance, that below a particular temperature water exists as ice, and above another it exists as the gas we call steam.

Table 2.2	Examples of Crystalline and Glassy Substances	
Crystalline	Glassy
Salt	Window glass
Sugar	Hard taffy
Diamond	Plastics
Ice	Rubber

2.2 **Energy**

After studying this section, you should be able to:

- Distinguish between kinetic and potential energy.
- Give examples of the conversion of energy from potential to kinetic and vice versa.

Energy is often defined as work or the ability to do work. Although an in-depth discussion of energy is beyond the scope of these chapters, we will explore different types of energy that will be useful in our understanding of chemistry.

When something is moving, it is said to have **velocity.** The velocity of an object is its speed in a particular direction. To change the velocity of a sample of matter, the object must undergo *acceleration*. Although you may ordinarily think of accelerating something as causing it to move faster, the term "acceleration" can be used in three senses: to increase velocity (positive acceleration), to decrease it (negative acceleration), or to change the direction of the motion.

An object in motion possesses **kinetic energy.** Kinetic energy, KE, is dependent on both the velocity and the mass of the object and is described by the equation

$$KE = \frac{1}{2}mv^2$$

where m represents the mass of the object and v represents the velocity.

Potential energy is the energy an object possesses due to its position in relation to objects exerting force on it. Suppose that you were to lift a brick from the floor and place it on a table. To lift it, you had to do some work on the brick to overcome the force of gravity, the force that the earth was exerting on the brick. That work was then stored in the brick because of its new position relative to the earth. Lifting the brick increased its potential energy. Energy that does not appear as kinetic energy is stored in the object as potential energy and can be recovered. As we will see, *chemical energy*, the energy released by chemical reaction, is a form of potential energy, as is nuclear energy.

Kinetic energy can be transformed into potential energy and vice versa. Look at the slingshot and rock in Figure 2.2. In the slingshot, potential energy is stored in the rubber bands as they are stretched. When the rubber bands are let go, they accelerate the rock upward, thus turning some of their potential energy into the kinetic energy of the rock in its vertical flight. As the rock rises, it continuously converts its kinetic energy into potential energy at its higher position. When the rock stops at the top of its flight, it has no kinetic energy left, but it soon gains some as the force of gravity accelerates it toward the earth's surface. The potential energy of the rock is converted to kinetic energy during its fall as the rock accelerates downward until it strikes the earth.

Although kinetic energy and the many kinds of potential energy are all associated with matter, there is also a form of pure energy that can exist in a vacuum. This is called *radiant energy,* one form of which is familiar to us as light.

acceleration

This climber is doing work against the force of gravity. The potential energy of the climber is being stored.

Potential and kinetic energy are easily converted, one to the other. Such conversions are happening constantly.

During the days of alchemy, energy and matter were confused with one another. The universe was thought to be made of four "elements," fire, air, earth, and water, but fire is really a chemical reaction involving energy.

2.3 Conservation of Mass and Energy

After studying this section, you should be able to:

- State the law of conservation of mass and the law of conservation of energy.

Two of the most important and widely known laws of science are the **law of conservation of mass** and the **law of conservation of energy.**

Occasionally we may think that mass has been destroyed. A large block of ice sitting on a sidewalk will melt, and eventually the water will evaporate. But the water has not been destroyed. It now exists in the form of a gas mixed with the air. It may some day be part of a rainstorm.

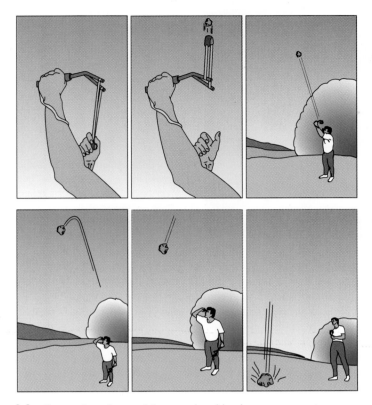

Figure 2.2   Conversion of potential energy into kinetic energy.

In the history of chemistry, chemical reaction was more or less a mystery until Antoine Lavoisier demonstrated that matter is conserved during reaction (Figure 2.3). The mass of the substances before reaction is always equal to the mass of the substances after reaction.

About 400 BC, Democritus said, "Nothing can be created out of nothing, nor can it be destroyed and returned to nothing." This was a statement of the law of conservation of mass, but it was pure speculation.

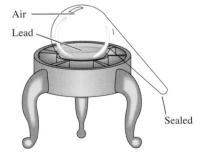

Air
Lead
Sealed

Figure 2.3   Lavoisier's experiment (about 1774 to 1785). In one of the earliest examples of quantitative chemistry, Antoine Lavoisier made a series of experiments demonstrating the law of conservation of mass and determining the proportion of oxygen in the air. In several experiments, he sealed samples of tin or lead into glass vessels, weighed the vessels, heated them strongly until no more reaction occurred between the metals and the air, and weighed them again. The results demonstrated that no weight gain accompanied the chemical reaction. When he opened the vessels, air rushed in, and the resulting weight gain was exactly the same as the weight increase of the metals when they turned to "calc" (their oxides). This indicated that the metals had combined with something in the air (what we now know as oxygen).

In 1905, Albert Einstein predicted that mass and energy could be interconverted and that they were related by the now famous equation $E = mc^2$. One such spectacular conversion occurs in the explosion of the nuclear bomb.

Our slingshot and rock system offers an example of the conservation of energy. Our muscles provided the energy to start the process. That energy next went into the rubber bands, then into the flying rock. When the rock hit the earth, it may have seemed as though the energy was lost, but it was not. The kinetic energy of the falling rock was converted mostly to heat both in the rock and in the earth near where it struck. A car moving down the street uses a large amount of energy, obtained from burning gasoline. Is that energy lost? No. Most of it heats the car, the atmosphere, and objects near it; a little is probably converted to various kinds of potential energy, but none of it is destroyed.

The law of conservation of energy was first proposed by Joule (see James Prescott Joule, p. 61) and was later firmly established by Lord Kelvin.

heat produced

We can combine the conservation laws into one statement:

✳ (*Neither mass nor energy can be created or destroyed.*)

Although under certain circumstances either mass or energy can be inter-converted, there are no known exceptions to the laws. Chemistry, as such, does not include processes in which mass and energy are detectably inter-converted; thus we will apply the laws as they are stated above.

The laws of conservation of mass and energy will form an important part of our studies to come. We rely on them in most chemical statements we make.

2.4 Temperature and Heat

After studying this section, you should be able to:

- Distinguish between heat and temperature.

As we discussed earlier, all matter is composed of individual particles (atoms) so small as to be invisible. In any sample of matter, these atoms constantly move in relation to one another, but that motion is neither visible to the eye, nor does it cause the sample itself to move. As we learned above, if particles are moving, they must possess kinetic energy.

In any sample of matter, some particles will have more or less kinetic energy than others. All the particles taken together, however, will have an *average* kinetic energy. The **temperature** of a sample is a measure of this average kinetic energy. The temperature of a sample rises when energy is added. When energy leaves a sample, the temperature drops.

Be careful not to confuse heat with temperature.

avg. KE = temp

56

Profiles in Science

Antoine Laurent Lavoisier (1743–1794)

Often called the Father of Modern Chemistry, Lavoisier was one of the first to realize the importance of careful analytical measurements in chemistry. Lavoisier's book, *New System of Chemistry*, established the modern concepts of element and compound, connected energy change with chemical reaction, and explained rusting of iron as a chemical reaction.

Lavoisier was a French aristocrat, and as such, during the reign of terror after the French Revolution, was automatically considered a traitor to the French people. He went to the guillotine in 1794, ending the career of the most promising scientist of the time. (One of Lavoisier's most important experiments is discussed on p. 55.)

Heat is equal to the amount of energy transferred from a sample of matter to another sample in contact with it. Heat will flow from a sample of higher temperature to one of lower temperature. For example, a piece of hot metal (a pan, for example) will transfer energy to cold water, not the other way around. Heat is measured in energy units.

2.5 **Measuring Energy**

After studying this section, you should be able to:

- Calculate the temperature change of water, given the mass of water and the number of calories.
- Calculate the number of calories, given the mass and temperature change of water.
- Convert from calories to joules and vice versa.

The calorie is a unit of energy that is easy to understand. <u>A **calorie** (cal) is</u> defined as the amount of energy required to <u>raise the temperature of one gram of water from a temperature of 14.5°C to 15.5°C.</u>

A kilocalorie is 1,000 calories. What nutritionists call a Calorie (with a capital C) is really a kilocalorie. A label on a chocolate bar states that it contains 290 Cal. However, the bar really contains 290,000 cal (small c), or enough energy to raise the temperature of 290,000 grams of water one degree Celsius. Starting at room temperature, that is enough energy to bring approximately three liters of water to the boiling point.

Temperature change is equal to $(T_f - T_i)$, where T_f is the final temperature at the end of the experiment and T_i is the initial temperature. For convenience, the temperature change is often written as $\triangle T$. That is, $\triangle T = (T_f - T_i)$. In the discussion and calculations that follow, the units on $\triangle T$ are given as "deg" rather than °C. Since kelvins are the same size as Celsius degrees, we can also use kelvins in the calculations.

Although the calorie is defined for a temperature change from 14.5°C to 15.5°C, the calorie is a good approximation for other temperature ranges.

EXAMPLE 2.1

Since one calorie is the amount of energy that changes the temperature of one gram of water by one degree, $\frac{cal}{g\ deg} = 1$. We can use this relation as a conversion factor in making calculations.

The average person burns about 2,000 Cal a day. If this amount of energy were used to raise the temperature of 50,000 grams of water, by how many Celsius degrees would the temperature increase?

Solution:

$$2,000\ \text{Cal} \times \frac{1,000\ \text{cal}}{\text{Cal}} \times \frac{\text{g deg}}{\text{cal}} \times \frac{1}{50,000\ \text{g}} = 40\ \text{deg}$$

Exercise 2.1

What will be the temperature change of 45 grams of water if 250 calories are used to raise its temperature?

EXAMPLE 2.2

An 82.6-gram sample of water increases in temperature from 4.0°C to 12°C. How many calories of energy must have been added?

Strategy and Solution:

Step 1.   Find the temperature change.

$$\triangle T = (T_f - T_i) = 12°C - 4.0°C = 8\ \text{deg}$$

The answer to Example 2.2 should have only one significant figure, therefore we round to 700 cal.

Step 2.   Set up and calculate.

$$8\ \text{deg} \times \frac{\text{cal}}{\text{g deg}} \times 82.6\ \text{g} = 700\ \text{cal}$$

Exercise 2.2

How many calories are required to raise 1,000 grams of water from room temperature, 25°C, to the boiling point, 100°C?

Although the calorie has been used for many years, the **joule,** J, is now used in scientific work. The relationship between the calorie and the joule is

$$4.184\ \text{J} = 1\ \text{cal}$$

Since the joule is so small, the **kilojoule,** 1,000 J, is the unit that is most handy. The kilojoule, kJ, is the SI unit of energy. Since 4.184 J = 1 cal, it is also true that 4.184 kJ must be the same as 1,000 cal. We use the conversion factor,

$$\frac{4.184\ \text{kJ}}{1,000\ \text{cal}} \quad \text{or} \quad \frac{\text{kJ}}{239\ \text{cal}}$$

EXAMPLE 2.3

One bite of a chocolate bar contains 27 Cal. How many kilojoules is this?

Solution:

$$27 \text{ Cal} \times \frac{1,000 \text{ cal}}{\text{Cal}} \times \frac{\text{kJ}}{239 \text{ cal}} = 1.1 \times 10^2 \text{ kJ}$$

Exercise 2.3

How many kilojoules will be necessary to raise 25.0 g of water from 0.0°C to 16.5°C? From 22°C to 63°C?

2.6 **Specific Heat**

After studying this section, you should be able to:

- Determine the specific heat of a substance, given the amount of energy lost or gained, the mass, and the temperature change.
- Calculate the third property, given any two of the following properties: temperature change, mass, and specific heat.

As we just learned, a calorie is defined as the amount of heat required to raise the temperature of one gram of water from 14.5°C to 15.5°C, i.e., one Celsius degree. If a calorie of energy is applied to one gram of aluminum, however, the temperature of the sample will rise 4.65 Celsius degrees. What's more, one calorie of energy will raise the temperature of one gram of uranium more than 36 Celsius degrees. As we can see, the same amount of energy produces different amounts of temperature change in samples of different substances.

Consider the following experiment. (You may want to try this.) Put an ounce of water (about 4 tablespoons) in a small, tightly closed plastic bag. Put it in the refrigerator. Beside it, put four quarters. The four coins have about the same mass as the water does. Let both substances get cold. Now, take out the coins and warm them in one closed hand. At the same time, warm the water bag in the other hand. Notice that warming the water takes much longer and leaves your hand cooler. Changing the temperature of an ounce of water takes far more energy than making the same change in an ounce of metal.

The amount of energy gained or lost by one gram of substance when its temperature changes by one Celsius degree is called the **specific heat** of the substance. Table 2.3 lists the specific heats of several substances. We will usually use the units kilojoules/(mass × deg) for specific heat.

The water is still cold after the quarters have become warm.

$$\text{specific heat} = \frac{\text{kilojoules added or taken away}}{\text{mass} \times \text{degrees of temperature lost or gained}}$$

Given any two of the following values, the specific heat definition can be used as a conversion factor to find the third: the number of grams of a substance that has its temperature changed, the amount of energy required, or the temperature change itself. The following examples show you how.

Table 2.3 Specific Heats of Some
Substances at 25°C

Substance	Specific Heat	
	kJ/(g deg)	cal/(g deg)
Aluminum	8.9×10^{-4}	0.21
Carbon (diamond)	5.0×10^{-4}	0.12
Magnesium	1.0×10^{-3}	0.24
Gold	1.3×10^{-4}	0.031
Water	4.18×10^{-3}	1.00
Ice*	2.1×10^{-3}	0.49
Benzene	1.8×10^{-3}	0.42
Ethyl alcohol	2.5×10^{-3}	0.59

*Taken at 0°C.

EXAMPLE 2.4

The specific heat of Al is

$$\frac{9.00 \times 10^{-4} \text{ kJ}}{\text{g deg}}$$

To a 13.1-gram sample of Al is added 3.08×10^{-2} kJ of energy. How many degrees does the temperature of the sample rise?

Solution Map:

$$\boxed{\text{kJ}} \longrightarrow \boxed{\text{g deg}} \longrightarrow \boxed{\text{deg}}$$

Solution: Invert the specific heat expression and use it as a conversion factor. This yields grams times degrees, which is then divided by grams (in this case multiply by 1/13.1 g) to give degrees.

$$3.08 \times 10^{-2} \text{ kJ} \times \frac{\text{g deg}}{9.00 \times 10^{-4} \text{ kJ}} \times \frac{1}{13.1 \text{ g}} = 2.61 \text{ deg}$$

Exercise 2.4

The specific heat of liquid water is 4.18×10^{-3} kJ/g deg. To a 13.1-gram sample of liquid water is added 3.08×10^{-2} kJ of energy. How many degrees does the temperature of the sample rise? (Compare your answer with that of the previous example. Why is there such a great difference?)

EXAMPLE 2.5

To a 19.5-gram sample of Li at 20.00°C is added 0.173 kJ of energy, resulting in a temperature rise to 22.50°C. What is the specific heat of Li?

Profiles in Science

James Prescott Joule (1818–1889)

James Joule was an English scientist who studied heat and its relation to matter. Heat was once thought to be some kind of substance and was called *caloric*. Heat (amounts of transferred energy) was measured by temperature changes produced in matter. Thus, the definition of the calorie was based on temperature changes.

Benjamin Thompson, Lord Rumford, by watching brass cannons being machined, discovered that mechanical energy and heat were related, that mechanical friction could raise the temperature of a body of matter. This effect was later studied quantitatively by James Joule, a member of a wealthy British brewing family. Joule had studied chemistry with the originator of the atomic theory, John Dalton, and had learned physics and mathematics on his own. Joule was interested in the relations of chemical, electrical, and mechanical energy and how transfers of that energy affected samples of matter. In 1843, he announced his findings, "the quantity of heat produced by the friction of bodies . . . is always proportional to the quantity of energy expended."

His measurements of the mechanical equivalent of heat (in today's language, that 4.18 Joules equal one calorie) were so important that our unit of mechanical energy is named in his honor.

Solution: Merely place the data in the specific heat equation.

$$\text{specific heat} = \frac{\text{kJ}}{\text{g} \times \triangle T}$$

$$\text{specific heat} = \frac{0.173 \text{ kJ}}{19.5 \text{ g} \times (22.50°\text{C} - 20.00°\text{C})} = 3.55 \times 10^{-3} \text{ kJ/g deg}$$

(Remember that we write $\triangle T$ as $T_f - T_i$.)

Exercise 2.5

If 4.00 kJ of energy is added to a 1.75-kilogram sample of the metal strontium, Sr, a rise in temperature from 18.51°C to 26.10°C results. What is the specific heat of Sr?

2.7 Classification of Matter

After studying this section, you should be able to:

- Give examples of heterogeneous and homogeneous mixtures.
- Explain the difference between physical processes and chemical reactions.
- Explain the difference between compounds and elements.

To help us understand it better, we'll look at a way of classifying matter. Figure 2.4 shows a way in which we can distinguish and describe different kinds of matter.

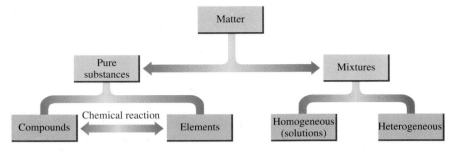

Figure 2.4   Classification of matter.

The Arab chemist Al-Razi (ninth century AD) classified all the known chemicals of that time. He used a system based on the origins of the substances—whether mineral, biological, and so on—giving characterizations that would not be very useful today.

First let's consider mixtures, of which there are two kinds, heterogeneous and homogeneous. **Heterogeneous** mixtures are composed of substances divided into distinct regions with definite boundaries. An example is a glass of ice cubes mixed with rock salt (Figure 2.5). In addition to the air between the pieces, the mixture contains two substances, ice and salt, and the boundaries of each piece are distinct. A heterogeneous mixture is composed of *phases,* each of which has a unique set of properties and distinct boundaries between it and the other phases. Although the mixture has many cubes of ice and of salt, it is said to have only two phases, the ice phase and the salt phase. (If the glass and air are to be included in what we define as the mixture, there are two more phases to consider.) Since phases are defined by sets of properties, a glass of ice and water (Figure 2.6) consists of two phases, even though both are water.

If we allow the ice in our glass of ice cubes and rock salt to melt and then we stir the mixture to dissolve the salt, the boundaries between the salt and the ice (now liquid water) disappear. Look at Figure 2.7. The resulting mixture is said to be **homogeneous,** or everywhere the same. Any sample taken from this mixture will have exactly the same composition as any other sample taken from the mixture. A homogeneous mixture is called a **solution.**

Figure 2.5   An example of a heterogeneous mixture. The ice cubes and rock salt are distinctly different regions of the mixture.

Figure 2.6   A heterogeneous mixture with two different phases (ice and liquid water) of the same substance.

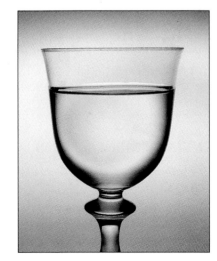

Figure 2.7   A glass of salt water is an example of a homogeneous mixture, a solution.

It is sometimes difficult to say whether a particular sample is heterogeneous or homogeneous. Milk, for instance, looks homogeneous but actually consists of two phases, one that is mostly water and one that is tiny globules of butterfat. The boundaries of the phases are distinct, but the fat particles are too small to be seen individually by the unaided eye.

Both heterogeneous and homogeneous mixtures can be separated into simpler substances by physical changes alone. Figure 2.8 shows that a mixture of sulfur and iron filings can be separated by using a magnet to attract the iron while leaving the sulfur. Or the mixture can be treated with carbon disulfide, a liquid that will dissolve the sulfur while leaving the iron behind. Both of these separations are simple physical processes. Pure salt and pure water can be obtained from a solution of salt and water by boiling the solution until all the water is converted to steam. Only solid salt will be left. The steam can be cooled to form water. In a **physical change,** matter may alter its shape, appearance, or state, but no substance is destroyed, nor is any new substance created.

Samples that cannot be separated into simpler substances by physical changes are called **pure substances** (or sometimes merely substances). A pure substance has a definite composition and a definite set of properties or characteristics.

A pure substance can be transformed into another substance by the process known as **chemical reaction.** The operational definition of chemical reaction is that substances (called **reactants**) change to form different substances (called **products**). For example, the gases hydrogen and oxygen combine during chemical reaction to produce water.

Energy changes often accompany chemical reaction. Frequently, energy is released. A chemical reaction that releases energy is said to be **exothermic.** A reaction that absorbs energy is called **endothermic.** An example of an exothermic reaction is the burning of gasoline with oxygen from the air, which produces the new substances water and carbon dioxide and releases energy.

Because substances are converted to other substances during chemical reaction, we must not imagine that any matter has been created or destroyed; it has merely changed arrangement. Moreover, because energy is released during reaction, we must not think that the energy was created; it was there all the time in the reactants as potential energy.

A pure substance that cannot be separated into simpler substances by either a physical process or a chemical reaction is called an **element.** A **compound** is a pure substance that is composed of elements. Chemical reactions can separate compounds into their elements, combine elements into compounds, or change compounds into other compounds. Table 2.4 gives examples of elements, compounds, and mixtures.

Figure 2.8 A heterogeneous mixture of sulfur and iron can be separated by using a magnet in a simple physical process.

Table 2.4 Examples of the Classes of Matter

Heterogeneous Mixtures	Solutions	Compounds	Elements
Salt and pepper	Gasoline	Water	Gold
Concrete	Wine	Sugar	Aluminum
Milk	Tea	Salt	Sulfur
This book	Brass	Alcohol	Oxygen
Mud	Air	Methane	Iron

2.8 The Elements

After studying this section, you should be able to:

- Explain the relation between elements and atoms.
- Name several elements and give their symbols.
- List typical properties of metals.
- List typical properties of nonmetals.
- Determine, from its properties, whether an element is a metal or a nonmetal.

The most abundant element in the universe is hydrogen.

A description of how "synthetic" elements are created appears in Chapter 3.

Atoms are made of smaller particles that we will discuss in a later chapter. With one exception the atoms of all the elements are composed of the same particles, none of which, themselves, have elemental properties.

Oxygen is the most abundant element in the earth's crust. Oxygen constitutes 20% of the earth's atmosphere and nearly half of the mass of the crust.

All substances in the universe, including those that make up the earth, the seas, the air, living creatures, the sun, and the most distant stars, are composed of elements or combinations of elements. In the entire history of science, only a few more than a hundred elements have been discovered; it is unlikely that many more will be found. The last few "discovered" were actually created in the laboratory.

If we take a sample of an element, such as gold, and divide it into smaller and smaller pieces, we would eventually come to a particle that we could no longer divide and still have gold. This very small particle is called an atom. An atom is the smallest particle of matter that can carry the distinguishing characteristics of an element; it is the smallest particle of any element that can enter into chemical reaction.

Atoms of nearly all the elements can combine with other atoms to form compounds held together by chemical bonds. There are only a few elements whose atoms do not form compounds. Many other elements, however, occur in nature only in the form of compounds. More than 99% of the earth and its contents is made from fewer than 20 elements, the other 85 elements being present either in small amounts or not at all. (See Table 2.5.) By far the most abundant element on earth is oxygen. The atmosphere is about 20% oxygen, but this is a small amount compared to the total amount of oxygen on the earth. Most oxygen exists in combination with other elements, in the rocks that form the earth's crust, and in the water of the oceans.

The elements were discovered one by one during many decades of experimentation and hard work. Because of their varied history, no system for naming the elements was ever established. A list of their names shows a colorful mix of Latin, Greek, German, English, and other languages; each name is meant to give some information about the element.

Table 2.5	Percentage of the Earth's Crust by Element (Percentages by Mass)
Element	Percentage
Oxygen	49.20
Silicon	25.67
Aluminum	7.50
Iron	4.71
Calcium	3.39
Sodium	2.63
Potassium	2.40
Magnesium	1.93
Hydrogen	0.87
Chlorine	0.19
Phosphorus	0.11
Manganese	0.09
Carbon	0.09
Sulfur	0.06
Barium	0.04
Fluorine	0.03
Nitrogen	0.03
Titanium	0.58
All others	0.47

Figure 2.9 Samples of two elements, mercury (left) and sulfur (right).

Profiles in Science

Joseph Priestley (1733–1804)

Priestley, a reclusive English chemist, was the first scientist to isolate oxygen. He obtained the oxygen by heating and decomposing mercuric oxide, HgO. This major discovery was only one of Priestley's contributions to chemistry, not the least of which were the careful and absolutely candid journals he kept. He wrote detailed descriptions of his experiments and their findings, not neglecting to discuss his mistakes and failures. In three of his six large volumes of research results, he discussed "Experiments and Observations on Different Kinds of Air" and reported valuable data from his experiments with many kinds of gases (he called them "airs"), including ammonia, nitric acid, hydrogen chloride, carbon dioxide, and nitrous oxide. Priestley was also the first to discover the connection between oxygen and breathing and the first to know that plants release oxygen when exposed to light.

When heated, red HgO decomposes to yield Hg and O_2.

Each element has its own symbol. Sometimes the symbols fit the names. For example, C for carbon, H for hydrogen, Cl for chlorine. Other symbols refer to the ancient Latin names and bear no resemblance to names used today; examples are Na for sodium (from natrium), Fe for iron (from ferrum), and Ag for silver (from argentum). Both the names and the symbols are used constantly in chemistry. Now is a good time to begin memorizing the names and symbols of elements most commonly encountered in chemistry. The list given in Table 2.6 is a good place to start.

Copper is named for Cyprus, where it was mined in ancient times.

Table 2.6 Names and Symbols of Some Important Elements

Metals		Nonmetals	
Name	*Symbol*	*Name*	*Symbol*
Aluminum	Al	Bromine	Br
Calcium	Ca	Carbon	C
Chromium	Cr	Chlorine	Cl
Copper	Cu	Hydrogen	H
Gold	Au	Helium	He
Iron	Fe	Iodine	I
Lead	Pb	Nitrogen	N
Magnesium	Mg	Neon	Ne
Potassium	K	Oxygen	O
Sodium	Na	Phosphorus	P
Silver	Ag	Sulfur	S

Exercise 2.6

Write the full names for the elements whose symbols are Mg, Mo, Mn, Ta, Ti, K, Na, Ag, Fe. (See the table inside the back cover of this textbook.)

Exercise 2.7

What are the symbols for barium, fluorine, silicon, and arsenic?

For reasons having to do with their atomic structure and their physical and chemical properties, the elements are often listed in an arrangement called the **periodic table** of the elements, shown in Figure 2.10. (See also the inside front cover of this textbook.)

Although we will later study the periodic table in depth, we can begin now to learn a few things about the elements and their properties. The periodic table can be divided into several areas, each of which contains elements of similar properties. On the left of the dark stairstep line are the **metals,** which are substances distinguished by their shiny appearance and their ability to conduct electricity and to be bent and hammered into different shapes without breaking. Metals are often found in compounds with the nonmetals, particularly oxygen

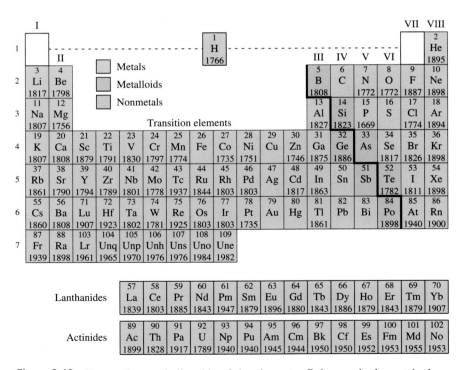

Figure 2.10 The modern periodic table of the elements. Below each element is the date of its discovery (elements with no dates have been known since ancient times). The color coding indicates whether the element is a metal, a nonmetal, or a metalloid.

Table 2.7   Properties of Metals and Nonmetals

Metals	Nonmetals
Shiny; reflect images	Dull surface
Conduct electricity	Nonconductors
Can be bent and hammered into different shapes	Brittle as solids
Conduct heat well	Poor conductors; many are insulators
React chemically mostly with nonmetals	React with metals and nonmetals

and sulfur. Much of the earth's crust is made of compounds called silicates, which are combinations of metals with silicon and oxygen. Most particularly, sodium and calcium, are found naturally only in the combined state, but the coinage metals, gold, silver, and copper, as well as a few others, do sometimes occur naturally in their elemental forms.

The **nonmetals,** such as fluorine, chlorine, oxygen, nitrogen, and carbon, are shown toward the upper right of the periodic table. The properties of nonmetals are distinctly different from those of metals. Nonmetals do not conduct electricity, shatter rather than bend, are dull in appearance when solid, or exist as gases or liquids. Most metals have high melting and boiling temperatures, whereas most nonmetals have relatively low melting and boiling temperatures (an exception is carbon, which has the highest melting point known). The nonmetals can also combine chemically with one another. Examples of such combinations are water, carbon dioxide, hydrogen chloride, and ammonia.

Not all metals are equally metallic, nor are all nonmetals alike. The extent to which an element displays the properties of its type depends roughly on its distance in the periodic table from fluorine, F, for the nonmetals, and from francium, Fr, for the metals. If we begin at the lower left of the periodic table, where the most metallic elements are, we find that the elements become generally less metallic toward the upper right corner.

Lying between metals and nonmetals in the periodic table are the **metalloids,** B, Si, Ge, As, Sb, Te, and Po, which combine many of the properties of both.

Some of the earliest samples of pure or nearly pure elements were metals. Gold may have been the first metal known to man and is known to have been in use in prehistoric times.

During the past few decades, silicon has become an important industrial metalloid because of its ability to conduct electricity. Integrated circuits, such as computer chips, are usually made of silicon.

Exercise 2.8

Classify the following elements as metals or nonmetals: Mg, S, Ne, Cs, V, C, P, I.

Most elements are solid at room temperature. Except for mercury, all metals are solids at room temperature, although a few are liquid at the temperature of

Science in Action

How the Element Sulfur Is Mined

Sulfur is a yellow solid.

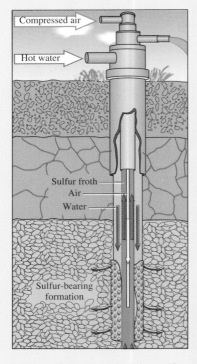

The Frasch process of mining sulfur.

The end of the nineteenth century saw a huge increase in the demand for sulfur, which had by then a variety of uses. At that time, most sulfur was mined by hand in Sicily and exported. A large deposit of sulfur-impregnated limestone was known to exist some 200 meters below a small island in the marshes of Louisiana, but repeated and expensive attempts to reach the sulfur had failed. Between the surface and the deposits were beds of quicksand saturated with poisonous hydrogen sulfide gas and layers of hard rock. Repeated attempts to penetrate this deadly barrier had resulted only in tragedy and lost resources. Consumers of sulfur were resigned to paying exorbitant prices for costly imports.

Late in the 19th century, Herman Frasch, an immigrant German chemist, was working for Standard Oil Company devising ways to make useful fuel out of oil contaminated with sulfur compounds that released corrosive gases when burned. Frasch became interested in the problem of the Louisiana sulfur deposits and began making theoretical calculations concerning a way to melt the sulfur below the ground and bring it up as a liquid. He devised, on paper, what is now known as the **Frasch process,** but there was no way to test the method in a laboratory. Only a full-scale attempt to mine sulfur would tell whether the process was usable.

Frasch proposed that the sulfur be melted in place by a mixture of superheated water and steam, which would be brought through a pipe installed in a well reaching the sulfur bed. At 119 C, the sulfur would melt and be pumped to the surface by a pump installed at the bottom of the pipe. Assigning to some backers a part of the rights to his process, he convinced them to gamble substantial funds on a test well. In 1894, when everything was finally ready, Frasch waited expectantly at the wellhead as the steam was turned on and the experiment begun. After a few minutes, the pump was started, and a deep brown stream of melted sulfur poured from the discharge pipe. This soon hardened into bright yellow crystals of nearly-pure sulfur, and Frasch was congratulated on his achievement. But the celebration was premature. After about four hours, the pump stopped, its shaft eaten away by the corrosive mixture. Then the sulfur solidified in the shaft. The well was plugged.

After the first temporary success, which essentially proved the method, it took six more years of repeated disappointments and increasing costs before Frasch finally achieved commercial production. Now Frasch wells produce a foamy mixture of air, steam, and sulfur. Much of the sulfur in the world, millions of tons, is now extracted by this process, but Frasch himself lost most of his gains in lawsuits involving his patents. He was nonetheless honored worldwide for his invention, without which much of the world's sulfur would be unobtainable.

Table 2.8	Elements that are Gases at Room Temperature
Hydrogen	Neon
Nitrogen	Argon
Oxygen	Krypton
Fluorine	Xenon
Chlorine	Radon
Helium	

boiling water. Only one other element, a nonmetal, is liquid under ordinary conditions: bromine, a brown, corrosive, poisonous substance. Eleven of the elements are gases at room temperature. These are listed in Table 2.8.

Exercise 2.9

Classify the elements listed here according to whether each is solid (s), liquid (l), or gas (g) in the natural state: Na, Sc, He, Kr, Br, I, Hg, H.

Mercury is the only metallic element that is a liquid at room temperature.

In some lists of elements, gallium, Ga, appears as a liquid. The melting point of gallium is about 30°C, so it will melt in your hand. In hot countries it is often a liquid.

2.9 The Law of Definite Proportions

After studying this section, you should be able to:

• State and explain the law of definite proportions.

When elements combine chemically to form compounds, they do not necessarily do so in the proportions in which they happen to be present, but, instead, according to specific natural laws that were discovered near the beginning of the nineteenth century. One such law, the **law of definite proportions,** was discovered experimentally in 1797. This law, also called the **law of constant composition,** states that *a pure sample of any compound, no matter what its origin, always contains its elements in a fixed proportion by mass*. In a sample of cupric oxide, for example, the ratio of the mass of oxygen to that of copper is always found to be 1.0 to 4.0. According to the law of definite proportions, every sample of pure cupric oxide, no matter what its history, will contain that same ratio of oxygen to copper. Zinc iodide, another compound, is made from the elements zinc and iodine. On careful analysis, it is found that zinc iodide contains its elements in the mass ratio

The first experimentally supported statement of the law of definite proportions was made by Joseph Louis Proust (1754 to 1826). In his pioneering analytical measurements, Proust also developed data that later led John Dalton to discover the law of multiple proportions.

fixed proportion

$$\frac{3.882 \text{ g iodine}}{1.000 \text{ g zinc}}$$

This means that in zinc iodide there are 3.882 grams of iodine for every 1.000 gram of zinc. The law of definite proportions tells us that all samples of zinc iodide will exhibit exactly this same mass ratio. Think about this. Any sample of zinc iodide from any source whatever has the same mass ratio, here given to four significant figures. A substance that has a different zinc and iodine mass ratio is not zinc iodide. The law holds not only for zinc iodide, but for every other compound. Every compound holds its elements in a particular, characteristic, fixed proportion by mass.

2.10 Compounds and Their Formulas

After studying this section, you should be able to:

- Distinguish between a compound and an element.
- Name the elements that are diatomic in the natural state.
- Name the elements and determine the number of atoms of each element, given a chemical formula.
- Determine the simplest formula of a compound, given the molecular formula.

We will be able to define the nature of compounds even better after we have studied the atomic theory in Chapter 3.

Because the constant composition of compounds is the inevitable result of chemical combination, we can use it to refine our definition of the term "compound." A compound is a chemical combination of two or more elements in fixed proportions by mass.

Some compounds are composed of individual units that are called **molecules.** With a few important exceptions, the word "molecule" is used to mean a particle capable of independent existence, composed of two or more atoms held together by a chemical bond, or bonds. A molecule can have atoms of the same element; each nitrogen molecule, for instance, is made up of two nitrogen atoms. Or a molecule can include atoms of two or more elements. For example, each molecule of ethyl alcohol contains six hydrogen atoms, two carbon atoms, and one oxygen atom.

As can be seen in Table 2.9, several elements exist in their natural states as **diatomic molecules,** molecules having two atoms. Except for I_2, a solid, and Br_2, a liquid, the diatomic elements are all gases. A few other elements, such as S_8 and P_4, are also molecular in their natural states.

Not all compounds, however, are necessarily molecular. A sample of sodium chloride (table salt), for instance, contains no individually existing particles whatever—no molecules. Sodium chloride pairs have strong attractions for other sodiums and chlorides, so that a single sodium chloride "molecule" can exist independently only under extremely unusual conditions and for a very short time. In any single piece of sodium chloride, large numbers of what are called *ions* of sodium and of chloride are stuck together in an organized pattern forming the solid compound. An ion may be an atom or a group of atoms, but

Table 2.9   The Diatomic Elements

Element	Molecular Formula	Natural State
Hydrogen	H_2	Colorless gas
Nitrogen	N_2	Colorless gas
Oxygen	O_2	Colorless gas
Fluorine	F_2	Pale yellow gas
Chlorine	Cl_2	Pale green gas
Bromine	Br_2	Red-brown liquid
Iodine	I_2	Purple-black solid

Chlorine, a gas (left). Bromine, a liquid (middle). Iodine, a solid (right).

it has electrical properties that ordinary atoms and molecules do not have. We will study ions in detail in Section 6.1.

Whether they are molecular or not, compounds always contain two or more elements in fixed proportion by mass.

We have learned that there are only about a hundred elements, and that only a third of those are plentiful. Even so, as we look around us, we see enormous numbers of different kinds of substances, far more than a thousand different kinds. Because only a few of these can be elements, it is obvious that most of the matter in the world consists of compounds. There are, in fact, more than six million different compounds known and no obvious limit on those yet to be discovered.

To identify the millions of known compounds, chemists use formulas. The **chemical formula** of a compound is a combination of symbols which indicate (1) which elements are in the compound and (2) the relative numbers of atoms of each element in the compound. Other kinds of formulas are used to show how many atoms of each element there are in an individual molecule, how the atoms are attached to one another, and even the physical positions of the atoms in space. Here is a set of rules that will help you write chemical formulas

1. *The formula includes the symbols of all the elements that appear in the substance.* Example: Water is made of hydrogen and oxygen. The symbols for both these elements appear in the formula for water, H_2O.

2. *The numbers of atoms of each kind are indicated by subscripts following the symbols for those atoms.* If the formula contains only one atom of an element, no subscript is used. Example: The formula for ammonia, NH_3, tells us that there are three hydrogen atoms and one nitrogen atom in the molecule.

3. *Sometimes atoms appear in groups.* If more than one such group appears in a formula, the group is shown in parentheses and the number of times it appears

$C_{12}H_{22}O_{11}$

This is the formula of sucrose, ordinary table sugar. It contains the symbols for its elements, carbon, hydrogen, and oxygen, along with the number of atoms of each in the sugar molecule.

is shown as a subscript to the right of the end parenthesis. Example: The formula for barium nitrate, $Ba(NO_3)_2$, contains two of the groups known as a nitrate ion. (As we continue, we will gradually learn the names and characteristics of various groups.)

EXAMPLE 2.6

In the formula $(NH_4)_3PO_4$, name the elements and the number of atoms of each element.

Solution: No subscript appears after P, so only one phosphorus atom is present. The subscript 4 indicates that there are four atoms of oxygen in the formula. The subscript 3 beside the parenthesis means that we must multiply all the numbers within the parentheses by three. We know then that there are 3 nitrogen atoms and 12 hydrogen atoms in the formula.

Exercise 2.10

For the following formulas, name the elements and say how many atoms of each appear: MgO, Na_2SO_4, $Al_2(SO_4)_3$.

Let's investigate formulas a little further. As we know, some compounds form substances that have no individual molecules. The formula for such a compound takes no notice of that; it merely tells us the *relative* numbers of the different kinds of atoms. For example, a sample of barium chloride contains twice as many atoms of chlorine as of barium. That ratio is reflected in the formula $BaCl_2$. There are, however, no individual molecules of $BaCl_2$, nor need there be for the formula to be written. Even if molecules exist in a compound, formulas are sometimes written to indicate only the relative numbers of atoms present, not the numbers of atoms actually present in the molecules. Such formulas are called **empirical formulas** or **simplest formulas.**

Many formulas, however, give more information. If a compound happens to be made up of molecules, its formula can tell us how many atoms of each element are actually in each molecule. For example, CH is the empirical formula of a compound called acetylene, but the acetylene molecule actually contains two carbon atoms and two hydrogen atoms; its molecular formula is C_2H_2. The **molecular formula** is always a multiple of the empirical formula, the multiplying factor being one, two, three, or another whole number, usually small.

Exercise 2.11

Give the empirical formula for the compounds with the following molecular formulas: H_2O_2, C_3H_6, and $C_4H_6O_2$.

reduce to simplest ratio

2.11 **Physical and Chemical Properties**

After studying this section, you should be able to:

- Name and describe some common methods of separating mixtures and suggest appropriate purification methods for particular kinds of mixtures.
- Explain how chemical reactions differ from physical processes and give examples of each.

Any pure substance, compound or element, can be identified by its properties, its characteristics. Every substance has a set of properties that differ from the set of properties of another substance.

Certain kinds of properties, called **physical properties,** can be measured without chemically destroying any of the sample. Some familiar physical properties are color, hardness, and density. Other physical properties may not be as easy to observe but are still characteristic. Compressibility and electrical conductivity are examples.

Other kinds of properties, those that determine how a substance will react chemically, are called **chemical properties.** Whether a substance will burn or whether bleach will fade it are examples of chemical properties.

Different pure substances may have some properties in common, but not all; each compound and element has its own unique set. Table 2.10 gives some examples of the properties by which some common substances can be identified. As your study of chemistry progresses, such properties will become familiar to you.

Table 2.10 Some Physical and Chemical Properties*

Substance	Physical Properties	Chemical Properties
Iron	Hard, high melting point, conducts electricity and heat, shiny	Liberates hydrogen from acids, rusts, forms colored ions
Sulfur	Soft, low melting point, nonconductor, dull yellow color	Burns readily, forms gaseous acidic oxide
Oxygen	Gaseous, odorless, tasteless, colorless	Supports combustion, combines with most other elements
Water	Colorless liquid, low melting point, good solvent	Can be decomposed to oxygen and hydrogen
Table salt	Colorless solid, high melting point, nonconductor as solid (but liquid salt conducts), brittle	Fairly inert, can be decomposed to metallic sodium and gaseous chlorine

* For the purposes of our discussion, *low melting point* refers to melting temperatures one or two hundred degrees Celsius above room temperature. Many substances with high melting points must be heated to such high temperatures that the substance glows before melting.

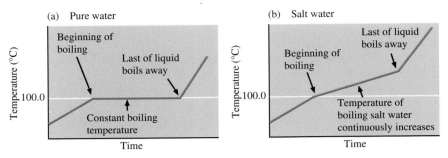

Figure 2.11   The effect of dissolved substances on boiling temperature. While water is boiling, the temperature remains constant. When a solution of salt and water is boiled (b), the temperature of the liquid rises steadily as it boils away.

One of the first compounds to be systematically characterized was antimony trichloride, described by Paracelsus and others about 1600 AD.

The properties of mixtures are usually combinations of the properties of the pure substances in them. To characterize any particular sample, you must first determine whether it is a mixture or a pure substance, and, if it is a mixture, separate the components. *extraction of elements*

Physical methods can be used to distinguish pure substances from mixtures. Pure substances will go through complete changes of phase at constant temperature; few mixtures will do so. A sample of pure water, once boiling, will boil entirely away without changing temperature. Once water has started freezing, it will freeze completely at the same temperature. The temperature of salt water, however, will rise continuously as the water changes to steam during the boiling process, as illustrated in Figure 2.11. Similarly, during freezing, the temperature of a solution will continuously decrease. Although there exist a few solutions that can maintain constant temperature during boiling, even these will change temperature while freezing.

Apart from simple mechanical methods such as sifting, the separation of substances from one another (that is, their purification) is usually accomplished by a change of state. The components of liquid solutions can usually be separated by means of **vaporization** in a process called **distillation,** shown in Figure 2.12. (Vaporization simply changes a liquid into a gas.) Distillation takes advantage of the fact that at the distillation temperature, one of the substances in the solution vaporizes more readily than the others, thus bringing about a separation.

A classic example of distillation is the separation of alcohol from other substances during the preparation of certain beverages. If a solution of alcohol in water is boiled, the first samples of gas will be richer in alcohol than was the original solution because alcohol vaporizes more readily than water. The gas is then captured and condensed into a liquid richer in alcohol than the original liquid. The process can be repeated until alcohol of 95% purity (190 proof) is obtained, but few beverage makers go beyond about 50% (100 proof).

Distillation is used on a large scale to remove salt from seawater. Making this process sufficiently economical to provide water for domestic use and irrigation in desert countries has required considerable engineering, but large plants are presently in operation in several parts of the world. Small-scale, sun-powered distillation apparatus can now be found in lifeboats, where it can supply drinking water to people who might otherwise perish.

In the ninth century AD, Arab alchemists purified several elements and compounds, unaware of their exact nature. The techniques used were distillation and redistillation, filtration, roasting, and sublimation, among several others.

Sublimation is similar to distillation, except that the change is from solid to vapor rather than from liquid to vapor. Several substances, sulfur and iodine for instance, are purified industrially by sublimation (Figure 2.13).

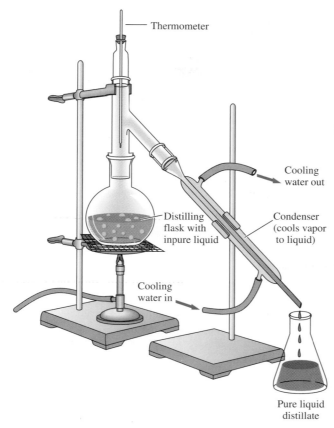

Thermometer

Cooling
water out

Distilling
flask with
inpure liquid

Condenser
(cools vapor
to liquid)

Cooling
water in

Pure liquid
distillate

Figure 2.12 A distillation apparatus. The liquid is vaporized and then condensed and collected in the flask.

Commercially, sugar is separated from water and impurities by means of *fractional crystallization*. A hot, concentrated sugar solution is allowed to cool slowly; crystals of pure sugar are formed, leaving the other substances behind. The preparation of colorless table sugar from ordinary brown sugar is a simple crystallization experiment often performed in beginning chemistry laboratories. This is an experiment that is easy to do at home; ask your instructor how to do it.

In other separation procedures, the change of state may be less obvious. A method called *solution chromatography* takes advantage of the fact that different materials, such as paper, glass, and clay, attract different substances with varying degrees of strength. When a solution is poured through the chromatography material, the various solution components move through at different rates and come out separately at the other side. A simple demonstration of this method may be done even without a laboratory. Ordinary black ink usually consists of colored dyes mixed together. You can separate the dyes if you place a tiny drop of ink near the bottom of a piece of filter paper (perhaps from a coffee maker) or blotting paper, put 3 or 4 mm of water in a pan, then prop the paper up so that its bottom edge (but not the spot) is in the water. As the water rises up the paper drawing the ink with it, the different dye components move at different

Figure 2.13 Sublimation. A mixture of sand and crystals of the element iodine is heated. The iodine sublimes (changes directly into a gas), forming crystals of the pure element when cooled.

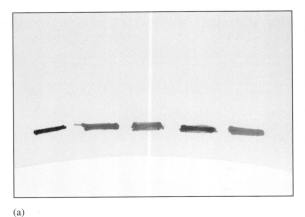
(a)

(b)

(c)

Figure 2.14 Chromatography can separate dyes from a colored felt-tip pen. (a) Ink marks on absorbent paper. (b) Water in the bottom of the beaker rises through the paper, carrying the dyes with it. (c) The finished chromatogram.

rates, causing them to appear as separate bands of color. Figure 2.14 illustrates the process.

Laboratory applications of chromatography use not only paper or coated glass plates but also columns filled with various materials such as aluminum oxide or even powdered brick. Chromatography is now used routinely to purify many substances once thought impossible to separate.

Separations are often made by taking advantage of the fact that one substance will dissolve in a particular liquid while another will not. As shown in Figure 2.15, a solid mixture of blue copper nitrate and yellow cadmium sulfide is added to water. The copper nitrate dissolves entirely, making a blue solution, while

(a)

(b)

(c)

Figure 2.15 (a) A solid mixture of copper nitrate (blue) and cadmium sulfide (yellow) is added to water. The copper nitrate dissolves entirely, making a blue solution, while the cadmium sulfide remains almost entirely undissolved as the yellow solid. (b) The mixture is filtered, leaving the cadmium sulfide on the filter paper. (c) The water is ...lowed to evaporate, giving nearly pure crystals of copper nitrate.

the cadmium sulfide remains almost entirely undissolved as the yellow solid. The mixture is filtered, and the water is allowed to evaporate, giving nearly pure crystals of copper nitrate.

Summary

Chemistry is concerned with matter, something that has mass and occupies space. Mass is what causes matter to experience gravitational attraction and to resist changes in motion. A sample of matter can be accelerated; when this happens, the sample gains or loses kinetic energy. Energy is work or the ability to do work.

Matter can possess potential energy (stored energy) due to its position. Although neither mass nor energy can be created or destroyed, each can be transformed into the other. Such transformation does not occur to a noticeable extent in chemical processes; what is more likely to occur in chemical reaction is the transformation of energy from one form to another and the transformation of matter from one kind to another.

Matter can be classified as pure substances, called elements and compounds, or mixtures of these. Mixtures can be separated into their component substances by physical processes such as distillation; compounds can be separated into elements by chemical reaction. Elements, however, cannot be separated into simpler substances by either physical processes or chemical reaction.

Matter is composed either of elements or of combinations of elements called compounds. As illustrated by the periodic table, the elements are classified as metals, nonmetals, or metalloids. Compounds are characterized by their formulas, which give the relative numbers of the atoms of the different elements in the compound.

Key Terms

matter	*(p. 51)*	homogeneous	*(p. 62)*
atom	*(p. 51)*	solution	*(p. 62)*
solid	*(p. 52)*	physical change	*(p. 63)*
liquid	*(p. 52)*	pure substance	*(p. 63)*
gas	*(p. 52)*	chemical reaction	*(p. 63)*
crystal	*(p. 53)*	reactant	*(p. 63)*
glass	*(p. 53)*	product	*(p. 63)*
energy	*(p. 54)*	exothermic	*(p. 63)*
velocity	*(p. 54)*	endothermic	*(p. 63)*
kinetic energy	*(p. 54)*	element	*(p. 63)*
potential energy	*(p. 54)*	compound	*(p. 63)*
law of conservation of mass	*(p. 55)*	periodic table	*(p. 66)*
law of conservation of energy	*(p. 55)*	metal	*(p. 66)*
temperature	*(p. 56)*	nonmetal	*(p. 67)*
heat	*(p. 56)*	metalloid	*(p. 67)*
calorie	*(p. 57)*	law of definite proportions	*(p. 69)*
joule	*(p. 58)*	molecule	*(p. 70)*
kilojoule	*(p. 58)*	diatomic molecule	*(p. 70)*
specific heat	*(p. 59)*	chemical formula	*(p. 71)*
heterogeneous	*(p. 62)*	empirical formula	*(p. 72)*
phase	*(p. 62)*	molecular formula	*(p. 72)*

physical property   *(p. 73)*
chemical property   *(p. 73)*
vaporization   *(p. 74)*

sublimation   *(p. 74)*
distillation   *(p. 74)*

Questions and Problems

Section 2.1 Matter

Questions

1. Distinguish between the terms "liquid," "solid," and "gas."

2. It is said that in certain old buildings, there are window-panes that are slightly thicker at the bottom than at the top, although they were presumably uniform when installed. Is window glass a solid or a liquid? How could you find out?

3. Why can gases be compressed easily, whereas liquids and solids cannot?

Section 2.3 Conservation of Mass and Energy

Questions

4. A rock is thrown into the air, falls back down onto a flat roof, and stays there. Discuss the fate of all of the energy involved in the process from the moment the energy initially leaves the thrower's muscle.

5. A train makes an emergency stop and melts part of its wheels. Where did the energy to melt the metal come from? What finally happened to it?

6. The combination of gasoline and air has stored energy (a subject we will study in detail somewhat later). A car is put into motion using that energy. It goes up a hill; then the brakes are applied and the car stops. Has the energy been destroyed? If not, describe in detail what happened to it from the time it ran the engine until the car was stopped.

Section 2.4 Temperature and Heat

Questions

7. What is the difference between heat and temperature?

8. What is the difference between heat and energy?

9. Exactly what is it that temperature measures?

10. Explain, if you can, why heat flows from places of higher temperature to places of lower temperature.

Section 2.5 Measuring Energy

Problems

11. Calculate the number of calories and the number of kilojoules in 39 Cal.

12. How many calories are there in 1.2×10^{-2} kJ?

13. How many kilojoules will heat 250 mL of water from room temperature ($25°C$) to the boiling point?

14. How many calories will be necessary to raise 2.45 kg of water from $14.5°C$ to $15.5°C$?

15. A certain reaction produces enough heat to raise 55 mL of water from $14.5°C$ to $15.5°C$. Calculate the number of kilojoules produced by the reaction.

16. How many mg of water will rise in temperature from $14.5°C$ to $15.5°C$ on receiving 723 kJ of heat?

17. How many dL of water can you heat from room temperature ($25°C$) to the boiling point with 4.5×10^4 cal?

Section 2.6 Specific Heat

Questions

18. How does the definition of the calorie involve specific heat? Would the value of the calorie be different if a substance other than water were used in the definition? Why?

19. A pan made of copper is put on a warm stove and allowed to get hot. The pan is then cooled. Water is put into the pan until the mass of the pan with the water in it is twice that of the empty pan. The pan with water is then put on the same stove and heated to the same temperature as before. Does the process take about twice as long? Explain.

20. A battery transfers electric energy to a flashlight bulb. The light leaves the flashlight and falls on your book. What becomes of the energy, both the part that is reflected from the book and the part that is not?

Problems

21. The specific heat of the element phosphorus, P, is 7.94×10^{-4} kJ/(g deg). How many kilojoules of energy are

required to raise the temperature of a 3.75-g sample of P from 16°C to 29°C?

22. A sample of silver, Ag, weighing 10.0 g was heated from 25°C to 40°C. The energy required to accomplish this temperature change was 3.5×10^{-2} kJ. What is the specific heat of silver in this temperature range?

23. To a sample of benzene, C_6H_6, with a mass of 257 g and a temperature of 25°C, 7.15×10^{-2} kJ of energy is added. What is the new temperature of the sample? The specific heat of benzene is 1.8×10^{-5} kJ/(g deg).

24. A sample of lithium weighing 129 g and at a temperature of 0°C was allowed to warm to 23°C. In this process, it absorbed 0.105 kJ of energy. What is the specific heat of lithium?

Section 2.7 Classification of Matter

Questions

25. Expand Table 2.4 by giving three more examples of each of the four classes listed.

26. Remembering that milk is not a solution, name the phases present in a system consisting of a glass of milk with ice in it.

27. How many phases are in (a) a glass of iced tea; (b) a cloud that contains a hailstorm; (c) a steel cylinder filled with steel nuts and bolts covered with oil?

28. What are the differences between a solution and a pure substance? Between an element and a compound?

29. Matter can't be created or destroyed. Can phases? What happens to the number of phases when the ice melts in a glass of ice water?

Section 2.8 The Elements

Questions

30. Go through the list of elements inside the front cover of the book. List those whose symbols start with the first letters of their names, those whose symbols are the first two letters of their names, and those whose symbols bear little resemblance to their names. In a source book such as the *Handbook of Chemistry and Physics* (try a library), look up the background of the symbols belonging to the third group.

31. Although oxygen is the most abundant element on earth, the atmosphere, which is 20% oxygen, contains only a small fraction of the earth's oxygen. Where and in what form is most of the oxygen on earth?

32. List the properties of metals and nonmetals.

33. List the elements that exist as liquids and those that exist as gases.

Problems

34. Give the names of elements whose symbols follow: Mn, Ta, Ti, K.

35. Give the names of elements whose symbols follow: P, Ba, Pd, Sn.

36. Write the symbols of the following elements: strontium, rhodium, phosphorus, cerium, cesium.

37. Write the symbols of the following elements: gold, silver, iron, lead, cobalt.

38. Of the elements that follow, which are metals and which are nonmetals: Cr, Ba, Te, O, Ar, Kr, K, As, P?

39. Classify the following elements according to whether each is solid, liquid, or gas in the natural state: Na, Ti, B, C, Br, P, Rh, Cs, Cl, I, Si, As.

Section 2.9 The Law of Definite Proportions

Questions

40. Explain the law of definite proportions.

41. The law of definite proportions is true for compounds. Does it also hold for solutions? Explain.

Section 2.10 Compounds and Their Formulas

Questions

42. Write the names and formulas of the elements that exist as diatomic molecules.

43. Which of the following are diatomic: CO, He, H_2O, N_2O_5, Ar, HBr, BrCl, P_2O_5?

44. What is the primary purpose of a chemical formula? What is the difference between a symbol and a formula?

45. Explain the difference between an empirical formula and a molecular formula. Can a formula be both?

Problems

46. Name the elements present in each of the following compounds, and say how many atoms of each element appear in the formula for the compound: H_2SO_4, $NaNO_3$, $CaCl_2$, $COCl_2$, $AlBr_3$.

47. Name the elements present in each of the following compounds, and say how many atoms of each element appear in the formula for the compound: $(NH_4)_2SO_4$, CS_2, Cs_2S, $(NH_4)_3Fe(CN)_6$, C_2H_7SN.

48. Give the empirical formulas for the compounds that have the following molecular formulas: P_4O_{10}, N_2H_4, $C_4H_{12}N_2$, Hg_2Cl_2, $C_6H_{12}O_6$.

Section 2.11 Physical and Chemical Properties

Questions

49. Explain the difference between a physical and a chemical property.

50. Expand Table 2.10 by giving physical properties of three more substances. If you can, do this for the chemical properties as well.

51. In Lavoisier's experiment with mercuric oxide, Figure 2.3, were there phase changes? Explain. 、 ~ there a chemical reaction or merely a physical process? How do you know?

52. A glass of water is left on a table and the water evaporates. Has the water been destroyed? If not, exactly what has happened to it? Could it be recovered somehow? Give two examples of similar disappearances and discuss the possibility of recovery.

53. If water and ethyl alcohol are mixed and then distilled over and over, the ethyl alcohol may be obtained with less and less water in it until the solution contains only 5% water. Further distillation will make no further separation. Is the alcohol now a pure substance? How might you go about finding out?

54. From everyday life or industry, give three examples of separations of heterogeneous mixtures and three examples of separations of solutions.

55. In the process of making the alcoholic beverages called spirits or liquors, two methods of separation are used. Try to name them.

56. What method of separation would you use to obtain pure antifreeze, ethylene glycol, from the dilute solution in a car radiator?

57. What method of separation does an automobile engine use to remove particles of dirt from motor oil? What kind of a mixture is the dirty oil?

Solutions to Exercises

2.1
$$250 \text{ cal} \times \frac{g \text{ deg}}{\text{cal}} \times \frac{1}{45 \text{ g}} = 5.6 \text{ deg}$$

2.2
$$100°C - 25°C = 75 \text{ deg}$$

$$1{,}000 \text{ g} \times 75 \text{ deg} \times \frac{\text{cal}}{g \text{ deg}} = 75{,}000 \text{ cal}$$

2.3
$$25.0 \text{ g} \times \frac{1 \text{ cal}}{g \text{ deg}} \times \frac{kJ}{239 \text{ cal}} \times 16.5 \text{ deg} = 1.73 \text{ kJ}$$

$$25.0 \text{ g} \times \frac{1 \text{ cal}}{g \text{ deg}} \times \frac{kJ}{239 \text{ cal}} \times 41 \text{ deg} = 4.3 \text{ kJ}$$

2.4
$$3.08 \times 10^{-2} \text{ kJ} \times \frac{g \text{ deg}}{4.184 \times 10^{-3} \text{ kJ}} \times \frac{1}{13.1 \text{ g}} = 5.62 \times 10^{-1} \text{ deg}$$

The difference arises because the specific heat of water is much higher than that of Al.

2.5 Specific heat $= \dfrac{4.00 \text{ kJ}}{(1.75 \times 10^3 \text{ g}) \times (26.10°C - 18.51°C)} = 3.01 \times 10^{-4} \text{ kJ/g deg}$

2.6 magnesium, molybdenum, manganese, tantalum, titanium, potassium, sodium, silver, iron

2.7 Ba, F, Si, As

2.8 metal, nonmetal, nonmetal, metal, metal, nonmetal, nonmetal, nonmetal

2.9 S, S, G, G, L, S, L, G

2.10 1 magnesium and 1 oxygen; 2 sodium, 1 sulfur and 4 oxygen; 2 aluminum, 3 sulfur and 12 oxygen.

2.11 HO, CH_2, C_2H_3O

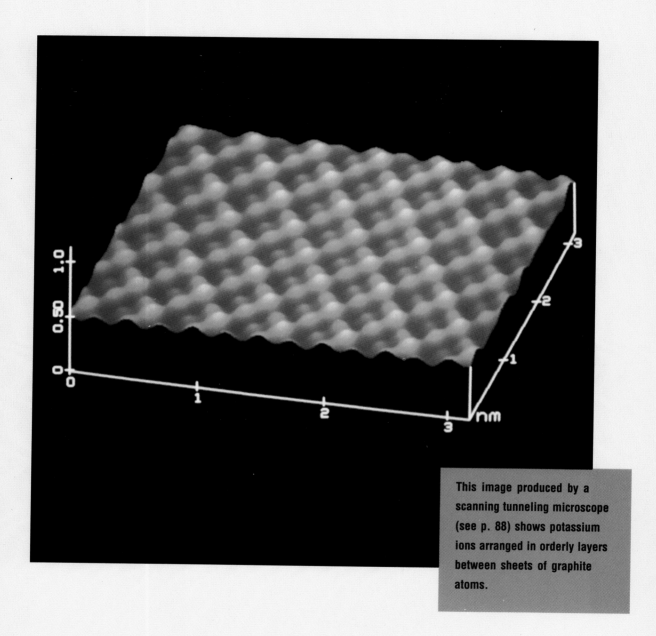

This image produced by a scanning tunneling microscope (see p. 88) shows potassium ions arranged in orderly layers between sheets of graphite atoms.

The Atomic Theory

Although we are learning numerous facts and theories about matter, nothing we have seen thus far can explain why different kinds of matter behave in different ways. Limestone is hard, acid corrodes, gasoline burns, sugar is sweet and so on. How is it possible that only slightly more than a hundred elements form millions of different substances displaying an enormous range of properties? In Chapter 3 we begin to explore this question by focusing on the nature of atoms.

What are atoms and what are they made of? For more than 2,000 years we have tried to answer these questions. No one has directly "seen" an atom; atoms are too small. But within the last 200 years we have accumulated enough experimental evidence that chemists unhesitatingly accept the existence of atoms and have begun to understand how atoms behave. Furthermore, scientists have discovered that atoms are made of smaller particles which determine many of the properties that atoms display. In this chapter we study the concept of atoms and the internal composition of atoms.

3.1 Early Views of the Nature of Atoms

Speculation about the ultimate basis of matter dates back to several Greek philosophers who lived about 400 BC. This early view of matter was a simple one. If you cut a sample of a pure substance such as gold into smaller and smaller pieces, you will eventually reach a particle that can no longer be divided and still be gold. To this smallest particle these philosophers applied the

keep cutting until can't

Although Democritus (460–370 BC) is credited with the first use of the word "atomos," Aristotle credits Leucippus (fifth century BC) with the original concept.

Greek word "atomos," meaning uncuttable. The idea was surprisingly correct; however, for lack of any experimental evidence, it remained for centuries only a matter for speculation.

Interest in the study of matter declined during the Middle Ages and was not revived again until the time of Newton, Boyle, and Bernoulli, scientists of the late seventeenth and early eighteenth centuries. By 1770–1800, carefully controlled experiments by Priestley, Proust, and Lavoisier made possible the discovery of several laws of chemical combination. One such is the law of definite proportions (a compound always contains its elements in a fixed proportion by mass).

If the discussion of this law is not fresh in your mind, go back to Section 2.9 and review it before continuing.

At the beginning of the nineteenth century John Dalton, an English school teacher and scientist, made quantitative studies of compounds and elements. He knew about the newly discovered law of definite proportions, and his work disclosed a similar pattern of chemical behavior, the law of multiple proportions. This law states that when elements form more than one compound, the ratio of the weight of one element to that of a second element will be small whole numbers such as two to one, three to one, or three to two. To explain these chemical laws, John Dalton resurrected the idea of atoms and proposed the **atomic theory.**

3.2 Dalton's Atomic Theory

After studying this section, you should be able to:

- Describe the main features of Dalton's atomic theory.
- Identify the features of Dalton's atomic theory that are no longer accepted.

Based on the results of his laboratory experiments, John Dalton became curious. He wanted to understand why such things happened. How can inanimate substances combine in the same ratio every time? His asking "How can it be?" led eventually to the atomic theory.

theory

Dalton theorized that the matter in a sample of an element existed in individual packages (atoms), each of which had a mass characteristic of the element itself. He believed that atoms somehow attached themselves to one another to form larger packages of which compounds were made. Dalton called the larger packages "compound atoms"; we now call them molecules. A sample of compound always contains its elements in a fixed proportion by mass simply because of the masses of the *atoms* of those elements. In Dalton's view, a sample of a compound was just a large number of molecules of that compound, all alike and all having the same number of atoms of each kind.

To help clarify this idea, pretend that you are going into the hardware business. You buy a supply of nuts and bolts (your sample of compound). These come to you assembled, one nut on every bolt. Each bolt happens to weigh 26 grams and each nut 13 grams. Consider one nut–bolt pair. The mass of the bolt (26 grams) is twice that of the nut (13 grams). Now pick up two pairs. The mass of the two bolts (52 grams) is twice that of the two nuts (26 grams). Take a sample containing any number of pairs, ten, a thousand, a hundred million. The mass of all the bolts in the pile is still twice the mass of all the nuts in the pile because of two physical facts: (1) A bolt weighs twice as much as a nut, and (2) there is always one bolt for each nut (Figure 3.1).

Profiles in Science

John Dalton (1766–1844)

Ten years before the American Declaration of Independence, John Dalton was born in rural England. Because his family was poor, he had to leave school at age 11. However, he continued his education on his own. Dalton is said to have begun to teach in his village school at age 12.

At 27, Dalton joined the faculty at New College, Manchester, but found that his teaching duties were taking too much time from his growing interest in scientific investigation. He quit his job to devote full time to his research, making his living by private tutoring.

In 1803, Dalton presented two important papers, "Chemical Atomic Theory" and "First Table of Relative Weights of the Ultimate Particles of Bodies," to his local scientific society.

Recognition was long in coming. Not until 1816, 13 years later, was he first honored in a major way by election to the French Academy of Sciences. Later still, he was elected to the English Royal Society. Now he is considered the father of our present atomic theory of matter.

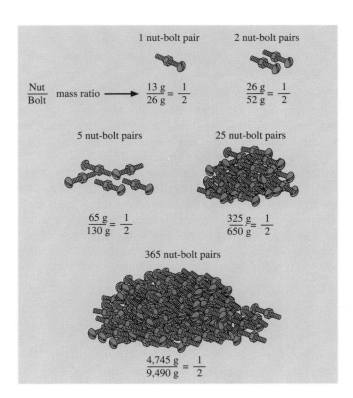

Figure 3.1 As long as there is one nut to every bolt, the mass ratio of nuts to bolts is 1/2, regardless of the number of pairs.

Dalton's reasoning was similar to that used in our example, except that the nuts and bolts (atoms of different kinds) were so small as to be invisible, and the nut–bolt pairs were what we now call molecules. In the compound hydrogen fluoride, the molecules (the nut–bolt pairs) each contain one hydrogen atom and one fluorine atom (one nut and one bolt). Because of this ratio and because one fluorine atom weighs 19 times as much as one hydrogen atom, the fluorine in one molecule of hydrogen fluoride weighs 19 times as much as the hydrogen. This mass relation is true for a sample containing two, a hundred, or any number of hydrogen fluoride molecules. Hydrogen fluoride obeys the law of definite proportions: The ratio of the masses of hydrogen and fluorine is the same in all samples of the compound.

Here is Dalton's theory, in detail, most of which is still accepted:

1. *All matter is made of indestructible, discrete particles.* Dalton called these atoms.

2. *All atoms of a given element are exactly alike, particularly in mass and size.*

3. *Atoms of different elements have different masses and sizes.*

4. *Chemical reactions are simply the joining or separation of atoms, of which none is created or destroyed in the process.*

5. *Compounds are made of molecules ("compound atoms"), which are combinations of atoms joined to one another.*

6. *Different compounds of the same two elements can result from combinations of the atoms in different ratios.* Such ratios, however, are always in the form of small whole numbers, such as one to one, one to two, two to three, and so on.

7. *The identity and properties of a compound are determined by the numbers and kinds of atoms composing it.*

Dalton's theory did fit the facts known at that time. The theory has since been tested by thousands of different kinds of experiments. The details have been changed, but the main features of the theory remain intact and are now far stronger than in Dalton's time. Let's consider further what Dalton proposed and then see how his theory explains the physical facts.

Using Dalton's reasoning and remembering our nut and bolt analogy, let's examine another slightly more complicated example. The compound water contains two hydrogen atoms and one oxygen atom in each molecule (like having two nuts on one bolt). In round numbers, one oxygen atom weighs 16 times as much as one hydrogen atom. But in one molecule of water, all the oxygen (one atom) weighs only eight times as much as all the hydrogen. Why? Because there are two hydrogen atoms, and the mass ratio in the molecule is

$$\frac{\text{mass of O atom}}{\text{mass of 2 H atoms}} = \frac{16}{1+1} = \frac{16}{2} = \frac{8}{1}$$

This mass ratio also holds for two molecules of water,

$$\frac{2 \times 16}{2(1+1)} = \frac{32}{4} = \frac{8}{1}$$

or for any number of water molecules, as shown in Figure 3.2.

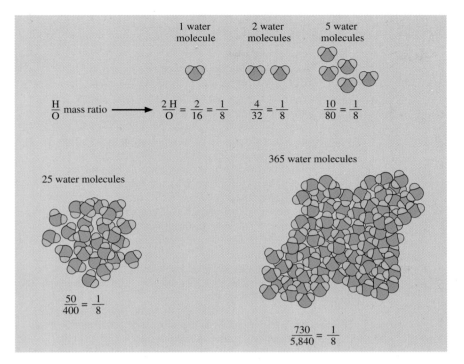

Figure 3.2 As long as there are two H atoms for one O atom, as in water molecules, the mass ratio O/H is always 8/1, no matter how many molecules are in the sample.

By considering the nuts and bolts and going through the example with water, you can see that the atomic theory does indeed explain why compounds always have their elements in constant proportion by mass (law of definite proportions). If the atoms of any given kind always have the same mass, and if the molecules always have the same numbers of atoms of each kind, the mass ratios of elements in a compound must then be constant.

Of the changes that have been made in Dalton's theory, one is particularly important. We now know that atoms of the same element do not necessarily all have the same mass. Also, Dalton believed that atoms were simple structures that could not be subdivided in any way; however, atoms are neither simple nor indivisible. Nonetheless, it is still considered true that atoms are neither created nor destroyed in chemical reactions—although their complex structures may exchange small amounts of matter with other atoms in processes called electron-transfer reactions. The essence of Dalton's atomic theory holds true, the idea that matter is composed of atoms.

The atomic theory remains a *theory* because no one has directly seen an individual atom. Modern instruments, however, have so convincingly demonstrated the atomic nature of matter that chemists use the theory as the cornerstone of their studies. It is unlikely that this concept will ever be proved incorrect. In fact, the theory may be the single most important concept in all of science; certainly, it is essential to understanding chemistry.

Electron-transfer reactions are discussed in detail in Chapter 16.

Science in Action

The Scanning Tunneling Microscope

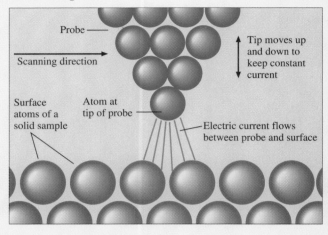

(a)

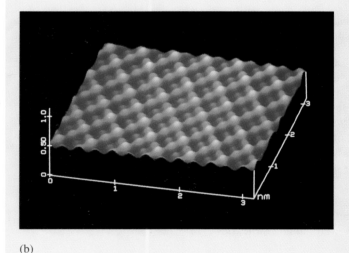

(b)

Atoms are so small that even the most powerful optical microscope is unable to distinguish them. The reason for this is not that the instruments are not good enough but that the light waves themselves are much too large. The smallest object that an optical microscope can distinguish is thousands of times larger than an atom.

Developed in the 1930's, the electron microscope, which "sees" objects by bouncing very high-energy electrons off them, can make dim images of large molecules containing many atoms (electrons themselves are explained in Section 3.4). Electron microscopes, however, work under the disadvantage that high-energy electrons damage the samples being viewed. Using another method, called X-ray crystallography, scientists can make extremely accurate maps of the positions of atoms in crystals, but the X-rays do not make images of the atoms themselves.

Only within the past few years has it become possible to make reasonably faithful pictures of actual atoms. The scanning tunneling microscope, developed in the 1980's, makes use of a tiny "tunneling" electric current that flows between a probe with an atom-sized tip and the atoms in the surface of a solid sample. As the probe is moved along, passing each atom, the tunneling current is kept constant by varying the distance from the probe to the surface, just as your finger would move up and down while you passed it over a series of bumps. As the probe sweeps in its predetermined pattern, its movements toward and away from the surface are recorded and mapped. The result is an image of the surface, showing the contours of each atom as it relates to its neighbors.

(a) How a scanning tunneling microscope works.
(b) An image of the surface of a silicon crystal as seen by a scanning tunneling microscope.

3.3 Electricity and Electric Charge

After studying this sectio , you should be able to:

- Describe a property of electric charges.

For more than a hundred years, it has been possible to demonstrate experimentally that electric charge is intimately involved in the inner workings of atoms. Although much study has been devoted to the topic, the fundamental nature

(a)   (b)

Figure 3.3 These two lightweight balls hang from fine threads. (a) If the two balls both have the same charge, ++ or −−, they repel each other and hang apart. (b) If they are charged differently, +− or −+, they are attracted to one another.

of electric charge is hardly better understood than it was in the early 1800's. Although we need not review the experiments from which the knowledge came, there are several things we should know about electric charge before we go on.

An **electric charge** is something that exerts force on other electric charges. Such force is either attractive, tending to pull charges together, or repulsive, tending to separate them. There are two kinds of charge, named positive (+) and negative (−). Two charges of the same kind, either two positive or two negative, repel each other; charges of different kinds attract each other (Figure 3.3). Forces between electric charges are the fundamental cause of most chemical phenomena.

If the amount of positive charge in a location is the same as the amount of negative charge, the negative charge cancels the positive, and the location seems to have no charge at all. Such an arrangement is said to be electrically neutral.

Electricity is the flow of electric charge, often called electric current.

Electric charges exert no forces on electrically neutral bodies.

3.4 The Electron

After studying this section, you should be able to:

- Give the charge and approximate mass of an electron.
- Describe how the electron was discovered.

Not long after Dalton proposed the atomic theory, Michael Faraday, an English chemist, discovered that passing an electric current through certain solutions caused chemical reaction. Faraday discovered that the amount of reaction was directly proportional to the amount of electric current applied. To explain this observation, he suggested that electric charge exists in individual packages (much as matter exists in the units called atoms) and that each package has the

Profiles in Science

Michael Faraday (1791–1867)

Faraday was born in a small English town, the son of a blacksmith and a farmer's daughter. As a child, he became an errand boy for the local bookbinder, who soon took him on as an apprentice. Exposed to the constant flow of books, Faraday read extensively, especially in science, a subject that fascinated him. He became particularly excited by the ***Encyclopedia Britannica*** series on electricity.

In 1812, Faraday attended a series of lectures by the famous English chemist Sir Humphry Davy. The youth was so impressed that he immediately applied to Davy for a job, sending a copy of some of his science notes. Davy became interested and hired Faraday as a laboratory helper and valet. In 1816, working in Davy's laboratory, Faraday published his first original paper, an analysis of lime.

During the next ten years, he studied steel making and steel analysis, organic chemistry (he discovered the important compound benzene), glass making, and the purification and liquefaction of gases at low temperatures. But it was Faraday's later research in magnetism, electricity, and the chemical effects of electric currents (1833–34) that made him famous. His work in these fields formed the foundation for the study of subatomic structure. Our present terms having to do with electrochemistry—***electrolysis, electrode, electrolyte, ion, cathode, anode, cation,*** and ***anion***—were all introduced by Faraday, after whom an electric charge unit, the *Faraday,* is named. Faraday was an example of the nineteenth century lone worker. Overcoming a scanty education, he tutored himself. In his productive years he worked without assistants, making discovery after discovery alone in his primitive laboratory.

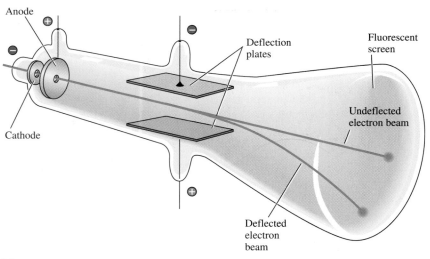

(a)

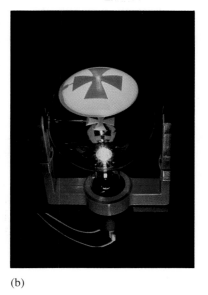

(b)

Figure 3.4   In a cathode ray tube, an electron beam is formed when electrons leave the cathode, and, at high speed, bombard the anode. Some electrons pass through the hole in the anode, forming a beam. When the electrons in the beam strike the fluorescent screen at the end of the tube, they produce a brightly glowing spot. If charged plates are installed in the cathode ray tube, the beam is deflected, showing that the electrons have mass and negative charge. A television tube is just an elaborate cathode ray tube.

Profiles in Science

John J. Thomson (1856–1940)

One of the great scientists of the late nineteenth and early twentieth centuries, John J. Thomson presided for nearly a generation over the famous Cavendish Laboratory at Cambridge University in England. Like many others, Thomson was fascinated by the effects of passing electric current through evacuated glass vessels fitted with metal plates and conducting wires.

Although scientists of the time held the general opinion that glowing beams in "cathode ray tubes" were a form of light, Thomson demonstrated that they were streams of moving particles (electrons) and measured the properties of the particles. In 1906, Thomson received the Nobel Prize for this work.

Thomson's contributions in science encompassed fields extending from electronics, through X-rays, to radioactivity, positive ion beams (mass spectrometry), mathematical theory of electric phenomena, isotopes, and much more. He was the enthusiastic sponsor of Ernest Rutherford, later Lord Rutherford, who discovered the nuclear structure of the atom (p. 93). Thomson was buried in Westminster Abbey near his protégé, who had predeceased him by three years.

same amount of charge. Somewhat later (1891), these packages of negative charge were named **electrons.**

The actual charge of the electron was determined experimentally by Robert Millikan (1868–1953) as 1.6×10^{-19} coulombs, where the coulomb is a unit of charge. Scientists find it convenient now, however, to assign charges to electrons and other subatomic particles using a relative scale in which the electron charge is -1.

In the late 1800's, J. J. Thomson and others performed several important experiments with electricity. These scientists used partially evacuated tubes, called cathode ray tubes. A stream of electricity was made to jump across gaps in wires, creating a "ray" (see Figure 3.4). Thomson's experiments revealed that the ray of electric charge was actually a stream of negatively charged particles of matter. Thomson's results, combined with Millikan's data on the electron charge, made it possible to calculate the mass of an individual electron, which is 9.1×10^{-28} grams.

Thomson found that all atoms contain electrons. Because atoms are known to be electrically neutral, it follows that atoms must also contain some positive charge. Thomson suggested that the negative electrons in an atom were evenly distributed in a pool of positive charge.

An electron has a charge of -1 and a mass of 9.1×10^{-28} grams.

1×10^{23} electrons weigh less than a fingerprint.

3.5 **The Proton**

After studying this section, you should be able to:

- Identify the charge and approximate mass of the proton.

During Thomson's time, scientists already knew (from Faraday's work) that individual atoms could lose or gain one or more electrons, thus becoming what

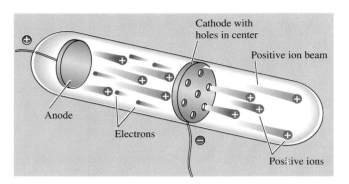

Figure 3.5   In this tube, atoms of a gas at very low pressure are struck by electrons and become positive ions. The positive ion beam can be studied in the same way that cathode rays are, yielding information about the masses of the positive ions.

are called **ions.** Several experiments similar to Thomson's were made using streams of positive ions in place of electrons (Figure 3.5). It was found that the mass and charge of the ions were smallest when the ions came from hydrogen. These positive hydrogen ions were named **protons.** The amount of positive charge associated with the proton is equal to that of the electron but is opposite in sign, $+1$.

We now know that a hydrogen atom consists of one proton and one electron.[*] Removing the electron leaves only the positively charged proton, usually represented by the symbol H^+.

In a way similar to that used to determine the mass of an electron, the mass of a proton was found to be 1.673×10^{-24} grams, or about 1,840 times that of an electron. Figure 3.6 gives a schematic version of the apparatus used to measure the mass of positive ions. In its refined form, this apparatus is called a mass spectrometer.

Now we can say what the mass of a single hydrogen atom is. The mass of the proton is 1.673×10^{-24} grams and that of the electron is 9.1×10^{-28} grams. Since a hydrogen atom consists of one proton and one electron, we merely need to take the sum of these two masses to have the mass of the atom itself.

$$\text{mass of H atom} = 1.673 \times 10^{-24} \text{ g} + 0.00091 \times 10^{-24} \text{ g} = 1.674 \times 10^{-24} \text{ g}$$

Clearly, because the mass of the electron is so small compared to the proton mass, we can say for most purposes that an H atom and a proton have the same mass.

Before we continue, let's review what we have covered. Electrons have extremely light masses and carry negative charges. Protons are much heavier than electrons and have positive charges. Atoms are electrically neutral; therefore atoms contain protons and electrons in equal numbers.

proton weighs more

A proton has a charge of +1 and a mass 1,840 times that of an electron, or 1.673×10^{-24} grams.

---

[*]Actually, there are two other rare varieties of hydrogen—isotopes—that include other particles. We will discuss these in Section 3.8.

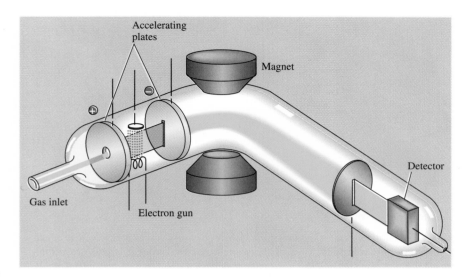

Figure 3.6 The mass spectrometer is an elaborate version of the cathode ray tube equipped with a magnet to bend the ion beam and a detector for the ions. Ion masses are easily measured to high precision in a mass spectrometer.

3.6 The Nuclear Atom

After studying this section, you should be able to:

- Describe and explain the experiment by which Rutherford discovered the nuclear atom.
- Identify the main features of the nuclear model of the atom.

Considering what we have learned so far about electric charges and atoms, it seems reasonable to expect that the charged particles in atoms would arrange themselves so that charges of the same sign would be as far apart as possible and charges of different sign as close together as possible. Perhaps the atom has a uniform distribution in which the protons and the electrons are uniformly mixed together in a kind of atomic pudding. At the end of the nineteenth century, this seemingly logical and attractive "pudding model" was universally accepted.

~mixed together~

Shortly after the beginning of the present century, the English scientist Ernest Rutherford was investigating the nature of matter by bombarding thin sheets of metal foil with alpha particles from radioactive substances. (An alpha particle is a high-speed helium atom with its two electrons stripped off.) Rutherford directed a stream of alpha particles at a thin metal foil to see whether their collisions with the atoms in the foil would alter their direction.

Because he accepted the pudding model of the atom, Rutherford expected alpha particles to penetrate the foil but to be slowed down as they plowed through the atoms of the metal. He reasoned that if the atoms of the metal foil were made of uniformly distributed particles, no single particle in the foil would be heavy enough to make the alpha particles stray far from their paths.

To make the situation more clear, consider the following imaginary experiment. You are given a stack of oranges. In each orange there are ten BB's

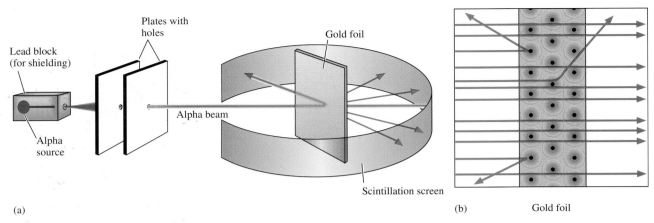

(a)

(b)   Gold foil

Figure 3.7 Rutherford's experiment. A narrow beam of alpha particles was aimed at the very thin gold foil. Most of the particles passed through the foil with little change, but a few (red tracks) were widely deflected. Rutherford interpreted these results as showing that the gold atoms were mostly empty space, with most of the mass concentrated in tiny nuclei.

evenly dispersed. You plan to fire 22-caliber rifle bullets through the oranges. If the BB's are scattered randomly inside the oranges, no one BB will substantially deflect your bullet, and the bullets will all hit the target beyond the oranges.

In Rutherford's experiment (Figure 3.7), most of the alpha particles indeed went directly through the foil, as expected. Amazingly, however, a few of them were *greatly* deflected. These results were a tremendous surprise to Rutherford, who said that the experiment, planned as a "small research," resulted in "quite the most incredible event that has happened to me in my life . . . almost as incredible as if you fired a 16-inch shell at a piece of tissue paper and it came back and hit you."

How can Rutherford's results be explained? Consider the oranges and BB's again. Suppose that in each orange, the ten BB's are not evenly distributed. Instead, all ten BB's are welded together into a solid mass bigger than a 22-caliber bullet and placed in the center of the orange. Not many bullets will hit these solid chunks, but when one does, the bullet will bounce widely and land nowhere near the target.

In 1911, Rutherford published his findings. His calculations from repeated experiments showed that nearly all the mass of the atom (the protons and other particles) was concentrated in one tiny dense unit, which he named the **nucleus.** He found that the nucleus had a diameter only 1/10,000 to 1/100,000 that of the atom itself. The electrons, of which there were enough to balance the charge of the protons, had negligible mass but occupied most of the remaining volume of the atom. Rutherford's experiments demolished the pudding model of the atom forever.

Many important scientific discoveries have been made by accident. Experiments designed for a particular purpose have sometimes produced totally unexpected results. What is important in scientific work is not that any particular experiment succeed as planned but that significant experiments be performed by people who understand the experimental systems and who have enough knowledge and imagination to interpret their results, regardless of what they may be. Rutherford was astonished at what his experiment revealed, yet he continued

Profiles in Science

Ernest Rutherford (1871–1937)

Ernest Rutherford was born in rural New Zealand, the fourth child of a couple who had migrated from Scotland to the new country. Interested in science from early boyhood, Rutherford excelled in school and was given a scholarship to the University of New Zealand. His accomplishments in science at the university won him a grant for continued work at Cambridge, where he became the protege of J. J. Thomson (p. 91).

Although Rutherford made substantial scientific contributions in the fields of electricity, magnetism, isotopes, atomic numbers, element transmutation, and more, his fame arose from his work in radioactivity and its use in discovering the atomic nucleus.

Rutherford received the Nobel Prize in chemistry in 1908. He was knighted in 1914 and made a Baron in 1931, taking his seat in the House of Lords. He died after a short illness in 1937 and was buried in Westminster Abbey near Newton, Kelvin, Darwin, and other giants of science.

with it, modified it to display the results even better, and brilliantly interpreted what he saw. The result was one of the most famous experimental discoveries in scientific history.

Rutherford's model, which required that the atom consist mostly of empty space and which demanded that the mutually repulsive positive charges stay grouped close together without the gluing effect of negative electrons, was at first regarded as a sophisticated joke. When it became clear, however, that Rutherford was serious, scientists in many laboratories began feverish research to determine the characteristics of the nucleus and electrons.

The diameter of a hydrogen atom is about 10,000 times that of its nucleus. Let's try to visualize this. Imagine yourself standing in the center of a racetrack two miles across. You are holding an atomic nucleus the size of a basketball. Now, the edge of the atom belonging to that nucleus is a mile away in every direction. The intervening space is occupied only by one tiny electron (Figure 3.8).

Now imagine that you drive an automobile off a cliff 50 stories high. When it hits the bottom, you'll think that both the car and the ground are pretty solid, but

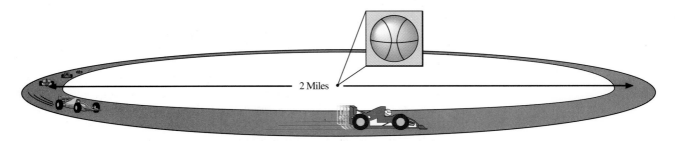

2 Miles

Figure 3.8 The relation of the size of a basketball to the size of a racetrack is the same as relation of the size of a nucleus to the size of an atom.

they are not. Both are made of atoms, and atoms are mostly empty space. Solid matter—truly solid, with no empty space—is unbelievably dense. A basketball would have a mass of about 10^{12} pounds, or about a billion tons, if it were "solid" matter.

3.7 Atomic Number

After studying this section, you should be able to:

- Define atomic number.
- Explain the characteristic that identifies an atom as being of a particular element.

Shortly after the discovery of the nuclear atom, Rutherford's experiment with alpha particles yielded another important result. From studies of the amount of alpha particle deflection, Rutherford found that (1) he could estimate the number of positive charges in the atoms of different elements and that (2) this number changed in a regular way from one element to the next. Rutherford's finding suggested that the number of positive charges in the nucleus of an atom characterizes that which is unique and unchanging about all atoms of that element. The number of protons in its atoms is the defining characteristic of an element.

Each atom of an element has the same number of protons.

In the early years of the twentieth century, experiments with ions showed that more than a single electron could be stripped from some kinds of atoms, and it soon became possible to strip and count all the electrons from the atoms of the lighter elements. In these experiments, it was found that the total number of electrons in an atom, and therefore the number of protons in its nucleus, was characteristic of the element to which the atom belonged. Hydrogen atoms were found to have one electron; helium, two electrons; lithium, three; and so on. Such findings were consistent with Rutherford's discovery that a specific number of protons was characteristic of a given element.

The atomic number is equal to the number of protons in an atom of an element.

Around 1913, experiments with X-rays confirmed all these findings, and it is now accepted that each element is characterized by the number of protons in its atoms. This number is called the **atomic number** (usually denoted by Z). An element is identified by its atomic number as much as by its name or symbol. The element having atomic number 5, for instance, is boron; number 10 is neon; number 80 is mercury; and so on. The periodic table shows each element with its symbol and with its atomic number printed above the symbol.

6 —— Atomic number / protons

C

—— Symbol

12.011

3.8 Neutrons, Isotopes, and Atomic Mass Numbers

After studying this section, you should be able to:

- Give the approximate mass of the neutron and its location in an atom.
- Explain what isotopes are and how they differ.
- Define atomic mass number.
- Calculate the number of neutrons in the nucleus of an atom.

The Neutron

Using apparatus like that described in Figure 3.6, scientists have been able to determine the masses of individual atoms. It was found that not all the atoms of any particular element necessarily have the same mass. At this point, you may wonder how two atoms can have the same number of protons and the same number of electrons and still have widely different masses. It must be that atoms include particles other than protons and electrons that account for these differences. Such particles, however, must be electrically neutral; otherwise they would upset the electrical charge balance of the protons and electrons in the atoms.

The same conclusion was reached by scientists around the turn of the century. It wasn't until 1932, however, that experiments by James Chadwick (1891–1974) finally revealed the particle that accounted for these mass differences. It was given the name **neutron.** A neutron has approximately the same mass as a proton, 1.675×10^{-24} grams, but it has no charge. Neutrons are found in the nuclei of atoms.

Chadwick, a close associate of Rutherford and J. J. Thomson, was another of the great English scientists who opened the door to modern experimental chemistry and physics. In 1935 he was awarded the Nobel Prize for his discovery of the neutron.

Isotopes and Mass Number

Samples of atoms belonging to the same element and having different numbers of neutrons (and therefore different masses) are said to be different **isotopes** of that element. Nearly all the elements have two or more isotopes. Hydrogen, for example, has three isotopes. The most common H isotope is the lightest, which has no neutrons. Each atom of the second isotope has only one neutron in the nucleus. Atoms of the third isotope each have two neutrons.

The atomic number of an element identifies that element and gives the number of protons in the nucleus. The **mass number** (usually represented by *A*) identifies an isotope and gives the total number of protons and neutrons in an atom of that isotope. The symbol for an isotope is shown by writing the symbol for the element preceded by the mass number as a superscript (Figure 3.9). The three isotopes of hydrogen are identified as ^{1}H, ^{2}H, and ^{3}H, in order of increasing mass. Two isotopes of oxygen are ^{16}O and ^{18}O; two isotopes of helium are ^{3}He and ^{4}He.

The atomic number is sometimes shown as a left subscript, an example being $^4_2$He for the isotope 4 of helium. However, because an element can be identified either by its symbol or by its atomic number, writing the atomic number beside

protons + neutrons

The mass number of an isotope is shown by a left superscript.

^{60}Co

Figure 3.9 The nuclear particles in H and O. The purple particles are neutrons. The yellow particles are protons.

Science in Action

Transmutation: Changing One Element into Another

Medieval alchemists were interested in miracle cures of illnesses and in finding a way to achieve eternal life, but their chief interest lay in finding a way to turn base metals, such as iron, into gold, in ***transmuting*** one element into another. Of course, they were unsuccessful, although some gained riches and power by claiming to have found the answer.

In 1898, Marie and Pierre Curie announced the discovery of radioactivity and the new element radium. A few years later, they showed that, in the radioactive process, the atoms of radium changed themselves, eventually turning into atoms of lead. Radium transmuted into lead. Then, in 1911, Lord Rutherford (p. 95) was able to show that bombardment with alpha particles (a type of radiation) could change nitrogen into oxygen and hydrogen.

$$^{14}_{7}N + ^{4}_{2}He \longrightarrow ^{17}_{8}O + ^{1}_{1}H$$

In the 1930's, E. O. Lawrence and M. S. Livingston at the University of California invented the cyclotron, a machine that could accelerate ions such as H^+ or He^{2+} to extremely high speeds, creating the equivalent of artificial radioactivity. Using the cyclotron, Lawrence created substantial quantities of radioactive sodium, some of which was used in medical experiments, inaugurating what is now a multimillion-dollar medical isotope industry.

In 1940, E. M. McMillan and Phillip Abelson created the first sample of a synthetic element, one that does not naturally exist on earth. They bombarded $^{238}_{92}U$ with deuterium ions, $^{2}_{1}H^+$, creating the first man-made element, neptunium, $^{239}_{93}Np$.

A modern cyclotron, showing a proton beam (in blue).

Since that time, other synthetic elements have been prepared, primarily by the bombardment of lighter elements by high-speed particles. The periodic table now contains 109 elements.

Atoms of the elements of atomic mass greater than 92, uranium, are unstable; they explode, creating smaller atoms and releasing energy. The lifetimes of some are far shorter than a second. Because of this instability, no natural samples of those elements exist on earth.

The first cyclotron was small enough to be held in the palm of the hand. Since then, in the quest for higher and higher ion speeds, larger and larger cyclotron-type machines have been built. This accelerator at Batavia, Illinois, is four miles across. In 1995, a similar accelerator 50 miles across is planned to be operating in Texas.

(a)

(b)

Particle accelerator at the Fermi National Laboratory in Illinois. (a) The main accelerator is 4 mi in circumference (1.27 mi in diameter). (b) An aerial view of the main accelerator.

Table 3.1   Calculations of the Number of Neutrons in Nuclei

Element	Hydrogen	Carbon	Oxygen	Sodium	Chromium
Symbol	^1H	^{12}C	^{16}O	^{23}Na	^{52}Cr
Mass number	1	12	16	23	52
Atomic number	1	6	8	11	24
Number of neutrons	0	6	8	12	28

the symbol is redundant. For example, chlorine is the element with atomic number 17 and symbol Cl; to write $_{17}Cl$ is repetitious. Moreover, writing the atomic number beside the symbol can be confusing in a formula; the two-atom molecule of hydrogen 1 would be written like this:

$$_1^1H_2$$

Although it is sometimes useful to show the atomic number this way, we will not often find it necessary to do so.

Because the mass number is equal to the sum of the number of protons and neutrons in an atom, we can find the number of neutrons in the nucleus of an isotope by subtracting the atomic number from the mass number.

$$\text{number of neutrons} = A \text{ (mass number)} - Z \text{ (atomic number)}$$

EXAMPLE 3.1

How many neutrons are there in the nucleus of an atom of ^{15}N? Of ^{18}O?

Solution: The mass number of ^{15}N is 15, and the atomic number is 7. The number of neutrons in the nucleus of a ^{15}N atom is $15 - 7 = 8$. The mass number of ^{18}O is 18, and the atomic number is 8. There are ten neutrons, $18 - 8 = 10$, in a nucleus of an ^{18}O atom.

Table 3.1 gives several more examples.

Exercise 3.1

How many neutrons are there in the nucleus of a ^{28}Si atom? Of a ^{29}Si atom? Of a ^{30}Si atom?

Exercise 3.2

How many neutrons are there in the nucleus of each atom of ^{35}Cl? Of ^{14}C?

In the atoms of the first 15 or so elements, the number of neutrons is about the same as the number of protons. The mass number, therefore, is about twice the atomic number. The common isotope of oxygen, for example, is ^{16}O and that of neon is ^{20}Ne. For the heavier elements, however, the number of neutrons increases with atomic number, growing to about one and a half times the atomic number in the heaviest elements. In the common isotope of gold, $^{197}_{79}\text{Au}$, for example, there are 1.49 neutrons for every proton.

Let's review:

1. *An atom consists of an extremely small, dense nucleus surrounded by one or more electrons.*

2. *The two major constituents of a nucleus are protons and neutrons. The mass of a proton is approximately equal to the mass of a neutron. Protons have a charge of +1; neutrons have no charge.*

3. *Electrons have a charge of −1. The mass of an electron is about $\frac{1}{1,840}$ that of a proton and, consequently, is negligible compared to that of protons and neutrons.*

4. *The atomic number (the number of protons in the nucleus) determines the element to which the atom belongs.*

5. *The mass number gives the total number of protons and neutrons in an atom of an isotope.*

6. *An atom has as many electrons as it has protons.*

Summary

Dalton's atomic theory, which is a cornerstone of science, was first proposed to explain certain laws of chemical combination which, in turn, represented the combined results of quantitative analytical measurements. The theory states that matter is composed of atoms and that atoms of different elements combine in fixed-number ratios to give compounds. Although there were certain faults in Dalton's atomic theory, the fundamental concept is still considered valid, and our understanding of the nature of matter is largely based on it.

Atoms are made up of subatomic particles. To chemists, the most important subatomic particles are protons, neutrons, and electrons. Each proton carries a positive electric charge, and each electron carries an equivalent negative charge. Neutrons do not have a charge.

A proton or neutron has a mass of about 1.6×10^{-24} grams. This mass is about 1,840 times that of an electron. Compared to the mass of a proton or neutron, the mass of an electron is negligible.

Every atom contains a nucleus, which, in turn, contains all the protons and neutrons. The nucleus, then, contains all the positive charges and nearly all the mass of an atom. The protons and neutrons are crowded into the tiny nucleus, which typically has a diameter only about 1/10,000 that of the atom itself. The nucleus is surrounded by electrons. These electrons occupy all the remaining space of an atom.

Atoms are neutral; therefore the number of electrons in an atom must equal the number of protons. Each element is characterized by its atomic number; that is, by the number of protons in the nucleus. An element can be identified by its atomic number, its name, or its symbol.

Two or more atoms that have the same atomic number but different mass numbers are different isotopes of the same element. The mass number is defined as the total number of protons and neutrons in the nucleus of an atom. To determine the number of neutrons in the nucleus of any isotope, subtract the atomic number from the mass number.

Key Terms

atomic theory	*(p. 84)*	proton	*(p. 92)*	neutron	*(p. 97)*
electric charge	*(p. 89)*	nucleus	*(p. 94)*	isotope	*(p. 97)*
electron	*(p. 91)*	atomic number	*(p. 96)*	mass number	*(p. 97)*
ion	*(p. 92)*				

Questions and Problems

Section 3.2 Dalton's Atomic Theory

Questions

1. Give a brief description of Dalton's atomic theory.

2. Briefly define the term "compound atoms."

3. List the deficiencies of Dalton's atomic theory.

4. Give an example from technology showing that Dalton was incorrect in assuming that atoms could not be subdivided.

5. How are atoms, elements, and compounds related to each other?

Section 3.4 The Electron

Questions

6. What observation led Faraday to conclude that electric charge exists in individual units?

7. What is an electron? What is its mass in grams?

8. What does it mean to say that an atom is electrically neutral?

9. Atoms are known to be electrically neutral. If electrons in atoms are negatively charged, what must also be present in atoms?

Section 3.5 The Proton

Questions

10. Explain the difference between an ion and an atom.

11. What is a proton?

12. What is the mass of a proton in grams?

13. Explain why the mass of a hydrogen atom is almost the same as that of its proton.

Section 3.6 The Nuclear Atom

Questions

14. Why was it so easy for scientists before Rutherford to accept the "pudding model" of the atom? What was found to be incorrect about this model?

15. What is an alpha particle?

16. Describe the nuclear model of the atom. What are the relative sizes of an atom and its nucleus?

17. Explain the experiment, including its results, that led Rutherford to conclude that atoms have a nucleus.

Section 3.7 Atomic Number

Questions

18. Describe a second important discovery made possible by Rutherford's experiment.

19. What is an atomic number? How does the atomic number of an element relate to its name?

20. Describe one experimental method of finding the atomic number of one of the lighter elements.

Problems

21. Give the name and the atomic number of each of the following elements: Na, Co, Au, Li, Po, K, Mg, Hg, Si, Ag, Ti, Sn, F, Fe.

22. Write the symbols for the elements with the following atomic numbers: 3, 6, 8, 9, 10, 18, 36.

Section 3.8 Neutrons, Isotopes, and Atomic Mass Numbers

Questions

23. Explain how two atoms can belong to the same element and yet have different masses. Define the words "isotope" and "neutron."

24. What is a mass number?

25. Is it possible for the atomic number of an element to be greater than its mass number? Explain.

Problems

26. The element carbon has three isotopes with mass numbers of 12, 13, and 14, respectively. Write the symbols for these three isotopes. How many neutrons are there in the nuclei of the atoms of each isotope?

27. The element cadmium has many isotopes. ^{106}Cd, ^{108}Cd, ^{110}Cd, ^{111}Cd, ^{112}Cd, ^{113}Cd, ^{114}Cd, and ^{116}Cd are a few of the naturally occurring ones. How many neutrons are in an atom of each isotope?

28. How many neutrons are there in the nuclei of the atoms of the following isotopes: ^{27}Al, ^{3}He, ^{24}Mg, ^{86}Rb, ^{23}Na?

29. The following is a list of isotopes, most of which exist. The element symbols are all represented by E.

$^{16}_{8}E$ $^{14}_{7}E$ $^{18}_{8}E$ $^{13}_{7}E$
$^{15}_{7}E$ $^{15}_{8}E$ $^{14}_{6}E$

 (a) Which are isotopes of the same element? Name the elements.
 (b) Which have the same mass numbers?
 (c) In which are the numbers of neutrons in the nuclei the same?
 (d) In which are the numbers of protons in the nuclei the same?
 (e) In which are the atomic numbers the same?

30. Complete the following table. (The first line is an example.)

Atomic Number	Mass Number	Symbol	Number of Protons	Number of Neutrons	Number of Electrons
17	35	Cl	17	18	17
___	___	C	6	6	___
8	18	___	___	___	8
___	___	Ni	___	30	___
___	199	___	80	___	___
___	___	Na	___	12	___
___	55	Fe	___	___	___
___	___	S	___	16	___
___	15	N	___	___	___

Solutions to Exercises

3.1   14; 15; 16

3.2   18; 8

Using a special device, we can create a three-dimensional wave.

Electron Structure of the Atom

E very part of an atom contributes directly or indirectly to the properties of the element to which the atom belongs. As we have seen, the nucleus supplies most of the mass and all of the positive charge. In turn, the number of protons in the nucleus determines the number of electrons in the atom.

The part of the atom that interacts with the rest of the world is the outside part, the electrons. Except for mass, nearly all the elemental properties that concern us are directly related to the electrons and only indirectly to the nucleus. For this reason, we will direct our attention to electrons in atoms, leaving the details of nuclear structure to physicists.

In Chapter 4, we learn about the behavior of electrons, the puzzling little particles that refuse to obey the laws we so confidently apply to larger particles of matter. Even though electrons behave in unusual ways, we can learn enough about them to understand how their behavior determines the properties of atoms. This will lay the foundation for an understanding of how elements form compounds.

4.1 Electron Energy Levels

After studying this section, you should be able to:

- Explain what is meant by the term *quantized*.
- Explain how the spectroscope gives information about energies of electrons in atoms.

Rutherford's experiments have shown us that the outer part of an atom consists only of electrons and empty space. Except for some exotic particles that can

A better name for the uncertainty principle is the indeterminancy principle, which is a literal translation of Heisenberg's German word. Uncertainty, however, is the word most used now.

electrons
uncertain

In light, color is determined by energy. The individual packages, photons, of blue light have more energy than those of green light. Red photons have the least energy of any in the visible spectrum.

emit

absorb

sometimes be temporarily coaxed out of a nucleus, electrons are the smallest particles of matter. Because they are so small, electrons exhibit properties unlike those that we see in larger particles. This makes it difficult for us to learn much about what electrons do. You can find out a little about how they behave by shooting fast particles at them and observing what happens, much as Rutherford did in his experiment with alpha particles and nuclei. But when you hit an electron with a fast particle, you knock it out of the atom entirely. You can say only where it was—not what it was doing. Was it circling the nucleus? Was it vibrating back and forth? Was it standing still? You can't tell. The properties of matter are such that, with very small particles, it is impossible to tell simultaneously where the particles are and what they are doing. During the first quarter of this century, this fact was stated in a theory called the **Heisenberg uncertainty principle.**

Fortunately, there are a few ways in which electrons in atoms can make themselves known, and from these we can make some good guesses about their behavior. Electrons have energy, just as any other body of matter can have energy. Because electrons move, they have kinetic energy. They also change position with or against the forces binding them to the nucleus, which gives them energy of position, or potential energy. When an electron in an atom *loses* energy, the lost energy can *leave* the atom as light. (Light is pure energy not associated with matter.) The energy of the emitted light, which exactly equals the amount of energy change made by the electron, can be measured. Light can also *enter* an atom, be totally absorbed, and *increase* the energy of electrons in the atom. This kind of energy change can also be measured. A field of study called **spectroscopy** measures and characterizes the energies of light absorbed and emitted by atoms.

Spectroscopy, then, lets us find out the energies that electrons in atoms can have, not the locations, unfortunately, only the energies. Even this much, however, allows us to build a solid theory.

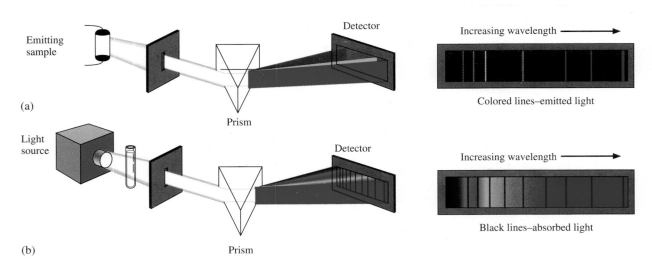

(a) When electrons in an atom lose energy, the energy can leave as emitted light. (b) The atoms in a sample can absorb light, increasing the energy of its electrons.

Profiles in Science

Lyle Lovett
~~Werner Heisenberg~~ *(1901–1976)*

When he was only 21 years old, the German student Werner Heisenberg attended a lecture by Niels Bohr. In the lecture hall, which was crowded with admiring scientists, Heisenberg boldly objected to one of Bohr's statements. Rather than resenting the interruption, Bohr was pleased to find a bright student willing to speak out. He invited Heisenberg to walk with him and discuss science. Soon after, he asked the young man to join the Bohr research group. Flattered by the invitation, Heisenberg joined the group in 1927. Three years later, still working with Bohr, he first proposed what is now called the Heisenberg uncertainty principle, one of the foundations of the understanding of subatomic particles.

Heisenberg was a patriotic German. During the Second World War, he was involved with the abortive and short-lived German effort to develop an atomic bomb but also continued to make contributions to basic scientific knowledge, primarily in mathematical theories of physics.

Researchers have found that an atom can absorb or emit light of particular energies only, not just *any* energy. This has led to the conclusion that electrons in atoms can *have* only certain amounts of energy. In other words, electron energy is **quantized,** restricted to certain quantities. To clarify the situation, *specific quantities* consider a brick and a bookcase like the one shown in Figure 4.1.

To raise a brick from a lower to a higher shelf, you must do work on it. That is, energy must be added to the brick to raise it. The energy of the brick

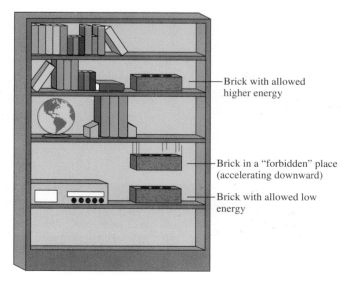

Brick with allowed higher energy

Brick in a "forbidden" place (accelerating downward)

Brick with allowed low energy

Figure 4.1 A brick on a bookshelf can have only certain energies, corresponding to positions on any of the shelves, that is, the energy of the brick is quantized. If the brick is placed between shelves, it will fall to a shelf and assume an allowed energy.

no in
b/w "

on the right, then, is greater than the energy of the brick on the left. A brick can have only certain energies, corresponding to its position on one of the shelves. If it is placed between shelves, it will fall to a shelf and assume an allowed energy. We see that the energy of the brick is quantized, because the brick can rest only on a shelf, not between shelves. Remember that the picture of the bookshelf shows the *physical* location of the bricks and only indirectly represents the *energies* of the bricks. With electrons, however, the **energy levels** strictly represent amounts of energy; they have only an indefinite relation to the electron location. We know electrons not by their positions, but by their energy levels.

Just as you can, in a way, characterize a bookcase by measuring the energies of bricks placed on various shelves, so too can you characterize an atom by the set of energy levels possible for its electrons. Spectroscopy has shown that the pattern of light absorption and emission (called a spectrum) is unique for each element. The emission spectra of several elements are shown in Figure 4.2. Because each element has a unique spectrum, it follows then that each element has a set of electron energy levels unlike that of any other element. Accordingly, it is possible to use spectra to identify the presence of a given element much as a fingerprint is used to identify people.

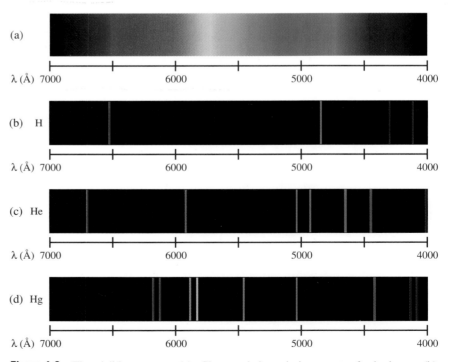

Figure 4.2   The visible spectrum (a). Characteristic emission spectra for hydrogen (b), helium (c), and mercury (d).

Science in Action

Early in the twentieth century, spectroscopy, the study of the energies of light absorbed or emitted by atoms or molecules, contributed significantly to our knowledge about electrons. Now it plays a large role in many kinds of present-day technological processes.

When atoms are given energy (are "excited") by heating to high temperatures or by being bombarded with electrons, they often immediately lose the energy in the form of light. Only certain energy changes can occur in atoms of a given element; therefore the atoms of that element can emit light only of those same energies, identifiable by wavelength. The instrument used to measure the wavelengths of light is called a spectrometer.

The wavelengths of light emitted by excited atoms are routinely used to identify the presence of particular elements in samples. For example, a chip of paint found at a fatal auto accident can be analyzed in an atomic emission spectrometer to indicate what kinds of atoms are present and in what proportions. The sample is placed in a hot flame, and the emitted light is studied. By spectroscopic comparison with known samples, the chip can be identified with a brand, perhaps even with an individual batch of paint. Steel makers use emission spectra to analyze samples of their metal, even as it is being manufactured.

In a similar technique, called atomic absorption spectrometry, samples are heated, but instead of looking for emission of light, the spectrometer passes light through the flame to see which wavelengths are absorbed. This technique is good for samples in solution and, among other things, is used to analyze for traces of heavy atoms, such as lead or arsenic, in animal tissues.

Spectroscopy is used for studying molecules as well as atoms. Molecules can gain or lose energy in ways that are characteristic of their structures. In particular, molecular patterns of absorption of infrared light are highly individual and can be used both to indicate the presence and the amounts of various kinds of molecules, as well as to study the actual structures of molecules. Infrared spectroscopy is used widely in research and is employed universally to test biological samples, as in the test for alcohol concentration in blood or in urinalysis to check for the presence of drugs.

4.2 The Theory of Electron Behavior in Atoms

After studying this section, you should be able to:

- State the principal concepts of the Bohr theory of the atom.
- Explain the basis for the wave theory of the electron and state its consequences.
- Describe the differences between the Bohr theory and the wave mechanical theory of the electron.

By now we should be accustomed to seeing how theories are devised to explain observed phenomena. Let's list the facts that a useful theory should explain about electrons in atoms, then examine the early approaches to a satisfactory theory.

1. *Negatively charged electrons occupy most of the space in an atom and stay away from the positively charged nucleus.* Why doesn't the atom collapse instead?

2. *From spectroscopy, we know that the energies of electrons in atoms are quantized.* Why can't electrons have any amount of energy?

3. *Although different elements have different physical and chemical properties, many elements have similar sets of properties that allow us to group them into "families."* Characteristic properties are presumably determined by the electrons in the atoms of the elements. If all electrons are the same, how can they cause such different sets of properties, and how can electron behavior create families of elements?

Bohr's Planetary Theory

We know that there is a strong gravitational attraction between the sun and the earth. What keeps the earth from falling into the sun? It is the fact that the earth is moving rapidly and would, in fact, leave the solar system entirely unless the sun's gravity were present to keep it in its orbit. In an atom, there is a strong attraction between the positively charged nucleus and the negatively charged electrons, but the electrons do not fall into the nucleus. In 1912, a Danish physicist named Niels Bohr proposed that electrons in atoms behave much like planets. He proposed that electrons occupy orbits around the nucleus and are kept in those orbits by the force of the attraction between their negative charges and the positive charge of the nucleus. Bohr explained that the emission and absorption of light results from the transitions of electrons from one orbit to another. Because he could not explain why the electron energies were quantized, he declared, arbitrarily and without explanation, that the electrons could occupy only certain orbits at particular distances from the nucleus. Apart from the arbitrary features, the **Bohr atom** fit the available experimental facts quite well.

like planets

Bohr's theory was attractive because it was easy to visualize. It did not, however, explain the *reason* for quantization. What is more, at that time the Heisenberg uncertainty principle was beginning to be understood and seemed to imply that we can never know the exact locations of electrons. Thus, it appeared that electrons in atoms could not be in orbits because orbits are predictable and therefore would allow exact knowledge of electron locations. Further, the Bohr picture worked only for atoms and ions containing single electrons, although most atomic spectra arise from multi-electron atoms.

Even so, Bohr's theory was useful because it caused scientists to begin thinking about the quantization of electron energies and led to the theories now considered correct. Although it is now realized that Bohr's theory was far from reality, we often see representations of it, with electrons shown whirling around a nucleus in orbits.

The Wave Theory of the Electron

The 1920's and 1930's were a time of intense change in science. They marked the boundary between "classical" and "modern" physics and changed all the theories about the structure of matter.

During the second and third decades of the twentieth century, several lines of research contributed to an understanding of electron behavior. Because of insights furnished by the uncertainty principle, scientists began to suspect that tiny particles like electrons could not be expected to behave like other matter.

Profiles in Science

Niels Bohr (1885–1962)

Niels Bohr, the second son of a physiology professor, was born and grew up in Copenhagen, Denmark. An exceptional athlete, he played championship soccer for the Akademsk Boldklub and excelled in many other sports.

Finishing his early schooling in Copenhagen, Bohr attended the University of Copenhagen, where he received his doctorate in 1911. He then traveled to England for postdoctoral work, where he joined the laboratory of J. J. Thomson (Section 3.4) in Cambridge and, somewhat later, Rutherford's laboratory (Section 3.6) in Manchester. While at Manchester, and in the months immediately after, he conceived of the "Bohr atom," the theory that made him famous and won him the Nobel Prize in 1922. Soon after, Bohr became chief of the Institute for Advanced Studies in Copenhagen, where he remained for nearly a generation. During his career as a theoretical physicist, in the decades in which modern physics was born, he made solid contributions, working with other great scientists such as Heisenberg, de Broglie, Planck, Schrödinger, Fermi, and Einstein.

During 1939, while in the United States for a conference, Bohr brought the news that two Germans, Hahn and Strasberg, had discovered nuclear fission in uranium. This immediately caught the interest of other scientists at the meeting, and much speculation took place. Somewhat later in the United States, work began on the development of a nuclear bomb, a project in which Bohr played a substantial part. In that arena, however, his greatest influence was exerted in trying to establish controls for the applications of the awesome power of the nucleus. For this work, in 1957 Bohr was awarded the first Atoms for Peace prize.

Researchers already knew that sometimes light acts like waves, such as waves in water, and that under other circumstances light behaves as though it were composed of fast-moving particles. Louis de Broglie used this information to advance the theory that electrons in motion might be considered to do the same, that is, electrons might sometimes behave as waves do and sometimes as particles do. In 1924, de Broglie proposed that moving electrons be assigned wavelengths; until then, wavelength had been used only to describe light and wave phenomena in water, strings, and similar systems. Shortly thereafter, Erwin Schrödinger took the final step and applied a mathematical description to electrons in atoms. Schrödinger's approach led to a theory that successfully avoided the disadvantages of the Bohr atom. This new theory was given the name **wave mechanics** or **quantum mechanics.**

The wave mechanics theory of electron behavior in atoms includes the following concepts:

1. *The electrons in an atom can in no way be regarded as particles of matter that obey the laws of classical physics.*

2. *We must consider that electrons in atoms behave as waves do.* Just as the wave created by a pebble dropped in a quiet pond exists in all parts of its

In 1932 Louis de Broglie (1892–1987) received the Nobel Prize in physics and published a comprehensive book on the wave mechanics of the electron.

Erwin Schrödinger (1887–1961)

Erwin Schrödinger was born in Austria and received his early schooling in Vienna. During World War I, he served as an artillery officer in the Austrian army. Soon after the war, he studied particle physics and thereafter taught physics at several European and English universities. Excited by de Broglie's wave concept of the electron, he began work on an explicit mathematical description of electron waves, beginning with the complex equations that describe waves in vibrating objects such as guitar strings.

Using this approach, he worked out the famous Schrödinger equation, the nearest thing we have to a description of an electron in an atom. This work, published around 1926, made him famous among physicists and in 1934 won him the Nobel Prize.

Unfortunately, easily visualized models of the atom, such as the Bohr atom, cannot be correct. Electrons can be thought of only in an abstract way.

circular pattern at once (Figure 4.3), so too *the wave representing an electron in an atom exists in many parts of the atom at once.*

3. *No possibility exists of assigning an exact location to an electron in an atom.* The best we can do is to know where the electron is most likely to be found at any particular moment, not where it actually is.

4. *The only precise things that can be known about electrons in an atom are their energies.*

In return for accepting the idea that electrons in atoms behave as waves rather than particles, we are rewarded by finding that wave mechanics easily explains why the atom does not collapse and why electron energy levels exist. If an electron is a wave, then it is not a particle that can fall into the nucleus. Just as the movement of a plucked guitar string, a wave motion, can create only certain tones, depending on the length of the string, so can an electron wave in an atom have only certain amounts of energy (exist only in quantized energy levels). Wave mechanics answers most of our questions satisfactorily and neatly fits all the known data about the various properties of elements.

Figure 4.3 In these circular waves, the wave, as such, exists in all parts of the circle at the same time.

4.3 Quantum Numbers

After studying this section, you should be able to:

- Explain what a quantum number is and does.
- List the four kinds of quantum numbers and state their relationships to each other.

Because we do not know the locations of electrons, we cannot use locations to distinguish one electron from another. Although we could keep track of electrons by reporting their actual individual energies, this would be quite cumbersome. As a consequence, a system of numbers and letters has been developed to label electron energy levels. These number–letter combinations are called **quantum**

Figure 4.4 A circular wave showing seven full wavelengths.

(n)

numbers. For our purposes, the quantum number system can be compared to the address system in a country. The designation of a certain house in the United States, for instance, proceeds from state to city to street and finally to street number. When all the levels are given, that house is uniquely specified. The address of an electron is similarly expressed by four quantum numbers, the difference being that the set of quantum numbers identifies the precise energy level of the electron, *not* its physical location in the atom. Although there is a relation between energy level and location, any information about the location of an electron in an atom is sketchy at best, whereas its energy is definite, precise, and useful. Just as no two houses can have the same address, no two electrons in an atom have the same set of quantum numbers. This last statement is often referred to as the **Pauli exclusion principle.**

We will frequently use quantum numbers in our future discussions, but we will be primarily interested in only two of the four kinds of quantum numbers that are needed to specify the energy level of an electron. The first is called the **principal quantum number,** which is represented by n. The principal quantum number, n, can have values of 1, 2, 3, or any greater positive whole number. Only seven values of n, given in order of increasing energy as $n = 1$, $n = 2$, $n = 3$, and so on up to $n = 7$, are needed to accommodate the electrons of all the elements now known. The value of n designates the main, or principal, energy level.

For each principal quantum number or value of n, there are sublevels designated by a second quantum number (usually represented by a letter). In order of increasing energy, the sublevels are given by s, p, d, and f. The letters s, p, d, and f represent labels that early researchers attached to spectral data from their experiments, and the labels have stuck. For each principal energy level, the number of sublevels equals the value of n. For $n = 1$, there is only one sublevel, s. For $n = 2$, there are two sublevels, s and p. $n = 3$ contains three sublevels, s, p, and d. Each of the other principal energy levels contains all four sublevels, s, p, d, and f. (See Table 4.1 and Figure 4.5.)

Theoretically, for energy levels greater than $n = 4$, there are sublevels of higher energy than s, p, d, and f. Only these four sublevel designations, however, are needed for all the electrons of the elements now known.

The third quantum number designates the sub-sublevels, or **orbitals.** Each s sublevel has one orbital. Each p sublevel has three orbitals. Each d sublevel

✳ all estimates

When we give values for all four quantum numbers of an electron in an atom, we have identified it uniquely.

Historically, sets of spectra were named according to some property seen or imagined by their discoverer. Thus s meant sharp, p meant principal, d meant diffuse, and f meant fundamental.

The term "orbitals" dates from the time when electrons were thought to travel in precise orbits. The "-al" now added to the end of the word avoids the undesirable preciseness of the word "orbit." An orbital is not an orbit.

Table 4.1 Sublevels Present for Each Principal Quantum Number

Principal Energy Level, n	Sublevels
1	s
2	s, p
3	s, p, d
4	s, p, d, f
5	$s, p, d, f, -$
6	$s, p, d, f, -, -$

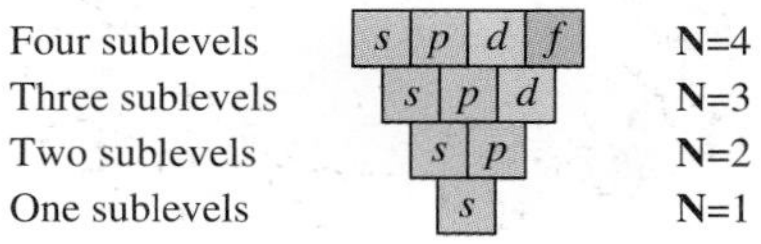

Four sublevels	s p d f	**N=4**
Three sublevels	s p d	**N=3**
Two sublevels	s p	**N=2**
One sublevels	s	**N=1**

Figure 4.5 The four main levels and their sublevels.

In each sublevel, the separate orbitals have labels. For example, p_x, p_y, and p_z are the three different p orbitals. In our study, we will not need to make such distinctions.

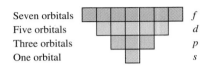

Seven orbitals	f
Five orbitals	d
Three orbitals	p
One orbital	s

Figure 4.6 The s sublevels each have one orbital, each p has three, each d has five, and each f has seven.

has five orbitals, while each f sublevel has seven orbitals (Figure 4.6). For our purposes we need to know only how many orbitals exist for each sublevel. Table 4.2 provides such a list.

The fourth quantum number is commonly called the *spin quantum number*. Although electrons don't actually *spin,* they do behave in certain situations as though they were spinning on an axis as a top does. The symbols used to draw electrons, arrows pointing up ↑ and down ↓, represent the fact that electrons can "spin" in opposite directions. Each orbital can contain a maximum of two electrons. When an orbital contains two electrons, the electrons must have opposite spin and are said to be *paired*. Table 4.3 shows the total number of electrons that can occupy each kind of sublevel.

Theoretically, each main energy level has enough sublevels and orbitals to hold $2n^2$ electrons if all its orbitals are filled to capacity with two electrons each. No atom of any known element, however, has enough electrons to fill completely the energy levels when $n = 5$ or greater. Despite not being filled, the sublevels do exist. Look at Table 4.4.

Table 4.5 summarizes the quantum numbers for you.

Table 4.2   Number of Orbitals in Each Kind of Sublevel

Sublevel	Number of Orbitals
s	1
p	3
d	5
f	7

EXAMPLE 4.1

What is the electron capacity of the $n = 2$ main level? In which sublevels are the electrons located?

Solution: The second principal energy level, $n = 2$, can hold eight electrons, because $2n^2 = 2 \times 2^2 = 8$. The s sublevel has one orbital with two electrons, and the p sublevel has three orbitals with six electrons.

Exercise 4.1

What is the electron capacity of the $n = 4$ energy level? In what sublevels are the electrons located?

Table 4.3   Number of Electrons That Can Occupy Each Orbital

Sublevel Type	Number of Orbitals	Number of Electrons in Sublevel
s	1	2
p	3	6
d	5	10
f	7	14

Table 4.4   Maximum Number of Electrons in Each Principal Energy Level

Principal Energy Level, n	Maximum Number of Electrons	$2N^2$
1	2	$2 \times (1)^2$
2	8	$2 \times (2)^2$
3	18	$2 \times (3)^2$
4	32	$2 \times (4)^2$
5	32 (theoretically 50)	$2 \times (5)^2$
6	32 (theoretically 72)	$2 \times (6)^2$
7	32 (theoretically 98)	$2 \times (7)^2$

Table 4.5 Quantum Numbers for All Known Elements

Principal Energy Level, n	Sublevels	Number of Orbitals	Maximum Number of Electrons
1	s	1	2
2	s	1	2
	p	3	6
3	s	1	2
	p	3	6
	d	5	10
4	s	1	2
	p	3	6
	d	5	10
	f	7	14
5	s	1	2
	p	3	6
	d	5	10
	f	7	14
6	s	1	2
	p	3	6
	d	5	10
	$(f)^*$	(7)	(14)
7	s	1	2
	(p)	(3)	(6)
	(d)	(5)	(10)
	(f)	(7)	(14)

$2n^2$

*Numbers and letters in parentheses represent energy levels in which no elements yet discovered have electrons.

Scientists have devised a shorthand notation to indicate a specific energy sublevel using quantum numbers. First, write the number n that designates the principal energy level and follow it by the letter designating the sublevel; for example, $3p$ refers to the p sublevel of the principal energy level 3. The number of electrons populating a sublevel is then given as a superscript to the right of the sublevel designation. For example, $4d^6$ says that there are six electrons in the d sublevel of the fourth main level.

EXAMPLE 4.2

Give the shorthand notation for four electrons in the $n = 3$ main level and the p sublevel.

Solution: The level and sublevel designation is $3p$. Put the four electrons in as a superscript, giving $3p^4$.

Exercise 4.2

What is the shorthand notation for three electrons in the d sublevel of $n = 4$?

Exercise 4.3

What is the value of *n*, and which sublevel is occupied by how many electrons, if the shorthand notation is $3d^6$?

Let's summarize what we have learned. We can specify the energy level of an electron using four quantum numbers: (1) *n*, the principal quantum number; (2) *s, p, d,* and *f* for the sublevels; (3) a quantum number for the orbitals; and (4) a quantum number for the electron spin. Each principal level has *n* sublevels, and each sublevel has a specific number of orbitals. Each orbital can contain a maximum of two electrons.

4.4 Quantum Numbers and the Order of Electron Energies

After studying this section, you should be able to:

- List, in increasing order of energy, the principal energy levels and sublevels of electrons in atoms.
- List the sublevels contained in each principal level and the maximum number of electrons possible in each.
- Explain why, for example, the 4*s* sublevel begins filling before the 3*d* sublevel begins to fill.
- Use the notation that gives the actual electron population in any principal level.

Quantum numbers correspond to energy levels. The energies of sublevels, however, do not follow the simple sequence of increasing principal quantum number. In certain cases, the difference in energy between the lowest and higher sublevels *within* a main energy level is larger than the difference in energy *between* that main level and the next higher main level. For example, the 3*d* sublevel lies at higher energy than the 4*s* sublevel. The result is that the sublevels of some main levels overlap those of the next main level, an effect that can be more easily understood from an illustration. Figure 4.7 shows both the main levels and their sublevels.

We just saw that the 3*d* sublevel is higher in energy than the 4*s* sublevel. Notice also that both the 4*d* and 4*f* sublevels lie higher than the 5*s* sublevel. Look at Figure 4.7 and find other cases of overlap.

You can use a simple memory device to remember the sublevels in order of increasing energy. To set up the memory device, simply write the designations of the sublevels in the order *s, p, d,* and *f,* beginning a new line for each new value of *n*. Then, starting at the top right side of the paper, draw a diagonal arrow through 1*s*, another parallel one through 2*s*, another through 2*p* and 3*s*, and so on through the entire set.

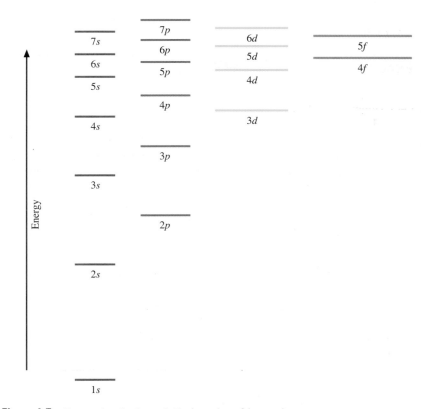

Figure 4.7   Energy levels through $7p$ in order of increasing energy.

As you draw the arrows one after another, your pencil will pass through the sublevels in order of increasing energy, $1s$, $2p$, $3s$, $3p$, $4s$, $3d$, $4p$, $5s$, and so on. Write down the order in which the points of the arrows pass through the various sublevels; then compare the result with Figure 4.7. It is far easier to remember this pattern than to memorize the entire sequence of energy levels.

The periodic table, which we will study in the next chapter, has a form that follows the sublevel energy sequence. This will also be useful to us in learning the overlapping pattern in which electrons fill energy levels in atoms.

4.5 **Electron Configurations**

After studying this section, you should be able to:

- Draw electron orbital distribution diagrams for the first 20 elements.
- Write the electron configurations of the first several dozen elements.

Every atom has a very large set of energy levels reaching far beyond any levels that are normally occupied by electrons. Ordinarily, however, the electrons of an atom occupy the energy levels of lowest possible energy. The effect can be compared to the behavior of water in a bowl. Naturally the water will lie in the bottom of the bowl. (The lowest part of the bowl is the part from which the largest amount of energy is required to lift the water out.) If more water is added to the bowl, the water level rises, because the lower part of the bowl has already been taken. Electrons in atoms behave in a similar manner; they fill the lowest energy levels first. An atom in which all the electrons occupy the lowest energy levels is said to be in the **ground state.**

Let's take a look at the rules that electrons follow when filling the energy levels of atoms.

1. *Every atom possesses a full set of main levels, sublevels, and orbitals. Some of these may be occupied by electrons, depending on how many electrons are present in the atom.*

2. *The number of electrons in an atom is the same as the atomic number of the element to which that atom belongs.*

3. *In the ground state of an atom, energy levels and sublevels fill in order of increasing energy.* This is known as the **aufbau principle.**

4. *It is possible for an orbital to contain zero electrons, one electron, or two electrons, but it cannot contain more than two.*

5. *Each orbital in a sublevel receives a single electron before any orbital in that same sublevel receives a second electron.* This is called **Hund's rule.**

Let's investigate Hund's rule by looking at diagrams that represent the orbitals within a sublevel as a set of boxes grouped together. For instance, the set of three orbitals for 2*p* would appear as

The electrons occupying the orbitals are then shown as vertical arrows. A single arrow represents one electron. A pair of arrows—one pointing up and one pointing down—represents a pair of electrons. One electron would occupy one of the orbitals:

According to Hund's rule, a second electron must occupy a different orbital:

A third electron would occupy the third orbital:

A fourth electron must now pair with one of the other electrons because there is no longer an empty orbital of the same sublevel available:

And so on until each of the three orbitals is filled with a pair of electrons.

It is possible for one or more electrons in an atom not to be in the ground state, but such a situation is unstable and soon "decays" to the ground state.

There is a physical basis for Hund's rule. Because electrons have negative charges, they repel one another. When they occupy separate orbitals, they can remain farther apart than when they are together in the same orbital. Two electrons will not occupy the same orbital as long as there is an empty orbital available in the same sublevel.

We are now ready to put the entire picture together. Figure 4.8 shows the electron distribution diagrams for the first 11 elements. Notice that in the diagram for nitrogen, the orbitals of the $2p$ sublevel receive one electron each. The first pairing of electrons in the $2p$ sublevel occurs when we reach oxygen. Because oxygen has four electrons in $2p$, one more than there are orbitals, it is necessary for two electrons to pair. It does not matter to us at this point which of the orbitals receives the additional electron.

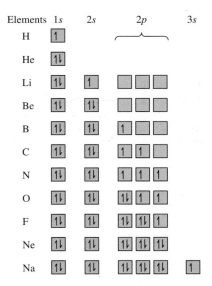

Figure 4.8 Orbital distribution diagrams for the first 11 elements.

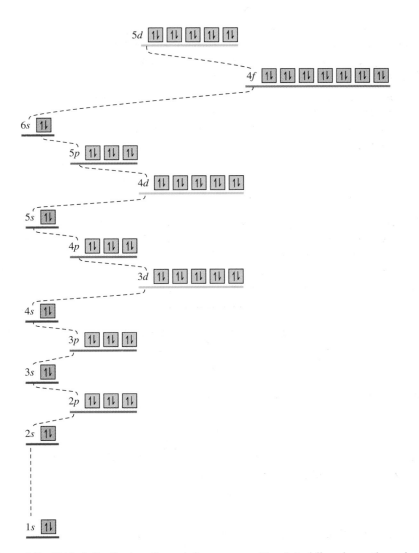

Figure 4.9 Orbital distribution diagram for mercury. The dotted line shows the order in which the energy levels fill with electrons.

Figure 4.9 shows the energies of the main levels and their sublevels for an atom of 80 electrons. It also shows the orbitals in each sublevel and the full population of electrons, two electrons in each orbital. Take the time to count the total number of electrons in some of the main levels. Check to see if the number in each case is $2n^2$. For example, in the s and p sublevels of the $n = 2$ main level, are there 2×2^2 electrons?

EXAMPLE 4.3

Draw the orbital distribution diagram of iron, Fe.

Strategy and Solution:

Step 1. Determine the number of electrons in an atom of Fe. There will be 26 of them because Fe has an atomic number of 26.

Step 2. Write the energy levels in order of increasing energy. Fe will require all of the energy levels through 4s, including the 3d's, although 3d will not be entirely filled. Using boxes to represent the orbitals, draw and label the orbitals that will be needed:

$1s$ $2s$ $2p$ $3s$ $3p$ $4s$ $3d$

Step 3. Place 26 arrows in the orbitals. First put a single arrow in each orbital, not pairing until you need to:

$1s$ $2s$ $2p$ $3s$ $3p$ $4s$ $3d$

> **Exercise 4.4**
>
> Draw the orbital distribution diagram of sulfur, S.

> **Exercise 4.5**
>
> What element has the following orbital distribution diagram?
>
>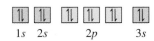
>
> $1s$ $2s$ $2p$ $3s$

We now have the information required to determine which energy levels will be occupied by the electrons in an atom of any element. The electron arrangement, called the **electron configuration,** is determined by following the rules we've discussed. The electron configuration of an atom is written as a list of the occupied main levels and sublevels of that atom, showing the number of electrons in each. In an electron configuration, no distinction is made between orbitals. The number of electrons in a sublevel is represented as a superscript to the right of the symbol for the sublevel. Let's look at the element carbon.

You will be needing to write electron configurations again and again as you continue in the chapters to come.

Carbon has two electrons in 1*s,* two electrons in 2*s,* and 2 electrons in 2*p*. The electron configuration for carbon, then, is $1s^2 2s^2 2p^2$.

EXAMPLE 4.4

What is the electron configuration of nitrogen?

Strategy and Solution:

Step 1. The atomic number of nitrogen is 7, so there are seven electrons in the nitrogen atom.

Step 2. Adding electrons one by one, we find that the first two electrons go into the 1*s* sublevel, giving us $1s^2$. The next two electrons go into the 2*s* sublevel, giving $2s^2$. The last three electrons go into 2*p,* giving $2p^3$. The entire electron configuration, then, is $1s^2 2s^2 2p^3$.

Step 3. Now check the work. The superscripts should add up to seven.

EXAMPLE 4.5

What is the electron configuration of selenium, Se?

Strategy and Solution:

Step 1. Selenium has atomic number 34.

Step 2. Proceed as before. The finished configuration is

$$1s^2 2s^2 2p^6 3s^2 3p^6 4s^2 3d^{10} 4p^4$$

In writing electron configurations, it is often useful to abbreviate sets of filled energy levels by using the symbols of noble gases that have those configurations. The electron configuration of argon is $1s^2 2s^2 2p^6 3s^2 3p^6$. We can substitute [Ar] for these energy levels in the electron configuration of selenium. The electron configuration of selenium becomes $[Ar]4s^2 3d^{10} 4p^4$.

> ### Exercise 4.6
>
> What is the electron configuration of zirconium, Zr (Z = 40). (The last sublevel with electrons in it is 4*d*.)

> ### Exercise 4.7
>
> What is the electron configuration of V (vanadium) (Z = 23)? Of Br (Z = 35)?

Table 4.6 is a list of the electron configurations of all the elements.

There are a few elements in which the energy levels are not populated according to the usual sequence discussed above. For our purposes, it is not necessary to know which elements these are or why these exceptions occur. A table of electron configurations is usually available if we need to look up unusual cases. We should remember, however, that some irregularities do exist. As an example, consider the element Cr, which has the configuration $1s^2 2s^2 2p^6 3s^2 3p^6 4s^1 3d^5$. If the usual sequence were followed, the 4*s* sublevel would contain two electrons

Table 4.6 Electron Populations of the Elements

Atomic Number	Element	1	2		3			4				5				6				7
		s	s	p	s	p	d	s	p	d	f	s	p	d	f	s	p	d	f	s
1	H	1																		
2	He	2																		
3	Li	2	1																	
4	Be	2	2																	
5	B	2	2	1																
6	C	2	2	2																
7	N	2	2	3																
8	O	2	2	4																
9	F	2	2	5																
10	Ne	2	2	6																
11	Na	2	2	6	1															
12	Mg	2	2	6	2															
13	Al	2	2	6	2	1														
14	Si	2	2	6	2	2														
15	P	2	2	6	2	3														
16	S	2	2	6	2	4														
17	Cl	2	2	6	2	5														
18	Ar	2	2	6	2	6														
19	K	2	2	6	2	6		1												
20	Ca	2	2	6	2	6		2												
21	Sc	2	2	6	2	6	1	2												
22	Ti	2	2	6	2	6	2	2												
23	V	2	2	6	2	6	3	2												
24	Cr	2	2	6	2	6	5	1												
25	Mn	2	2	6	2	6	5	2												
26	Fe	2	2	6	2	6	6	2												
27	Co	2	2	6	2	6	7	2												
28	Ni	2	2	6	2	6	8	2												
29	Cu	2	2	6	2	6	10	1												
30	Zn	2	2	6	2	6	10	2												
31	Ga	2	2	6	2	6	10	2	1											
32	Ge	2	2	6	2	6	10	2	2											
33	As	2	2	6	2	6	10	2	3											
34	Se	2	2	6	2	6	10	2	4											
35	Br	2	2	6	2	6	10	2	5											
36	Kr	2	2	6	2	6	10	2	6											
37	Rb	2	2	6	2	6	10	2	6			1								
38	Sr	2	2	6	2	6	10	2	6			2								
39	Y	2	2	6	2	6	10	2	6	1		2								
40	Zr	2	2	6	2	6	10	2	6	2		2								
41	Nb	2	2	6	2	6	10	2	6	4		1								
42	Mo	2	2	6	2	6	10	2	6	5		1								
43	Tc	2	2	6	2	6	10	2	6	5		2								
44	Ru	2	2	6	2	6	10	2	6	7		1								
45	Rh	2	2	6	2	6	10	2	6	8		1								
46	Pd	2	2	6	2	6	10	2	6	10										
47	Ag	2	2	6	2	6	10	2	6	10		1								

Table 4.6 Electron Populations of the Elements *(con't)*

Atomic Number	Element	1	2		3			4				5				6				7
		s	*s*	*p*	*s*	*p*	*d*	*s*	*p*	*d*	*f*	*s*	*p*	*d*	*f*	*s*	*p*	*d*	*f*	*s*
48	Cd	2	2	6	2	6	10	2	6	10		2								
49	In	2	2	6	2	6	10	2	6	10		2	1							
50	Sn	2	2	6	2	6	10	2	6	10		2	2							
51	Sb	2	2	6	2	6	10	2	6	10		2	3							
52	Te	2	2	6	2	6	10	2	6	10		2	4							
53	I	2	2	6	2	6	10	2	6	10		2	5							
54	Xe	2	2	6	2	6	10	2	6	10		2	6							
55	Cs	2	2	6	2	6	10	2	6	10		2	6			1				
56	Ba	2	2	6	2	6	10	2	6	10		2	6			2				
57	La	2	2	6	2	6	10	2	6	10		2	6	1		2				
58	Ce	2	2	6	2	6	10	2	6	10	1	2	6	1		2				
59	Pr	2	2	6	2	6	10	2	6	10	3	2	6			2				
60	Nd	2	2	6	2	6	10	2	6	10	4	2	6			2				
61	Pm	2	2	6	2	6	10	2	6	10	5	2	6			2				
62	Sm	2	2	6	2	6	10	2	6	10	6	2	6			2				
63	Eu	2	2	6	2	6	10	2	6	10	7	2	6			2				
64	Gd	2	2	6	2	6	10	2	6	10	7	2	6	1		2				
65	Tb	2	2	6	2	6	10	2	6	10	9	2	6			2				
66	Dy	2	2	6	2	6	10	2	6	10	10	2	6			2				
67	Ho	2	2	6	2	6	10	2	6	10	11	2	6			2				
68	Er	2	2	6	2	6	10	2	6	10	12	2	6			2				
69	Tm	2	2	6	2	6	10	2	6	10	13	2	6			2				
70	Yb	2	2	6	2	6	10	2	6	10	14	2	6			2				
71	Lu	2	2	6	2	6	10	2	6	10	14	2	6	1		2				
72	Hf	2	2	6	2	6	10	2	6	10	14	2	6	2		2				
73	Ta	2	2	6	2	6	10	2	6	10	14	2	6	3		2				
74	W	2	2	6	2	6	10	2	6	10	14	2	6	4		2				
75	Re	2	2	6	2	6	10	2	6	10	14	2	6	5		2				
76	Os	2	2	6	2	6	10	2	6	10	14	2	6	6		2				
77	Ir	2	2	6	2	6	10	2	6	10	14	2	6	7		2				
78	Pt	2	2	6	2	6	10	2	6	10	14	2	6	9		1				
79	Au	2	2	6	2	6	10	2	6	10	14	2	6	10		1				
80	Hg	2	2	6	2	6	10	2	6	10	14	2	6	10		2				
81	Tl	2	2	6	2	6	10	2	6	10	14	2	6	10		2	1			
82	Pb	2	2	6	2	6	10	2	6	10	14	2	6	10		2	2			
83	Bi	2	2	6	2	6	10	2	6	10	14	2	6	10		2	3			
84	Po	2	2	6	2	6	10	2	6	10	14	2	6	10		2	4			
85	At	2	2	6	2	6	10	2	6	10	14	2	6	10		2	5			
86	Rn	2	2	6	2	6	10	2	6	10	14	2	6	10		2	6			
87	Fr	2	2	6	2	6	10	2	6	10	14	2	6	10		2	6			1
88	Ra	2	2	6	2	6	10	2	6	10	14	2	6	10		2	6			2
89	Ac	2	2	6	2	6	10	2	6	10	14	2	6	10		2	6	1		2
90	Th	2	2	6	2	6	10	2	6	10	14	2	6	10		2	6	2		2
91	Pa	2	2	6	2	6	10	2	6	10	14	2	6	10	2	2	6	1		2
92	U	2	2	6	2	6	10	2	6	10	14	2	6	10	3	2	6	1		2
93	Np	2	2	6	2	6	10	2	6	10	14	2	6	10	4	2	6	1		2
94	Pu	2	2	6	2	6	10	2	6	10	14	2	6	10	6	2	6			2

Table 4.6 Electron Populations of the Elements *(con't)*

Atomic Number	Element	Main Levels and Sublevels																
		1	2		3			4				5				6			7
		s	*s*	*p*	*s*	*p*	*d*	*s*	*p*	*d*	*f*	*s*	*p*	*d*	*f*	*s*	*p*	*d* *f*	*s*
95	Am	2	2	6	2	6	10	2	6	10	14	2	6	10	7	2	6		2
96	Cm	2	2	6	2	6	10	2	6	10	14	2	6	10	7	2	6	1	2
97	Bk	2	2	6	2	6	10	2	6	10	14	2	6	10	9	2	6		2
98	Cf	2	2	6	2	6	10	2	6	10	14	2	6	10	10	2	6		2
99	Es	2	2	6	2	6	10	2	6	10	14	2	6	10	11	2	6		2
100	Fm	2	2	6	2	6	10	2	6	10	14	2	6	10	12	2	6		2
101	Md	2	2	6	2	6	10	2	6	10	14	2	6	10	13	2	6		2
102	No	2	2	6	2	6	10	2	6	10	14	2	6	10	14	2	6		2
103	Lr	2	2	6	2	6	10	2	6	10	14	2	6	10	14	2	6	1	2
104	Unq	2	2	6	2	6	10	2	6	10	14	2	6	10	14	2	6	2	2
105	Unp	2	2	6	2	6	10	2	6	10	14	2	6	10	14	2	6	3	2
106	Unh	2	2	6	2	6	10	2	6	10	14	2	6	10	14	2	6	4	2
107	Uns	2	2	6	2	6	10	2	6	10	14	2	6	10	14	2	6	5	2
108	Uno	2	2	6	2	6	10	2	6	10	14	2	6	10	14	2	6	6	2
109	Une	2	2	6	2	6	10	2	6	10	14	2	6	10	14	2	6	7	2

and the 3*d* sublevel would contain four electrons. You can find other exceptions by studying Table 4.6.

4.6 Electron Energy Level and Electron Location

After studying this section, you should be able to:

* Identify the probability density plots of 1*s*, 2*s*, and 2*p* energy levels.

According to the Heisenberg uncertainty principle, it is impossible to know the exact location of an electron occupying an energy level in an atom. The movements of electrons in atoms are entirely random, and the paths they take from place to place in the atom (so far as they can be said to take *paths*) are unknowable. All we can know about the position of an electron is its probable location. Depending on what energy level it occupies, an electron spends most of its time only in certain parts of an atom. There are also some parts of an atom to which electrons never go. If we know the energy level of an electron, we can say what the chances are of finding that electron in a particular part of an atom, although we can't determine precisely *where* it actually is located at any particular time.

Let's look at an analogy. Suppose you are asked to photograph a baseball game so that you can determine where each player spent his or her time on the diamond. One way to do this would be to climb a tall tower and use a time exposure to photograph the game. Although the images of the players would be blurred, and the photo would not reveal exactly where any player had been at any given time, it would show, for example, that the concentrated blur around the pitcher's box meant that you could almost always find the pitcher there.

Anything energetic enough to locate an electron in a particular place in an atom would knock the electron completely out of the atom. (This is another way of stating Heisenberg's uncertainty principle.)

The blur for the center fielder would be in a different place and would be more smeared out because fielders move around more than pitchers do. By studying the picture, you could come to some valid conclusions about the most likely locations of the players during the game.

If we could, by magic, take a time-exposure photograph of an electron in an atom, we would get a smeared blur telling us in what part of the atom the electron would most probably be found at any particular time. Electron probability plots, calculated mathematically from wave theory, give us such pictures. Studying probability plots yields much useful information about the locations of electrons in atoms and consequently about important properties of atoms.

Probability plots have fuzzy boundaries because, with certain exceptions, there is no limit to where the electron can be; the plots show only where it is most likely to be. As they are usually shown, the most heavily shaded parts of the pictures represent the regions of the atom within which there is about a 90% chance of finding the electron at any given time. In the baseball photo, you might find the pitcher near first base, but about 90% of the time you would find him within a few feet of the pitcher's mound.

The shapes of the probability plots are determined primarily by the orbitals that the electrons occupy. A plot of a $1s$ orbital, for example, has the same general shape as a plot of a $2s$ orbital or of a $3s$ orbital. Figure 4.10 shows that the plot of an s orbital has a spherical shape. The plots for a particular kind of orbital in different main levels (say, the s orbitals in the first four main levels) vary mostly in their average distance from the nucleus, not in their shape. In other words, as the principal quantum number increases, the electrons have a higher probability of being farther from the nucleus. Looking at Figure 4.10, we see, for example, that the $2s$ plot is larger than the $1s$ plot.

The probability plots of p orbitals have dumbbell shapes. Under ordinary circumstances, p orbitals in any particular main level have the same energy, but as Figure 4.11 shows, within a main level, the orbitals are oriented in different directions.

Because the plots for the d and the f orbitals are much more complex in their shapes, we will leave their discussion to more advance study.

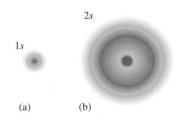

(a)   (b)

Figure 4.10 (a) Probability density plot for a $1s$ orbital. (b) Probability density plot for a $2s$ orbital.

We will discuss the importance of this fact in the next chapter.

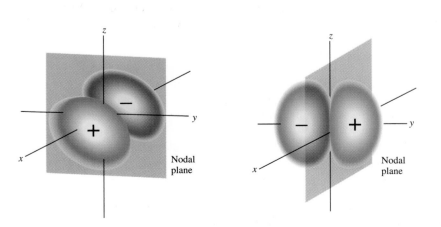

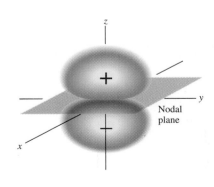

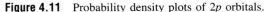

Figure 4.11 Probability density plots of $2p$ orbitals.

4.7 Lewis Diagrams of Atoms

After studying this section, you should be able to:

• Determine the number of valence electrons from electron configurations.
• Draw Lewis diagrams of atoms.

The valence electrons of an atom are the electrons having the highest principal quantum number.

As discussed in the previous section, the electrons having the highest principal quantum number are those spending their time farthest from the nucleus. Because the outer electrons are the part of an atom that interacts most with other matter, most properties of the elements result primarily from the actions of these electrons. These important electrons are called **valence electrons.** For example, the electron configuration of sodium is $1s^2 2s^2 2p^6 3s^1$, and the valence electron of sodium is the one 3s electron. *outer shell*

A method devised by G. N. Lewis of the University of California in 1902 (see Section 6.4) shows the valence electrons of an element pictorially. The symbol for the element is written and surrounded by as many dots as there are electrons with the highest principal quantum number. The dots are often shown in pairs. This kind of picture is called a **Lewis diagram** of an atom. In a Lewis diagram, the symbol of the element stands for the nucleus as well as for all the electrons that do not have the highest principal quantum number (the "inner" electrons). To write a Lewis diagram for an element, follow these steps:

Lewis diagrams show only the valence electrons of an atom.

1. *Write the symbol for the element.*

2. *Determine how many valence electrons the element has.* You can do this by writing the electron configuration for the element and seeing how many electrons have the highest principal quantum number. Soon we will learn how to use the periodic table to do this.

3. *Draw as many dots around the symbol as there are valence electrons.* Paired dots represent paired electrons. Single dots represent unpaired electrons.

EXAMPLE 4.6

Draw the Lewis diagram for fluorine.

one on each then pick one to start again

Strategy and Solution:

Step 1.   Write the symbol for fluorine.

$$F$$

Step 2.   The electron configuration is $1s^2 2s^2 2p^5$. Therefore, there are seven valence electrons. The two 2s electrons are paired. Since there are five electrons in the 2p level, there must be two pairs and a single electron.

Step 3.   Draw the valence electrons. Put in a pair of dots on each of three sides of the symbol and a single dot on the remaining side.

Drawing Lewis diagrams of atoms and molecules will be one of our most useful tools. Practice doing it.

$$: \overset{\displaystyle ..}{\underset{\displaystyle ..}{F}} \cdot$$

Exercise 4.8

Draw the Lewis diagram of phosphorus.

Exercise 4.9

Name three elements that have a Lewis diagram that looks like this one, where "X" stands for the symbol of the element.

4.8 Octets of Electrons

After studying this section, you should be able to:

- Understand why no atom has more than eight valence electrons.

Now that we know about electron energy levels, we are ready to learn why sublevel overlap has such a great effect on the properties of elements. Look at the electron configurations in Table 4.6. Notice that the total number of *valence* electrons never exceeds eight for any element.

Sublevel energy overlap causes new main levels to begin filling after the *s* and *p* sublevels of the previous main level are full, regardless of whether there is a *d* sublevel in the previous main level. In any main level, the *s* and *p* sublevels can hold eight electrons. *After eight electrons go into any main level of any atom, the next electron starts a new main level.* Any remaining sublevels of the previous level then fill sometime afterward.

As an illustration, let's consider the row of elements in the periodic table from K to Kr.

K	$1s^2\ 2s^2\ 2p^6\ 3s^2\ 3p^6\ 4s^1$
Ca	$1s^2\ 2s^2\ 2p^6\ 3s^2\ 3p^6\ 4s^2$
Sc	$1s^2\ 2s^2\ 2p^6\ 3s^2\ 3p^6\ 4s^2\ 3d^1$
Zn	$1s^2\ 2s^2\ 2p^6\ 3s^2\ 3p^6\ 4s^2\ 3d^{10}$
Ga	$1s^2\ 2s^2\ 2p^6\ 3s^2\ 3p^6\ 4s^2\ 3d^{10}\ 4p^1$
Ge	$1s^2\ 2s^2\ 2p^6\ 3s^2\ 3p^6\ 4s^2\ 3d^{10}\ 4p^2$
As	$1s^2\ 2s^2\ 2p^6\ 3s^2\ 3p^6\ 4s^2\ 3d^{10}\ 4p^3$
Se	$1s^2\ 2s^2\ 2p^6\ 3s^2\ 3p^6\ 4s^2\ 3d^{10}\ 4p^4$
Br	$1s^2\ 2s^2\ 2p^6\ 3s^2\ 3p^6\ 4s^2\ 3d^{10}\ 4p^5$
Kr	$1s^2\ 2s^2\ 2p^6\ 3s^2\ 3p^6\ 4s^2\ 3d^{10}\ 4p^6$

The electron configuration of K is $1s^2 2s^2 2p^6 3s^2 3p^6 4s^1$. For K, the highest principal quantum number is 4. There is one electron in the 4s orbital. Therefore, K has a single valence electron. Ca has an additional electron in the 4s sublevel, for a total of two valence electrons. Then in the series Sc ($1s^2 2s^2 2p^6 3s^2 3p^6 4s^2 3d^1$) through Zn ($1s^2 2s^2 2p^6 3s^2 3p^6 4s^2 3d^{10}$) the 3d sublevel fills. Zn has ten more electrons than Ca, all in the *d* sublevel, but even so, Zn has only two valence electrons, both in the 4s.

Although none of the atoms of the transition elements formally have more than two valence electrons, their chemistry sometimes involves d sublevel electrons which then behave as valence electrons.

Now the $4p$ sublevel begins to fill, beginning with the element Ga, $1s^2 2s^2 2p^6 3s^2 3p^6 4s^2 3d^{10} 4p^1$. In Ga, there are three electrons with the highest principal quantum number, two in the $4s$ and one in the $4p$; thus Ga has three valence electrons. The $4p$ sublevel continues to fill, so that Ge has four valence electrons, As has five, Se has six, Br has seven, and Kr has eight.

When another electron is added for the element Rb, however, the new electron must occupy the $5s$ orbital. The highest principal quantum number is now 5. Rb has only one valence electron, $5s^1$.

For none of these elements, nor for that matter for any element, will *the number of valence electrons in an atom ever exceed eight.* Eight valence electrons are often referred to as an **octet.**

Octets of electrons have particular importance in chemical bonding, a subject we will discuss in detail in Chapter 6.

Summary

We study electrons in atoms because they form the part of the atom that interacts most with the world. Because electrons are so small, they do not behave the way that larger particles do. Because electrons in atoms have definite, fixed energies, the uncertainty principle tells us that they can't be said to have any measurable or predictable location. It is not correct to think of an electron as a tiny particle orbiting a nucleus.

We can best understand electrons in atoms by treating them as though they were waves instead of tiny particles. Wave theory and spectroscopic measurements have enabled us to learn much about electrons. Electrons in atoms can have only certain amounts or levels of energy; that is, their energies are quantized.

Energy levels exist in a pattern described by the quantum number system. The main levels characterized by the principal quantum number, n, are divided into sublevels. The quantum numbers of the sublevels are usually designated by the letters s, p, d, and f. Sublevels are divided into orbitals. Each orbital can accommodate a maximum of two electrons, which are said to be paired. No two electrons can have the same set of four quantum numbers.

Atoms are usually in the ground state, in which case the electrons occupy the lowest set of energy levels available. When the s and p sublevels of any main level become completely filled, the next higher principal level begins filling. Because of this pattern of filling energy levels, no atom can have more than eight valence electrons.

Key Terms

Heisenberg uncertainty principle	*(p. 106)*	orbital	*(p. 113)*
spectroscopy	*(p. 106)*	ground state	*(p. 118)*
quantized	*(p. 107)*	aufbau principle	*(p. 118)*
energy level	*(p. 108)*	Hund's rule	*(p. 118)*
Bohr atom	*(p. 110)*	electron configuration	*(p. 120)*
wave mechanics	*(p. 111)*	valence electrons	*(p. 126)*
quantum number	*(p. 112)*	Lewis diagram	*(p. 126)*
Pauli exclusion principle	*(p. 113)*	octet	*(p. 128)*
principal quantum number	*(p. 113)*		

Questions and Problems

Section 4.1 Electron Energy Levels

Questions

1. Why is it more important to a chemist to learn about the electrons in atoms than to discover the details of the nucleus, where most of the mass of the atom resides?

2. If it were possible to build a sufficiently powerful microscope, would it then be possible to discover directly the characteristics of electrons in atoms? Explain your answer.

3. What is the best source of information about the energies of electrons in atoms?

4. What do we mean when we say that the energies of electrons in atoms are quantized? Use an analogy to explain your answer.

Section 4.2 The Theory of Electron Behavior in Atoms

Questions

5. Describe the Bohr atom.

6. Discuss one of the difficulties of the conceptually attractive Bohr theory.

7. Explain what is meant by the de Broglie wavelength.

8. For decades, the electron was regarded only as a particle. Describe the principal characteristic of an electron as viewed in terms of wave mechanics.

Section 4.3 Quantum Numbers

Questions

9. What is a sublevel? What is an orbital? Why not just say "orbit," instead of "orbital?"

10. Does a set of four quantum numbers for an electron in an atom give the location of the electron? Explain.

11. We say that four quantum numbers can describe an electron in an atom. What information is it that the quantum numbers actually give?

Problems

12. Which principal quantum number ($n = ?$) would contain a maximum of 50 electrons?

13. What is the maximum number of electrons that can occupy the $n = 6$ main level? The $n = 9$ main level?

14. What is the electron capacity of the $n = 1$ main level? In what sublevels are the electrons located?

15. What is the electron capacity of the $n = 3$ main level? In what sublevels are the electrons located?

16. Which principal quantum number contains eight electrons? 72 electrons?

17. Identify the principal quantum number and the sublevel, along with the number of electrons, in the following: $2s^2$, $4p^5$, $7s^1$, $4d^7$.

18. What is n for the electrons described as $4f^8$? How many of these electrons are there? In which sublevel are they?

19. Give the shorthand notation for the following.
    (a) five electrons in the $n = 4$ principal level and the p sublevel
    (b) two electrons in the $n = 3$ principal level and and d sublevel
    (c) one electron in the $n = 2$ principal level and s sublevel.
    (d) ten electrons in the $n = 6$ principal level and f sublevel.

20. Give the shorthand notation for the following:

n	Number of Electrons	Sublevel
2	2	p
3	1	s
3	4	d
4	10	f

21. Write, in order of increasing energy, the shorthand notations of the energy levels through principal quantum number 6.

22. If the next energy sublevel after f were g, what would be the maximum number of electrons that the g sublevel could contain?

Section 4.4 Quantum Numbers and the Order of Electron Energies

Questions

23. Explain what is meant by the overlap of electron energy levels.

24. Could the $n = 4$ main level have more than four sublevels? (Use the memory device given in this section.) Could the $n = 5$ main level have more than four sublevels?

25. Could the $n = 6$ main level have more than six possible sublevels? Explain.

26. Section 4.4 presents a memory device for the filling of energy levels. Complete the memory device, assuming that the g and h sublevels have the next higher energy after the f sublevels.

27. Develop another memory device for the filling of energy levels.

Section 4.5 Electron Configurations

Questions

28. Why is the nineteenth electron in K located in the fourth main level, rather than in the third?

29. What is the aufbau principle?

30. State Hund's rule in your own words.

31. What is the physical basis for Hund's rule?

32. Can an electron join an orbital that is already occupied by another electron? If so, under what conditions?

33. In an orbital distribution diagram, what information is given that is not given in an electron configuration?

34. When it is said that two electrons are "paired," what does this mean with respect to the sets of quantum numbers of those electrons?

35. What might be the reason for the irregular electron configurations of Cr, Cu, and Mo?

Problems

36. Draw the orbital distribution diagrams of Na, C, P, Se, and Ar.

37. Draw the orbital distribution diagrams for the following elements: F, Mg, Ga, V.

38. What element has the following orbital distribution diagram?

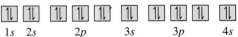

$1s$ $2s$ $2p$ $3s$ $3p$ $4s$

39. Identify the element that has this orbital distribution diagram.

$1s$ $2s$ $2p$ $3s$ $3p$

40. Write the electron configurations for the following elements: Mg, P, B, Sc.

41. Give the electron configurations for the following elements: Zn, K, Br, Te.

42. Seven electrons are in the $3d$ sublevel of a neutral atom. What element does the atom belong to? How many of these electrons are paired?

43. Another atom has four electrons in the $4p$ sublevel. What element does it belong to? How many of these electrons are paired?

44. Identify each element having the following in its electron configuration: $3s^1$, $2p^5$, $3d^5$, $4d^2$, $5p^1$.

45. Identify each element that has the following in its electron configuration: $3p^4$, $4d^6$, $4s^1$, $2p^2$, $5s^1$.

46. Assuming that [J] represents any noble gas, what elements have the following electron configurations?
    (a) $[J]s^2p^2$
    (b) $[J]s^2d^3$
    (c) $[J]s^1d^5$
    (d) $[J]s^2p^4$
    (e) $[J]s^2d^2$?

47. Assuming that [J] represents any noble gas, what elements have the following outer electron configurations?
    (a) $[J]s^2d^{10}p^5$
    (b) $[J]s^2d^7$
    (c) $[J]s^2p^3$
    (d) $[J]s^2p^6$

48. On Xanthu, the 112th element has been found. How many electrons are in its outermost main level? What is n for that level?

Section 4.6 Electron Energy Level and Electron Location

Questions

49. What can we know about the location of electrons in atoms?

50. How does the location of an electron depend on the principal quantum number?

51. What is an electron probability plot? Make rough sketches of the probability plots for s and p electrons.

52. Would the plot for a $1s$ electron be different than that for a $3s$ electron? Explain.

Section 4.7 Lewis Diagrams of Atoms

Questions

53. What are valence electrons?

54. Why are valence electrons important?

55. Describe how to draw a Lewis diagram of an atom.

Problems

56. Draw Lewis diagrams of the following elements: N, O, C, Ca.

57. Draw Lewis diagrams for Cl, Kr, Al, Sn, and As.

58. Which elements have Lewis diagrams like the following? ("A" stands for the symbol of the element.)

59. Name the element that has a Lewis diagram like the following ("X" stands for the symbol of the element).

Section 4.8 Octets of Electrons

Questions

60. Explain what is meant by an octet of electrons.

61. Main levels of n greater than 2 can accommodate 18 or more electrons, but atoms having electrons in those main levels never have more than eight valence electrons. Explain why this is.

62. Demonstrate that atoms never have more than eight valence electrons by explaining the sequence in which main levels and sublevels fill. Use the elements Cs to Rn as examples.

63. How does the fact that there are no atoms with more than eight valence electrons relate to Lewis diagrams of atoms?

Solutions to Exercises

4.1   32, s, p, d, f

4.2 $4d^3$

4.3 $n = 3$; the d sublevel has 6 electrons.

4.4
| ⇅ | ⇅ | ⇅ | ⇅ | ⇅ | ⇅ | ⇅ | ↑ | ↑ |
| 1s | 2s | | 2p | | 3s | | 3p | |

4.5   Mg

4.6 $1s^2 2s^2 2p^6 3s^2 3p^6 4s^2 3d^{10} 4p^6 5s^2 4d^2$.

4.7 $1s^2 2s^2 2p^6 3s^2 3p^6 4s^2 3d^3$ $1s^2 2s^2 2p^6 3s^2 3p^6 4s^2 3d^{10} 4p^5$

4.8

· **P** ·

4.9   N, P, As

Three of the group IV elements, tin, silicon, and lead.

Groups of Elements and the Periodic Table

B ecause electrons form the outer parts of atoms, it is the electrons that interact with the rest of the world. In Chapter 4, we learned that electrons in atoms occupy energy levels. We might reasonably expect, then, that patterns in electron energy levels would produce patterns in the physical and chemical properties of atoms. Such patterns of properties do indeed exist and are organized and displayed in the *periodic table,* an arrangement of the elements according to their physical and chemical properties.

The simple classification of elements into groups having common sets of properties has been practiced for more than a hundred years, but until fairly recently no one knew why such relationships existed. The most satisfying achievement of the wave theory is the detailed explanation of these well-known regularities in the behavior of the elements.

We begin Chapter 5 with a review of the historical development of the periodic table. We will learn how the form of the periodic table follows the sequence of electron energy levels. We can then begin to relate the properties of the elements to the structures of their atoms, a relation that is central to the study of chemistry.

5.1 Historical Development of the Periodic Table

After studying this section, you should be able to:

- Explain how elemental properties were used as the basis for the construction of the first periodic tables.
- Draw a periodic table in outline, filling in only the atomic numbers of the elements.

The importance of elemental substances began to be appreciated at about the time of Dalton. By 1830, more than 50 elements had been discovered and isolated in the pure state. Although chemists had been trying for many decades to find systems to classify the substances they studied, little progress was made until Dalton's atomic theory was accepted and the masses of atoms were measured.

The measurement of atomic weight is discussed in Chapter 8.

In 1863, the English chemist John Newlands made an arrangement of the known elements in order of increasing atomic weight. He found that, in many cases, certain properties of a given element closely resembled those of the element seven spaces away from it in the list. Newlands noticed, for instance, that lithium, sodium, and potassium (second, ninth, and sixteenth in his list) had similar chemical properties. At this time, however, atomic weight was considered to be of so little importance that Newlands was ridiculed.

Mendeleev based his periodic table on atomic weight and then rearranged it according to the chemical properties of the known elements.

In 1869, Lothar Meyer, in Germany, and Dmitrii Mendeleev, in Russia, made similar and somewhat improved listings. They arranged the elements into columns or groups based on elemental properties. Meyer was working with physical properties and Mendeleev mostly with chemical properties.

Although several other chemists also deserve credit, Mendeleev's work shows the greatest insight. While organizing material for a soon-to-be successful chemistry textbook, he listed the properties of the elements and attempted to classify them in some fundamental way. Then, arranging the elements in order by atomic weight, he noticed, as others had before, that at every seventh and seventeenth element there was a repetition of properties. Where the order of atomic weight did not agree with the physical and chemical properties, he changed the order or left a blank space. In the case of tellurium and iodine, for example, he listed the elements out of order with respect to atomic weight because tellurium resembled selenium and iodine resembled bromine, chlorine, and fluorine. The resulting table (Table 5.1) contained vertical groupings of elements with similar physical and chemical properties. The blank spaces in the table represented elements that had not yet been discovered (these spaces are shown by dashes in Table 5.1).

Until the era of Mendeleev, most chemists thought that the elements had no relation to one another. That is why Newlands received no attention and Mendeleev found difficulty convincing other chemists of the value of his findings.

Mendeleev's table, unlike that of Newlands, contained no serious exceptions to the similarities of properties. Although Mendeleev believed that the arrangement of the table reflected natural law, his belief received no actual verification for several years.

hypothesis confirmed later

In a spectacular piece of detective work, Mendeleev published a second paper predicting the properties of three elements that had not yet even been discovered. He based his predictions on the known properties of the elements nearby the undiscovered elements in his table. His predictions for one of these elements are shown in Table 5.2.

When these three elements (whose names today are gallium, scandium, and germanium) were discovered a few years later and their properties confirmed, Mendeleev's work became widely celebrated by the scientific world. The importance of systematic classification of the elements began to be appreciated. It is ironic that the greatest honors came from England, where the original work of Newlands had been dismissed.

Table 5.1 Mendeleev's Version of the Periodic Table (Produced in 1872)

TABELLE II

REIHEN	GRUPPE I. — R²O	GRUPPE II. — RO	GRUPPE III. — R²O³	GRUPPE IV. RH⁴ RO²	GRUPPE V. RH³ R²O⁵	GRUPPE VI. RH² RO³	GRUPPE VII. RH R²O⁷	GRUPPE VIII. — RO⁴
1	H = 1							
2	Li = 7	Be = 9,4	B = 11	C = 12	N = 14	O = 16	F = 19	
3	Na = 23	Mg = 24	Al = 27,3	Si = 28	P = 31	S = 32	Cl = 35,5	
4	K = 39	Ca = 40	— = 44	Ti = 48	V = 51	Cr = 52	Mn = 55	Fe = 56, Co = 59, Ni = 59, Cu = 63.
5	(Cu = 63)	Zn = 65	— = 68	— = 72	As = 75	Se = 78	Br = 80	
6	Rb = 85	Sr = 87	?Yt = 88	Zr = 90	Nb = 94	Mo = 96	— = 100	Ru = 104, Rh = 104, Pd = 106, Ag = 108.
7	(Ag = 108)	Cd = 112	In = 113	Sn = 118	Sb = 122	Te = 125	J = 127	
8	Cs = 133	Ba = 137	?Di = 138	?Ce = 140	—	—	—	— — — —
9	(—)	—	—	—	—	—	—	
10	—	—	?Er = 178	?La = 180	Ta = 182	W = 184	—	Os = 195, Ir = 197, Pt = 198, Au = 199.
11	(Au = 199)	Hg = 200	Tl = 204	Pb = 207	Bi = 208	—	—	
12	—	—	—	Th = 231	—	U = 240	—	— — — —

The arrangement of elements begun by Mendeleev is called the periodic table of the elements, or simply the **periodic table.** Because of its consistency with known chemical and physical properties, the periodic table became generally accepted by about 1890.

Over the next several decades, elements were discovered and placed in the appropriate groups in the periodic table. When argon was discovered, another discrepancy, like that of tellurium and iodine, was observed. Argon has an atomic weight greater than that of potassium but was placed before potassium in the periodic table because the physical and chemical properties of potassium resemble those of sodium and lithium, not of argon.

In the early 1900's, the concept of atomic number was developed and experimentally confirmed. It then became clear that the sequence of elements in the

Table 5.2 Mendeleev's Predictions for the Element Germanium*

Property	Mendeleev's Predictions in 1871	Observed in 1885
Atomic weight	72	72.60
Density (g/mL)	5.5	5.47
Specific heat (cal/g°C)	0.073	0.076
Formula of oxide	EsO₂	GeO₂
Density of oxide (g/mL)	4.7	4.703
Formula of chloride	EsCl₄	GeCl₄
Density of chloride (g/mL)	1.9	1.887
Boiling point of chloride (°C)	Under 100°C	86°C

*What Mendeleev called ekasilicon, Es, we now call germanium, Ge.

Profiles in Science

Dmitrii Ivanovich Mendeleev (1834–1907)

Dmitrii Mendeleev was born in Tobolsk, Siberia, where his father was a teacher. Mendeleev received his early schooling in Tobolsk, graduating from the gymnasium (high school) with a concentration in science and mathematics. At about this time, his father died, leaving his mother to carry on. Knowing of her son's talents, she was determined that he would get a good education. Moving the great distance to Moscow, she tried to get help from her brother for Mendeleev's education, but to no avail. Then, moving again, this time to St. Petersburg, she prevailed on friends to get Mendeleev into the Chief Pedagogical Institute. Soon afterwards she died from overexertion. Now alone, Mendeleev worked night and day, graduating finally with a Master's Degree and high honors. He was immediately appointed to the faculty of the institute.

Not long afterwards, Mendeleev was invited to go to Heidelberg, a center of German science, where he worked for several years. At Heidelberg, Mendeleev began working on the properties of liquids in equilibrium with their vapors, an investigation that he would follow for many decades and to which he would make substantial contributions.

In 1861, Mendeleev returned to the faculty at St. Petersburg. Soon thereafter, he began work on what was to be a phenomenally successful chemistry textbook. An outgrowth of this work was the arrangement of elements that brought him fame. In 1869, his first paper on the classification of elements, the periodic table, was read in Moscow, with a second paper following soon after. The work received little attention and no praise because the subject had previously been studied by several other researchers. But Mendeleev boldly added something; he noted vacant spaces in the table, predicted that elements would be discovered to fill the spaces, and even predicted the properties of these unknown elements. When his predictions were fulfilled a few years later, he became famous. Mendeleev now is probably more greatly recognized than any other Russian chemist.

Although he is best known for the periodic table discoveries, Mendeleev worked on that subject for only a few years. In his long and fruitful career in science, he made contributions to petroleum chemistry, the development of natural resources, inorganic chemistry, the properties of gases and liquids, barometry, alloys, and smokeless powder. He was even interested in ballooning and made many ascensions. He died of pneumonia at the age of 73.

The modern periodic table is based on atomic number.

periodic table was determined by atomic number, rather than by atomic weight, thus solving the mysteries surrounding I/Te and Ar/K. A modern version of the periodic table is shown in Figure 5.1 and inside the front cover of this book.

5.2 The Periodic Table

After studying this section, you should be able to:

- Identify the groups and periods in the periodic table.
- Locate the alkali metals, noble gases, transition metals, lanthanides, and actinides in the periodic table.

The most common arrangement of the periodic table, Figure 5.1, consists of seven horizontal rows of elements called **periods.** The two yellow rows that

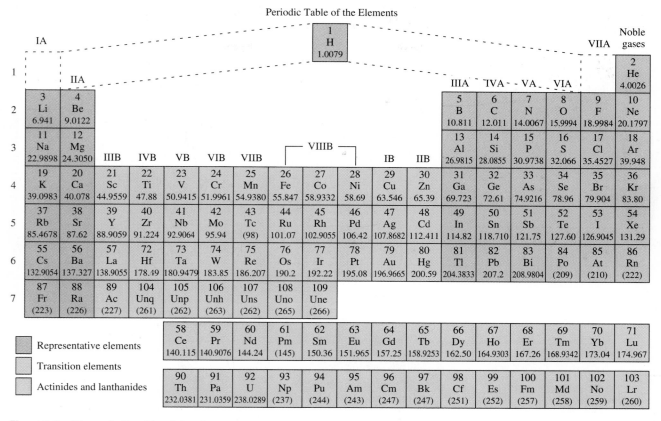

Figure 5.1 The periodic table of the elements.

appear below the table actually belong in the sixth and seventh periods, as shown in Figure 5.2. (These rows are normally shown separately for convenience.)

The periodic table is arranged in sets of **groups,** vertical families of elements vert. that have similar physical and chemical properties. Thus Li, Na, K, Rb, Cs, and Fr, the first column, constitute a group known as the **alkali metals.** The last column, He through Rn, forms a group called the **noble gases.** Except for the first period, each period begins on the left with one of the alkali metals and ends with one of the noble gases.

The colors in Figure 5.1 designate three categories into which the elements naturally fall. The elements in the pink areas are called the **representative elements.** The elements in the blue area in the center of the table, are known as the **transition elements,** and the two yellow rows form the **lanthanides** and **actinides.** Most of our interest will be focused on the representative elements.

Occasionally, the representative elements are called main group elements.

Exercise 5.1

Referring to the periodic table, give the symbol, atomic number, and period for silicon.

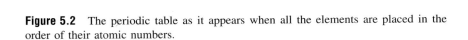

	IA								1 H										Noble VIIA gases
1		IIA											IIIA IVA VA VIA						2 He
2	3 Li	4 Be											5 B	6 C	7 N	8 O	9 F	10 Ne
3	11 Na	12 Mg				IIIB IVB VB VIB VIIB			⌐VIIIB⌐		IB IIB		13 Al	14 Si	15 P	16 S	17 Cl	18 Ar
4	19 K	20 Ca					21 Sc	22 Ti	23 V	24 Cr	25 Mn	26 Fe	27 Co	28 Ni	29 Cu	30 Zn	31 Ga	32 Ge	33 As	34 Se	35 Br	36 Kr
5	37 Rb	38 Sr					39 Y	40 Zr	41 Nb	42 Mo	43 Te	44 Ru	45 Rh	46 Pd	47 Ag	48 Cd	49 In	50 Sn	51 Sb	52 Te	53 I	54 Xe
6	55 Cs	56 Ba	57 La	58 Ce	59 Pr	60 Nd	61 Pm	62 Sm	63 Eu	64 Gd	65 Tb	66 Dy	67 Ho	68 Er	69 Tm	70 Yb	71 Lu	72 Hf	73 Ta	74 W	75 Re	76 Os	77 Ir	78 Pt	79 Au	80 Hg	81 Tl	82 Pb	83 Bi	84 Po	85 At	86 Rn
7	87 Fr	88 Ra	89 Ac	90 Th	91 Pa	92 U	93 Np	94 Pu	95 Am	96 Cm	97 Bk	98 Cf	99 Es	100 Fm	101 Md	102 No	103 Lr	104 Unq	105 Unp	106 Unh	107 Uns	108 Uno	109 Une

Lanthanides and actinides   Transition metals

Representative elements

Figure 5.2 The periodic table as it appears when all the elements are placed in the order of their atomic numbers.

5.3 Electron Energy Levels and the Periodic Table

After studying this section, you should be able to:

- Describe how the form of the periodic table follows the changes in the electron energy levels of the elements.
- Use the periodic table to determine the electron configuration of most elements.

The arrangement of the periodic table was originally based on the physical and chemical properties of the elements. But not until 40 years after the first tables were accepted did scientists begin to understand why periodicity, the regular, periodic repetition of similar properties, of elements occurs. Periodicity was finally explained in detail only after electron wave theory was developed.

To understand the relationship between the periodic table and wave theory, we will study the electron configurations of several elements. First, look at hydrogen and helium and their electron configurations.

H $\quad 1s^1$

He $\quad 1s^2$

The electrons in hydrogen and helium are in the $n = 1$ energy level. Looking at the periodic table, you can see that hydrogen and helium are also the *only* elements in the first period.

Next consider the electron configuration of lithium.

Li $1s^2 2s^1$

Notice that one of the three electrons of lithium occupies the $n = 2$ energy level. Lithium is the first element in the second period. Now look at the electron configurations of the other elements in the second period.

Be $1s^2 2s^2$
B $1s^2 2s^2 2p^1$
C $1s^2 2s^2 2p^2$
N $1s^2 2s^2 2p^3$
O $1s^2 2s^2 2p^4$
F $1s^2 2s^2 2p^5$
Ne $1s^2 2s^2 2p^6$

All of these elements have electrons in the $n = 2$ energy level. The next element, sodium, is the first element of the third period and has one electron in the $n = 3$ energy level.

Na $1s^2 2s^2 2p^6 3s^1$

Each of the elements H, Li, Na, K, Rb, Cs, and Fr starts a new period and also begins filling a new principal energy level.

Now notice that lithium and sodium atoms each have one valence electron in an s sublevel. As a matter of fact, atoms of all members of the alkali metal group each have one valence electron in an s sublevel.

Li $1s^2 2s^1$
Na $1s^2 2s^2 2p^6 3s^1$
K $1s^2 2s^2 2p^6 3s^2 3p^6 4s^1$
Rb $1s^2 2s^2 2p^6 3s^2 3p^6 4s^2 3d^{10} 4p^6 5s^1$
Cs $1s^2 2s^2 2p^6 3s^2 3p^6 4s^2 3d^{10} 4p^6 5s^2 4d^{10} 5p^6 6s^1$
Fr $1s^2 2s^2 2p^6 3s^2 3p^6 4s^2 3d^{10} 4p^6 5s^2 4d^{10} 5p^6 6s^2 4f^{14} 5d^{10} 6p^6 7s^1$

Now look at the electron configurations of the next family, the **alkaline earth metals.**

Be $1s^2 2s^2$
Mg $1s^2 2s^2 2p^6 3s^2$
Ca $1s^2 2s^2 2p^6 3s^2 3p^6 4s^2$
Sr $1s^2 2s^2 2p^6 3s^2 3p^6 4s^2 3d^{10} 4p^6 5s^2$
Ba $1s^2 2s^2 2p^6 3s^2 3p^6 4s^2 3d^{10} 4p^6 5s^2 4d^{10} 5p^6 6s^2$
Ra $1s^2 2s^2 2p^6 3s^2 3p^6 4s^2 3d^{10} 4p^6 5s^2 4d^{10} 5p^6 6s^2 4f^{14} 5d^{10} 6p^6 7s^2$

The atoms of the alkaline earth metals each have two valence electrons in an s sublevel. In fact, all the elements in the first two columns or groups have their highest energy electrons in s sublevels. These elements are the members of what is called the **s block.** The s block is shown in pink in Figure 5.3.

All s block elements have their highest energy electron(s) in an s sublevel.

If an element appears in the first column, or family, its atoms each have one valence electron, which lies in an s sublevel. If an element appears in the

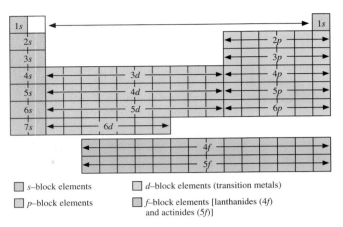

Figure 5.3   Electron energy levels and the periodic table.

second column, its atoms each have two valence electrons in an s sublevel. The principal quantum number of the valence electrons in the s sublevel corresponds to the period in which we find the element. For example, potassium, K, is located in the fourth period and the first column of the periodic table. Its atoms each have one valence electron in the $4s$ energy level.

Now let's look at boron, B. Because the $1s$ and $2s$ energy levels are filled, the fifth electron of boron must enter the $2p$ energy level.

B $1s^2 2s^2 2p^1$

The atoms of the elements from boron to neon, Ne, all have electrons in the $2p$ energy level. Carbon atoms each have two electrons in $2p$. Nitrogen atoms have three, and so on. The blue set of columns in Figure 5.3, headed by boron through neon, make up the p **block.** The number of electrons in the highest p energy level corresponds to the column in which the element is located. For example, the **halogens,** members of the family in the fifth column of the p block, beginning with fluorine, have atoms each containing five valence electrons in a p sublevel.

All p block elements have their highest energy electron(s) in a p sublevel.

F $1s^2 2s^2 2p^5$
Cl $1s^2 2s^2 2p^6 3s^2 3p^5$
Br $1s^2 2s^2 2p^6 3s^2 3p^6 4s^2 3d^{10} 4p^5$
I $1s^2 2s^2 2p^6 3s^2 3p^6 4s^2 3d^{10} 4p^6 5s^2 4d^{10} 5p^5$
At $1s^2 2s^2 2p^6 3s^2 3p^6 4s^2 3d^{10} 4p^6 5s^2 4d^{10} 5p^6 6s^2 4f^{14} 5d^{10} 6p^5$

The atoms of the noble gas group—the last column, neon to radon—each have six electrons in the highest energy p sublevel.

Ne $1s^2 2s^2 2p^6$
Ar $1s^2 2s^2 2p^6 3s^2 3p^6$
Kr $1s^2 2s^2 2p^6 3s^2 3p^6 4s^2 3d^{10} 4p^6$
Xe $1s^2 2s^2 2p^6 3s^2 3p^6 4s^2 3d^{10} 4p^6 5s^2 4d^{10} 5p^6$
Rn $1s^2 2s^2 2p^6 3s^2 3p^6 4s^2 3d^{10} 4p^6 5s^2 4d^{10} 5p^6 6s^2 4f^{14} 5d^{10} 6p^6$

In atoms of elements in the s and p blocks, the highest energy electrons have principal quantum numbers, n, that correspond to the period number. For example, Ar is in the *p* block and the third period. The highest energy electrons of Ar, $3p^6$, have *n* of 3.

From the preceding examples, we can see that *elements in a given group all have atoms containing the same number of valence electrons in the same type of sublevels.*

The **d block,** indicated in yellow in Figure 5.3, contains elements that have their highest-energy electrons in a *d* sublevel. The principal quantum number of electrons in the *d* sublevel corresponds to the period number minus one. Consider the electron configuration of iron. Iron is in the *d* block of the fourth period.

The principal quantum number, n, for the highest-energy d *electrons equals the period number less one.*

−1

Fe $1s^2 2s^2 2p^6 3s^2 3p^6 4s^2 3d^6$ *d level subtract one*

There are six electrons in the $3d$ energy level of iron. The principal quantum number, $n = 3$, of these *d* electrons, is given by $4 - 1 = 3$.

The **f block,** indicated by green in Figure 5.3, contains the inner transition elements. The inner transition series contains elements whose highest energy electrons are in *f* sublevels, making these elements in many ways similar to those of the transition metals. The series is called *inner* transition because the electrons in *f* sublevels on the average spend their time closer to the nucleus than do those of the *d* sublevels. The principal quantum number of the electrons in the *f* sublevel of the lanthanides and actinides is the same as the period number minus two. Europium, Eu, located in the sixth period, for example, has its highest-energy electrons in the $4f$ sublevel, $n = 6 - 2 = 4$.

The principal quantum number, n, for the highest-energy f *electrons equals the period number less two.*

−2

Armed with this information, we can now use the periodic table to help us determine the electron configuration of any element. Electrons fill energy levels in order from $1s$ to $2s$ to $2p$ and so on according to Figure 5.3.

EXAMPLE 5.1

Use the periodic table in Figure 5.3 to determine the electron configuration of antimony, Sb.

Strategy and Solution:

Step 1. Determine the number of electrons. Antimony has 51 electrons.

Step 2. These electrons will fill energy levels in the order given in Figure 5.3. Two electrons fill the $1s$ energy level. Two electrons fill the $2s$ energy level. The fifth through tenth electrons occupy the $2p$ sublevel. The eleventh and twelfth electrons must occupy the $3s$ energy level. The $3p$ sublevel fills with six more electrons. The $4s$ sublevel fills next with two electrons. The next ten electrons fill the $3d$ sublevel. Six more electrons will occupy the $4p$ sublevel. The $5s$ energy level fills with the next two electrons. Ten more electrons occupy the $4d$ sublevel. Because antimony is in the third column of the *p* block, there are three electrons in the $5p$ sublevel.

Sb $1s^2 2s^2 2p^6 3s^2 3p^6 4s^2 3d^{10} 4p^6 5s^2 4d^{10} 5p^3$

Step 3. Check by adding the superscripts. The total must be 51 for antimony.

Exercise 5.2

Using the periodic table, give the period, block, and electron configuration for magnesium, Mg.

As you study the periodic table, keep in mind that it was originally arranged according to the properties of the elements, not to the order of the energy levels. Our knowledge of electron energy levels comes mostly from electron wave theory and spectroscopy. The fact that the periodic table agrees so well with wave theory might lead you think that one came from the other, but the two were actually discovered independently. The remarkable agreement between them gives powerful support to the wave theory of the electron.

5.4 Valence Electrons and the Periodic Table

After studying this section, you should be able to:

- Use the periodic table to determine the number of valence electrons in the atoms of most elements.

The elemental families are given group numbers, which are printed above the vertical columns. There are various systems for numbering groups, the most common of which are current American usage and new notation. Both of these are shown in Figure 5.1. In current American usage, the A groups (IA, IIA, and so on) are the representative elements, and the B groups are the transition elements. Presently, there is considerable discussion and effort being expended on establishing an improved method of notation for groups in the periodic table.

In Section 4.5 we learned how to find the number of valence electrons in an atom by writing the electron configuration. Now we can see that it is possible to say how many valence electrons there are by looking at the position of an element in the periodic table. *For the representative elements (the A groups), the number of valence electrons is the same as the group number* (examples: one for Rb, four for Si, seven for Br).

The elements of groups IIIB through VIIIB and those of groups IB and IIB are the transition and the inner transition elements. The concept of valence electrons is not entirely applicable to the transition and inner transition elements because the reactions of these elements can involve the *d* or *f* electrons.

Exercise 5.3

(a) What is the number of valence electrons in the atoms of each of the following elements: Al, Se, Xe, F, He (be careful), and Ge?

(b) Why is He an exception to the general rule for counting valence electrons?

5.5 Metallic and Nonmetallic Character

After studying this section, you should be able to:

- Give rough comparisons of the extent of metallic character of several elements.
- Describe the relation between metallic character and the positions of elements in the periodic table.

In Section 2.8 we learned about the properties of metallic and nonmetallic elements. We can now use our knowledge of those characteristics to find examples of periodicity in the periodic table. First we consider the 17 elements beginning with Li and ending with K. Li is metallic, but as we follow the period from Be to Ne, we find that the elements become less and less metallic as we approach fluorine, F, which is one of the least metallic of all the elements. Next to fluorine is the noble gas, neon, Ne. Sodium, Na, which starts a new period, is strongly metallic, like Li. As we continue across this row, element by element, we find the metallic character again diminishing until we reach Cl, a strong nonmetal followed by the almost inert Ar. A new row starts with the highly metallic K, and the cycle is repeated.

Metallic behavior is strongest in the elements on the left and lower parts of the periodic table and weakest in the upper and right portions. A step-shaped boundary beginning on the left and bottom sides of the boron square and stepping down and right to At (see Figure 5.1) separates the metals and nonmetals. The few elements lying close to the boundary, the metalloids, display both metallic and nonmetallic properties. They cannot properly be included in either class.

Metallic character decreases from left to right across a period.
increases from up to down
Metals are found on the left and lower parts of the periodic table.

Three elements: (left to right) tin, germanium, and silicon. Although they resemble tin, silicon and germanium are metalloids and have some nonmetallic properties.

5.6 Atomic Radius

After studying this section, you should be able to:

- Describe the relative sizes of atoms within a period and within a group.
- Explain the reasons for the general trends of atomic radii in the periodic table.

Another example of periodicity can be seen by studying atomic sizes. Let's look at the second period in Figure 5.4. Of the elements in this period, Li has the largest atoms. Those of Be are smaller, those of B smaller still, and so on to F, the smallest. Similarly, the atoms of the elements in the third period decrease in size as the elements are taken from left to right. Although there are interruptions in the trend in the *d* and *f* blocks, the general pattern of decreasing atomic size from left to right repeats itself through the entire periodic table.

The reason for this periodic behavior can, of course, be seen in the structure of the atoms. Taking the elements from left to right in a period, we note that the atomic number, the charge on the nucleus, increases. However, the number of "inner" electrons (those having *n* less than the *n* of the period) remains constant across the period. Inner electrons tend to shield valence electrons from the charge of the nucleus. From left to right across a period, then, shielding remains relatively constant, while nuclear charge increases. Increasing nuclear charge increases the attraction between the valence electrons and the nucleus, resulting in smaller atoms.

As we can see from Figure 5.4, when we take the elements from top to bottom in a column, we find that the atoms of any given element are smaller than those of the element below. Also, from top to bottom in a column, the highest principal quantum number, *n*, becomes larger, element by element. Recall that in Section 4.6 we learned that electrons of higher principal quantum number, *n*, compared to those of lower *n*, are more likely to be a greater distance from the nucleus. Thus valence electrons of the atoms of elements at the bottom of the periodic table spend their time, on the average, farther from the nucleus than valence electrons of the lighter elements. In general, within a column of elements,

Because the sizes of noble gas atoms are measured differently from those of other atoms, noble gases appear to have atomic radii that do not always follow the trend.

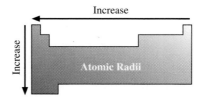

General trends in atomic radii in the periodic table.

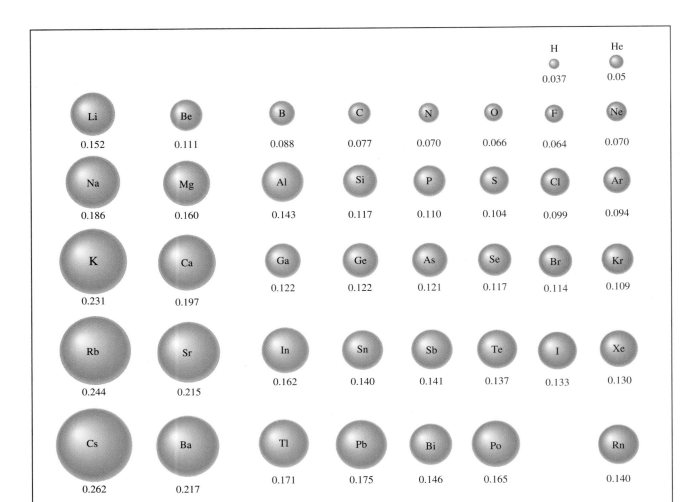

Figure 5.4   Atomic radii of the representative elements.

atomic size increases with an increase in the principal quantum number of the valence electrons. Even so, the size increase is not proportional to the increase in number of electrons. As the atomic number increases, the increasing charge of the nucleus exerts greater and greater force on the electrons, pulling them closer to the nucleus. So the increase in size of atoms, as you take elements down a column, is not as great as you might at first think.

5.7 Ionization Energy

After studying this section, you should be able to:

- Give rough comparisons of the ionization energies of several elements.
- Describe the relation between ionization energy and the positions of elements in the periodic table.
- Explain the reasons for the general trends of ionization energies in the periodic table.

Table 5.3   Table of Ionization Energies*

1	2	3	4	5	6	7	8
						H 1312	He 2372
Li 520	Be 900	B 801	C 1086	N 1402	O 1314	F 1681	Ne 2081
Na 496	Mg 738	Al 578	Si 786	P 1012	S 1000	Cl 1251	Ar 1520
K 419	Ca 590	Ga 579	Ge 762	As 944	Se 941	Br 1140	Kr 1351
Rb 403	Sr 550	In 558	Sn 709	Sb 832	Te 869	I 1009	Xe 1170
Cs 376	Ba 503	Tl 589	Pb 716	Bi 703	Po 812	At	Rn 1037

*The units are kilojoules per mole.

Every atom has substantial attraction for its own electrons. Thus energy is required to remove an electron from an atom. The energy required to take an electron completely away from a neutral atom is called the first ionization energy, or just the **ionization energy** (Table 5.3). We see an example of ionization energy when an electron is removed from a sodium atom.

| Na | + | ionization energy | $\longrightarrow$ | Na^+ | + | e^- |
| atom | | | | ion | | electron |

The second ionization energy is the amount of energy required to remove a *second* electron from the atom. For successive electrons, there is a third ionization energy, a fourth, and so on. As more electrons are removed, the ionization energy increases. How can this be explained? When an electron is removed, a positive ion results. The nucleus of this positive ion attracts its electrons more strongly than does that of a neutral atom (one fewer valence electron, less repulsion between electrons), increasing the ionization energy. In general, as attractive forces between electrons and nuclei increase, the ionization energy increases.

This effect can be seen in another way. Look at the curve for the ionization energies of the alkali metals, Li to Cs in Figure 5.5. The ionization energy of Na is slightly less than that of Li. Why? Recall what we learned in Section 4.6. Valence electrons of higher principal quantum number, n, spend their time, on the average, farther from the nucleus than do those of lower n. Correspondingly, there is a decrease in attraction between valence electrons and their nuclei. Because the single $3s$ electron of sodium spends more of its time farther from the nucleus than does the $2s$ electron of lithium, the sodium valence electron feels less attraction to its nucleus than does that of lithium. We see, then, that the ionization energies of the alkali metals decrease as we take the elements from top to bottom in the group. This trend is generally followed in all of the columns of the periodic table.

We know from our previous discussion of atomic radii that atomic size de-

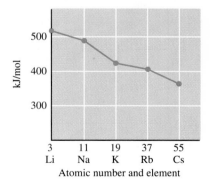

Figure 5.5   Ionization energies of the alkali metals, lithium to cesium.

increse across left to right

Generally, ionization energy decreases from top to bottom in a column.

Science in Action

Silicon: Uncommon Uses for a Common Element

The tweezers hold a single computer chip made of silicon.

ilicon, the second most abundant element on earth, has been recognized as an element since the early 1800's. It was first isolated and purified in 1824 by Jacob Berzelius of Sweden. For more than a century, however, its only industrial use was in making glass, which begins with sand, SiO_2. But in the mid-twentieth century, silicon suddenly became an exciting element with a large number of essential uses.

One such use is in silicones. Silicones are a family of long-chain compounds mostly of silicon, carbon, oxygen, and hydrogen. They were first discovered in the General Electric laboratories in 1945 and became industrially important during the next three decades. Nowadays, silicones are found everywhere and used for all manner of sealants, lubricants, stain repellents, and special kinds of rubber. Because of the strengths of the bonds in the long-chain molecules, silicones maintain their properties even at high temperatures. Silicone greases and rubbers are used in furnaces and space shuttles. Their unusual molecular structures make silicones unattractive to other substances; this makes them useful as stain repellents and coatings for raincoats. More than 3 million pounds of silicones were produced in the United States in 1991. Silly Putty is a silicone product. The first footprints on the moon were made by boots with silicone soles.

As a metalloid, silicon displays both nonmetallic and metallic properties; it can conduct electricity. Ever since the discovery of transistors in 1948, germanium, silicon, and a few other elements have become essential to the electronic industry. The now-famous "computer chips" are made from thin wafers of extremely pure silicon. Such a "chip," no bigger than a postage stamp, can store millions of bits of information. The region in Northern California in which the most advanced electronic components are made often is called "Silicon Valley," not because silicon is mined there but because the solid-state revolution in electronics was nurtured there.

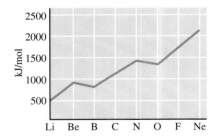

Figure 5.6 First ionization energies of the elements lithium through neon.

2p

The 2p sublevel electron distribution of an oxygen atom.

Ionization energy increases from left to right across a period.

creases from left to right across a period. The atoms on the right are smaller because of larger attractive forces between their nuclei and electrons. Since electrons that are more strongly held are also more difficult to remove, ionization energy *increases* from left to right across a period. Look at the ionization energies of the second period shown in Figure 5.6, for example.

Although the trend is for ionization energy to increase from left to right across the periodic table, there are two dips in the ionization energy curve shown in Figure 5.6. The first dip between Be and B results from the difference in energies of the *s* and *p* sublevels. Because the electrons in a *p* sublevel are higher in energy than electrons in an *s* sublevel, a *p* electron is more easily removed from an atom. A second dip between nitrogen and oxygen is caused by the fact that oxygen is the first element in which two electrons must occupy the same *p* orbital. The resulting electron–electron repulsion makes it easier to remove one of the electrons from the atom. This effect outweighs the increase in nuclear charge, causing oxygen to have a slightly smaller ionization energy than does nitrogen.

Finally, we look at the pattern of ionization energies as it relates to the entire periodic table. The curve in Figure 5.7 repeatedly rises to the large ionization energy of a noble gas, then plunges dramatically to the low ionization energy of an alkali metal. Notice that the curve tends to level out when it comes to a transition metal series or to the lanthanides or actinides, an effect consistent

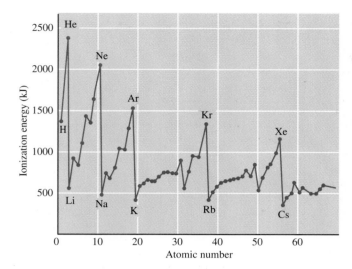

Figure 5.7 A graph of the ionization energies of some of the elements.

with the many other close similarities in the properties of these sets of metals. The ionization energy curve gives us one of the best displays of periodicity and group similarities.

5.8 Electron Affinity

After studying this section, you should be able to:

- Give rough comparisons of the electron affinities of several elements.
- Describe the relation between electron affinity and the positions of elements in the periodic table.
- Explain the reasons for the general trends of electron affinities in the periodic table.

Atoms of many of the nonmetals readily gain one or more extra electrons to become negative ions. **Electron affinity** is a measure of the attraction of an atom for an extra electron brought from far away and added to the atom. When an atom gains an extra electron, energy is released as the electron responds to the attractive force (much as a brick loses energy in falling to earth). For example, energy is released when a bromine atom accepts an electron into its $4p$ sublevel. In this process, a bromine ion, Br^-, is produced.

$$Br + e^- \longrightarrow Br^- + energy$$

Bromine atoms have one of the highest electron affinities. (Look at Table 5.4.)

In general, the values of electron affinity tend to follow those of ionization energy. That is, electron affinity increases from left to right and from bottom to top in the periodic table. If an atom attracts its own electrons strongly, it will also have a strong attraction for an extra electron, provided that the s and p sublevels are not both full. The noble gas atoms with their full s and p sublevels have no attraction for extra electrons and correspondingly have little or no electron affinity.

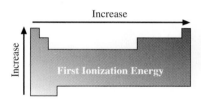

General trends in first ionization energies in the periodic table.

same ionization & affinity

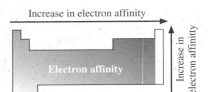

Increase in electron affinity

Electron affinity

Increase in electron affinity

General trends in electron affinities in the periodic table.

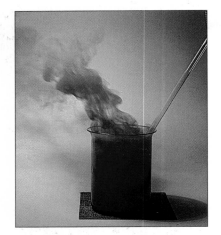

Chlorine reacts vigorously with iron to produce iron(III) chloride.

Only recently have any noble gas compounds, such as XeF_6, been successfully made. None are very stable.

Sodium can easily be cut with a knife.

Table 5.4   Table of Electron Affinities

H −72.8							He +21
Li −59.6	Be +241	B −26.7	C −122	N 0	O −141	F −328	Ne +29
Na −52.9	Mg +230	Al −42.5	Si −134	P −72.0	S −200	Cl −349	Ar +34
K −48.4	Ca +156	Ga −28.9	Ge −119	As −78.2	Se −195	Br −325	Kr +39
Rb −46.9	Sr +167	In −28.9	Sn −107	Sb −103	Te −190	I −295	Xe +40
Cs −45.5	Ba +52.0	Tl −19.3	Pb −35.1	Bi −91.3	Po −183	At −270	Rn +41

(up & down)

5.9 Characteristics of Groups of Elements

After studying this section, you should be able to:

* List similar properties for elements of several groups.

Although there are many trends in the periodic table, the strong physical and chemical similarities that elements within the same group bear to one another are of most interest to us. As we discussed earlier, the properties of the elements are determined primarily by the outer, or valence, electrons of their atoms. Elements in the same group have the same number of valence electrons in the same type of energy sublevel. This sameness of electron configuration causes the properties to be similar.

The halogens, F to At, for example, are all strongly nonmetallic. Except for iodine, they all have fairly low melting and boiling temperatures, and their solids are soft and easily deformed. The halogens are also highly reactive chemically.

Next to the halogens in the periodic table are the noble gases (He to Rn), which until recently were known as the inert gases. Because their atoms have full *s* and *p* sublevels, the noble gases rarely form compounds. The members of this group have no metallic characteristics whatever and must be classified as nonmetals; however, they lack the chemical reactivity of the other nonmetals.

At the far left side of the table are the alkali metals. These elements have melting and boiling points that are, for metals, quite low. The alkali metals can easily be hammered or bent into new shapes and can easily be cut with a knife. They are among the most chemically reactive of all the elements.

The alkaline earth elements are also quite metallic. They are, however, less reactive, harder, and have higher melting temperatures than do the alkali metals.

These four groups of elements—halogens, alkali metals, alkaline earth metals, and noble gases—exhibit the most distinct sets of group characteristics. The elements lying in other vertical columns share some properties but often to a lesser extent.

The physical and chemical properties of the transition elements are very similar because their properties are strongly influenced by *d* electrons. The dif-

Table 5.5   Formulas of Some Oxides (Arranged According to Periodic Law)*

H_2O				
Li_2O	BeO	B_2O_3	CO_2	N_2O_5
Na_2O	MgO	Al_2O_3	SiO_2	P_2O_5
K_2O	CaO			
Rb_2O	SrO			
Cs_2O	BaO			

*This list does not necessarily give all of the oxides formed by these elements; these compounds have been chosen to illustrate periodicity. The elements of many groups, however, form only one type of oxide, BeO and MgO, for example.

Like other alkali metals, potassium reacts violently with water.

ferences that do occur, some of which are significant, cause us to consider these elements individually rather than in groups.

A good example of periodicity is given by the similarities in formulas of compounds containing the elements within groups. Table 5.5 shows the formulas of the oxides of some of the elements, arranged by group. Notice that the same atom number ratio appears in all the formulas in each vertical column.

Suppose we had been given the formulas for the oxides of Li, Na, and K and had been asked to make a prediction as to what the formula would be for the oxide of Rb. We would undoubtedly have predicted Rb_2O, which is correct. Similarities in formulas occur not only in the oxides of the alkali metals, but also in the compounds of many groups. Often, knowing the formula for one compound of one element in a group will allow you to make an accurate prediction about compounds formed by other members.

Exercise 5.4

The formulas for some of the chlorides of the alkali metals are NaCl, RbCl, and CsCl, while those for some of the sodium halides are NaCl, NaBr, and NaF. What would you guess to be the formula for the compound of Rb and I?

Calcium, an alkaline earth metal, reacts with water less vigorously than do the alkali metals.

5.10 Changes in Properties Within Elemental Groups

After studying this section, you should be able to:

- Explain why there are differences in the properties of the elements in several groups of the periodic table.

Although groups of elements are often characterized by strong resemblances, there are also differences in properties among the elements of most groups.

Toward the left and the right sides of the periodic table, where elements are either very metallic or very nonmetallic, changes in group characteristics are small.

Because all the elements in a column have the *same number* of valence electrons, the differences in properties of those elements must be caused by some other factor. That factor is the distance of the valence electrons from the nucleus. Figure 5.8 shows the sizes of the atoms of the alkali metals and of the halogens. As the sizes of the atoms increase, so do the distances of the valence electrons from the nuclei. accounts for diff. in properties

Changes of properties within columns are most easily seen in the center groups of the periodic table, where the columns cross the boundary between the metals and the nonmetals. In group VA, for instance, the top elements, N and P, are classic nonmetals, but As and Sb display some metallic character, and Sn and Pb must be classed as metals. The metallic character of an element depends on how loosely the atom holds its valence electrons. In any given kind of sublevel, the farther the valence electrons are from the nucleus, the less the attraction between the nucleus and the electrons. Metallic properties, therefore, *increase* from top to bottom in a column. Look at Figure 5.9.

Exercise 5.5

Using the periodic table and the known properties of groups, describe the element helium in terms of size, reactivity, and metallic character. Do the same for aluminum.

Even in the groups whose elements are highly similar, differences in properties can be seen from top to bottom in a column. For instance, both Cl and I are nonmetals, but I begins to display a few metallic characteristics, while Cl has none whatever. Look at Figure 5.10.

Similarly, in the most metallic elements, changes in properties can be seen as we take the elements from top to bottom in a column. For example, in the alkali metals, Li to Cs, reactivity increases markedly with an increase in atomic radius. Li reacts moderately well with water, Na readily, and K explosively (Figure 5.11). And in its other characteristics, K is clearly more metallic than Li.

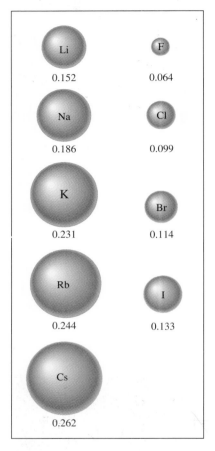

Figure 5.8 Relative sizes of the atoms of the alkali metals and the halogens.

Figure 5.9 Although both are in group VIA, sulfur is a yellow powder and selenium resembles a metal.

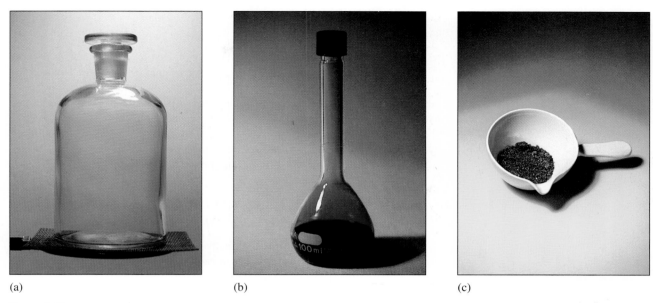

(a)   (b)   (c)

Figure 5.10   (a) Chlorine, a yellow gas. (b) Bromine, a brown liquid. (c) Iodine, a purple solid.

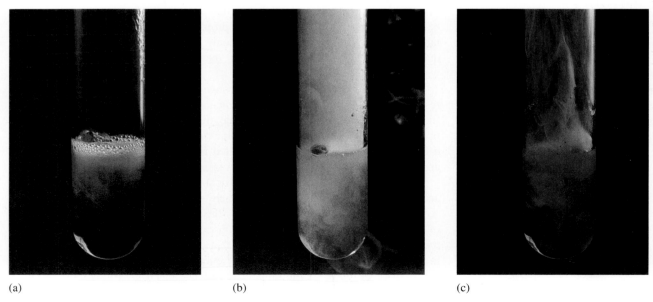

(a)   (b)   (c)

Figure 5.11   The reactions of (a) Li with water, (b) Na with water, and (c) K with water.

5.11 The Special Nature of Hydrogen

After studying this section, you should be able to:

- List the chemical and physical properties of hydrogen and explain the differences between this element and others that have the same number of valence electrons.

Although hydrogen and the alkali metals have similar electron configurations, hydrogen is not an alkali metal. Most of the physical and chemical properties of H are those of nonmetals, even though solid hydrogen is metallic. In addition, hydrogen has many other properties not shared by any other element.

Hydrogen is a solid only below −259°C; it is one of the lowest-melting substances.

There are two reasons for the unusual properties of hydrogen:

1. *Hydrogen has the smallest atoms.*
2. *The hydrogen electron lies close to the nucleus without electrons of lower principal quantum number to shield it from the nuclear charge.*

When a hydrogen atom loses its electron to become the ion, H^+, only the bare nucleus is left, and, since He is unreactive, H is the only element whose atoms exist in ionic form as only nuclei. Because of the unique properties given H by these characteristics, it really cannot be classified in any group but must be studied separately. We will again encounter situations in which H displays its unique behavior.

Summary

Long before the concept of atomic number was developed, chemists had begun to arrange the elements according to mass and to group together elements of similar properties. What is now known as the periodic table of the elements was the result of these efforts. The subsequent discovery of atomic numbers led to improvements in the table and provided a new basis for the arrangement that had previously reflected only observed properties. A full explanation of why the table takes the form it does did not become available, however, until the second quarter of the twentieth century, when the wave theory of electrons in atoms was developed.

The periodic table can be divided conveniently into three sections: the representative elements, the transition elements, and the lanthanides and actinides. In the standard form of the periodic table, the representative elements are shown in two blocks (the *s* block and the *p* block), separated by the three rows of the *d* block. The *f* block is usually separated from the other elements in the table and is shown as two rows of 14 elements each.

The table is further divided into vertical columns, called groups or families, and horizontal rows, called periods. There are seven periods and 18 groups.

The progression of the table follows the electron energy levels of the atoms of the elements. For example, the alkali metals (the column that is farthest to the left) contain elements with atoms that each have a single valence electron in an *s* sublevel. If we take the representative elements from left to right of any period, we find that the number of valence electrons follows the group A number. For example, in the third period, Na to Ar, Na atoms each have one valence electron, Mg atoms each have two, Al atoms three, and so on.

Because the properties of the elements are determined by the electron energy levels of the atoms, groups of elements exhibit strong resemblances in physical and chemical properties. Although there are differences in properties within elemental groups, especially in groups IIIA through VIA, it is nonetheless possible to predict many properties of an element simply by knowing the group to which it belongs.

Although hydrogen and the alkali metals have similar electron configurations, hydrogen does not share the properties of the alkali metals but has its own individual characteristics. The distinctive physical and chemical properties of hydrogen arise from its structure: The single electron of a hydrogen atom is directly exposed to the nucleus without being shielded by electrons of lower principal quantum number. Moreover, when an atom of H loses an electron, it becomes a bare nucleus; no other atom shares that characteristic.

Key Terms

periodic table	*(p. 135)*	alkaline earth metals	*(p. 139)*
periods	*(p. 136)*	*s* block	*(p. 139)*
groups	*(p. 137)*	*p* block	*(p. 140)*
alkali metals	*(p. 137)*	halogens	*(p. 140)*
noble gases	*(p. 137)*	*d* block	*(p. 141)*
representative elements	*(p. 137)*	*f* block	*(p. 141)*
transition elements	*(p. 137)*	ionization energy	*(p. 145)*
lanthanides	*(p. 137)*	electron affinity	*(p. 147)*
actinides	*(p. 137)*		

Questions and Problems

Section 5.1 Historical Development of the Periodic Table

Questions

1. What did Newlands use as a basis for his proposed table?

2. What did Mendeleev use as a basis for the periodic table he designed? Explain in detail.

3. Mendeleev published a paper that predicted the properties of then undiscovered elements. Why was this prediction important?

Section 5.2 The Periodic Table

Questions

4. Explain how groups and periods are arranged in the periodic table.

5. To what period do the actinides belong? The lanthanides?

6. Which elements are in group IIA? Period 2?

7. In which groups would the undiscovered elements with the following atomic numbers belong: 110, 117?

Problems

8. Give the symbols of the elements that fit these descriptions.
   (a) Period 2, group VIIA
   (b) Period 4, group VIB
   (c) Period 1
   (d) Period 6, group IIA
   (e) Period 4, representative elements
   (f) Group VA
   (g) Group IIB
   (h) Period 7, group IIIB

9. Give the period and group for each of the following elements: Ar, As, Sn, Ir, Sr.

10. Without looking at another table, draw the periodic table through element 86. Use atomic numbers to identify the elements. Fill in the symbols for the elements you already know.

Section 5.3 Electron Energy Levels and the Periodic Table

Questions

11. Explain what is meant by the statement that "each period represents the beginning of a new main electron energy level."

12. Why does the first period have only two elements, while periods two and three each contain eight elements?

13. Distinguish between the terms "*s* block," "*p* block," "*d* block," and "*f* block." Which of these blocks corresponds to the representative elements?

14. What is the structural factor that relates the elements in any group of the periodic table? In any period? In any block?

Section 5.4 Valence Electrons and the Periodic Table

Questions

15. What do groups of elements have in common, besides similar properties?

16. In which groups do the atoms of the elements each have four valence electrons? In which groups do atoms have six?

Section 5.5 Metallic and Nonmetallic Character

Questions

17. In what part of the periodic table are the elements of greatest metallic character located?

18. In what part of the periodic table are the elements of least metallic character located?

19. Which of the elements, O, S, Se, Te, and Po, shows the greatest metallic character?

20. The metallic properties of the elements increase from the top to the bottom of the periodic table. Why is the increase in metallic properties not obvious in group IA?

Problems

21. Classify the following elements as metals, nonmetals, or metalloids: Cs, C, S, Cl, As, Ne, Si.

22. Complete the following table

Atomic Number	Name of Element	Electron Configuration	Metal or Nonmetal
12			
25			
14			
9			
19			
34			

23. Draw Lewis diagrams for atoms of the following elements: Rb, Se, S, I, As, B, H, He, Ne.

24. Using the periodic table, give the block, group, period, and number of valence electrons for each of the following elements. Classify each as a metal or nonmetal.
    (a) magnesium
    (b) phosphorus
    (c) potassium
    (d) bromine
    (e) krypton
    (f) copper
    (g) vanadium
    (h) nitrogen
    (i) cesium

25. An element with the atomic number M is a noble gas. What kind of element is the one with atomic number $M - 2$? $M + 2$?

Section 5.6 Atomic Radius

Questions

26. In a given group of elements, why do those at the top of the periodic table have smaller atoms than those at the bottom?

27. Even though atomic weight increases in any period from left to right in the periodic table, the atoms of those elements become progressively smaller. Why?

28. What two factors have the greatest effect on the attraction of an atom for its valence electrons?

Problems

29. List the symbols of the elements in each of the following groups in order of increasing size of the atoms.
    (a) K, As, Ga, F
    (b) Cs, Rb, Br, Cl
    (c) Cs, Al, Na, Ar
    (d) O, Sn, P
    (e) Ta, V, U, Cu

30. It is said that the element 118 exists on Xanthu. Do you believe that the element could exist anywhere? Why? If it did exist, what would be its period, group, size, block, and number of valence electrons ? What other information can you estimate about it?

Section 5.7 Ionization Energy

Questions

31. Define the term "ionization energy."

32. Why does it require energy to remove an electron from an atom?

33. How do ionization energies vary across the periodic table and down the periodic table? Why do they vary in this way?

34. Explain why the ionization energy of helium is greater than that of lithium.

35. Why is the ionization energy of boron less than that of beryllium?

36. Why is the ionization energy of Al less than that of Mg?

37. Explain why the ionization energy of oxygen is less than that of nitrogen.

38. Explain why the ionization energy of sulfur is less than that of phosphorus.

39. Plot the second ionization energies for the elements B through Ne. Explain why the first break in the curve (the notch) comes at C, rather than at B, as it does in the plot

of Figure 5.6. (Data: second ionization energies: Be, 417; B, 575; C, 558; N, 678; O, 803; F, 801; Ne, 1143.)

Section 5.8 Electron Affinity

Questions

40. Define the term "electron affinity."

41. Why is the electron affinity of iodine less than that of fluorine?

42. Predict which will have the higher electron affinity, oxygen or nitrogen. Why?

43. Why does the change in electron affinities generally follow the same pattern as the changes in ionization energies from one place to another in the periodic table? Why are there notable exceptions?

44. Can an atom have both a low ionization energy and a high electron affinity? Do the atoms of greatest ionization energy always have the greatest electron affinity? Explain.

Section 5.9 Characteristics of Groups of Elements

Questions

45. Name some elemental physical and chemical properties that demonstrate periodicity. In each case, give examples to support your answer.

46. What physical and chemical properties do the alkali metals have in common?

47. What physical and chemical properties do the halogens have in common?

48. Why are the physical and chemical properties of He, Ne, and Ar similar?

49. The compounds NaOH, NaCl, and $MgCl_2$ are well known. What is the formula for the compound containing hydroxide ion, OH^-, and calcium?

50. The following compounds are well known: $MgBr_2$, NaCl, and NH_3. What are likely formulas for the compounds of the following pairs of elements: Ba and I, Rb and Br, As and H?

51. Carbon dioxide has the formula CO_2. Write the formula for an oxide of each of the other elements in group IVA.

52. Each atom of the elements of group IIB has two valence electrons. Atoms of the elements of group IIA also have two valence electrons. Knowing that $MgCl_2$ and $CaCl_2$ exist, would you have predicted that $ZnCl_2$ and $CdCl_2$ exist? Explain fully.

Section 5.10 Changes in Properties Within Elemental Groups

Questions

53. What main factor is responsible for the differences in properties within a group?

54. Phosphorus is a nonmetal and Bi is a metal, yet both are in the same group. Explain.

Section 5.11 The Special Nature of Hydrogen

Questions

55. The alkali metals are the most metallic group known. Hydrogen and the alkali metals have similar electron configurations, yet hydrogen is not considered an alkali metal. Why are the properties of hydrogen different from those of the alkali metals?

56. In many periodic tables H appears twice, once above Li and once above F. Discuss a probable reason for this unique listing.

57. Why doesn't He exhibit the same unusual properties as H?

More Difficult Questions

58. Use the periodic table to make your choices.
    (a) Which has the larger atom size, Na or K? Se or S? O or N?
    (b) Which is the more likely compound, PH_2 or PH_3? TeH_2 or TeH?
    (c) Which is the better conductor of electricity, P or As?
    (d) Which has the greater density, Li or B?

59. The properties of the five elements B through F vary greatly; those of the five elements Ti through Fe are more similar, and the properties of the five elements Nd through Gd are almost identical. Why are there differences in the similarities of properties among these three horizontal series of elements?

Solutions to Exercises

5.1 Si, 14, 3

5.2 3, *s* block, $1s^2 2s^2 2p^6 3s^2$

5.3 a. 3, 6, 8, 7, 2, 4. b. The $n = 1$ energy level can contain only two electrons.

5.4 RbI

5.5 He is the second smallest atom; only H is smaller. He is unreactive, entirely nonmetallic. Al is a metal; it is smaller than Mg, larger than Si, and it is rather reactive as a metal.

White phosphorus reacts violently with chlorine to form the compound PCl_5.

How Elements Form Compounds

Just as a brick falls to earth in response to the attractive force of gravity or a compass needle moves under a magnetic field, so do electrons respond to the forces acting on them by undergoing chemical reaction. By far the most important requirement for an initial understanding of chemical reaction and its results is an understanding of the forces exerted on electrons by nuclei, both the nuclei of their own atoms and those of other atoms. In the previous chapter we studied ionization energy and electron affinity, both of which give us information about the forces on electrons. We apply that knowledge now to learn about the strong forces called chemical bonds that hold atoms together.

Chemical bonds are forces that hold atoms together. The structures that result are stable and can be identified by formulas, either of elements (as in N_2 or S_8) or of compounds (as in H_2O or NaCl). Chemical bonds are the principal object of much of the study of chemistry.

For our purposes, this definition of a chemical bond will be sufficient, although there are certain exceptions, such as the bonds in diamonds.

To begin our study of chemical bonds, we investigate the formation of ions during chemical reaction. We can then learn about the chemical bonds in both ionic and covalent compounds. What we learn will eventually enable us to predict for many elements in the periodic table the kinds of compounds that they form, as well as the formulas and some of the properties of those compounds.

Although our present discussion of chemical bonding focuses primarily on the chemical reactions between elements, we should keep in mind that reactions can also occur between compounds and between compounds and elements, topics that we will explore somewhat later. The fundamental facts we will learn by examining the reactions of elements, however, are also true for reactions of other species.

6.1 Ion Formation

After studying this section, you should be able to:

- Predict how many electrons the atoms of any particular element are likely to gain or lose in an electron transfer reaction.
- List several properties of ions.

As we will learn, one of the ways atoms form compounds is by losing or gaining electrons. When atoms lose or gain electrons, they become electrically charged; that is, they become ions. If a neutral atom *gains* one or more electrons, it thereby adds the same number of negative charges and becomes a negative ion. Negative ions are called **anions.** When an atom *loses* one or more electrons, it no longer has enough electrons to equal the number of protons in its nucleus; it becomes a positive ion. Positive ions are called **cations.** The amount of charge on either a positive or negative ion is exactly the same as the number of electrons gained or lost, since each electron carries one unit of charge. To designate ions and their charges, we use the symbol of the element and a superscript on the right-hand side showing the sign and amount of charge the ion carries. For example, magnesium atoms can lose two electrons; the charge on a Mg ion is +2, and the ion is written as Mg^{2+}. Occasionally, if the charge on an ion is +1 or −1, the 1 is omitted, as in F^-.

EXAMPLE 6.1

Write the symbol of the ion formed when an aluminum atom loses three electrons.

Solution: When an aluminum atom loses three electrons, it forms an ion that has three more protons than electrons. The resulting ion has a +3 charge. The symbol for this ion is Al^{3+}.

> ### Exercise 6.1
>
> Write the symbol for the ion formed when an oxygen atom gains two electrons.

Large amounts of energy are required to remove electrons from atoms of some elements, those that have high ionization energy. With certain exceptions (the noble gases, for example) atoms having high ionization energy also have strong attractions for extra electrons; they have high electron affinities. Atoms of this kind, then, do not easily lose electrons, but they do readily add enough electrons to fill their *s* and *p* sublevels. Once any atom has eight valence electrons, however, it will then have little attraction for more electrons. Fluorine, which has a high ionization energy and a high electron affinity, is a good example.

A fluorine atom, which has seven valence electrons, will attract one additional electron, which fills the $2p$ sublevel.

$$:\overset{..}{\underset{..}{F}}\cdot \quad + e^- \longrightarrow \quad :\overset{..}{\underset{..}{F}}:^-$$
$$1s^2 2s^2 2p^5 \qquad\qquad\qquad 1s^2 2s^2 2p^6$$

The resulting anion has a total of eight valence electrons. Because its $2s$ and $2p$ sublevels are full and especially because it now has a negative charge, F^- has no attraction for more electrons.

Atoms of metals hold their own electrons weakly and have little or no attraction for extra electrons. That is, these atoms have relatively low ionization energies and low electron affinities. Consequently, little energy is required to remove the loosely held valence electrons from atoms of this kind. Magnesium is a good example. Magnesium atoms have relatively low first and second ionization energies as well as low electron affinities. As a result, the two valence electrons of a magnesium atom are easily removed.

$$Mg: \quad \longrightarrow \quad Mg^{2+} \quad + 2\,e^-$$
$$1s^2 2s^2 2p^6 3s^2 \qquad 1s^2 2s^2 2p^6$$

Notice that in each of the reactions the resulting ion, F^- or Mg^{2+}, has the electron configuration of a noble gas. *In losing or gaining electrons, atoms tend to form ions that have noble gas configurations, that is, ions that have eight valence electrons.* This is a general statement of the (octet rule.) Using the octet rule, we can predict for many elements the number of electrons that will be lost or gained when an ion is formed and also what the formulas of the resulting compounds will be.

8 like noble gases

In Section 6.4 we will broaden this statement of the octet rule still further.

EXAMPLE 6.2

How many electrons is an oxygen atom likely to gain or lose when an ion is formed?

Solution: Oxygen has the following electron configuration and six valence electrons.

$$O \qquad 1s^2 2s^2 2p^4$$

If an oxygen atom gains two electrons, it will have a total of eight valence electrons and an electron configuration identical to that of neon.

Exercise 6.2

How many electrons would a Cl atom ordinarily add when it forms an ion?

Exercise 6.3

How many electrons is an Al atom likely to lose when it forms an ion? Use Lewis diagrams to show an atom of Al and its ion.

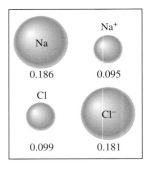

Figure 6.1 Relative sizes of a sodium atom and its ion. Relative sizes of a chlorine atom and its ion.

Ions are structurally different from their atoms. In general, when atoms lose electrons, they decrease in size. Let's consider the element sodium. The valence electron of a sodium atom lies in the $n = 3$ energy level, $1s^2 2s^2 2p^6 3s^1$. The outer electrons of Na^+, however, lie in the $n = 2$ energy level, $1s^2 2s^2 2p^6$. Because the $n = 2$ electrons have a higher probablity of being closer to the nucleus than the $n = 3$ electrons, the ion is smaller than the neutral atom. That is, when an atom of Na loses its valence electron to become Na^+, it decreases in size (Figure 6.1).

On the other hand, adding electrons to an atom tends to increase its size, even if the added electrons do not occupy a new energy level. A Cl^- ion, for example, is larger than a Cl atom. When an electron is added to the seven valence electrons in the $n = 3$ energy level of Cl, the increased repulsion created by the larger number of electrons causes the ion to expand a little. The increase in size caused by the formation of negative ions is usually not as great as the decrease caused by the loss of electrons in the formation of positive ions. Figure 6.2 shows the sizes of some atoms and their corresponding ions.

Let's consider the noble gases Ne to Rn. The atoms of each of these elements have eight valence electrons. As a result, they have the highest possible

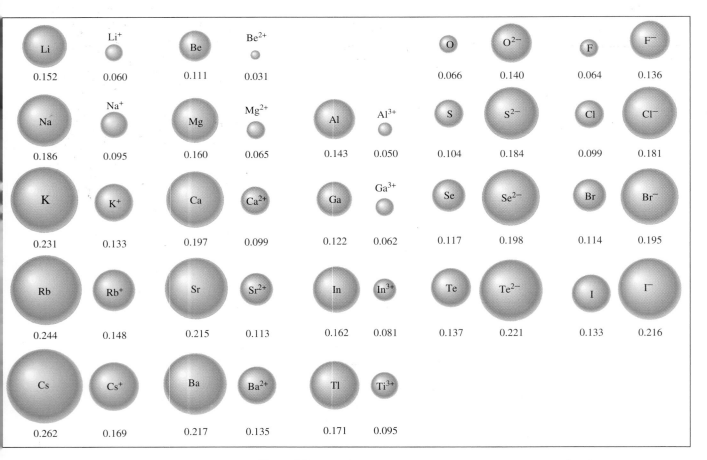

Figure 6.2 Relative sizes of some atoms and their corresponding ions.

ionization energies and, therefore, have no tendency to lose electrons to other atoms. Since they have eight valence electrons, the atoms of the noble gases don't attract extra electrons.

6.2 **Ionic Compounds**

After studying this section, you should be able to:

● Explain how ionic compounds form.

Ions are formed in chemical reactions that involve the transfer of electrons between atoms. Chemical reactions of this kind are known as **electron transfer** or **oxidation-reduction** reactions.

Now we'll consider the reaction between a potassium atom and a chlorine atom.

$$K \cdot \; + \; : \overset{\cdot\cdot}{\underset{\cdot\cdot}{Cl}} \cdot \; \longrightarrow \; K^+ \; + \; : \overset{\cdot\cdot}{\underset{\cdot\cdot}{Cl}} : ^-$$

Because chlorine atoms attract electrons strongly, a chlorine atom can accept a loosely held electron from a potassium atom. The acceptance of an electron is a process known as **reduction.** At the same time, the potassium atom loses an electron. Loss of an electron is **oxidation.** During chemical reaction, oxidation and reduction occur simultaneously.

Positive potassium ions and negative chloride ions result from electron transfer. Because of their opposite charges, the ions exert strong attractive forces on each other. That is, they form **ionic bonds.** Ionic bonds hold the ions together in an **ionic compound.** The electron transfer process does not, however, create *specific* bonds between individual pairs of ions. Instead, all the positive and negative ions exert forces on one another. Under ordinary circumstances, there are no isolated pairs or molecules of the sort K^+Cl^-. Two such sets always join to form

$$K^+Cl^-$$
$$Cl^-K^+$$

and then another joins to form

$$K^+Cl^-K^+$$
$$Cl^-K^+Cl^-$$

and so on, in a process limited only by the number of ions present. In the solids formed by ionic bonds, very large numbers of ions join together in such a way that each ion is surrounded by ions of opposite charge and is equally attracted to all of these (Figure 6.3). For this reason, *it is not correct to speak of molecules of ionic compounds.*

Atoms of the noble gases do not need to gain or lose electrons to have eight valence electrons. That is why noble gases are so unreactive.

We will study oxidation-reduction reactions in detail in Chapter 16.

Chlorine occurs in the natural state as Cl_2, but for our purposes we will consider a chlorine atom.

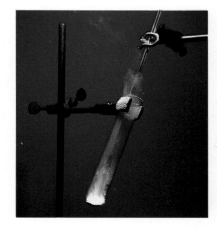

In this vigorous reaction, potassium is oxidized and chlorine is reduced. The product is KCl.

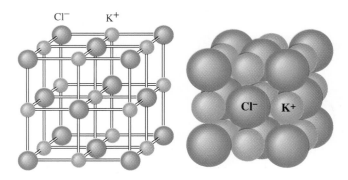

Figure 6.3 Two models of a KCl crystal.

Of the very large number of ionic compounds, almost all form characteristic crystals, some of which are quite beautiful. The photograph shows crystals of NaCl, $MnCO_3$, and As_4S_4.

Because atoms of many nonmetals are attracted to electrons, and atoms of many metals lose their electrons relatively easily, electron transfer reactions frequently occur between metal–nonmetal pairs, sometimes violently. For example, chlorine and sodium will react rapidly to give NaCl. The more metallic an element (low and left in the periodic table), the more likely it is to transfer electrons to nonmetals. The more nonmetallic an element (high and right on the periodic table), the more likely it is to accept electrons. Tellurium and germanium do not react under normal conditions, even though one element lies to the right of the metal–nonmetal boundary of the periodic table and one to the left.

It is important to understand that when atoms lose or gain electrons to become ionic compounds, their properties change dramatically. For example, we think nothing of eating the compound NaCl, table salt, in which sodium and chlorine are in ionic form, but both sodium and chlorine in their elemental forms are extremely harmful to animal tissues. A piece of Na metal would react rapidly with the moisture in your mouth (not to mention reacting with your tongue) and form poisonous sodium hydroxide. Elemental chlorine was used as a poisonous gas in the First World War.

Actually, both germanium and tellurium are so close to the boundary that neither can be classed as entirely metal or nonmetal.

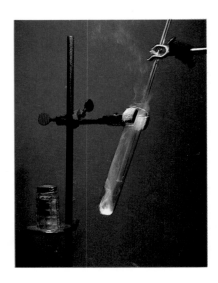

Sodium reacts violently with chlorine to produce NaCl (common table salt).

Since ions themselves are individual particles, the properties of ionic compounds are combinations of the properties of their ions. Because Cu^{2+} ion is blue, for example, all the compounds containing Cu^{2+} ion are blue unless the blue color is altered by the addition of color from other ions. Because cyanide ions, CN^-, are violently poisonous, the compounds NaCN, KCN, and $Mg(CN)_2$ are poisonous, although the cations Na^+, K^+, and Mg^{2+} are not.

Ionic compounds have another interesting property with which we will become quite familiar. With few exceptions, when ionic compounds dissolve in water, the ions separate, a process called **dissociation.** The solution then contains the individual ions rather than the compound itself. For example, when sodium chloride dissolves in water, it dissociates into sodium and chloride ions.

$$\text{(in water)} \quad NaCl \longrightarrow Na^+(aq) + \overset{..}{:}\overset{..}{\underset{..}{Cl}}:^-(aq)$$

Dissociation in solution creates characteristic physical, electrical, and chemical properties. Nonionic compounds do not dissociate and therefore do not show these properties in solution, although certain nonionic compounds react chemically with water to produce ions. We will study such reactions in a later chapter.

Crystals of $CuSO_4 \cdot 5\,H_2O$ are blue.

We will study dissociation in more detail in Chapter 13.

The symbol "(aq)" indicates that the ion is surrounded by water molecules in a solution made with water.

6.3 Formulas of Ionic Compounds

After studying this section, you should be able to:

- Predict the formulas of ionic compounds.

In electron transfer reactions, the atoms involved usually react to form ions with eight valence electrons, that is, to follow the octet rule. Metals tend to lose their valence electrons, and nonmetals tend to gain electrons, although there are exceptions. Nonetheless, we can make surprisingly accurate predictions about the formulas of many compounds of metals with nonmetals if we merely know which elements are involved.

First, we look at the periodic table to find the electron configurations of the elements. By remembering the octet rule, we then decide how many electrons are gained or lost by the atoms of each element and then how many atoms of each element belong in the formula. Here's how it's done.

Consider the difference in the reactions of K and Mg, each with Br. Remember that all the electrons involved must be transferred from the metal atoms to Br atoms. In the reaction of K atoms and Br atoms, each K atom loses its single valence electron, while each Br atom adds a single electron.

Bromine occurs in the natural state as Br_2, but for our present purposes we will use only bromine atoms.

$$K\cdot \; + \; :\overset{..}{\underset{..}{Br}}\cdot \; \longrightarrow \; K^+ :\overset{..}{\underset{..}{Br}}:^-$$

Therefore, one K atom reacts for every one Br atom, resulting in a compound with the formula KBr. The balanced equation, using naturally occurring Br_2, is

$$2\,K + Br_2 \longrightarrow 2\,KBr$$

We learn how to balance equations in Chapter 10.

For the reaction of Mg and Br, however, the proportions are not one to one. In this reaction, each Br atom gains a single electron as before, but because each Mg atom has two valence electrons, each Mg will lose two electrons. Because no electrons can be left over, there must be twice as many Br atoms taking part in the reaction as there are Mg atoms.

$$
\text{Mg:} \;+\; \begin{array}{c} :\overset{\cdot\cdot}{\underset{\cdot\cdot}{Br}}\cdot \\[1em] :\overset{\cdot\cdot}{\underset{\cdot\cdot}{Br}}\cdot \end{array} \;\longrightarrow\; \text{Mg}^{2+} \quad \begin{array}{c} :\overset{\cdot\cdot}{\underset{\cdot\cdot}{Br}}:^{-} \\[1em] :\overset{\cdot\cdot}{\underset{\cdot\cdot}{Br}}:^{-} \end{array}
$$

By extension, the numbers of atoms we have found in the formula for Mg and Br can be applied to any compound of an alkaline earth metal and a halogen.

Therefore, one Mg atom reacts for every two Br atoms, resulting in a compound with the formula $MgBr_2$.

$$
Mg + Br_2 \longrightarrow MgBr_2
$$

Let's look at another example, aluminum and oxygen. If we consider only a single Al atom, we find that it loses three electrons, while a single O atom gains two electrons. This yields an extra electron. If we use two O atoms for one Al atom, we are now one electron shy. If we start with two Al atoms, however, their six electrons can be transferred handily to three O atoms. The proportions in the compound must then be two Al^{3+} ions for every three O^{2-} ions. Thus, we find that the formula is Al_2O_3.

$$
\begin{array}{ccc}
\overset{\cdot}{\underset{\cdot}{Al}}\cdot & :\overset{\cdot\cdot}{O}\cdot & Al^{3+} \quad :\overset{\cdot\cdot}{\underset{\cdot\cdot}{O}}:^{-} \\[1em]
& +\;:\overset{\cdot\cdot}{O}\cdot \longrightarrow & :\overset{\cdot\cdot}{\underset{\cdot\cdot}{O}}:^{-} \\[1em]
\overset{\cdot}{\underset{\cdot}{Al}}\cdot & :\overset{\cdot\cdot}{O}\cdot & Al^{3+} \quad :\overset{\cdot\cdot}{\underset{\cdot\cdot}{O}}:^{-}
\end{array}
$$

In reactions of oxygen, sulfur, or any of the halogens with members of the alkali or alkaline earth families, you should be able confidently to predict the formulas of the resulting compounds. With more experience, you will be able to do the same for several other sets of elements. However, the atoms of some elements react in ways that do not lead to ions with eight valence electrons. Copper is an example. A copper atom can lose either one electron or two electrons, forming either a Cu^+ or Cu^{2+} ion. Many other metal and nonmetal atoms also behave in unpredictable ways. To the extent that it will be useful to us, we will learn about some of these reactions as we go along.

Exercise 6.4

Work out the electron transfer reactions that would occur between the atoms listed in the following pairs: Li and O, Sr and Cl, Al and F. For each pair, show the number of electrons gained or lost by the atoms of each element and predict the formulas of the resulting compounds.

In Section 6.10 we will learn about ions that contain more than one atom. Such ions also form ionic compounds. In Section 6.11 we learn to write formulas for those ionic compounds.

6.4 Covalent Bonding

After studying this section, you should be able to:

• Explain what covalent bonds are and how they are formed.

You may have noticed that the information in the last three sections has been confined to electron transfer reactions between metals and nonmetals. Based on what we know, however, we might guess that there must be another way in which chemical bonds are formed. Consider hydrogen, for example. Elemental hydrogen exists in the form of individual molecules each of which contains two H atoms. In H_2, the atoms are held together by a chemical bond, but, clearly, that bond is not ionic. In the first place, the molecules are individual, and we know that ions cannot form individual molecules. Furthermore, it is not possible for two H atoms to transfer an electron because the atoms are exactly alike; neither is more attracted to an electron than the other. Even so, the atoms of H are held tightly together in the H_2 molecule, and it has been shown experimentally that separating H_2 molecules into individual atoms requires considerable energy.

Let's examine a hydrogen atom more closely. The electron configuration for hydrogen is $1s^1$; its $1s$ energy level is half filled. Either hydrogen atom could accept a second electron from a willing giver, but neither will give up its own electron to another hydrogen atom. The attraction for another electron can be easily satisfied, however, if the two atoms join so that their two electrons are shared as a pair. A shared pair of electrons is called a **covalent bond.**

Covalent bonds can be looked at another way. Suppose we place two hydrogen atoms near each other. The most attractive place for their two electrons to be is as close as possible to both of the positive-charged nuclei; that is, between the nuclei. In molecules with covalent bonds, the electrons in the bond spend most of their time between the nuclei. (See Figure 6.4.) Since the nuclei are attracted to this concentration of negative charge, they tend to remain close to it, that is, together.

A simple diagram showing the shared pair of electrons between the atoms makes the concept easy to understand. Such a diagram is called a *Lewis diagram of a molecule* similar to the Lewis diagrams of atoms we met in Section 4.7. It was G. N. Lewis who, in 1916, first proposed the concept of covalent bonds. To draw a Lewis diagram of a molecule, use Lewis diagrams of the atoms and show shared pairs lying between the atoms. Although it is not necessary, you may, if you wish, draw a circle around each atom and its electrons, shared and otherwise, to help you keep track.

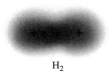

H_2

Figure 6.4 Electron distribution in the H_2 covalent bond. The density of the orange color represents the probability of finding an electron.

Take another example, that of elemental chlorine. An atom of Cl has seven valence electrons.

$$: \overset{..}{\underset{..}{Cl}} \cdot$$

$$1s^2 2s^2 2p^6 3s^2 3p^5$$

One more electron is needed to create the electron configuration of the noble gas argon, Ar. Two Cl atoms can each contribute one electron to a shared pair.

Profiles in Science

Gilbert Newton Lewis (1875–1946)

Although G. N. Lewis grew up in Nebraska, he received his education in the East, at Harvard University. After finishing his training, Lewis became an instructor at Harvard but soon left to join the faculty at the Massachusetts Institute of Technology, where he was promoted to a full professor in a remarkably short time. Somewhat later he was invited to become dean of the College of Chemistry at the University of California at Berkeley, where he remained for the rest of his life. His contributions as an educator were substantial; under his leadership, the college became one of the most prestigious in the nation.

During his years as an instructor at Harvard, Lewis became interested in the mechanism of chemical bonding, which at that time was thought to be entirely ionic. For a decade and a half, he pursued this study and in 1916 published his theory of the shared electron pair bond. His views also included the octet rule, electronegativity, and what we now call Lewis diagrams, all of which are ideas that laid the cornerstone of molecular chemistry. Lewis was interested also in reaction, especially that between electron-rich atoms and electron-deficient ones. He originated the concept of Lewis acids, which greatly broadened the understanding of acid-base reactions in nonaqueous solutions (Section 14.1). He made contributions in the fields of thermodynamics, isotope studies, and photochemistry.

After his formal retirement, he continued performing experiments in photochemistry, the interaction of molecules and light. He died in 1946.

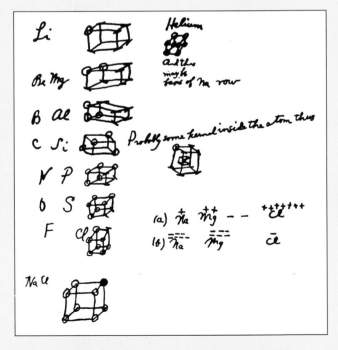

A page from Lewis' laboratory notebook.

Sharing the pair holds the atoms together as a molecule. Elemental chlorine is diatomic, that is, it exists as molecules, each of which contains two atoms. Its formula is Cl_2.

We can consider the shared electrons in a covalent bond as belonging to both atoms. Thus, in a Lewis diagram, each atom (except H, of course) can be shown as having eight valence electrons, a noble gas electron configuration, as in Cl_2 above. We can now state the octet rule to include electron sharing. *In chemical reaction, many elements transfer or share electrons so as to attain eight valence electrons.*

6.5 Polar Covalent Bonds and Electronegativity

After studying this section, you should be able to:

- Use electronegativities to predict the kind of bond formed between two atoms.

So far, we have considered only covalent bonds between atoms of the same element. Let's look at a molecule containing atoms of different elements, hydrogen and chlorine, for instance. If a hydrogen atom gains one electron, it will have its full complement of two valence electrons, and if a chlorine atom gains an electron, it will have its full complement of eight.

When a covalent bond is formed by two atoms of the same element, the two electrons forming the bond are shared equally by the two atoms, since both atoms have the same attraction for electrons. Because different elements have different ionization energies and different electron affinities, however, different elements attract a shared pair unequally. A covalent bond in which the electron pair is not equally shared is called a **polar covalent bond.** The bond between hydrogen and chlorine atoms is a polar covalent bond.   stronger pull

The attraction that an atom has for a shared pair of electrons is called the **electronegativity** of that atom. Electronegativity is expressed on a relative scale that ranges from somewhat less than one to a value of four (see Table 6.1). The least electronegative element is francium, Fr, 0.8 on the scale, and the most electronegative is fluorine, F, 4.0 on the scale. Electronegativity is greatest in the nonmetals that lie high and to the right in the periodic table (excluding the rare gases) and least in the elements that are low and to the left.

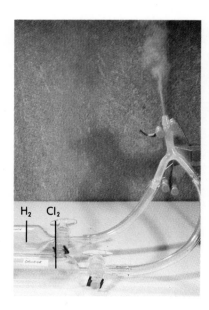

Hydrogen and chlorine gases react to produce covalently bonded HCl molecules. The requirement of each atom for another electron can be accomplished by sharing a pair of electrons, forming a covalent bond.

Table 6.1   The Electronegativities of the Elements

						1 **H**											
Li 1.0	Be 1.5	■ <1.0			▢ 2.0 – 2.4							B 2.0	C 2.5	N 3.0	O 3.5	F 4.0
Na 1.0	Mg 1.2	▢ 1.0 – 1.4			▢ 2.5 – 2.9							Al 1.5	Si 1.8	P 2.1	S 2.5	Cl 3.0
		▢ 1.5 – 1.9			▢ 3.0 – 4.0											
K 0.9	Ca 1.0	Sc 1.3	Ti 1.4	V 1.5	Cr 1.6	Mn 1.6	Fe 1.7	Co 1.7	Ni 1.8	Cu 1.8	Zn 1.6	Ga 1.7	Ge 1.9	As 2.1	Se 2.4	Br 2.8
Rb 0.9	Sr 1.0	Y 1.2	Zr 1.3	Nb 1.5	Mo 1.6	Tc 1.7	Ru 1.8	Rh 1.8	Pd 1.8	Ag 1.6	Cd 1.6	In 1.6	Sn 1.8	Sb 1.9	Te 2.1	I 2.5
Cs 0.8	Ba 1.0	La 1.1	Hf 1.3	Ta 1.4	W 1.5	Re 1.7	Os 1.9	Ir 1.9	Pt 1.8	Au 1.9	Hg 1.7	Tl 1.6	Pb 1.7	Bi 1.8	Po 1.9	At 2.1
Fr 0.8	Ra 1.0	Ac 1.1														

Ce	Pr	Nd	Pm	Sm	Eu	Gd	Tb	Dy	Ho	Er	Tm	Yb	Lu
1.1	1.1	1.1	1.1	1.1	1.1	1.1	1.1	1.1	1.1	1.1	1.1	1.0	1.2

Th	Pa	U	Np	Pu	Am	Cm	Bk	Cf	Es	Fm	Md	Mo	Lr
1.2	1.3	1.5	1.3	1.3	1.3	1.3	1.3	1.3	1.3	1.3	1.3	1.3	1.5

Exercise 6.5

Which element will more readily attract the shared electron pair in a CO molecule? In BrF?

When there is a difference in the electronegativity of two covalently bonded atoms, the shared electron pair spends its time on the average closer to the more electronegative atom. <u>If there is no difference, the sharing is equal, and the electron pair forms a pure covalent bond.</u> If the difference in the electronega-

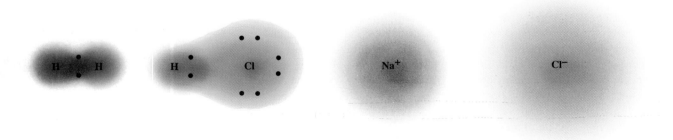

$$2.1–2.1 = 0 \qquad\qquad 3.0–2.1 = 0.9 \qquad\qquad\qquad 3.0–1.0 = 2.0$$

Figure 6.5   Electronegativity difference between two H molecules is $2.1 - 2.1 = 0$. Electronegativity difference between Cl and H is $3.0 - 2.1 = 0.9$. Electronegativity difference between Cl and Na is $3.0 - 1.0 = 2.0$.

tivities is greater than about 1.8 to 2, however, electron transfer usually occurs, and an ionic, not covalent, bond is formed.

Spend a few minutes studying the electronegativity values printed under the element symbols in Figure 6.5. These numbers enable you to decide what kind of bond will be formed between the atoms of any combination of elements. Along with information that we will be discussing later, the numbers will also allow you to predict many of the chemical and physical properties of compounds formed by pairs of elements.

Using electronegativity values enables you to predict the type of bond present in a particular compound.

$\triangle EN \geq 2$ ionic

$\triangle EN < 2$ polar covalent

$\triangle EN \approx 0$ covalent

Exercise 6.6

Fill in the table that follows.

Reacting Elements	Electronegativities of These Elements	Differences in Electronegativities	Kind of Bond Formed
Cs + F	Cs, 0.7; F, 4.0	3.3	Ionic
O + Na			
C + H			
Si + Cl			

6.6 Drawing Lewis Diagrams of Covalent Molecules

After studying this section, you should be able to:

- Draw Lewis diagrams of simple covalent molecules.

Two kinds of drawings are commonly used to show covalently bonded molecules. As we saw in the preceding section, our Lewis diagrams can be easily adapted to covalent molecules. We include all the valence electrons of both atoms and show the shared pair or pairs lying between the atoms.

Although electrons are all alike, it is occasionally convenient to use different symbols for electrons from different atoms. The shared pair is shown with one symbol from each atom.

$$: Br \underset{\cdot \cdot}{\overset{x\ x}{\underset{x\ x}{x}} Cl} \, \overset{x}{\underset{x}{}}$$

If, as we shall see, both electrons in the pair are furnished by the same atom, then both should be shown with the same symbol.

A covalent bond is often shown simply as a line connecting the two atoms. In this method, the line represents a pair of shared electrons. Using this approach, we can represent any pair of electrons with a short line. If we use a line to show unshared as well as shared pairs, we draw the HCl molecule like this:

$$H-\overline{\underline{Cl}}\,|$$

Such a diagram is sometimes called a bond line drawing. Although this is a faster method of drawing covalent bonds, be sure you have had plenty of practice in counting electrons before you use this simple abbreviation. For emphasis, we will sometimes show unshared electron pairs as dots in bond line diagrams,

$$H-\overset{\displaystyle \cdot\cdot}{\underset{\displaystyle \cdot\cdot}{Cl}}:$$

After we have had more practice, we will draw bond line diagrams in which the unshared electrons are assumed to be present but are not shown,

$$H-Cl$$

In what follows, we will use dots (or crosses, x's, or small circles) when we wish to keep track of where electrons come from, but when we are interested only in the total number of electron pairs, we will use the bond line method.

Exercise 6.7

Construct both Lewis and bond line diagrams for the following molecules: HF, I_2, IBr. Show all electrons or electron pairs.

6.7 Atoms Forming More Than One Covalent Bond

After studying this section, you should be able to:

- Determine how many covalent bonds the atoms of any particular element are likely to make.

Atoms of many of the elements can each share more than one pair of electrons, either with one other atom or with more than one atom. A water molecule, H_2O, is a good example. An oxygen atom has six valence electrons, needing two more to have its full complement of eight. When an oxygen atom combines with two hydrogen atoms to form a molecule of water, it shares two pairs of electrons, one with each hydrogen atom. Each of the shared pairs becomes a covalent bond, thus bonding the oxygen atom to both of the hydrogens.

A similar situation occurs in the case of ammonia, NH_3. Nitrogen, which has five valence electrons, can share three pairs, one with each of three hydrogen atoms. In an NH_3 molecule, nitrogen has eight valence electrons, and each hydrogen has two valence electrons.

Exercise 6.8

In every molecule of phosphorus trichloride, PCl_3, there are three chlorine atoms, each of which is joined by a covalent bond to a central phosphorus atom. Draw a Lewis diagram and a bond line diagram for PCl_3.

Exercise 6.9

Draw Lewis and bond line diagrams for CH_4. Each of the H atoms is bonded to the C atom.

6.8 Multiple Bonds and Coordinate Covalent Bonds

After studying this section, you should be able to:

- Draw electron dot diagrams of a single bond, a double bond, and a triple bond.
- Identify a coordinate covalent bond.

A shared pair of electrons forms a **single bond** between two atoms, but in many cases, a single pair of atoms can share more than one pair of electrons and form more than one bond. An oxygen atom has six valence electrons. It can share two pairs of electrons to obtain eight valence electrons. If two oxygen atoms combine, each atom can share two pairs of electrons and form two covalent bonds, as you see in the diagram below. When two atoms share two pairs of electrons, the result is called a **double bond.**

Another good example of multiple bonding is the N_2 molecule. Each N atom has five valence electrons, and by sharing three pairs, each N atom can have eight valence electrons. Three pairs of electrons shared between two atoms make a **triple bond.**

A shared pair of electrons creates a strong force between two atoms. Two shared pairs create a stronger force, and three pairs, stronger yet.

Multiple bonds also occur in molecules composed of atoms of several different elements. We will see examples in the next section.

It is not always the case that each of two covalently bonded atoms donates one electron to the shared electron pair. In some cases, both electrons are furnished by only one of the atoms. A shared pair that has originated from only one of the bonded atoms is called a **coordinate covalent bond.** Take, for example, the compound BF_3NH_3. Because BF_3 is deficient in electrons (even counting shared pairs, the boron atom has only six valence electrons), it is attractive to electrons. Each molecule of ammonia, NH_3, however, has an *unshared* pair of electrons. BF_3 and NH_3 react, forming a covalent B—N bond in which both electrons in the bond formerly belonged to the N atom.

$$:\overset{\cdot\cdot}{\underset{\cdot\cdot}{F}}\overset{\cdot\cdot}{\underset{\cdot\cdot}{:}}\overset{:\overset{\cdot\cdot}{F}:H}{\underset{:\overset{\cdot\cdot}{F}:H}{B}}\overset{x}{\underset{x}{N}}\overset{x}{H}$$

Coordinate covalent bonds are characteristic of several compounds such as H_2SO_4 and HNO_3.

6.9 Lewis Diagrams of Larger Molecules

After studying this section, you should be able to:

- Draw Lewis diagrams for many different molecules.

Now that we have drawn Lewis diagrams for simple molecules and have learned more about the characteristics of covalent bonds, we can take an organized approach to making Lewis diagrams for more complex molecules. Here are a few helpful hints and steps to follow.

1. *Draw a preliminary structure in which the atoms will later be connected by bonds.*

To draw a preliminary structure, you must first know which atom is bonded to which. Experience and practice help most here, but there are useful guidelines. It is helpful to know how many bonds the atoms of some common elements usually make; for example, a hydrogen atom makes only one bond and an oxygen atom usually makes two. It is important to know which atom in the molecule is the central atom (the one with the most atoms bonded to it) and what bonding patterns are possible in the rest of the molecule. In compounds that have more than one atom of one element and only a single atom of another, the single atom is ordinarily the central atom, with the others bonded to it instead of to each other. Examples are H_2O, SO_3, and PCl_3, in which the central atoms are, respectively, O, S, and P.

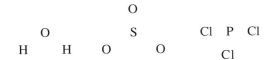

2. *Determine how many bonds must exist in the molecule.*
   To do this, follow steps (a) through (c).

   a. According to the octet rule, each atom (except H) is to have 8 electrons. Therefore, multiply the number of atoms by 8 (for each H atom, multiply only by 2). In PCl_3, the number is $4 \times 8\ e^- = 32\ e^-$.
   b. Count the total number of valence electrons in all the atoms that will be used to make the molecule. The molecule can have neither more nor fewer electrons than the sum of the electrons of all the atoms in the molecule.

Science in Action

Of all the elements, carbon is special, so special that it has an entire branch of chemistry named after it, organic chemistry. Serious study and use of carbon compounds began in the mid 1800's, but the discoveries in organic chemistry that have perhaps made the greatest changes in our lives did not occur until well into this century.

Carbon lies at the horizontal center of the representative elements; therefore it has an average electron affinity and an average ionization energy. Because it has four valence electrons, a carbon atom can form four covalent bonds with other atoms (—C—), including other carbon atoms. Consequently, there can be molecules containing very large numbers of carbon atoms.

Large molecules of molecular weights in the hundreds of thousands are called polymers, and polymers can now be synthesized. By far, the most useful synthetic polymers are carbon compounds.

Before 1935, humans knew of and employed several natural polymeric materials such as rubber, leather, cotton, and silk. In fact, humans and all living things are themselves mostly made of organic polymers, the two most important of which are proteins and genetic polymers such as DNA. The structures of natural polymers had been partly understood during the first quarter of this century. Some natural polymers were chemically modified to make them more useful, polymers like rayon (modified cotton) and celluloid (also made of cotton). But the first organic polymer specifically designed for a particular purpose and synthesized from simple materials was discovered in 1935. Its name was nylon.

Today synthetic polymers have supplanted natural materials in almost every kind of use. Suitcases, telephones, shoes, clothing, dinnerware, paints of all kinds, artificial hearts, airplanes, and literally hundreds of millions of other items are made of polymeric materials. By designing the molecules of which polymers are made, it is easy to make the properties of the finished polymers be exactly what is needed. And all this is possible because of the special nature of carbon.

A laboratory demonstration of nylon being made.

We will confine ourselves to valence electrons because the other electrons remain unaffected by the bonding.

Each Cl atom has 7 valence electrons. A P atom has 5 valence electrons. PCl_3, then, has a total of 26 valence electrons, $(3 \times 7\ e^-) + 5\ e^- = 26\ e^-$. (As we will see soon, this step must be modified for ions.)

c. Subtract the number in step (b) from the number in (a) to obtain the number of electrons that must be shared.

For PCl_3, $32\ e^- - 26\ e^- = 6\ e^-$. The atoms in PCl_3 share 6 electrons (or three pairs) by making three bonds.

3. *Draw in the bonds.*

Place a pair of electrons representing a bond between each pair of atoms in the molecule. If the molecule still needs to share more electrons, draw more pairs of electrons between the atoms (keeping in mind how many bonds atoms of certain elements usually form; H, one bond, O, two bonds; etc.) until as many are shared as you calculated in step 2(c).

PCl_3 has three bonds. Draw a pair of electrons between the P and each Cl.

$$Cl-P-Cl$$
$$\overset{|}{Cl}$$

4. *Draw the other electrons around the atoms of the molecule, making sure that each atom except H has eight electrons.*

The finished molecule should obey the octet rule, if possible. (There are a few exceptions, such as NO_2 and BF_3, but we will not be concerned with these.)

In the drawing of PCl_3 above, six electrons are located around P. To obey the octet rule, P requires eight valence electrons. Add two more electrons to P. Each Cl has only two electrons. Add six electrons to each Cl.

$$: \overset{..}{\underset{..}{Cl}} - \overset{..}{\underset{|}{P}} - \overset{..}{\underset{..}{Cl}} :$$
$$: \overset{..}{\underset{..}{Cl}} :$$

5. *Check and count electrons.*

Count the electrons in the structure we have just finished. Confirm that for PCl_3 there is a total of 26 valence electrons, that each atom has an octet, and that 6 electrons lie between atoms (three pairs are shared).

EXAMPLE 6.3

As we will see in later chapters, sulfuric acid, H_2SO_4, is an essential industrial chemical. It is made from SO_3. Draw a Lewis diagram of SO_3.

Strategy and Solution:

Step 1. Draw a preliminary structure in which the atoms will later be connected by bonds. The S atom is the central atom with three oxygen atoms bonded to it.

$$\begin{matrix} O & & O \\ & S & \\ & O & \end{matrix}$$

Step 2. Determine how many bonds must exist in the molecule.
a. There are four atoms that obey the octet rule; $4 \times 8\ e^- = 32\ e^-$.
b. A sulfur atom has 6 valence electrons, and each oxygen atom has 6 valence electrons. Therefore, SO_3 has a total of 24 valence electrons.
c. Subtracting, we get $32\ e^- - 24\ e^- = 8$ electrons (or four pairs) that are shared. An SO_3 molecule must have four bonds.

Step 3. Draw in the bonds. To get four bonds in the structure, we will have to make one double bond. It makes no difference which oxygen shares the double bond with sulfur.

$$\begin{matrix} O \\ \ \diagdown \\ \quad S = O \\ \ \diagup \\ O \end{matrix}$$

Step 4. Draw the other electrons around the oxygen atoms so that each atom has eight electrons.

$$\overset{\cdot\cdot}{\underset{\cdot\cdot}{O}}\diagdown_{\diagup}\underset{\overset{\cdot\cdot}{\underset{\cdot\cdot}{O}}}{}\!\!\!\!\!\!\!\!\!\! S=\overset{\cdot\cdot}{\underset{\cdot\cdot}{O}}$$

Step 5. Check and count electrons. There should be 24 electrons in the diagram.

EXAMPLE 6.4

Oxidation by nitric acid, a process we will study in detail in Chapter 16, is common in chemistry. Draw a Lewis diagram for nitric acid, HNO_3.

Strategy and Solution:

Step 1. Draw a preliminary structure in which the atoms will later be connected by bonds. N is the central atom. Each O is attached to the N. H is attached to one of the O's. (Why not have the H attached to N? Try it. You can't draw an octet-rule structure like that.)

$$\begin{array}{ccc} O & & O \\ & N & \quad H \\ & O & \end{array}$$

Step 2. Determine how many bonds must exist in the molecule.

a. In addition to the H atom, there are four atoms that will obey the octet rule, $(4 \times 8 \; e^-) + 2 \; e^- = 34 \; e^-$.

b. A nitrogen atom has 5 valence electrons, each oxygen atom has 6 valence electrons, and a hydrogen atom has 1 valence electron. Therefore, HNO_3 has a total of 24 valence electrons, $5 \; e^- + (3 \times 6 \; e^-) + 1 \; e^- = 24 \; e^-$.

c. Subtracting, we get $34 \; e^- - 24 \; e^- = 10$ electrons, or five pairs that are shared. An HNO_3 molecule must have five bonds.

Step 3. Draw in the bonds. To get five bonds in the structure, we will have to make one double bond. The double bond is between the nitrogen and one of the oxygen atoms that is not bonded to the hydrogen.

$$\overset{O}{\diagdown}\!\!\!\!\underset{\underset{O}{|}}{N}\!\!\!\!\overset{O}{\diagup}\!\!\! H$$

Step 4. Draw the other electrons around the oxygen atoms, making sure that each has eight electrons.

$$\overset{\cdot\cdot}{\underset{\cdot\cdot}{O}}\!\!\!\!\diagdown\underset{\underset{:\overset{\cdot\cdot}{O}:}{|}}{N}\!\!\!\!\diagup\!\!\overset{\cdot\cdot}{\overset{\cdot\cdot}{O}}\!\!\! H$$

Step 5. Check and count electrons. There should be 24 electrons in the diagram.

Exercise 6.10

Draw a Lewis diagram for hydrogen peroxide, H_2O_2.

Exercise 6.11

Draw a Lewis structure for carbon dioxide, CO_2.

Exercise 6.12

Draw a Lewis structure for phosphoric acid, H_3PO_4.

6.10 Lewis Diagrams of Polyatomic Ions

After studying this section, you should be able to:

- Draw Lewis diagrams for several different polyatomic ions.

Many chemical compounds contain both covalent and ionic bonds. Such compounds are composed of ions, but some of the ions contain more than one atom. These ions are called **polyatomic ions**. The atoms within polyatomic ions are joined by covalent bonds. A simple example of such a compound is sodium hydroxide, NaOH, which is composed of Na^+ ions and OH^- ions. A hydroxide ion, OH^-, contains an oxygen atom and a hydrogen atom joined by a single covalent bond. In addition to the seven valence electrons that the neutral O and H atoms would have, this pair has an extra electron and thus carries a negative charge.

$$\left[:\overset{..}{\underset{..}{O}}-H \right]^-$$

In drawing the structure of a polyatomic ion, we begin as though we were drawing a molecule. When we count the number of valence electrons in step 2(b), however, we must allow for the charge on the ion. For ions, step 2(b) should read as follows:

Count the number of valence electrons in all the atoms. If the ion is negatively charged, add one electron for every charge. If the ion is positively charged, subtract one electron for every charge.

Finish the procedure just as you would the diagram of a molecule. When the structure is complete, draw brackets around it and indicate the charge.

EXAMPLE 6.5

In Chapter 14 we will learn that hydroxide ion, OH^-, is one of several substances called bases. Draw a Lewis diagram for an OH^- ion.

Strategy and Solution:

Step 1. Draw a preliminary structure in which the atoms will later be connected by bonds.

O   H

Step 2.   Determine how many bonds must exist in the molecule.
a. The total number of electrons needed for the octet rule equals $8\ e^- + 2\ e^- = 10$ electrons.
b. The number of actual valence electrons equals $6\ e^- + 1\ e^- = 7\ e^-$. Add one electron for the one negative charge, $7\ e^- + 1\ e^- = 8$ electrons.
c. Subtracting, we get $10\ e^- - 8\ e^- = 2$ electrons, or one bond.

Step 3.   Draw in the bond.

$$O{-}H$$

Step 4.   Draw the other six electrons around the oxygen atom. Draw brackets around the ion and indicate the negative charge.

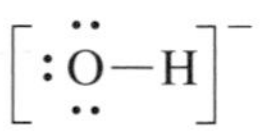

$$\left[\ :\overset{..}{\underset{..}{O}}{-}H\ \right]^{-}$$

Step 5.   Check and count electrons. There should be eight electrons in the diagram.

For an example of a polyatomic ion that carries a positive charge, consider the ammonium ion, NH_4^+, which is a covalently bonded combination of an ammonia molecule, NH_3, and a hydrogen ion, H^+. In the ammonium ion, N is the central atom; it is bonded to each of the four H atoms.

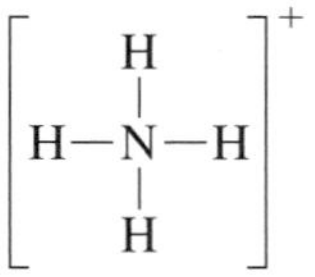

$$\left[\ \begin{array}{c} H \\ | \\ H{-}N{-}H \\ | \\ H \end{array}\ \right]^{+}$$

EXAMPLE 6.6

Worldwide, more sulfuric acid is produced than any other chemical. In solution, sulfuric acid produces SO_4^{2-}. Draw a Lewis diagram of SO_4^{2-}.

Strategy and Solution:

Step 1.   Draw a preliminary structure in which the atoms will later be connected by bonds.

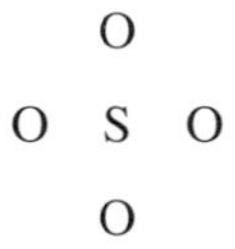

$$\begin{array}{c} O \\ O\ \ S\ \ O \\ O \end{array}$$

Step 2.   Determine how many bonds must exist in the molecule.
a. The total number of electrons needed for the octet rule equals $5 \times 8\ e^- = 40$ electrons.
b. The number of valence electrons equals $6\ e^- + (4 \times 6\ e^-) = 30\ e^-$. Add 2 electrons for the negative-two charge, $30\ e^- + 2\ e^- = 32$ electrons.
c. Subtracting, we get $40\ e^- - 32\ e^- = 8$ electrons, or four bonds.

Step 3.   Draw in the bonds.

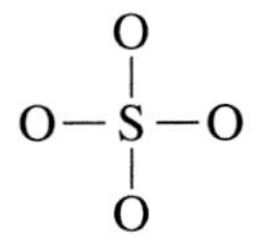

$$
\begin{array}{c}
\text{O} \\
| \\
\text{O}-\text{S}-\text{O} \\
| \\
\text{O}
\end{array}
$$

Step 4.   Draw the other electrons around the oxygen atoms. Draw brackets and indicate the negative charge.

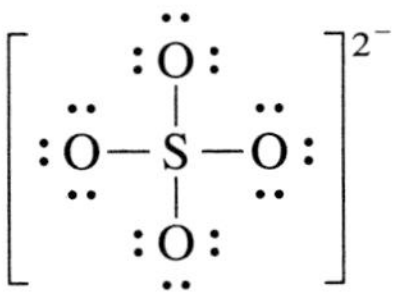

$$
\left[
\begin{array}{c}
.. \\
:\ddot{\text{O}}: \\
| \\
:\ddot{\text{O}}-\text{S}-\ddot{\text{O}}: \\
| \\
:\ddot{\text{O}}: \\
..
\end{array}
\right]^{2-}
$$

Step 5.   Check and count electrons. There should be 32 electrons in the diagram.

> **Exercise 6.13**
>
> Draw Lewis diagrams for a ClO_4^- ion and for an NO_3^- ion.

Other common polyatomic ions are carbonate, CO_3^{2-}, and phosphate, PO_4^{3-}. See Section 7.4 for the formulas and names of more polyatomic ions.

6.11 Formulas of Compounds Containing Polyatomic Ions

After studying this section, you should be able to:

- Give formulas of compounds containing monatomic and polyatomic ions.

In Section 6.3, we learned how to determine the formulas of some ionic compounds. We counted how many electrons each metal atom lost and how many each nonmetal atom gained. Then, by remembering that in chemical reaction no electrons are created or destroyed, we determined the ratio between numbers of metal ions and nonmetal ions and wrote a formula.

A sample of any compound is electrically neutral; that is, there are as many positive charges as negative charges. To determine the formula of an ionic compound, whether it contains polyatomic ions or not, we merely set the ratio between the number of positive ions and negative ions so that the overall charge of the compound will be zero. If the ions are all singly charged, then the formula will have one positive ion for each negative ion. Na^+ and Cl^- form NaCl, for example. Compounds with polyatomic ions form in a similar manner. Na^+ combines with OH^- to form NaOH, and K^+ combines with NO_3^- to form KNO_3. If one of the ions is singly charged and the other doubly or triply charged, there will always be as many singly charged ions as the value of the charge on the multiply charged ion. To understand this better, look at Example 6.7.

EXAMPLE 6.7

What is the formula of the compound containing Cs^+ and PO_4^{3-}?

Solution: For every one PO_4^{3-} ion, there must be three Cs^+ ions. The formula, then, is Cs_3PO_4.

Exercise 6.14

What is the formula of the compound containing Al^{3+} and BO_3^-. Of the compound containing Rb^+ and SO_4^{2-} ions?

Finally, if both kinds of ions have multiple charges, merely take as many of each ion as the numerical value of the charge on the other ion. Simplify the subscripts if necessary.

EXAMPLE 6.8

What is the formula of the compound containing Al^{3+} and CO_3^{2-}?

Solution: CO_3^{2-} has a charge of -2, so take two Al^{3+} ions. Al^{3+} has a charge of $3+$; therefore, take three CO_3^{2-} ions. The formula is $Al_2(CO_3)_2$.

EXAMPLE 6.9

What is the formula of the compound containing Ba^{2+} and SO_4^{2-}?

Solution: Because the charges are the same on both ions, take one Ba^{2+} ion for every one SO_4^{2-} ion, giving $BaSO_4$.

Exercise 6.15

What is the formula of the compound containing Ca^{2+} and PO_4^{3-}? Of that containing Sr^{2+} and SO_4^{2-}?

6.12 **Shapes of Molecules**

The molecular shape, that is, the arrangement of the atoms in space, is important because it often contributes to the chemical and physical properties of molecules. As you know, a Lewis diagram shows the bond structure (which atom is bonded to which) of a molecule. For many formulas, only one correct Lewis diagram can be drawn, and that drawing usually represents a molecule that actually exists. Whenever we can draw more than one Lewis diagram, there is often more than one compound. A Lewis diagram alone cannot tell us what *shape* a molecule has, but there are other guidelines that can help.

The shapes of molecules have distinct effects on their properties.

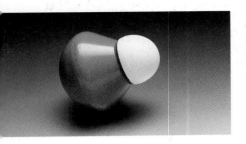

Molecules with two atoms, for example, must form a linear shape: A straight line can be drawn through the nuclei. The HCl molecule, for instance, is linear.

$$H-\ddot{\underset{..}{Cl}}:$$

If there are more than two atoms, the nuclei may or may not lie in a straight line. The CO_2 molecule is linear, that is, all three atoms lie in a straight line.

$$\ddot{\underset{..}{O}}=C=\ddot{\underset{..}{O}}$$

This model of HCl shows that the molecule is linear.

Water molecules, however, are bent. Lines drawn along the bonds from each H nucleus through the O nucleus intersect at an angle of about 105°. When two atoms are bonded to a central atom, the angle between their bonds is called the **bond angle.**

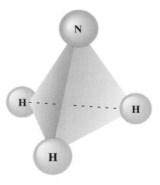

Most molecules containing more than three atoms are not planar. For example, the molecules of ammonia, NH_3, form three-sided pyramids with the nitrogen at one of the corners.

In the methane pyramid, the three sides and the bottom are all identical. Such a pyramid is called a regular tetrahedron.

Most large molecules can change their shapes as atoms rotate around bonds. Tinkertoy models can similarly change, even though the balls are firmly held together by their sticks.

Molecules of methane, CH_4, also have shapes like pyramids, but in methane the carbon atom is in the center and the four corners are occupied by hydrogen. Depending on the numbers of atoms and the bonding structures, larger molecules often have complex three-dimensional shapes.

To study and understand the properties of molecules, chemists make drawings and models that display molecular architecture. What are affectionately called "Tinkertoy models" show the atoms connected by long rods that represent bonds. Figure 6.6 shows Tinkertoy models of H_2O, CO_2, and NH_3.

Although the Tinkertoy models indicate the bond angles and bond structure of molecules, they cannot show what a molecule actually looks like. Other molecular models, called space-filling models, are more realistic. Notice that the atoms actually blend into one another in the space-filling models of H_2O, CO_2, and NH_3 in Figure 6.7.

The properties of compounds are often decisively affected by the shapes of the molecules themselves. We will examine the relationship between molecular shape and physical properties in the next section and in later chapters.

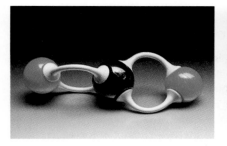

Figure 6.6 Tinkertoy type models of CO_2, H_2O, and NH_3.

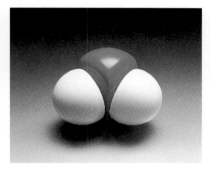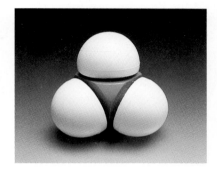

Figure 6.7 Space-filling models of CO_2, H_2O, and NH_3.

6.13 Dipoles and Polar Molecules

After studying this section, you should be able to:

- Explain what causes polarity in a molecule.
- Determine whether a molecule is polar or nonpolar.

Molecular shape frequently exerts its influence on the properties of substances by means of polar covalent bonds. As we saw in Section 6.5, polar covalent bonds are formed when the bonded atoms have different electronegativities. When two bonded atoms share a pair of electrons unequally, the molecule acquires a partial negative charge (δ^-) toward the more electronegative atom and a partial positive charge (δ^+) toward the other atom. That is, the molecule as a whole has a charge difference between its two ends. In such a case, the two bonded atoms have what is called an **electric dipole,** or just **dipole.** Even though the charge difference is not as great as it would be if an electron had been completely transferred, it can still have a substantial effect on the properties of the molecule. Polar covalent bonds are so named because of the dipole, and we say that a molecule with an electric dipole, such as that in HCl, is **polar.**

Electric dipoles in molecules cause the molecules to have properties that non dipolar molecules do not have.

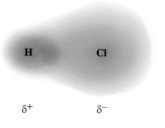

δ^+ δ^-

In HCl the electrons of the shared pair spend their time on the average closer to the chlorine than to the hydrogen. This gives chlorine a slight excess of negative charge and hydrogen a slight excess of positive charge.

Larger molecules, however, may or may not be polar, even though some of the bonds in the molecule are polar covalent bonds. In carbon dioxide, for example, both the C—O bonds are polar covalent because oxygen is more electronegative than carbon. The linear carbon dioxide molecule, however, is not polar because the two opposite dipoles cancel one another. (Imagine that two people are pulling with equal force on the two ends of a rope. The rope doesn't move because the opposing forces on it cancel each other.)

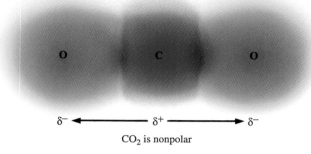

$$CO_2 \text{ is nonpolar}$$

Because of the symmetric pyramid shape of the CH_4 molecule, all the C—H dipoles cancel each other.

Because the water molecule is bent, the dipoles do not cancel; therefore, a water molecule is polar. If the water molecule were linear, it would not be polar.

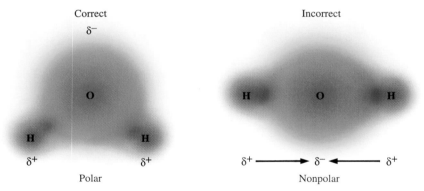

In methane, CH_4, all four of the dipoles cancel, but in CH_3Cl, they do not.

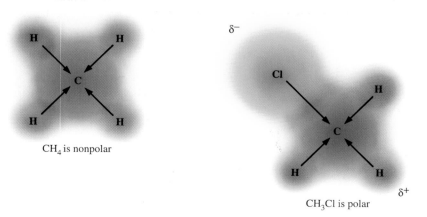

EXAMPLE 6.10

Determine whether NH_2Cl is polar.

Strategy and Solution:

Step 1. Draw a Lewis diagram of the molecule.

$$H—\overset{\cdot\cdot}{\underset{|}{N}}—\overset{\cdot\cdot}{\underset{\cdot\cdot}{Cl}}:$$
$$H$$

Step 2. Determine which element is more electronegative. In this case, N and Cl are more electronegative than H. The electron density will be greater near the N—Cl bond than the two N—H bonds.

$$H—\overset{\cdot\cdot}{\underset{\underset{\delta+\ H}{|}}{N}}—\overset{\overset{\delta-}{\cdot\cdot}}{\underset{\cdot\cdot}{Cl}}:$$

The molecule is polar.

Exercise 6.16

Determine whether HNO_3 is polar.

Whether a molecule is polar or not strongly influences its properties. We will study the effects of polarity in later chapters.

Summary

The understanding we have gained in this chapter forms the core of our study of chemistry. Let's now pull together the basic facts about chemical reaction and the formation of chemical compounds.

A compound is a stable combination of two or more elements whose atoms or ions are joined by chemical bonds. Ionic solids are united by electric attractions, called ionic bonds, between cations and anions. The formation of cations and anions results from electron transfer (oxidation-reduction) reactions.

Covalent bonds are formed when pairs of electrons are shared by pairs of atoms. There are several types of covalent bonds: pure, polar, and coordinate. Polar covalent bonds are created when atoms of different electronegativities share electrons. The atoms in individual molecules are covalently bonded.

The properties of compounds arise not only from the numbers and kinds of atoms in molecules, but also from the ways in which the atoms are bonded, and even from the shape of the molecule itself. Polarity is one such property.

Key Terms

chemical bond	*(p. 157)*	electron transfer	*(p. 161)*
anion	*(p. 158)*	oxidation-reduction reaction	*(p. 161)*
cation	*(p. 158)*	reduction	*(p. 161)*
octet rule	*(p. 159)*	oxidation	*(p. 161)*

ionic bond	*(p. 161)*	double bond	*(p. 171)*
ionic compounds	*(p. 161)*	triple bond	*(p. 171)*
dissociation	*(p. 163)*	coordinate covalent bond	*(p. 171)*
covalent bond	*(p. 165)*	polyatomic ion	*(p. 176)*
Lewis diagram	*(p. 165)*	bond angle	*(p. 180)*
polar covalent bond	*(p. 167)*	electric dipole	*(p. 181)*
electronegativity	*(p. 167)*	dipole	*(p. 181)*
single bond	*(p. 171)*	polar molecule	*(p. 181)*

Questions and Problems

Section 6.1 Ionic Compounds

Questions

1. Explain in terms of ionization energy and electron affinity why metals form positive ions and nonmetals form negative ions.

2. Why is it possible for elements to form doubly or triply charged ions despite the fact that larger amounts of energy are required to form those ions than are required to form singly charged ions?

3. Oxygen is often found as the ion O^{2-}. Why never as O^{3-}?

4. Explain why Al^{3+} is common, but B^{3+} is rarely found.

5. Atoms and ions that have the same electron configuration are called isoelectronic. Give an example of
   (a) two cations that are isoelectronic.
   (b) two anions that are isoelectronic.
   (c) one cation and one anion that are isoelectronic.
   (d) an atom and an ion that are isoelectronic.

6. Why is a magnesium atom smaller than a sodium atom, even though the Mg ion has more electrons and protons?

7. Given two isoelectronic ions, one positive and one negative, which will be larger. Explain.

8. What is the smallest negative ion? The smallest positive ion?

Problems

9. How many electrons will each of the following atoms most likely add in an electron transfer reaction: N, Cl, Br, O, S, P?

10. How many electrons will each of the following atoms most likely lose in an electron transfer reaction: K, Al, Ca, Na, Mg, Sr, Be?

11. Write the electron configurations for the atoms and the ions in Problem 9.

12. Write the electron configurations for the atoms and the ions in Problem 10.

13. Draw Lewis diagrams for the atoms and the corresponding ions in Problems 9 and 10.

14. Write the symbols, including charge, for the ions of the atoms in Problem 9.

15. Write the symbols, including charge, for the ions of the atoms in Problem 10.

16. Give the symbols for the ions formed from atoms of Li, Ca, Cs, S, I, B.

17. Write the symbols for the ions formed from atoms of Ba, Ne, Br, Be, Bi.

18. Give the electron configurations of the following ions: Mg^{2+}, Al^{3+}, O^{2-}, Cl^{-}, Li^{+}.

19. Arrange the ions and atoms in each part in order of increasing size.
    (a) Na, Na^+, P, S^{2-}, Cl
    (b) Li, Be^{2+}, C^{4-}, C^{4+}, O^{2-}
    (c) Mg^{2+}, Al^{3+}, P^{3-}, S, Ar
    (d) N^{3-}, N^{3+}, O^{2-}, Ne, F^-

20. Arrange the ions and atoms in each part in order of decreasing size.
    (a) Sr^{2+}, Mg^{2+}, Ba, Mg
    (b) Li^+, Be^{2+}, B, C, N
    (c) K, Ca, Ca^{2+}, Ag, Br^-
    (d) O^{2-}, C^{4-}, C^{4+}, F, Li

Section 6.2 Ionic Compounds

Questions

21. Explain what happens, and why, during an electron transfer reaction.

22. Explain two ways by which ions can be formed.

23. Define "ionic compound" and "ionic bond."

24. Why are there no ionic molecules?

25. What is meant by dissociation?

Section 6.3 Formulas of Ionic Compounds

Questions

26. Based on the explanation of how KBr is formed, describe the formation of MgO.

27. Will one aluminum atom combine ionically with two chloride atoms? Explain.

28. Compounds containing Cu^{2+} are usually blue. Why is Cu_2O not blue?

Problems

29. For each of the following pairs of atoms, work out all electron transfer reactions that could occur, and predict the formulas of the resulting compounds. (Assume that the elements ordinarily occurring as diatomic gases have already been split into atoms.)
    (a) Mg and O (b) Cl and Cs
    (c) Al and H (d) K and S

30. For each of the following pairs of atoms, work out all electron transfer reactions that could occur, and predict the formulas of the resulting compounds. (Assume that the elements ordinarily occurring as diatomic gases have already been split into atoms.)
    (a) K and Se (b) S and Rb
    (c) F and Ba (d) Ba and S

31. What might be the formula of a molecule resulting from elements X and Y when
    (a) X has six valence electrons and Y has three valence electrons?
    (b) X has seven valence electrons and Y has four valence electrons?
    (c) X has five valence electrons and Y has one valence electron?
    (d) X has six valence electrons and Y has two valence electrons?

32. For each of the following compounds, determine the charge on the ions that form the compounds: $BaBr_2$, Al_2S_3, MgO, Ca_3N_2.

Section 6.4 Covalent Bonding

Questions

33. Explain how two hydrogen atoms form a molecule.

34. Explain the difference between an ionic bond and a covalent bond.

35. In a molecule of H_2, where do the electrons of the shared pair have the highest probability of being found? What consequences does this arrangement have?

Section 6.5 Polar Covalent Bonds and Electronegativity

Questions

36. What is the difference between a pure covalent bond and a polar covalent bond?

37. Define the term "electronegativity."

38. Using electronegativities as a guide, how can one determine whether a chemical bond is ionic or covalent?

Problems

39. In the following compounds, which element will attract a bonding electron pair more strongly? If the bonds are ionic instead of covalent, so state. Hint: Draw the Lewis diagram of each molecule before trying to answer.
    (a) SiS_2 (b) NI_3 (c) Cl_2O (d) PBr_3
    (e) InSb (f) KH (g) SiC

40. Which element in each pair has the highest electronegativity?
    (a) P and O (b) C and Cl
    (c) Al and I (d) H and O
    (e) F and I (f) B and H

41. For the following molecules, which element will attract the electron pair more strongly in each of the bonds?
    (a) HCN (b) CH_3NH_2 (c) CClHCBrF
    (d) CH_2O (e) CH_3OH (f) HSO_3F

42. What type of bond (ionic, polar covalent, or covalent) will form between the atoms of the following pairs?
    (a) Al and N (b) B and H
    (c) Cs and I (d) Mg and O
    (e) O and F (f) Si and C

Section 6.8 Multiple Bonds and Coordinate Covalent Bonds

Questions

43. Based on the bonding of N_2, would you predict the existence of P_2? Why, or why not?

44. How many bonds would there be in a molecule of C_2 if C_2 existed?

45. What is the difference between an ordinary covalent bond and a coordinate covalent bond? Can a molecule contain both types of bonds?

Section 6.9 Lewis Diagrams of Larger Molecules

Problems

46. Draw Lewis diagrams for the following molecules: NCl_3, Cl_2O_7, SiH_4.

47. Draw Lewis diagrams for the following molecules: CF_4, P_2O_4, S_2F_2.

48. Draw Lewis diagrams for the following molecules, all of which contain multiple bonds: C_2H_4, N_2, $HC_2H_3O_2$, CO_2.

49. Draw Lewis diagrams for the following molecules, all of which contain multiple bonds: HCN, C_2H_2, ONF, SO_2.

50. Draw Lewis diagrams for the following molecules: $HClO_3$, CCl_4, SO_2Cl_2.

51. Draw Lewis diagrams for the following molecules: H_2S, NF_3, N_2O_4.

52. Draw Lewis diagrams for the following molecules: C_2H_6, CH_3OH, CH_2O.

53. Draw Lewis diagrams for the following molecules: HNO_2, CH_3COH, NH_3BH_3.

54. Draw Lewis diagrams for the following molecules. There may be more than one possible diagram for some: C_2H_7N, C_3H_5Cl, C_5H_8.

55. Draw Lewis diagrams for the following molecules. There may be more than one possible diagram for some: C_4H_8O, C_3H_6, $C_3H_4O_2$, $C_4H_6O_2$.

Section 6.10 Lewis Structures of Polyatomic Ions

Problems

56. Draw Lewis diagrams for the following polyatomic ions: NO_2^-, CO_3^{2-}, ClO_4^-, HPO_4^{2-}.

57. Draw Lewis diagrams for the following polyatomic ions: PO_4^{3-}, BO_3^{3-}, BF_4^-, $S_2O_4^{2-}$.

Section 6.11 Formulas of Compounds Containing Polyatomic Ions

Problems

58. Complete the table by placing the proper formula in each box.

	Cl^-	O^{2-}	NO_2^-	PO_4^{3-}	$Cr_2O_7^{2-}$	OH^-
Rb^+						
Ca^{2+}			$Ca(NO_2)_2$			
Al^{3+}						
NH_4^+					$(NH_4)_2Cr_2O_7$	
H^+						
Cu^{2+}						

59. Complete the table by placing the proper formula in each box.

	AsO_4^{3-}	SO_4^{2-}	Br^-	CO_3^{2-}	$C_2O_4^{2-}$	HPO_4^{2-}
K^+						
Sr^{2+}	$Sr_3(AsO_4)_2$					
Fe^{3+}						
H^+						
Cu^+			CuBr			
Zn^{2+}						

Section 6.13 Dipoles and Polar Molecules

Questions

60. Why is it important to know the shapes of molecules?

61. Would water be polar or nonpolar if water molecules were linear?

62. What is an electric dipole?

63. What causes electric dipoles in molecules?

64. Explain how a molecule can have polar bonds yet not be a polar molecule.

Problems

65. Based on what you know about the shapes and bonding of the following molecules, predict whether the molecules are polar or nonpolar: CCl_4, H_2S, NOF, PCl_3, NF_3, CH_2O

66. Based on what you know about the shapes and bonding of the following molecules, predict whether the molecules are polar or nonpolar: CO_2, C_2H_4O, NH_3BH_3, N_2, C_2H_4, CH_3OH, CH_2O

Solutions to Exercises

6.1 O^{2-}

6.2 Cl is likely to gain one electron.

6.3 Al will ordinarily lose three electrons.

$$\overset{\cdot}{\underset{\cdot}{Al}}\cdot \longrightarrow Al^{3+} + 3\,e^-$$

6.4 Li loses 1, O gains 2: Li_2O
Sr loses 2, Cl gains 1: $SrCl_2$
Al loses 3, F gains 1: AlF_3

6.5 Oxygen will more readily attract the shared electron pair in CO. Fluorine attracts the shared electrons more readily in BrF.

6.6

Reacting Elements	Electronegativities of These Elements	Difference in Electronegativities	Kind of Bond Formed
Cs + F	Cs, 0.7; F, 4.0	3.3	Ionic
O + Na	O, 3.5; Na, 1.0	2.5	Ionic
C + H	C, 2.5; H, 2.1	0.4	Covalent
Si + Cl	Si, 1.8; Cl, 3.0	1.2	Covalent

6.7 H:F: H—F̄| :I:I: |Ī—Ī| :I:Br: |Ī—B̄r|

6.8 :Cl:P:Cl: |Cl—P̄—Cl|
      :Cl: |Cl|

6.9 H:C:H   H—C—H   (with H above and below each central C)

6.10 H—O—O—H

6.11 :O=C=O:

6.12 H—O—P—O—H   (with H—O: above and :O: below the P)

6.13

$$\left[\begin{array}{c} :O: \\ :O\!-\!Cl\!-\!O: \\ :O: \end{array} \right]^{-} \qquad \left[\begin{array}{c} :O \\[-2pt] \searrow \\ N\!-\!O: \\ :O: \end{array} \right]^{-}$$

6.14 $Al(BrO_3)_3$, Rb_2SO_4

6.15 $Ca_3(PO_4)_2$, $SrSO_4$

6.16 polar

When ammonium nitrate, NH_4NO_3, reacts, N_2, O_2, H_2O, NO, and NO_2 are produced.

An Introduction to Naming Compounds

There are several million known compounds, all of which have names, but fortunately it is not necessary to memorize a name for each compound. An international system has been established for naming all compounds based on their formulas and the families of compounds to which they belong. In Chapter 7, we learn how to name simple inorganic compounds using this system.

Less formal systems of naming compounds have been in common use for decades. We briefly review one such system for naming compounds that contain metals.

Many compounds have common names that arose historically and have nothing to do with formulas. For example, hardly anyone, even a chemist, calls water by its descriptive name, hydrogen oxide. We will learn the common names of several compounds.

The language of chemistry, then, is a combination of official and common names. We begin to deal with this array of names by learning about oxidation numbers and learning a few simple naming rules.

7.1 Oxidation Numbers

After studying this section, you should be able to:

- Determine the oxidation number of an atom in a molecule or ion.
- Determine the oxidation numbers of ions in ionic compounds.

Before learning how to name compounds, we will learn about oxidation numbers because they are needed for writing formulas and naming certain kinds of compounds. We were introduced to oxidation-reduction reactions in Chapter 6. The reaction of Na and Cl atoms is a simple example. An electron is transferred from a sodium atom to a chlorine atom, resulting in a cation and an anion. The cation has a +1 charge and the anion has a −1 charge.

$$\text{Na} \cdot + \; :\overset{..}{\underset{..}{\text{Cl}}} \cdot \; \longrightarrow \; \text{Na}^+ + \; :\overset{..}{\underset{..}{\text{Cl}}}^- :$$

We can keep track of how many electrons are transferred in a reaction by using the system of **oxidation numbers,** sometimes called **oxidation states.** An oxidation number is a positive, negative, or zero value that is assigned to each element in a compound. In ionic compounds, the oxidation numbers are the same as the charges on the ions. In NaCl, for example, the oxidation number of Na is +1 while that of Cl is −1.

# of ion charge

Although electrons are not actually transferred when covalent bonds are formed, it is often convenient to assign oxidation numbers to covalently bonded atoms as well as to ions. Oxidation numbers are assigned according to a set of rules. In many cases, the rules are somewhat arbitrary, but because the assignment of oxidation numbers is only a kind of bookkeeping, this causes no problem as long as the rules are consistently applied. The rules that follow are the most widely accepted.

Rules for Assigning Oxidation Numbers

1. *In an element, each atom has an oxidation number of zero.* In N_2, for example, each N atom has an oxidation number of zero, and the sum of the oxidation numbers in N_2 is zero. Other examples are Cl_2, H_2, O_2, and He.

2. *In compounds, the alkali metals, Na through Cs in the first column of the periodic table, have an oxidation number of +1.*

3. *In compounds, the alkaline earth metals, Be, Mg, Ca, Sr, and Ba, have an oxidation number of +2.*

4. *In compounds, Al has an oxidation number of +3, Ag of +1, and Cd and Zn of +2.*

5. *The oxidation number of hydrogen in compounds is +1, except for metal hydrides such as NaH.* (Hydrides, in which the oxidation number of H is −1, occur only in compounds with metals. Compounds containing hydride ions have the word "hydride" in their names.)

6. *The oxidation number of oxygen in compounds is −2, except in OF_2 and in peroxides.* (Peroxides contain O—O bonds and have the word "peroxide" in their names. An example is hydrogen peroxide, H_2O_2.)

7. *In compounds with metals, the halogens (F through I) have an oxidation number of −1. Fluorine has an oxidation number of −1 in all its compounds.*

8. *The sum of the oxidation numbers of all the atoms in a molecule is zero.* In H_2O, the oxidation number of each H is $+1$, and that of O is -2; thus the sum is zero: $(2 \times (+1)) + (-2) = 0$.

9. *In an ion, the sum of the oxidation numbers of all the atoms must equal the charge on that ion.* For example, in a sulfate ion, SO_4^{2-}, the oxidation numbers of the elements must add up to -2. since -2 ion

10. *To assign oxidation numbers to individual atoms in covalent compounds, assign the electrons in each bond to the more electronegative atom sharing those electrons.* (Remember that the sum of the oxidation numbers of neutral molecules must be zero.) In PCl_3, for example, Cl is more electronegative than P, therefore, assign each Cl a -1 oxidation number.

Although we need to be aware that there are exceptions, we can, in most cases, assign oxidation numbers by relying on our general knowledge of chemistry and by considering the positions of the elements in the periodic table; that is, we can avoid referring constantly to the rules.

We will usually be able to determine oxidation numbers of elements in compounds by relying on the fixed oxidation numbers of a few elements and doing a little arithmetic.

EXAMPLE 7.1

The compound H_2S is sometimes called "rotten egg gas." What are the oxidation numbers of H and S in H_2S?

Strategy and Solution:

Step 1. Since this compound is not a metal hydride, H has an oxidation number of $+1$, giving a total of $+2$ for the two H atoms.

Step 2. Assign an oxidation number to the other element. H_2S is neutral. S must, therefore, have an oxidation number of -2 so that the total will equal zero.

$$[2 \times (+1)] - 2 = 0$$

EXAMPLE 7.2

At one time carbon tetrachloride, CCl_4, was used in commercial dry cleaning. What are the oxidation numbers of carbon and chlorine in CCl_4?

Strategy and Solution:

Step 1. To assign oxidation numbers to individual atoms in covalent compounds, assign the electrons in each bond to the more electronegative atom sharing those electrons. Cl is significantly more electronegative than C, so we assign an oxidation number of -1 to each chlorine. This gives a total of -4 for the chlorine.

Step 2. Assign an oxidation number to the other element. CCl_4 is neutral. Therefore, C must have an oxidation number of $+4$ for the total to equal zero.

Exercise 7.1

What are the oxidation numbers of phosphorus and bromine in PBr_3? In PBr_5?

Exercise 7.2

Chloroform, $CHCl_3$, was once widely used as an anesthetic. What is the oxidation number of each element in $CHCl_3$?

EXAMPLE 7.3

What are the oxidation numbers of the atoms in sulfate ion, SO_4^{2-}?

Strategy and Solution:

Step 1. From the name, we know that sulfate ion has no O—O peroxide bonds, so each oxygen atom must have an oxidation number of -2.

Step 2. Assign an oxidation number to the other element. Sulfate ion has a -2 charge, and the sum of the oxidation numbers of all the atoms must equal -2. We can write this relation: Sum of oxidation numbers of four oxygen atoms + oxidation number of S $= -2$.
Therefore,

$$[4 \times (-2)] + \text{oxidation number of S} = -2$$

Oxidation number of S $= -2 + 8 = +6$.

EXAMPLE 7.4

What are the oxidation numbers of the atoms in sulfurous acid, H_2SO_3?

Strategy and Solution:

Step 1. The name tells us that there are neither hydride ions nor peroxides in H_2SO_3; therefore, we assign oxidation numbers of $+1$ to hydrogen and -2 to oxygen.

Step 2. Assign an oxidation number to the other element. Because H_2SO_3 is a neutral molecule, the sum of its oxidation numbers must be zero.

$$[2 \times (+1)] + [3 \times (-2)] + \text{oxidation number of S} = 0$$

The oxidation number of S $= +4$.

EXAMPLE 7.5

When mixed with water, sulfurous acid produces SO_3^{2-} ions. (We will study this phenomenon in Chapter 14.) What is the oxidation number of sulfur after this occurs?

Strategy and Solution:

Step 1. Assign oxidation number -2 to oxygen.

Step 2. This polyatomic ion has a -2 charge, and the sum of the oxidation numbers of all the atoms must equal -2.

$$[3 \times (-2)] + \text{oxidation number of S} = -2$$

The oxidation number of $S = -2 + 6 = +4$.
The ionization process has not changed the oxidation number of sulfur.

In three of these examples, the oxidation number for sulfur has three different values: -2, $+4$, and $+6$. Sulfur is one of the many elements that can have any of several oxidation states, depending on the nature of the compound it is in.

Table 7.1 gives oxidation numbers for a few elements.

Many elements can have any of several oxidation states. In most cases, you can determine the oxidation number of such elements in different compounds by using the rules for assigning oxidation numbers.

Exercise 7.3

What are the oxidation numbers of nitrogen and oxygen in NO_3^-?

Exercise 7.4

What is the oxidation number of nitrogen in NO_2^-?

Table 7.1	Oxidation Numbers of a Few Elements
Element	Oxidation Numbers
Au	$+1, +3$
Cr	$+2, +3, +6$
Cu	$+1, +2$
Co	$+2, +3$
Hg	$+1, +2$
Mn	$+2, +4, +6, +7$
Pb	$+2, +4$
Sn	$+2, +4$
C	$-4, +2, +4$
Cl	$-1, +1, +3, +5, +7$
Br	$-1, +1, +3, +5, +7$
I	$-1, +1, +3, +5, +7$
S	$-2, +4, +6$
N	$-3, +1, +2, +3, +4, +5$
P	$-3, +3, +5$

7.2 Naming Binary Compounds

After studying this section, you should be able to:

- Name binary compounds that contain a nonmetal and a metal with a fixed oxidation number.
- Use the Stock system to name binary compounds that contain a nonmetal and a metal with variable oxidation numbers.
- Use the "old" system to name binary compounds that contain a nonmetal and a metal with variable oxidation numbers.
- Name binary compounds that contain two nonmetals.
- Determine the formula of a binary compound, given its name.

Compounds containing only two elements are called **binary compounds.** Many binary compounds consist of a metal and a nonmetal, such as magnesium chloride, $MgCl_2$. Binary compounds containing two nonmetals are also common, such as sulfur trioxide, SO_3, for instance. The naming rule governing binary compounds requires that the name of the more metallic element appear first, followed by the stem of the name of the second element written with the ending *-ide*. Notice that in writing names and formulas, the *symbols* of the elements always begin with capital letters, but the *names* of the elements and compounds are not capitalized.

✳ The names of binary compounds always end in -ide.

Naming Binary Compounds Containing Metals with Fixed Oxidation Numbers

The binary compounds easiest to name contain nonmetals combined with metals that have fixed oxidation numbers. Recall from the previous section that alkali metals, alkaline earth metals, aluminum, silver, cadmium, and zinc, all have fixed oxidation numbers. A binary compound of any of these metals with a nonmetal is named by giving the name of the metal first, then the name of the nonmetal, which is changed by replacing the ending of the name with *-ide*. Let's look at $CaCl_2$. Drop *-ine* in the name "chlorine" and add *-ide*. The compound is named calcium chloride.

EXAMPLE 7.6

Name the compound MgF_2.

Strategy and Solution:

Step 1. Name the metal, magnesium.

Step 2. Drop the ending of the nonmetal and add *-ide*. For fluorine, the ending *-ine* is replaced with *-ide,* giving fluoride.

Step 3. Put the two together. The name is magnesium fluoride.

> **Exercise 7.5**
>
> Name these three compounds: BaO, K_3N, and Al_2S_3.

You will often need to determine the formula of a compound from its name. Here is an example of how to do this.

EXAMPLE 7.7

Write the formula for aluminum chloride.

Strategy and Solution:

Step 1. Write the symbol and determine the oxidation number for the metal. In this case, the metal is aluminum, Al, with an oxidation state of +3.

Step 2. Write the symbol for the nonmetal and determine its oxidation number. Chloride is Cl with a −1 oxidation number.

Step 3. Find the ratio of the number of metal ions to nonmetal ions. To make the sum of the oxidation numbers in the compound zero, there must be three Cl^-'s for every Al^{3+}; therefore, the formula of the compound is $AlCl_3$.

> **Exercise 7.6**
>
> Give the formulas for silver iodide and zinc oxide.

Naming Binary Compounds Containing Metals with Variable Oxidation Numbers

As they form different compounds, certain metals can appear in different oxidation states. Because the name of a binary compound must be fully descriptive, it must designate the oxidation number of the individual metal (unless the oxidation number is fixed). Binary compounds containing metals that have different oxidation states can be named using a method known as the **Stock system. In the Stock system, the name of the element is followed by a Roman numeral in parentheses, indicating the oxidation state of that metal.** For example, iron in its +3 oxidation state is named iron(III) and is pronounced *"iron three."* Fe_2O_3 is written as iron(III) oxide and is called *"iron three oxide."* CuCl and $CuCl_2$ are respectively copper(I) chloride and copper(II) chloride. This system is being used more and more as some of the older names are slowly being phased out.

Metals that do not have variable oxidation states are the alkali metals, alkaline earth metals, Cd, Ag, and Zn.

only with transition metals

In the Stock system, the oxidation number of the metal is shown as a Roman numeral in parentheses.

EXAMPLE 7.8

Name NiF_2.

Strategy and Solution:

Step 1. Determine the oxidation number of the metal. In this compound the oxidation number of nickel is +2.

Step 2. Drop the ending of the nonmetal and add -*ide*. For fluorine, the ending -*ine* is dropped and -*ide* is added, giving fluoride. The name is nickel(II) fluoride.

Chromium forms two different oxides, Cr_2O_3 (green) and CrO_3 (red).

Exercise 7.7

Use the Stock system to name the following compounds: CoSe, $CrBr_2$, Hg_2Cl_2.

EXAMPLE 7.9

Write the formula for chromium(III) oxide.

Strategy and Solution:

Step 1. Write the symbol and determine the oxidation number for the metal. In this case, the metal is chromium, Cr, with an oxidation state of +3.

Step 2. Write the symbol for the nonmetal and determine its oxidation number. The nonmetal in chromium(III) oxide is O, which has a −2 oxidation number.

Step 3. Find the ratio of metal to nonmetal. Summing the oxidation numbers to zero tells us that the ratio of Cr^{3+} to O^{2-} must be two to three. Accordingly, the formula of the compound is Cr_2O_3.

Exercise 7.8

Give the formulas for iron(II) oxide and tin(IV) fluoride.

The "old system" of naming compounds indicates only the *relative* oxidation state of the metal in a compound. In this system, two suffixes (*-ous* and *-ic*)

Table 7.2   Names of Some Common Metal Ions that Have Variable Oxidation Numbers

Element	Oxidation Number	Stock Name	Old Name
As	+3	Arsenic(III)	Arsenous
As	+5	Arsenic(V)	Arsenic
Cu	+1	Copper(I)	Cuprous
Cu	+2	Copper(II)	Cupric
Cr	+2	Chromium(II)	Chromic
Cr	+3	Chromium(III)	Chromous
Co	+2	Cobalt(II)	Cobaltous
Co	+3	Cobalt(III)	Cobaltic
Fe	+2	Iron(II)	Ferrous
Fe	+3	Iron(III)	Ferric
Hg	+1	Mercury(I)	Mercurous
Hg	+2	Mercury(II)	Mercuric
Pb	+2	Lead(II)	Plumbous
Pb	+4	Lead(IV)	Plumbic
Sn	+2	Tin(II)	Stannous
Sn	+4	Tin(IV)	Stannic

are used to indicate lower and higher oxidation states. To name compounds containing a metal with the lowest oxidation state, we use the suffix that indicates a lower state, *-ous,* as part of the name of the metal. If the metal has a higher oxidation number, the suffix *-ic* is used. In this system, the name of the metal is most often derived from its Latin name. Iron, for example, can assume an oxidation state of either +2 or +3. In the +2 compounds, the name *ferrous* (from the Latin, *ferrum,* for iron) is used, as in ferrous chloride, $FeCl_2$. In the +3 compounds, the name *ferric* is used, as in ferric chloride, $FeCl_3$.

Because it does not indicate actual oxidation numbers as the Stock system does, the "old system" is falling into disuse. Nonetheless, we need to know it because this system is still used on the labels of many bottles of chemicals.

Table 7.2 lists the names of some common metals that have more than one oxidation state.

You need to know the Latin names and common oxidation states of many of the metal ions (example: cupric and cuprous ions) to be able to use the "old system."

EXAMPLE 7.10

Several toothpastes contain SnF_2 as a source of fluoride ion. Use the "old system" to name SnF_2.

Strategy and Solution:

Step 1. Determine the oxidation number of the metal. In this compound, Sn is in the +2 state, called *stannous.*

Step 2. Drop the ending of the nonmetal and add *-ide*. For fluorine, the ending *-ine* is dropped and *-ide* is added, giving fluoride. The name of this compound is stannous fluoride.

Exercise 7.9

Using the "old system," name PbO_2, CuI_2, and Hg_2Cl_2.

EXAMPLE 7.11

Write the formula for cobaltous chloride.

Strategy and Solution:

Step 1. Write the symbol for the metal. In this case, the metal is Co with an oxidation state of +2.

Step 2. Write the symbol for the nonmetal. Chloride is Cl with a −1 oxidation number.

Step 3. From the oxidation numbers, we find that the ratio of the number of Co^{2+} ions to Cl^- ions must be one to two. The formula of the compound is $CoCl_2$.

Crystals of cobaltous chloride.

Exercise 7.10

Give the formulas for ferrous chloride and stannic fluoride.

Still
-ide

Hydrogen is a nonmetal that has some metallic characteristics as a solid.

You will need to memorize several of the prefixes and their meanings.

Naming Binary Compounds Containing Two Nonmetals

As with all binary compounds, the symbol and name of the more metallic element appears first, followed by the second element written with the ending *-ide.* Thus in HCl, hydrogen is the more metallic element and is followed by chlorine, the name of which is changed to end in *-ide,* giving hydrogen chloride.

When two nonmetals form more than one compound, it is necessary to distinguish between those compounds. There are two ways to accomplish this. The "old" method of naming nonmetal binary compounds uses two suffixes (*-ous* and *-ic*). Because it is clumsy and sometimes causes confusion, it is seldom used. Instead, we will study the preferred method of naming binary nonmetal compounds, which includes prefixes that indicate the numbers of atoms in the formula. A familiar example is given by carbon monoxide, CO, and carbon dioxide, CO_2. *Mono-* indicates that one oxygen atom appears in the formula, and *di-* signals that there are two oxygen atoms. Theoretically, CO would be named monocarbon monoxide; however, the prefix *mono-* is usually omitted for the first element in a compound. Thus, if there is no prefix with the name of an element, we can assume that there is only one atom of that element in the formula. In this system, *tri-* means three, as in nitrogen triiodide, NI_3. Table 7.3 lists the prefixes used.

EXAMPLE 7.12

Name P_2O_4.

Solution: In P_2O_4 there are two P atoms and four O atoms. We use the prefixes *di-* and *tetra-* in the name: diphosphorous tetroxide. (Notice that the "a" of *tetra-* is dropped when added to oxide, as is the "o" in *mono-* of carbon monoxide.)

Exercise 7.11

Name SO_2, SO_3, and S_2O_3.

(sreek)

Table 7.3   Prefixes That Denote Numbers of Atoms

Prefix	Number Designated	Example
mono-	one	Carbon monoxide, CO
di-	two	Carbon dioxide, CO_2
tri-	three	Phosphorus trichloride, PCl_3
tetra-	four	Carbon tetrabromide, CBr_4
penta-	five	Phosphorus pentachloride, PCl_5
hexa-	six	Selenium hexafluoride, SeF_6
hepta-	seven	Iodine heptafluoride, IF_7

EXAMPLE 7.13

Write the formula for phosphorus pentachloride.

Solution:

Step 1. Write the symbol for phosphorus.

Step 2. Write the symbol for five Cl. The formula is PCl_5.

Exercise 7.12

Give the formulas for selenium trioxide and silicon tetrafluoride.

7.3 Naming Acids Derived from Binary Compounds

After studying this section, you should be able to:

- Name acids derived from binary compounds.
- Determine the formula of an acid from a binary compound when given the name.

Certain binary compounds that contain hydrogen display acidic properties when they are dissolved in water. Because of these acidic properties, the names of the compounds include the word "acid." To name such a compound, the prefix *hydro-* is substituted for the word hydrogen and the name of the nonmetal ends in *-ic.* Hydrogen chloride gas, for example, is acidic when mixed with water. The acidic solution is known as hydrochloric acid.

It is important to remember that not all binary compounds that contain hydrogen are acidic when dissolved in water. To distinguish between those that are acidic and those that are not, the symbol for hydrogen appears first in the formula of an acid.

Phosphorus pentachloride is a white solid that forms when white phosphorus is burned in an atmosphere of chlorine gas. The yellow of Cl_2 disappears when Cl_2 makes the ion, Cl^-.

We will discuss acids in detail in Chapter 14.

hydro for hydrogen

EXAMPLE 7.14

Name the aqueous solution of H_2S.

Solution: The name of the compound is hydrogen sulfide. Replace hydrogen with the prefix *hydro-* and replace the ending *-ide* with *-ic.* Add the word "acid." The name is hydrosulfuric acid. (The *-ur-* is added to the name for phonetic reasons.)

Exercise 7.13

Name the aqueous solutions of HI and H_2Se.

EXAMPLE 7.15

Write the formula for hydrobromic acid.

200 CHAPTER 7 An Introduction to Naming Compounds

Strategy and Solution:

Step 1. We know that the formula begins with H.

Step 2. Write the symbol for the second element, in this case, Br with a −1 oxidation number.

Step 3. Determine the ratio of hydrogen to the other element. The ratio of H^+ to Br^- must be one to one; therefore, the formula of the compound is HBr.

Exercise 7.14

Give the formula for hydrofluoric acid.

7.4 Naming Compounds Containing Polyatomic Ions

After studying this section, you should be able to:

- Name compounds containing polyatomic ions.
- Determine the formula of a compound containing a polyatomic ion, given its name.
- Name oxyacids.
- Determine the formula of an oxyacid when given its name.

To name compounds containing polyatomic ions, you must know the names of the polyatomic ions themselves. Table 7.4 gives the names and symbols of some polyatomic ions. Most of these names need to be memorized. There are, however, some guidelines that can help you learn them. Many polyatomic ions have names ending with one of two suffixes, *-ite* and *-ate*. These suffixes indicate lower (*-ite*) and higher (*-ate*) oxidation states. For example, in the chlorite ion, ClO_2^-, chlorine has an oxidation state of +3, while in chlorate, ClO_3^-, chlorine has an oxidation state of +5. When there are more than two oxidation states, two prefixes, *hypo-* (meaning "lower than *-ite*") and *per-* (meaning "higher than *-ate*") are also used. Take hypochlorite, ClO^-, where chlorine is +1 and perchlorate, ClO_4^-, where chlorine is +7, for instance. Table 7.4 lists the names of several common polyatomic ions.

Several polyatomic ions are produced by adding H^+ to other polyatomic ions. The names of the resulting ions begin with the prefix *bi-*. For example, bicarbonate, HCO_3^-, is a combination of hydrogen and carbonate ion, CO_3^{2-}.

When you know the names of the polyatomic ions, it is easy to name compounds containing them. If a compound contains a metal and a polyatomic ion, name the metal according to Section 7.2 and then name the polyatomic ion. Lithium carbonate, Li_2CO_3, and copper(I) sulfate, Cu_2SO_4, are examples. If a compound contains a positive polyatomic ion, such as NH_4^+, and a simple non-metal ion, the ion appears first in the formula and is named first, followed by the name of the nonmetal with the ending *-ide.* Ammonium chloride, NH_4Cl, for example, is a common compound in the laboratory. If both ions are polyatomic, use both ion names, as in NH_4ClO_4, ammonium perchlorate.

Table 7.4 Common Polyatomic Ions

Acetate	$C_2H_3O_2^-$	Hydroxide	OH^-
Ammonium	NH_4^+	Hypochlorite	ClO^-
Arsenate	AsO_4^{3-}	Iodate	IO_3^-
Bicarbonate	HCO_3^-	Nitrate	NO_3^-
Bisulfate	HSO_4^{2-}	Nitrite	NO_2^-
Bisulfite	HSO_3^-	Oxalate	$C_2O_4^{2-}$
Bromate	BrO_3^-	Perchlorate	ClO_4^-
Borate	BO_3^{3-}	Permanganate	MnO_4^-
Carbonate	CO_3^{2-}	Phosphate	PO_4^{3-}
Chlorate	ClO_3^-	Phosphite	PO_3^{3-}
Chlorite	ClO_2^-	Silicate	SiO_3^{2-}
Chromate	CrO_4^{2-}	Sulfate	SO_4^{2-}
Cyanide	CN^-	Sulfite	SO_3^{2-}
Dichromate	$Cr_2O_7^{2-}$	Thiocyanate	SCN^-

EXAMPLE 7.16

Name $K_2Cr_2O_7$.

Solution:

Step 1. Name the metal. In this case, the metal is potassium.

Step 2. Name the polyatomic ion.
The name of the compound is potassium dichromate.

$K_2Cr_2O_7$ is an orange solid.

Exercise 7.15

Name $NaClO_2$ and CaC_2O_4.

EXAMPLE 7.17

Name $Fe(NO_3)_2$ using the Stock system.

Strategy and Solution:

Step 1. Determine the oxidation number of the metal. In this compound, the oxidation of iron is $+2$.

Step 2. Name the polyatomic ion.
The name of the compound is iron(II) nitrate.

Exercise 7.16

Using the Stock system, name Cu_2SO_4 and $Cr_2(SO_3)_3$.

$Fe(NO_3)_2$ crystals are light green because of the Fe^{2+} ion.

EXAMPLE 7.18

Write the formula for mercury(II) nitrate.

Strategy and Solution:

Step 1.   Write the symbol and determine the oxidation number for the metal. In this case, the metal is Hg with an oxidation state of +2.

Step 2.   Write the formula for the polyatomic ion. Nitrate is NO_3^-.

Step 3.   Find the ratio of metal ions to polyatomic ions by summing the oxidation numbers to zero. The ratio of Hg^{2+} to NO_3^- is one to two; therefore, the formula of the compound is $Hg(NO_3)_2$.

> **Exercise 7.17**
>
> Write the formula for cobalt(III) acetate.

EXAMPLE 7.19

Name $Fe(NO_3)_2$ using the "old system."

Strategy and Solution:

Step 1.   Determine the oxidation number of the metal and its name. In this compound, the oxidation number of iron is +2, and its name is ferrous.

Step 2.   Name the polyatomic ion.
The name of the compound is ferrous nitrate.

> **Exercise 7.18**
>
> Using the "old system," name Cu_2SO_4 and $SnSO_4$.

EXAMPLE 7.20

Write the formula for cobaltous chlorate.

Strategy and Solution:

Step 1.   Write the symbol for the metal. In this case, the metal is Co with an oxidation state of +2.

Step 2.   Write the formula for the polyatomic ion. Chlorate is ClO_3^-.

Step 3.   Determine the ratio of metal ions to polyatomic ions. The ratio of Co^{2+} to ClO_3^- must be one to two; therefore, the formula of the compound is $Co(ClO_3)_2$.

> **Exercise 7.19**
>
> Give the formulas for ferrous perchlorate and stannic chromate.

EXAMPLE 7.21

Write the formula for ammonium carbonate.

Strategy and Solution:

Step 1. This compound contains two polyatomic ions. Write the formula for each polyatomic ion. Ammonium ion is NH_4^+ and carbonate is CO_3^{2-}.

Step 2. Determine the ratio of ammonium ion to carbonate ion. The ratio is two to one.
The formula is $(NH_4)_2CO_3$.

> **Exercise 7.20**
>
> Write the formula for ammonium nitrate.

Naming Oxyacids

oxygen in the polyatomic

The **oxyacids** are compounds containing polyatomic ions in combination with hydrogen. Like acidic binary compounds, oxyacids produce acidic solutions when dissolved in water. To name these compounds, use only the name of the polyatomic ion, modify the ending of the name (*-ous* replaces *-ite,* and *-ic* replaces *-ate*) and add the word "acid." One of the more common acids is the aqueous solution of HNO_3. HNO_3 contains the nitrate group, NO_3^-. To name the acid, replace the *-ate* ending of nitrate with *-ic* and add "acid," giving *nitric acid*. Notice that, unlike the acids of the binary compounds, the oxyacids do not use the prefix *hydro-*.

—ous over -ite
-ic over -ate

no hydro

EXAMPLE 7.22

Name the aqueous solution of H_2SO_3.

Strategy and Solution:

Step 1. Name the polyatomic ion. The compound contains sulfite, SO_3^{2-}.

Step 2. Substitute *-ous* for *-ite* and add the word "acid." The name is sulfurous acid. The *-ur-* is added before adding the *-ous* to make the name sound better to the ear.

> **Exercise 7.21**
>
> Name the aqueous solutions of H_2CO_3 and H_3PO_4.

EXAMPLE 7.23

Write the formula for nitrous acid.

Strategy and Solution:

Step 1. Write the formula for the polyatomic ion. In this case, the polyatomic ion is nitrite, NO_2^-.

Step 2. Determine the ratio of hydrogen ion to polyatomic ion. The ratio of H^+ to NO_2^- is one to one. The formula is HNO_2.

Exercise 7.22

Give the formula for hypochlorous acid.

7.5 Naming Compounds with More Than One Kind of Positive Ion

After studying this section, you should be able to:

- Name compounds that have more than one kind of positive ion.

Some compounds contain more than one kind of positive ion. Usually such compounds are the result of replacing some of the hydrogen ions of oxyacids or of acids of binary compounds with metallic ions. To name these compounds, first name each metallic ion, then add the nonmetal or polyatomic ion. For example, $KNaCO_3$ is named potassium sodium carbonate. If there are two kinds of positive ions in a compound and more than one of any ion, a prefix, such as *di-*, can be used. Potassium disodium phosphate, KNa_2PO_4, is a good example.

If one of the hydrogen ions has not been replaced by a metallic ion, the prefix *bi-* is often added to the name of the nonmetal ion or polyatomic ion. For example, HSe^- is biselenide ion, and NH_4HSe is ammonium biselenide or ammonium hydrogen selenide.

[handwritten margin note: treat metal seperately]

[handwritten margin note: — bi instead of hydrogen]

EXAMPLE 7.24

Name $LiHCO_3$.

Solution: The name is lithium hydrogen carbonate or lithium bicarbonate.

Exercise 7.23

Name NH_4HS and KRb_2PO_4.

EXAMPLE 7.25

Write the formula for lithium sodium sulfite.

Strategy and Solution:

Step 1.   Write the symbols for the metals. In this case, the metals are Li and Na. Each has an oxidation state of $+1$.

Step 2.   Write the formula for the polyatomic ion. Sulfite is SO_3^{2-}.

Step 3.   Determine the ratio of the number of metal ions to that of the polyatomic ion. The ratio of Li^+ and Na^+ to SO_3^{2-} must be one to one to one; therefore, the formula of the compound is $LiNaSO_3$.

Exercise 7.24

A common antacid is sodium bicarbonate, also known as sodium hydrogen carbonate. Give the formula for this compound.

Sodium hydrogen carbonate (sodium bicarbonate), better known as baking soda, has the formula $NaHCO_3$.

7.6 **Common Names of Compounds**

After studying this section, you should be able to:

• Name several common compounds, given their formulas or chemical names.

Long before systematic methods of naming compounds were developed, many compounds had already been named. These names have a long history, and chemists continue to use them. Many of the best-known binary compounds are among those that have common names. Table 7.5 lists the formulas and names of several of the better-known ones.

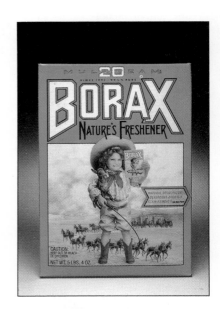

Sodium tetraborate decahydrate, $Na_2B_4O_7 \cdot 10\ H_2O$, is better known as Borax.

Table 7.5   Common Names of Some Compounds

Common Name	Chemical Name	Formula
Acetylene	Ethyne	C_2H_2
Ammonia	Hydrogen nitride	NH_3
Baking soda	Sodium bicarbonate	$NaHCO_3$
Borax	Sodium tetraborate decahydrate	$Na_2B_4O_7 \cdot 10\ H_2O$
Cream of tartar	Potassium bitartrate	$KHC_4H_4O_6$
Gypsum	Calcium sulfate dihydrate	$CaSO_4 \cdot 2\ H_2O$
Grain alcohol	Ethanol	C_2H_6O
Lime	Calcium oxide	CaO
Lye	Sodium hydroxide	$NaOH$
Milk of magnesia	Magnesium hydroxide	$Mg(OH)_2$
Sugar	Sucrose	$C_{12}H_{22}O_{11}$
Table salt	Sodium chloride	$NaCl$
Washing soda	Sodium carbonate	Na_2CO_3
Water	Hydrogen oxide	H_2O

Science in Action

Names of Complex Compounds

Structure of valium.

The chemical name of a compound must describe the compound completely, so that a person reading the name can write an accurate and complete formula. For simple compounds that have only a few atoms, the names are correspondingly simple; but there are large numbers of compounds, hundreds of thousands, whose molecules contain many tens or even many hundreds of atoms. To be properly descriptive, the names of these compounds must say how many atoms of each kind are present, which atoms are attached to which, and how the atoms lie in space relative to one another. Even using the special words invented for the purpose, the name of a compound containing, say, only a hundred atoms would take several lines of print. Although such names exist and are used routinely by chemists, they are too long and complex for everyday use, so shorter names take their place. A simple example is to say "water," meaning "hydrogen oxide."

Particularly in the pharmaceutical field, common names are needed for consumers of complex drugs. In many cases, a three-level name system is used: the true, chemical name for the pharmaceutical chemists, a shorter name for use by pharmacists, and the common or trade name by which consumers know the drug. An example is *Valium*, a name copyrighted by Roche Products for a tranquilizer. Doctors and druggists know that Valium is *diazepam*, but that name, too, is an abbreviation. The empirical formula for Valium is $C_{16}H_{13}N_2OCl$, and its chemical name is 7-chloro-1,3-dihydro-1-methyl-5-phenyl-2H-1,4-benzodiazepin-2-one. (The *-one* at the end does not refer to the numeral 1 but identifies the compound as one of a class called ketones.) The chemical name exactly describes the physical structure of the molecule, drawn to the left.

Valium is by no means the most complex pharmaceutical compound; many have names several times as long, indicating that the molecules are correspondingly large. Here are the names of some other pharmaceuticals.

Trade Name	Generic Name	Function	Chemical Name
Seldane	Terfenadine	Antihistamine	α-[4-(1,1-dimethyl) phenyl]-4-(dihydroxy- diphenylmethyl)-1- piperidine butanol
Zovirax	Acyclovir	Herpes treatment	9-[(2-hydroxy ethoxy) methyl guanine
Codeine	Codeine	Pain relief	7,8-didehydro-4,5α- epoxy-3-methoxy– methylmorphinan- 6α-ol phosphate
Tylenol	Acetaminophen	Pain relief	4'-hydroxyacetanilide
Cortisone	Hydrocortisone	Anti-inflammatory	11,17,21-trihydroxy- pregn-4-ene-3, 20-dione

Summary

Oxidation numbers can be assigned to atoms in covalent compounds as well as to those in ionic compounds. The rules are arbitrary, but the assignment of oxidation numbers is often essential, especially when naming certain compounds.

Compounds are named using a variety of methods. Binary compounds are systematically named. The name of the more metallic element comes first, followed by the name of the second element written with the ending -*ide*.

Some binary compounds contain metals with variable oxidation numbers. Such compounds can be named using a method known as the Stock system in which the name of the element is followed by a Roman numeral in parentheses, indicating the oxidation number of the metal. A less systematic method for naming these compounds uses metal ion names derived from the Latin names of the metals.

Compounds containing polyatomic ions can be named if the names of the polyatomic ions themselves are known. To cope with the names of polyatomic ions, it helps to know what the various prefixes and suffixes mean and something about how they are used. Even so, in the end it is necessary to resort to some memorization.

Key Terms

oxidation number (p. 190)
oxidation state (p. 190)
oxyacid (p. 203)

binary compound (p. 194)
Stock system (p. 195)

Questions and Problems

Section 7.1 Oxidation Numbers

Question

1. List the rules for assigning oxidation numbers to elements in compounds and ions.

Problems

2. Determine the oxidation number of chlorine in these compounds: $HClO$, $HClO_2$, $HClO_3$, $HClO_4$.

3. What is the oxidation number of carbon in each of the following compounds: CO, CO_2, CH_2O, $Al_2(C_2O_4)_3$, H_2CO_3.

4. Determine the oxidation number of manganese in each of the following compounds: K_2MnO_4, $KMnO_4$, MnO_2, $MnCl_2$.

5. Give the oxidation number of each element in the following compounds and ions: BaC_2O_4, HBO_2, I_2O_5, BrO_3^-, MnO_4^-, $Al(C_3H_7O)_3$.

6. Give the oxidation number of each element in the following compounds and ions: Al_2Se_3, CrO_4^{2-}, NH_4^+, NH_2OH, N_2H_4, $Na_2H_3IO_6$.

7. Give the oxidation number of each element in the compounds in Problem 20.

8. Give the oxidation number of each element in the compounds in Problem 15.

Section 7.2 Naming Binary Compounds

Questions

9. Briefly describe how to name a binary compound.

10. Briefly describe the Stock system of naming compounds.

11. What advantage does the Stock system have over the "old system" of naming compounds?

12. According to the "old system" of naming compounds, Cu^+ and Co^{2+} have the same suffix in their names but different oxidation numbers. Explain how this is possible.

13. When naming binary compounds containing more than one nonmetal, prefixes are used. List seven prefixes and the numbers of atoms they represent.

Problems

14. Name the following compounds: KBr, Li_2O, CaS, BeI_2, Al_2Se_3, BaO.

15. Name the following compounds: $CsBr$, SrI_2, Na_3P, CaC_2, Be_3N_2, Ag_2O.

16. Name the following compounds: $CaCl_2$, BaS, Li_3N, KF, Cd_3As_2, CaB_6.

17. Give formulas for the following binary compounds: potassium chloride, beryllium sulfide, potassium oxide, barium fluoride, calcium iodide, magnesium selenide.

18. Write formulas for the following compounds: lithium fluoride, beryllium oxide, potassium iodide, silver chloride, zinc oxide, cadmium bromide.

19. Using the Stock system, name the following compounds: SnO_2, PbO_2, Fe_2O_3, AsI_3, Hg_2I_2, Cu_3B_2.

20. Using the Stock system, name the following compounds: SnO_2, $HgBr_2$, $FeCl_2$, PbO_2, Cu_2O, As_2O_3.

21. Name the following compounds using the Stock system: FeF_3, Co_2O_3, $CoCl_3$, CuO, Hg_2O.

22. Using the Stock system, name the following compounds: Ni_2O_3, $CoBr_2$, MnO_2, CrI_3, TiO_2, VO.

23. Give formulas for the following binary compounds: lead(IV) chloride, copper(II) oxide, iron(III) iodide, tin(IV) oxide, arsenic(III) fluoride.

24. Write the formulas for the following compounds: mercury(II) oxide, cobalt(II) iodide, lead(II) bromide, manganese(IV) oxide, platinum(IV) fluoride, nickel(II) oxide.

25. Using the "old system" of naming binary compounds, name the compounds in Problem 19.

26. Using the "old system" of naming binary compounds, name the compounds in Problem 20.

27. Using the "old system" of naming binary compounds, name the compounds in Problem 21.

28. Write the formulas for the following compounds: cuprous chloride, ferric oxide, stannous bromide, mercuric oxide, plumbous iodide.

29. Write the formulas for the following compounds: ferric chloride, plumbic oxide, mercuric bromide, cupric oxide, cobaltous iodide.

30. Name the following compounds: NO, CO, ClO_2, H_2S_5, P_2Se_5, SF_4.

31. Name the following compounds: BrF_5, BBr_3, N_2S_5, H_2S_3, IF_7, CS_2.

32. Write formulas for the following compounds: bromine pentafluoride, carbon tetrachloride, iodine dioxide, sulfur trioxide, triboron silicide.

33. Give formulas for the following binary compounds: sulfur dioxide, tetrasulfur dinitride, arsenic pentaiodide, tetraphosphorus triselenide, boron nitride.

Section 7.3 Naming Acids Derived from Binary Compounds

Questions

34. Briefly describe how to name acids derived from binary compounds.

35. How do chemists distinguish hydrogen compounds that form acids from those that do not?

Problems

36. Name the aqueous solutions of these binary compounds: HI, H_2S, HF, HBr.

37. Give the formulas of these acids: hydrosulfuric acid, hydrochloric acid, hydroiodic acid.

Section 7.4 Naming Compounds Containing Polyatomic Ions

Questions

38. Briefly describe how to name compounds containing polyatomic ions.

39. What is an oxyacid?

40. Briefly describe how to name an oxyacid.

41. H_2SO_3 is sulfurous acid, but $HClO_3$ is chloric acid. The formulas are similar, but the suffixes in the names are different. How can this be?

Problems

42. Name the following compounds: K_2SO_4, $Ca(OH)_2$, $NaC_2H_3O_2$, $(NH_4)_2CO_3$, $Be(NO_3)_2$, $Mg(MnO_4)_2$.

43. Using the Stock system, name the following compounds: $Pb(SO_4)_2$, $Cu(ClO_2)_2$, $Sn(HCO_3)_2$, $FePO_4$, $Hg(NO_3)_2$, $PbCrO_4$.

44. Using the "old system," name the compounds in Problem 43.

45. Name the following compounds: $Ca(NO_2)_2$, $KClO_3$, Cs_2SO_3, $Mg(OH)_2$, $Ba(CN)_2$, $KBrO_3$.

46. Give formulas for the following compounds: barium bisulfite, potassium nitrite, lithium hydroxide, beryllium oxalate, calcium carbonate, sodium sulfate.

47. Give formulas for the following compounds: manganese(II) nitrite, cobalt(III) phosphate, copper(I) iodate, lead(VI) sulfate, iron(III) bicarbonate.

48. Using the "old system," name the compounds in Problem 47.

49. Give formulas for the following compounds: ferric chromate, cupric nitrite, plumbous acetate, stannic carbonate, cuprous sulfate, ferrous sulfite.

50. Name the following oxyacids: H_2SO_4, HNO_2, H_2CO_3, $HClO_2$, HNO_3, $HC_2H_3O_2$.

51. Name the following oxyacids of iodine: HIO, HIO_2, HIO_3, HIO_4.

52. Give formulas for the following oxyacids: hypochlorous acid, iodic acid, acetic acid, carbonic acid, sulfuric acid, chromic acid.

53. Give the formula for the following compounds: nitrous acid, oxalic acid, chlorous acid, phosphorus acid, perchloric acid.

Section 7.5 Naming Compounds with More Than One Kind of Positive Ion

Problems

54. Name the following compounds: $KNaCO_3$, NH_4HSO_4, CuK_2PO_4, KH_2PO_4, $LiHCO_3$, NaH_2PO_4.

55. Give formulas for the following compounds: lithium disodium phosphate, magnesium potassium phosphate, copper(I) bisulfite, ammonium hydrogen dichromate.

Solutions to Exercises

7.1 In PBr_3, Br is -1 and P is $+3$. In PBr_5, Br is -1 and P is $+5$.

7.2 In $CHCl_3$, H is $+1$, Cl is -1, and C is $+2$.

7.3 In NO_3^-, O is -2 and N is $+5$.

7.4 In NO_2^-, O is -2 and N is $+3$.

7.5 BaO is barium oxide. K_3N is potassium nitride. Al_2S_3 is aluminum sulfide.

7.6 Silver iodide is AgI and zinc oxide is ZnO.

7.7 CoSe is cobalt(II) selenide. $CrBr_2$ is chromium(II) bromide. Hg_2Cl_2 is mercury(I) chloride.

7.8 Iron(II) oxide is FeO and tin(IV) fluoride is SnF_4.

7.9 PbO_2 is plumbic oxide. CuI_2 is cupric iodide. Hg_2Cl_2 is mercurous chloride.

7.10 Ferrous chloride is $FeCl_2$ and stannic fluoride is SnF_4.

7.11 SO_2 is sulfur dioxide. SO_3 is sulfur trioxide. S_2O_3 is disulfur trioxide.

7.12 Selenium trioxide is SeO_3 and silicon tetrafluoride is SiF_4.

7.13 HI is hydroiodic acid and H_2Se is hydroselenic acid.

7.14 Hydrofluoric acid is HF.

7.15 $NaClO_2$ is sodium chlorite and CaC_2O_4 is calcium oxalate.

7.16 Cu_2SO_4 is copper(I) sulfate and $Cr_2(SO_3)_3$ is chromium(III) sulfite.

7.17 Cobalt(III) acetate is $Co(C_2H_3O_2)_3$.

7.18 Cu_2SO_4 is cuprous sulfate and $SnSO_4$ is stannous sulfate.

7.19 Ferrous perchlorate is $Fe(ClO_4)_2$ and stannic chromate is $Sn(CrO_4)_2$.

7.20 Ammonium nitrate is NH_4NO_3.

7.21 H_2CO_3 is carbonic acid. H_3PO_4 is phosphoric acid.

7.22 Hypochlorous acid is HClO.

7.23 NH_4HS is ammonium hydrogen sulfide or ammonium bisulfide. KRb_2PO_4 is potassium dirubidium phosphate.

7.24 Sodium bicarbonate is $NaHCO_3$.

Bromine (Br$_2$), an orange-brown liquid and aluminum metal react vigorously, forming a vapor consisting of vaporized Br$_2$ and some of the product of the reaction, Al$_2$Br$_6$.

Atomic Weight, Molecular Weight, The Mole

So far, we have learned much about atoms, molecules, ions, and chemical compounds. But in the laboratory we can't see individual atoms or molecules. We might see, for example, a small pile of powdered sulfur. How many atoms of sulfur are in the sample? Even if you could see the individual atoms, you certainly wouldn't want to count them physically one by one; there are just too many. What can you do? Well, you can certainly weigh the sample of sulfur. But how does the weight relate to the number of individual atoms? How can we put together what we know about atoms with what we can measure in the laboratory?

In Chapter 8, we learn how the masses of the atoms of different elements are determined. We see how to use mass data to find the number of atoms in a pure sample of any element or the number of molecules in any pure sample of a compound. We discover how these methods can be applied to isotopically mixed samples of elements and their compounds.

We are about to encounter one of the most important concepts in our study of chemistry: the mole. As merchants often think in dozens, chemists think in moles, but the mole is a number much larger than 12. With a full understanding of the mole concept, we will learn to perform many useful chemical calculations.

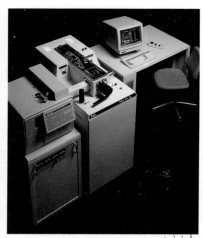

A modern mass spectrometer. *weighted avg.*

The atomic masses of other elements are measured relative to ^{12}C.

8.1 Isotopes and Their Atomic Masses

After studying this section, you should be able to:

- Explain the basis for the atomic mass scale.
- Define the atomic mass unit, amu.
- Convert from atomic mass units to grams and vice versa.

As you know from Chapter 3, not all the atoms of any particular element necessarily have the same mass. Atoms of the same element having different masses are different isotopes of that element. Nearly all the elements have two or more isotopes. We learned in Section 3.5 that a mass spectrometer can separate the isotopes of an element from one another and measure the masses of the atoms of any isotope.

Several years ago, an international committee of scientists agreed to establish a scale in which the masses of the atoms of all the elements would be compared to the mass of the atoms of the most common isotope of carbon, ^{12}C. This was then assigned what was defined as an **atomic mass** of exactly twelve. For example, the mass of an atom of ^{24}Mg is measured and found to be 1.998753 times as great as the mass of a ^{12}C atom. Compared to ^{12}C, then, a ^{24}Mg atom must have an atomic mass of $(1.998753 \times 12.00000\ldots) = 23.98504$. The masses of atoms of hundreds of different isotopes of different elements have been carefully measured in a similar manner and are reported in tables such as the one in the *Handbook of Chemistry and Physics* published by CRC Press. In this table, because the masses of atoms are directly related to the mass of ^{12}C atoms, they can be reported without mass units of any kind.

It is not unusual, however, to use a unit of mass known as the **atomic mass unit,** or **amu.** The amu is defined as being exactly one-twelfth of the mass of one ^{12}C atom. The mass of a ^{12}C atom has been experimentally determined to be 1.9927×10^{-23} grams; therefore, one amu is 1/12 of 1.9927×10^{-23} grams or 1.6606×10^{-24} grams.

 one amu $= 1.6606 \times 10^{-24}$ grams

Just as we can say that the mass of one atom of ^{24}Mg is 23.98504 on the atomic mass scale, so we can also say that the same atom has a mass of 23.98504 amu.

EXAMPLE 8.1

Calculate the mass in grams of one atom of ^{107}Ag, whose atomic mass is 106.91.

Solution Map:

$$\boxed{amu} \longrightarrow \boxed{grams}$$

Solution:

$$106.91 \text{ amu} \times \frac{1.6606 \times 10^{-24} \text{ g}}{\text{amu}} = 1.7753 \times 10^{-22} \text{ g}$$

Exercise 8.1

Calculate the mass in grams of one atom of ^{79}Br, the atomic mass of which is 78.9.

8.2 **Atomic Weight**

After studying this section, you should be able to:

- Explain what a natural sample of an element is.
- Calculate the atomic weight of an element, given its isotopic abundance and the atomic masses of its naturally occurring isotopes.

how often occur

A natural sample of an element is a mixture of its different isotopes. The proportions of isotopes in that mixture are referred to as the **isotopic abundance.** With few exceptions, the history of the elements on earth is such that the isotopic abundance of any natural sample of an element is independent of the origin of the sample. For example, a natural sample of copper contains 69.09% of ^{63}Cu and 30.91% of ^{65}Cu by weight no matter where the sample is found on earth. A sample of an element containing its isotopes in their natural abundances is sometimes called a geonormal sample; we will refer to it as a natural sample.

Although chemists can, with some difficulty, obtain pure samples of single isotopes of many of the elements, natural samples are used in most ordinary processes such as chemical reactions or purification procedures. Natural samples are more easily available and behave in the same way as pure samples of single isotopes. Most chemical calculations are therefore based on the *average* mass of the atoms in a natural sample.

For historical reasons, and to distinguish the average mass of the atoms of a natural sample of an element from the atomic mass of atoms of individual isotopes, chemists commonly use the term "atomic weight." The **atomic weight** of an element is the average *mass* of the atoms in a natural sample of an element. We will follow that usage. Atomic weights are read in amu or without units (relative to the mass of a ^{12}C atom). A table of atomic weights appears inside the front cover of this book. Atomic weights also appear below the symbols for the elements in the periodic table inside the front cover.

With a mass spectrometer, it is possible to determine the isotopic makeup of a natural sample of an element. For example, natural chlorine is made up of 75.5% ^{35}Cl and 24.5% ^{37}Cl. From this information and the atomic masses of the two isotopes, it is possible to calculate that the atomic weight of chlorine is 35.5.*

To understand the averaging process better, let's consider school grades. We want to calculate an average grade for a student that received two A's, four B's, and one C. An A grade is assigned 4.00. B and C grades are assigned 3.00 and

Natural sample of copper.

Atomic weight is really a mass.

---

*Contrasting with chlorine are many cases in which the naturally occurring isotopic mixture of an element is composed almost entirely of a single isotope. In such cases, the atomic weight of the element is quite close to the atomic mass of that isotope. For example, the atomic weight of naturally occurring oxygen is 15.9994, and the atomic mass of ^{16}O is 15.99914. There is very little of the other naturally occurring isotope, ^{18}O, in naturally occurring oxygen.

2.00, respectively. For each grade, multiply the abundance of a grade times its assignment. To determine the average grade, add the results and divide by the total number of grades.

$$\frac{(2 \times 4.00) + (4 \times 3.00) + (1 \times 2.00)}{7} = 3.14$$

The average grade is 3.14. However, there is no single grade that is assigned that value; 3.14 is an average (strictly speaking, a weighted average).

To calculate an atomic weight, the same averaging process is applied to isotopes of an element. Let's calculate the atomic weights of several elements.

EXAMPLE 8.2

A natural sample of copper contains 69.09% ^{63}Cu, atomic mass 62.9298, and 30.91% ^{65}Cu, atomic mass 64.9278. Calculate the atomic weight of copper.

Strategy and Solution:

Step 1. For each isotope, multiply the abundance or the percentage of the isotope times its atomic mass.

Step 2. To determine the average mass, add the results and divide the total by 100.

$$\frac{(69.09 \times 62.9298) + (30.91 \times 64.9278)}{100} = 63.55$$

The atomic weight of copper is 63.55. Remember, there is no one copper atom with a mass of 63.55.

EXAMPLE 8.3

Determine the atomic weight of magnesium. Magnesium is 78.70% ^{24}Mg, 10.13% ^{25}Mg, and 11.17% ^{26}Mg. ^{24}Mg has an atomic mass of 23.98504. ^{25}Mg has an atomic mass of 24.98584. ^{26}Mg has an atomic mass of 25.98259.

Strategy and Solution:

Step 1. Multiply the percentage of each isotope times its atomic mass.

Step 2. Add the results and divide by 100.

$$\frac{(78.70 \times 23.98504) + (10.13 \times 24.98584) + (11.17 \times 25.98259)}{100} = 24.31$$

Exercise 8.2

Natural bromine contains 50.54% ^{79}Br (atomic mass = 78.9183) and the rest ^{81}Br (atomic mass 80.9163). Calculate the atomic weight of bromine.

8.3 Measuring Atomic Weights

After studying this section, you should be able to:

- Explain how the atomic weight of an element can be measured using laboratory data.

Although it is now possible to work with samples of highly purified isotopes and measure the masses of individual atoms, the atomic weight table was well established long before these modern methods were available. How were the measurements made a century ago?

Because isotopically mixed samples of elements behave chemically the same way as pure samples of an isotope, early scientists never suspected that isotopes existed. To study the laws of chemical combination, they had only natural samples of elements, all of which had identical isotopic distributions. Consequently, they assumed that all atoms of the same element had the same mass.

Chemical analysis was used to determine the relations of the masses of the atoms of different elements. Oxygen was assigned an atomic weight of exactly sixteen and was used as the standard to which other atomic weights would be compared. Compounds of oxygen with other elements were analyzed, and the relative masses with which the other elements combined with oxygen were carefully measured. Given laboratory mass data and the formulas of the compounds, the average masses of atoms of other elements were compared to the average mass of oxygen atoms. For example, in an oxide of copper, CuO, it is known that there is one copper atom for every oxygen atom and that, in a sample of the oxide, the mass of the copper is about four times (3.97) the mass of the oxygen. Copper atoms, therefore, must have a mass about four times that of oxygen atoms. Taking the atomic weight of oxygen as 16, the atomic mass of copper, then, must be about 4 × 16, or 64. Atomic weights for elements that do not combine directly with oxygen were calculated using compounds with elements whose atomic weights had already been determined.

relative mass

Careful measurements of this kind show the atomic weight of Cu to be 63.546.

Atomic weights are measured to the best precision that the element and the method allow. Most atomic weights are known to at least four or five significant figures, a few to eight or nine.

8.4 The Mole

After studying this section, you should be able to:

- Explain why it is necessary to define a unit such as the mole.
- Define the mole.
- Convert from numbers of moles of particles to numbers of particles and vice versa.

People who deal in numbers of individual items often prefer to count in units, rather than by single items. Grocery shoppers rarely buy single eggs; the unit of a dozen is more convenient. Egg dealers cannot be bothered with dozens; they deal in grosses (12 dozen). Soda cans come in six-packs, and a case contains 24 cans.

Because atoms are extremely small, it is inconvenient to count them individually. We might use units such as a dozen, but the unit dozen is not large

Examples of two counting units in use, a dozen eggs and a mole of copper.

Lorenzo Romano Amadeo Carlo Avogadro di Quarequa e di Cereto (1766–1856)

Born in 1766 in Turin, Italy, Amadeo Avogadro practiced law in Turin for several years. In his mid-thirties, Avogadro became interested in science and tinkered with it as a hobby. Soon he gave up law to study physics full-time. In 1820, he became the first professor of mathematical physics to be appointed at the University of Turin. By that time, however, he had already proposed the hypothesis for which we now honor him.

In the first decade of the 1800's, it was known that when gases react chemically, the volumes in which they react are in the ratio of small whole numbers. For example, two volumes of hydrogen gas react with exactly one volume of oxygen gas to produce two volumes of gaseous water. Although reactions of gases were to provide the basis for great advances in atomic theory, they were almost ignored until Avogadro provided the key to their understanding. Avogadro concluded that *equal volumes of gases of whatever kind must contain equal numbers of molecules.* This is a statement of Avogadro's hypothesis (explained further in Section 11.5).

Avogadro's conclusions met resistance because their application leads inescapably to the downfall of John Dalton's view that the elements all existed as single atoms. Even though Avogadro's hypothesis is today accepted as correct, and even though it did open the way to unravel the relations among laboratory data, molecular formulas, and atomic weights, it was ignored or rejected until long after Avogadro's death.

Because his hypothesis is so fundamental to atomic theory, Avogadro is honored also by having the number of molecules in a mole named after him. Again, the concept of the mole, and certainly the numerical value of the number, were things that came long after Avogadro died. Avogadro never had the chance to know the value of his number.

enough to be of much use chemically; a dozen molecules is just as invisible to the eye as a single molecule. Chemists need a much larger unit. To meet that need, chemists have defined a quantity called the mole, abbreviated in calculations as "mol." The mole is used just as pair, dozen, gross, and six-pack are used—simply for convenience. Although the mole can be applied to any individual countable item, such as tennis balls or people, the number of items in a mole is so large that a mole of tennis balls would occupy a volume equal to about one fifth of the entire earth. The mole, however, is just right for working with atoms and molecules, and we will soon begin to use it constantly.

The number of individual units in a mole has been measured by several methods, and the value is known to be 6.0221367×10^{23}, or approximately 6 with 23 zeroes after it. This number is called **Avogadro's number,** N_A, after an early Italian chemist. For most calculations, Avogadro's number can be rounded to 6.022×10^{23}.

A **mole** *of something is Avogadro's number of individual units of that thing.* A mole of silver atoms contains 6.022×10^{23} silver atoms. A mole of water molecules contains 6.022×10^{23} molecules.

To understand just how large a number 6×10^{23} is, imagine counting that many marbles. Say that one person could count about 120 marbles in one minute. If that person were to count marbles every minute of every day and had everyone in the United States (about 250 million people) helping count at the same rate, every minute of every day, it would take more than 38 million years for all those people to count 6×10^{23} marbles!

We frequently meet problems that require converting from numbers of moles to numbers of atoms or molecules or vice versa. The following examples show how this is done.

EXAMPLE 8.4

How many atoms are in 0.43 moles of sodium?

Solution Map:

$$\boxed{\text{moles}} \longrightarrow \boxed{\text{atoms}}$$

Solution: There is Avogadro's number of atoms in a mole of Na. We can use this as a conversion factor.

$$\frac{6.022 \times 10^{23} \text{ Na atoms}}{\text{mol of Na}}$$

$$0.43 \text{ mol Na} \times \frac{6.022 \times 10^{23} \text{ Na atoms}}{\text{mol Na}} = 2.6 \times 10^{23} \text{ Na atoms}$$

> **Exercise 8.3**
>
> A sample of copper contains 4.92×10^{24} atoms. Calculate the number of moles of atoms.

EXAMPLE 8.5

How many moles equal 6.82×10^{24} molecules?

Solution Map:

$$\boxed{\text{molecules}} \longrightarrow \boxed{\text{moles}}$$

Solution:

$$6.82 \times 10^{24} \text{ molecules} \times \frac{\text{mol molecules}}{6.022 \times 10^{23} \text{ molecules}} = 11.3 \text{ mol molecules}$$

> **Exercise 8.4**
>
> How many molecules are in 22.7 moles of H_2O?

In our study of chemistry, we will usually think in terms of moles of elements and moles of compounds, rather than in terms of individual particles.

8.5 Atomic Weight and the Mole

After studying this section, you should be able to:

- Explain why a sample with a mass in grams that is equal to the atomic weight of an element contains a mole of atoms of that element.
- Convert from the number of grams of an element to the number of moles of the element and vice versa.
- Define *molar mass* for elements.

Why does a mole of particles contain 6.0221367×10^{23} particles? Why not a more convenient number such as 6×10^{23} or just 1×10^{23}? At the same time that scientists assigned a mass of 12 amu to the individual atom of ^{12}C, they defined the mole to be *the number of atoms of ^{12}C in exactly 12 grams of ^{12}C.* That is, exactly 12 grams of ^{12}C contain 6.0221367×10^{23} atoms of ^{12}C.

Because a mole is defined as the number of atoms in 12 grams of ^{12}C, it follows that one mole of any element has a mass numerically equal to its atomic weight. This is most easily understood through an example. We use round numbers and take C and H for our example. We know that one atom of C has a mass 12 times as great as one atom of H. The same relation holds between a sample of C containing two atoms and a sample of H containing two atoms. It is true for any pair of samples of C and H having the same number of atoms. Now, one mole of C has 12 times the mass of one mole of H. But one mole of C is defined as 12 grams. Then one mole of H must be one twelfth of that, or 1 gram, the same as the atomic weight of H taken in grams. We rounded to 1 and 12 for simplicity, but that has no effect on the actual relations. Actually, to have a mole of H, we must take a sample weighing 1.0079 grams. The equivalent relation exists for every element.

Because it's so fundamental, let's look at the statement worded in another way.

A natural sample of any element contains a mole of atoms if its mass, in grams, equals the atomic weight of the element.

To illustrate this statement, we will calculate an atomic weight in grams.

EXAMPLE 8.6

What is the mass in grams of Avogadro's number of aluminum atoms?
Solution Map:

$$ \boxed{\text{atoms}} \longrightarrow \boxed{\text{amu}} \longrightarrow \boxed{\text{grams}} \qquad \text{2 steps} $$

Strategy and Solution:

Step 1. Find the atomic weight. The atomic weight of aluminum is 26.9815.

Step 2. Find the mass in amu of a mole of Al atoms.

$$ 6.022 \times 10^{23} \text{ Al atoms} \times \frac{26.9815 \text{ amu}}{\text{Al atom}} = 1.625 \times 10^{25} \text{ amu} $$

Step 3. Using the conversion factor for converting amu to grams, find the number of grams.

$$ 1.625 \times 10^{25} \text{ amu} \times \frac{1.6606 \times 10^{-24} \text{ g}}{\text{amu}} = 26.98 \text{ g} $$

Figure 8.1 One mole each of several elements, (clockwise from lower left) sulfur, bromine, aluminum, mercury, and zinc.

The atomic weight of aluminum is 26.9815. A natural sample of aluminum containing a mole (6.022×10^{23}) of aluminum atoms weighs (rounded to four significant figures) 26.98 grams.

The mole is an essential tool. It allows us to count numbers of individual atoms or molecules by weighing samples in the laboratory. To understand this concept better, let's consider jelly beans. You want to buy 10,000 jelly beans. How will you and the seller count them out? You will certainly not count them physically, one by one. Most likely, you and the seller will agree that a small error in the number is acceptable to both of you. Then you'll count out perhaps 100 jelly beans, weigh them, and buy 100 times that weight. You have counted the 10,000 jelly beans by weighing them. When you count atoms by weighing, you will make an error in the count, but that error is not likely to be proportionately greater than the error in the count of the jelly beans. The number of atoms in a mole is known to a far better precision than, for example, the number of people in any large city.

The value of the mole is known to about 1 part in 60,000,000, roughly the equivalent of knowing the population of the United States to plus or minus four or five people.

To measure out a mole of an element, all you need to do is weigh out a sample with a weight in grams numerically equal to the atomic weight. As in Example 8.6, when we weigh out 26.98 grams of aluminum, we obtain 1.000 mole of aluminum. The mass in grams of one mole of an element is called its **molar mass.** As we will see, this term also applies to the mass of a mole of molecules.

EXAMPLE 8.7

What is the mass in grams of 0.23 moles of aluminum?

Solution Map:

$$\boxed{\text{moles}} \longrightarrow \boxed{\text{grams}}$$

Strategy and Solution:

Step 1. Find the molar mass.

Step 2. Write a conversion factor. We know that 26.9815 grams of Al contains one mole of Al atoms.

$$\frac{26.9815 \text{ g Al}}{\text{mol Al}}$$

Step 3. Set up and calculate.

$$0.23 \text{ mol Al} \times \frac{26.9815 \text{ g Al}}{\text{mol Al}} = 6.2 \text{ g Al}$$

> **Exercise 8.5**
>
> How many grams of silver are in 0.0562 moles of silver?

When performing chemical calculations, it is best not to round off the atomic weights of the elements, but to use all the digits provided in the periodic table.

don't round atm. masses

Rounding off before a calculation may lead to round-off errors. Instead round the answers as needed to obtain the correct number of significant figures.

EXAMPLE 8.8

A sample of magnesium has a mass of 52.8 grams. Calculate the number of moles.

Solution Map:

$$\boxed{\text{grams}} \longrightarrow \boxed{\text{moles}}$$

Strategy and Solution:

Step 1.   Find the molar mass.

Step 2.   Write a conversion factor. We know that a mole of Mg will have a mass of 24.305 grams.

$$\frac{\text{mol Mg}}{24.305 \text{ g Mg}}$$

Step 3.   Set up and calculate.

$$52.8 \text{ g Mg} \times \frac{\text{mol Mg}}{24.305 \text{ g Mg}} = 2.17 \text{ mol Mg}$$

Exercise 8.6

Calculate the number of moles in 32.6 grams of copper.

It is important to keep in mind the difference between the atomic weight and the molar mass of an element. The atomic weight of an element is the average mass of the atoms in a natural sample of that element, taken either in amu or without units. The *molar mass* is the actual mass in units of grams of a *mole* of an element. For example, the atomic weight of hydrogen is 1.0079 (no units). The molar mass of hydrogen is 1.0079 g/mol, meaning that one mole of naturally occurring H atoms has a mass of 1.0079 grams.

The molar mass is the mass in grams of one mole of any element or compound.

EXAMPLE 8.9

How many atoms are in a 3.0-gram sample of aluminum?

Solution Map:

$$\boxed{\text{grams}} \longrightarrow \boxed{\text{moles}} \longrightarrow \boxed{\text{atoms}}$$

Strategy and Solution:

Step 1.   Find the molar mass of aluminum. The molar mass of aluminum is 26.98 g/mol.

Step 2.   Write the molar mass as a conversion factor to convert grams to moles.

$$\frac{\text{mol Al}}{26.9815 \text{ g Al}}$$

Step 3. Set up and calculate. Be sure to convert moles of Al to atoms of Al using Avogadro's number.

$$3.0 \text{ g Al} \times \frac{\text{mol Al}}{26.9815 \text{ g Al}} \times \frac{6.022 \times 10^{23} \text{ Al atoms}}{\text{mol Al}} = 6.7 \times 10^{22} \text{ Al atoms}$$

Exercise 8.7

How many atoms are in 1.5 grams of sodium?

Exercise 8.8

Calculate the number of atoms in 0.729 grams of calcium.

EXAMPLE 8.10

What is the mass of 3.61×10^{24} magnesium atoms?

Solution Map:

$$\boxed{\text{atoms}} \longrightarrow \boxed{\text{moles}} \longrightarrow \boxed{\text{grams}}$$

Strategy and Solution:

Step 1. Find the molar mass of Mg. It is 24.305 g/mol.

Step 2. Write the molar mass as a conversion factor to convert moles to grams.

$$\frac{24.305 \text{ g Mg}}{\text{mol Mg}}$$

Step 3. Set up and calculate.

$$3.61 \times 10^{24} \text{ Mg atoms} \times \frac{\text{mol Mg}}{6.022 \times 10^{23} \text{ Mg atoms}} \times \frac{24.305 \text{ g Mg}}{\text{mol Mg}} = 146 \text{ g Mg}$$

Exercise 8.9

Find the mass in grams of a sample containing 4.62×10^{24} atoms of aluminum.

8.6 Molecular Weight

After studying this section, you should be able to:

- Calculate molecular weights, given atomic weights and compound formulas.
- Explain why the term "molecular weight" can be used for compounds in which there are no molecules.

Just as an individual atom has a characteristic mass, so does an individual molecule. The mass of a single molecule is the sum of the masses of the atoms it contains.

We've seen that samples of elements as they ordinarily occur on earth are mixtures of isotopes. Accordingly, compounds of those elements will also be isotopically mixed; that is, compounds contain the same proportions of isotopes as do the elements in them. The mass of a molecule of a compound composed of isotopically mixed elements is called the **molecular weight** of the compound.

To calculate a molecular weight, take the sum of the atomic weights of the elements in the compound, allowing for the fact that more than one atom of an element may be present in the molecule. Molecular weights are read in amu or without units, just as atomic weights are.

EXAMPLE 8.11

Calculate the molecular weight of nitric acid, HNO_3.

Strategy and Solution:

Step 1. The formula tells us that a molecule of HNO_3 contains one hydrogen atom, one nitrogen atom, and three oxygen atoms.

Step 2. We look up the atomic weights of the elements and multiply each atomic weight by the number of atoms of that element that appears in the formula.

Step 3. Then we add the results.

$$H : 1 \times 1.0079 = 1.0079$$
$$N : 1 \times 14.0067 = 14.0067$$
$$O : 3 \times 15.9994 = \underline{47.9982}$$
$$63.0128$$

The molecular weight of nitric acid is 63.0128.

Exercise 8.10

Calculate the molecular weight of sucrose (ordinary table sugar), $C_{12}H_{22}O_{11}$.

As we learned in Section 6.2, ionic compounds, NaCl for example, do not form individual molecules. Even so, those compounds have formulas that can be used to calculate a mass just as though the molecules really did exist. The **formula weight** of an ionic compound is the sum of the atomic weights of the elements in the compound, taking into account the fact that more than one atom of an element may appear in the formula.

EXAMPLE 8.12

Calculate the formula weight of calcium perchlorate, $Ca(ClO_4)_2$.

Strategy and Solution:

Step 1. Looking at the formula, we find that there are two chlorine atoms, eight oxygen atoms, and one calcium atom in the formula for $Ca(ClO_4)_2$.

Step 2. We look up the atomic weights of the elements and multiply each atomic weight by the number of atoms of that element that appears in the formula.

Step 3. Then we add the results.

$$
\begin{aligned}
Ca &: 1 \times 40.078 = 40.078 \\
Cl &: 2 \times 35.4527 = 70.9054 \\
O &: 8 \times 15.9994 = \underline{127.995} \\
& \qquad\qquad\qquad\quad 238.979
\end{aligned}
$$

Exercise 8.11

Calculate the formula weight of each of the following: BaF_2, $Fe_2(Cr_2O_7)_3$, and $KBrO_3$.

8.7 Molar Mass of Compounds

After studying this section, you should be able to:

- Define *molar mass* for compounds.
- Convert from grams to moles of compound and vice versa.

We have seen that if we take a sample of an element in which the number of grams in the sample is equal to the atomic weight of the element, that sample will contain one mole of atoms. In a similar way, for a compound, a sample whose weight in grams is equal to the molecular weight (or formula weight) will contain one mole of the compound. Let's demonstrate this to ourselves by calculating the mass of a mole of water, that is, of Avogadro's number of H_2O molecules.

EXAMPLE 8.13

Calculate the mass of Avogadro's number of H_2O molecules.

Solution Map:

$$\boxed{\text{molecules}} \longrightarrow \boxed{\text{amu}} \longrightarrow \boxed{\text{grams}}$$

Strategy and Solution:

Step 1. Find the molecular weight by adding the atomic weight of oxygen to twice the atomic weight of hydrogen. The molecular weight of H_2O is 18.0152.

Step 2. Find the mass in amu of Avogadro's number of H_2O molecules.

$$6.022 \times 10^{23} \text{ } H_2O \text{ molecules} \times \frac{18.0152 \text{ amu}}{\text{molecule}} = 1.085 \times 10^{25} \text{ amu}$$

Step 3. Using the conversion factor for converting amu to grams, find the number of grams.

$$1.085 \times 10^{25} \text{ amu} \times \frac{1.6606 \times 10^{-24} \text{ g}}{\text{amu}} = 18.02 \text{ g}$$

The mass of one mole of water is 18.02 grams.

The mass in grams of one mole of any compound is called the *molar mass.* The molar mass of any compound is numerically equal to the value of the molecular weight or formula weight of that compound taken in units of grams.

Exercise 8.12

Calculate the molar mass of NO_2 and of BF_3. Does this differ from the molecular weight? How?

EXAMPLE 8.14

How many grams of BF_3 are there in 5.50×10^{-2} moles of BF_3?

Solution Map:

$$\boxed{\text{moles}} \longrightarrow \boxed{\text{grams}}$$

Strategy and Solution:

Step 1. Find the molar mass of BF_3. The molar mass of BF_3 is 67.806 g/mol.

Step 2. Write the molar mass as a conversion factor. We know that 67.806 grams of BF_3 equals a mole.

$$\frac{67.806 \text{ g } BF_3}{\text{mol } BF_3}$$

Step 3. Set up and calculate.

$$5.50 \times 10^{-2} \text{ mol } BF_3 \times \frac{67.806 \text{ g } BF_3}{\text{mol } BF_3} = 3.73 \text{ g } BF_3$$

Exercise 8.13

Calculate the mass in grams of 4.8 moles of iodine, I_2.

EXAMPLE 8.15

How many moles are there in a 14.0-gram sample of NO_2?

Solution Map:

$$\boxed{\text{grams}} \longrightarrow \boxed{\text{moles}}$$

Strategy and Solution:

Step 1.   Find the molar mass. The molar mass of NO_2 is 46.0055 g/mol.

Step 2.   Set up and calculate.

$$14.0 \text{ g } NO_2 \times \frac{\text{mol } NO_2}{46.0055 \text{ g } NO_2} = 0.304 \text{ mol } NO_2$$

Exercise 8.14

Calculate the number of moles of compound in 45.32 grams of $K_2Cr_2O_7$.

EXAMPLE 8.16

How many molecules are there in a 4.00-gram sample of carbon dioxide, CO_2?

Solution Map:

$$\boxed{\text{grams}} \longrightarrow \boxed{\text{moles}} \longrightarrow \boxed{\text{molecules}}$$

Strategy and Solution:

Step 1.   Find the molar mass. The molar mass of CO_2 is 44.010 g/mol.

Step 2.   Write the molar mass as a conversion factor.

$$\frac{\text{mol } CO_2}{44.010 \text{ g } CO_2}$$

Step 3.   Set up and calculate.

$$4.00 \text{ g } CO_2 \times \frac{\text{mol } CO_2}{44.010 \text{ g } CO_2} \times \frac{6.022 \times 10^{23} \text{ } CO_2 \text{ molecules}}{\text{mol } CO_2} =$$

$$5.47 \times 10^{22} \text{ } CO_2 \text{ molecules}$$

Exercise 8.15

How many molecules are there in 34.8 grams of ethane, C_2H_6?

EXAMPLE 8.17

What is the mass of 8.63×10^{24} molecules of ammonia, NH_3?

Solution Map:

$$\boxed{\text{molecules}} \longrightarrow \boxed{\text{moles}} \longrightarrow \boxed{\text{grams}}$$

Strategy and Solution:

Step 1.   Find the molar mass. The molar mass of NH_3 is 17.0304 g/mol.

Step 2.   Write the molar mass as a conversion factor.

$$\frac{17.0304 \text{ g } NH_3}{\text{mol } NH_3}$$

Step 3.   Set up and calculate.

$$8.63 \times 10^{24} \text{ molecules of } NH_3 \times \frac{\text{mol } NH_3}{6.022 \times 10^{23} \ NH_3 \text{ molecules}} \times$$

$$\frac{17.0304 \text{ g } NH_3}{\text{mol } NH_3} = 244 \text{ g } NH_3$$

Exercise 8.16

Find the mass in grams of a sample containing $6.97 \times 10^{23} \ N_2$ molecules.

   You might object that ionic compounds have no individual molecules that can be counted, and therefore there can be no Avogadro's number of molecules and no mole of those compounds. Although it is true that many compounds have no countable molecules, the concept of the mole can still be applied as though there were molecules. NaCl, as we have seen before, has no molecules, but a mole of Na^+ ions and a mole of Cl^- ions form a sample of NaCl which, for all purposes of calculation, contains a mole of NaCl. Mass calculations using the mole are independent of whether there are individual molecules. As long as a formula can be written for the compound, the formula can be treated as though it represents a molecule.

EXAMPLE 8.18

What is the mass in grams of 0.67 moles of NaCl?

Solution Map:

$$\boxed{\text{moles}} \longrightarrow \boxed{\text{grams}}$$

Strategy and Solution:

Step 1.   Find the molar mass. The molar mass of NaCl is 58.4425 g/mol.

Step 2.   Write the molar mass as a conversion factor.

$$\frac{58.4425 \text{ g NaCl}}{\text{mol NaCl}}$$

Step 3.   Set up and calculate.

$$0.67 \text{ mol NaCl} \times \frac{58.4425 \text{ g NaCl}}{\text{mol NaCl}} = 39 \text{ g NaCl}$$

Exercise 8.17

How many grams of KBr are there in 0.491 moles of KBr?

Exercise 8.18

Calculate the number of moles in 9.0 grams of KI.

In calculating molecular and atomic weights, it is important to be sure what species the calculation is for. If we are asked the mass of a mole of nitrogen, for example, we don't know whether the question refers to nitrogen atoms or nitrogen molecules, N_2. The mass in grams of a mole of nitrogen *atoms* is 14 grams, but that of a mole of nitrogen *molecules* is 2×14 grams, or 28 grams. We do not know which is meant unless we are told. If we are asked for the *atomic weight* of nitrogen, we can confidently say that it is 14. If we are asked for the *molecular weight* of nitrogen, we can say it is 28. When only the word *mole* is used, it is important to specify the species, as in "a mole of nitrogen atoms," "a mole of nitrogen molecules," or "a mole of N_2."

Summary

Today, the mass of an individual atom is compared to the mass of ^{12}C atoms. The mass of a ^{12}C atom is defined as exactly 12, or 12 atomic mass units (amu).

Although the atomic mass of an isotope can be measured, the elements occur in their natural states as mixtures of isotopes. Most chemistry involves such natural samples rather than pure isotopes. The masses of the atoms of the isotopes and the isotopic abundances of all the elements have been measured. From this information, it is possible to calculate the average atomic mass of a natural sample, the atomic weight. Atomic weights can also be calculated using data from chemical reactions. Molecular weights and formula weights can be calculated from atomic weights and formulas.

Since most laboratory work is done with macroscopic samples, it is usually inconvenient to make calculations in terms of individual atoms and molecules. For that reason, the mole, the chemist's dozen, has been defined. A mole of something is Avogadro's number of that thing, namely 6.0221367×10^{23} of it. The mass in grams of a mole of a substance is called the molar mass of that substance. The concept of the mole is one of the most useful in chemistry.

In the language of chemistry, there are many terms, some of which sound much alike. Be sure you know how to distinguish among the similar terms that we have just learned. Here is a collected set of definitions for your review.

Atomic mass	On the ^{12}C scale, the mass of an individual atom of any isotope of any element. Atomic mass can be read in a relative way or in amu.
Atomic weight	Actually a term that refers to mass. On the ^{12}C scale, a weighted average mass of the atoms in a naturally occurring sample of an element. Atomic weight can be read in a relative way or in amu.

Molecular weight	A term that refers to mass. The sum of the atomic weights of the elements in a molecule of a compound, allowing for the fact that some elements appear more than once in the molecular formula.
Formula weight	For ionic compounds, the sum of the atomic weights of the elements in the compound as indicated by the formula, allowing for the fact that some elements appear more than once in the formula.
Mole	Avogadro's number (N_A) of individual items, such as atoms, molecules, or ions, where N_A is the number of atoms of ^{12}C in exactly 12 grams of ^{12}C, 6.0221367×10^{23}.
Molar mass	For an element, a sample whose mass in units of grams is equal to its atomic weight. The mass of a mole of an element. For a compound, a sample whose mass in units of grams is equal to its molecular weight. The mass of a mole of a compound.

Key Terms

atomic mass	(p. 212)	mole	(p. 216)
atomic mass unit, amu	(p. 212)	molar mass	(p. 219)
isotopic abundance	(p. 213)	molecular weight	(p. 222)
atomic weight	(p. 213)	formula weight	(p. 222)
Avogadro's number	(p. 216)		

Questions and Problems

Section 8.1 Isotopes and Their Atomic Masses

Questions

1. Explain the relationships between an atom, an element, and an isotope.

2. How would you go about developing a relative scale of any kind?

Section 8.2 Atomic Weight

Questions

3. Explain the basis for the atomic weight scale used by chemists.

4. Sometimes the values for atomic weight are written without units of any kind. How can it be correct to leave the units off these values?

5. Since it might be simpler to assign hydrogen a mass of 1.000... on the atomic weight scale, what might have been the reason for choosing carbon as the standard instead?

6. Does the existence of isotopes change the laws of chemical combination as they were discovered a century ago? Explain fully.

7. Explain the term "isotopic abundance."

8. Explain why there are no atoms of chlorine with a mass equal to 35.5 amu, even though that is the atomic weight of chlorine.

Problems

9. The isotopic abundances and isotopic atomic masses of three elements are listed in the table. What are the atomic weights of these elements?

		Percent Abundance	Atomic Mass
(a)	Silicon	92.2% ^{28}Si	27.98
		4.7% ^{29}Si	28.98
		3.1% ^{30}Si	29.97
(b)	Silver	51.82% ^{102}Ag	106.9
		48.18% ^{109}Ag	108.9
(c)	Gallium	60.4% ^{69}Ga	68.9
		39.6% ^{71}Ga	70.9

10. What is the atomic weight of antimony? There are two naturally occurring isotopes, ^{121}Sb, which has an atomic mass of 120.9038 and an abundance of 57.25%, and ^{123}Sb, with an atomic mass of 122.9041.

11. Europium has two naturally occurring isotopes, ^{151}Eu and ^{153}Eu. What is the atomic weight of europium? ^{151}Eu has an atomic mass of 150.9196 and an abundance of 47.82%, and ^{153}Eu has an atomic mass of 152.9209.

12. What is the atomic mass of ^{87}Rb? Rubidium has two naturally occurring isotopes: ^{85}Rb, which has an atomic mass of 84.9117, and ^{87}Rb. By weight, 27.85% of rubidium is ^{87}Rb.

13. What is the atomic mass of $^{65}_{29}$Cu? The atomic weight of Cu is 63.546, and the other naturally occurring isotope is ^{63}Cu, which has an abundance of 69.09% and atomic mass of 62.9298.

14. Calculate the atomic mass of $^{79}_{35}$Br if the atomic weight of Br is 79.904. The only other naturally occurring isotope, ^{81}Br, has an atomic mass of 80.9163 and an abundance of 49.46%.

15. What is the atomic mass of $^{191}_{77}$Ir? The only other naturally occurring isotope, ^{193}Ir, has an atomic mass of 192.9633 and an abundance of 62.6%.

Section 8.3 Measuring Atomic Weights

Question

16. How is it possible to measure the relative masses of the atoms of two elements if the mass of neither kind of atom is known?

Section 8.4 The Mole

Questions

17. Would a definition of the mole be just as satisfactory as a round number, say, 10^{22}? Why or why not?

18. Would it be practical to use the mole for other small objects such as rice grains? Explain.

Problems

19. How many atoms of silver are in 3.0 moles of silver?

20. How many atoms of copper are in 0.00027 moles of copper?

21. Calculate the number of molecules in 4.4 moles of molecules.

22. How many molecules are in 0.44 moles of molecules?

23. A sample that contains $3.01 \times 10_{23}$ atoms of carbon contains how many moles of carbon?

24. A sample contains 8.2×10^{21} molecules. How many moles of molecules are in the sample?

25. If a typewriter prints 75 characters per line and 27 lines per page, how many pages of typing will have a mole of characters? A house has five rooms, and each room holds ten million pages. How many houses will hold a mole of characters?

26. If one mm^3 of sand contains 88 grains, how long would a beach be if it contained a mole of sand grains and was 100 cm deep and 10 m wide?

27. If 52 tadpoles can live in one cubic meter of water, how large would a tadpole pond be if it could accommodate Avogadro's number of tadpoles? How does the volume of the tadpole pond compare with the volume of the earth, which is approximately 1.1×10^{21} m^3?

28. If a certain kind of molecule has a length of 2.70×10^{-8} cm, how many times will a chain containing one mole of such molecules, linked end-to-end, stretch around the earth? Take the circumference of the earth to be 24,800 mi. One ft contains 12 in., an inch equals 2.54 cm, and a mile is 5,280 ft.

Section 8.5 Atomic Weight and the Mole

Questions

29. How would you go about measuring out a sample containing a mole of an element?

30. A sample of an element contains a mole of atoms if its mass, measured in grams, is numerically the same as the atomic weight. Explain in detail why this is so.

Problems

31. What is the mass in grams of an average atom of each of the following elements: Ar, Ba, Ca, C, Cr, Co, Au?

32. What are the molar masses of the elements listed in Problem 31?

33. Give the molar masses for the elements B, H, Rb, N, Sc, Br, Sn, F.

34. Give the names and symbols of the elements that have the following molar masses: 26.9 g/mol, 10.8 g/mol, 55.8 g/mol, 207 g/mol, 197 g/mol.

35. What is the mass in grams of Avogadro's number of sulfur atoms?

36. How many atoms are in a 15.8-gram sample of boron?

37. How many atoms are in 125 grams of iron?

38. A sample of barium contains 5.82×10^{21} atoms. Find the mass in grams of the sample.

39. Calculate the mass in grams of 2.8×10^{18} atoms of lead.

40. How many moles of atoms are there in each of the following samples?
    (a) 72.8 g of Al   (b) 33.9 g of Ne
    (c) 0.22 g of Li   (d) 418 g of Si
    (e) 2.4×10^{-3} g of Au

41. Calculate the number of moles of atoms in each of the following samples.
    (a) 40.0 g of Mg   (b) 3.4×10^2 g of Pb
    (c) 79.6 g of Mn   (d) 4.3×10^{-2} g of He
    (e) 16.8 g of S

42. Calculate the number of grams in each of the following samples.
    (a) 6.30 mol of B   (b) 9.86×10^{-8} mol of Hg
    (c) 4.32×10^{-3} mol of Cr

43. What is the mass in grams of each of the following samples?
    (a) 0.0222 mol of He   (b) 4.56×10^{-4} mol of S
    (c) 2.89 mol of Hg   (d) 1.27 mol of Ag

44. What is the mass in grams of 0.88 moles of potassium? Of bismuth?

45. How many grams of lithium are in 56.8 moles?

46. How many grams of cobalt are in 0.14 moles?

47. Calculate the number of moles in 456 grams of lead; of mercury; of helium.

48. A sample of calcium has a mass of 23.7 grams. Find the number of moles.

49. Find the number of moles in 93 grams of neon; of silver; of arsenic.

50. A mole of ^{135}Ba has a mass of 135 g. What is the mass of one atom of ^{135}Ba in grams?

Section 8.6 Molecular Weight

Questions

51. The term "molecular weight" is sometimes used for compounds that do not form molecules. Explain why this practice works.

52. What does the term "formula weight" mean?

Problems

53. Calculate the molecular weights of CO_2, H_2O_2, NO, and CO.

54. Find the molecular weights of CCl_4, PH_3, SO_2, and BF_3.

55. What are the molecular weights of the following compounds: H_2SO_4, C_2H_6O, $Ba(NO_3)_2$, $Sr_3(PO_4)_2$?

56. Give the molecular weights of the following compounds: $HClO_4$, $C_6H_{12}O_5N$, $Al_2(SO_4)_3$.

Section 8.7 Molar Mass of Compounds

Questions

57. Distinguish between molecular weight and molar mass.

58. How do you calculate the molar mass of a compound?

59. Why is it ambiguous to speak of a mole of oxygen?

Problems

60. Calculate the molar mass of each of the following compounds: HIO_4, NH_4NO_3, Hg_2CO_3, $Cu_3(PO_4)_2$.

61. What is the molar mass of each of the following: $Mg(ClO_3)_2$, $C_9H_{16}N_4O_{13}P_2$, C_3H_7Cl?

62. What is the molecular weight of each of the compounds in Problem 61? How does this problem relate to that one?

63. Calculate the mass of Avogadro's number of CO molecules.

64. How many moles of compound are in each of the following samples?
    (a) 53.2 g of Fe_2O_3 (b) 48.7 g of Cl_2
    (c) 2.16 g of LiH   (d) 409 g of $Fe(H_2PO_2)_3$
    (e) 188 g of Al_2S_3

65. Calculate the number of moles of compound in each of the following samples.
    (a) 125 g of H_2SO_4
    (b) 252 g of NaOH
    (c) 7.9×10^{-2} g of K_2CrO_4
    (d) 8.54×10^{-1} g of CCl_4
    (e) 5.24 g of MnO_2

66. Calculate the number of grams of compound in each of the following samples.
    (a) 54.2 mol of LiOH
    (b) 0.0421 mol of H_3PO_4
    (c) 9.31×10^{-4} mol of $CaCl_2$
    (d) 27.2 mol of CO_2

67. What is the mass in grams of each of the following samples?
    (a) 9.27×10^{-3} mol $K_2Cr_2O_7$
    (b) 4.20 mol HCl
    (c) 1.1×10^4 mol CO
    (d) 0.0486 mol $BaBr_2$
    (e) 3.48×10^{-5} mol Al_2O_3

More Difficult Problems

68. If the atomic weight scale were based on the definition of ^{1}H exactly equal to one, what would be the atomic mass of ^{40}Ca, which is 39.96 on the present scale? On the present scale, the atomic mass of ^{1}H is 1.0078.

69. On the planet Xanthu, the atomic weight scale is based on the isotope 12 of carbon, just as on earth. Instead of assigning carbon an atomic mass of exactly twelve, the scientists of Xanthu have chosen a relative mass of one for carbon. On the Xanthu scale, what is the atomic weight of Ni?

70. If the atomic mass of ^{12}C were exactly 15.000 and a mole were defined as the number of atoms in exactly 15.000 kg of carbon 12, what would N_A be?

71. Again, we visit the planet Xanthu. The scientists of Xanthu have assigned ^{12}C a relative mass of one. In the Xanthu units, a mole of carbon is seven nuds of carbon, where a nud is the same as 20.5 of our grams. What is the Xanthu equivalent of Avogadro's number?

Solutions to Exercises

8.1 $78.9 \text{ amu} \times \dfrac{1.6606 \times 10^{-24} \text{ g}}{\text{amu}} = 1.31 \times 10^{-22} \text{ g}$

8.2 $100.00 - 50.54 = 49.46 \; ^{81}\text{Br}$

$\dfrac{(50.54 \times 78.9183) + (49.46 \times 80.9163)}{100} = 79.91$

8.3 $4.92 \times 10^{24} \text{ Cu atoms} \times \dfrac{\text{mol Cu}}{6.022 \times 10^{23} \text{ Cu atoms}} = 8.17 \text{ mol Cu}$

8.4 $22.7 \text{ mol H}_2\text{O} \times \dfrac{6.022 \times 10^{23} \text{ H}_2\text{O molecules}}{\text{mol H}_2\text{O}} = 1.37 \times 10^{25} \text{ H}_2\text{O molecules}$

8.5 $0.0562 \text{ mol Ag} \times \dfrac{107.8682 \text{ g Ag}}{\text{mol Ag}} = 6.06 \text{ g Ag}$

8.6 $32.6 \text{ g Cu} \times \dfrac{\text{mol Cu}}{63.546 \text{ g Cu}} = 0.513 \text{ mol Cu}$

8.7 $1.5 \text{ g Na} \times \dfrac{\text{mol Na}}{22.9898 \text{ g Na}} \times \dfrac{6.022 \times 10^{23} \text{ Na atoms}}{\text{mol Na}} = 3.9 \times 10^{22} \text{ Na atoms}$

8.8 $0.729 \text{ g Ca} \times \dfrac{\text{mol Ca}}{40.078 \text{ g Ca}} \times \dfrac{6.022 \times 10^{23} \text{ Ca atoms}}{\text{mol Ca}} = 1.10 \times 10^{22} \text{ Ca atoms}$

8.9 $4.62 \times 10^{24} \text{ atoms Al} \times \dfrac{\text{mol Al}}{6.022 \times 10^{23} \text{ Al atoms}} \times \dfrac{26.9815 \text{ g Al}}{\text{mol Al}} = 207 \text{ g Al}$

8.10 $(12.011 \times 12) + (1.0079 \times 22) + (15.9994 \times 11) = 342.299$

8.11 BaF_2: $137.327 + (2 \times 18.9984) = 175.323$
$\text{Fe}_2(\text{Cr}_2\text{O}_7)_3$: $(2 \times 55.847) + (6 \times 51.9961) + (21 \times 15.9994) = 759.658$
KBrO_3: $39.0983 + 79.904 + (3 \times 15.9994) = 167.000$

8.12 NO_2: $14.0067 + (2 \times 15.9994) = 46.0055 \text{ g/mol}$
BF_3: $10.811 + (3 \times 18.9984) = 67.806 \text{ g/mol}$
Molar mass has the units of g/mol. The molecular weight is in amu or without units. The numerical values are the same for both.

8.13 $4.8 \text{ mol I}_2 \times \dfrac{253.809 \text{ g I}_2}{\text{mol I}_2} = 1.2 \times 10^3 \text{ g I}_2$

8.14 $45.32 \text{ g K}_2\text{Cr}_2\text{O}_7 \times \dfrac{\text{mol K}_2\text{Cr}_2\text{O}_7}{294.185 \text{ g K}_2\text{Cr}_2\text{O}_7} = 0.1541 \text{ mol K}_2\text{Cr}_2\text{O}_7$

8.15 $34.8 \text{ g C}_2\text{H}_6 \times \dfrac{\text{mol C}_2\text{H}_6}{30.069 \text{ g C}_2\text{H}_6} \times \dfrac{6.022 \times 10^{23} \text{ C}_2\text{H}_6 \text{ molecules}}{\text{mol C}_2\text{H}_6} =$
$6.97 \times 10^{23} \text{ C}_2\text{H}_6 \text{ molecules}$

8.16 $6.97 \times 10^{23} \text{ N}_2 \text{ molecules} \times \dfrac{\text{mol N}_2}{6.022 \times 10^{23} \text{ N}_2 \text{ molecules}} \times$
$\dfrac{28.0134 \text{ g N}_2}{\text{mol N}_2} = 32.4 \text{ g N}_2$

8.17 $0.491 \text{ mol KBr} \times \dfrac{119.002 \text{ g KBr}}{\text{mol KBr}} = 58.4 \text{ g KBr}$

8.18 $9.0 \text{ g KI} \times \dfrac{\text{mol KI}}{166.003 \text{ g KI}} = 0.054 \text{ mol KI}$

The thermite reaction,
$Fe_2O_3(s) + 2\ Al(s) \longrightarrow$
$2\ Fe(s) + Al_2O_3(s)$, is so
exothermic that molten iron
results.

Mass Relations
in Compounds

S uppose that you are given a sample of an unknown pure compound. You are told that it contains only carbon and oxygen. How can you determine what the compound is? In Chapter 9, we learn how to do this. We explore the quantitative relations between atomic weight, molecular weight, chemical formula, and mass data derived from the analyses of compounds.

Chemistry, like many other sciences, owes much of its present state of advancement to the careful numerical interpretation of quantitative data. To appreciate chemistry as a science and to be able to perform significant experiments, we need to make chemical calculations. In Chapter 8, we learned about one of the most useful concepts in chemistry, the mole, and we began to make chemical calculations. Here we continue to make use of the mole in more advanced chemical calculations.

9.1 Review of Mass Calculations Using the Mole

Before continuing, let's review some of the chemical calculations we learned in Chapter 8. If atomic weights are known, laboratory mass measurements can be converted to numbers of moles and vice versa. We simply need to use the correct conversion factor, namely the molar mass of the element.

EXAMPLE 9.1

How many moles of argon atoms are there in 34.0 grams of argon?

Solution Map: | grams of Ar | $\longrightarrow$ | moles of Ar |

Strategy and Solution:

Step 1. A look at the periodic table shows us that argon has an atomic weight of 39.948.

Step 2. Use the molar mass as a conversion factor.

Step 3. Set up and calculate.

$$34.0 \text{ g Ar} \times \frac{\text{mol Ar}}{39.948 \text{ g Ar}} = 0.851 \text{ mol Ar}$$

There are 0.851 moles of Ar in 34.0 grams. Since one mole contains 39.948 grams of Ar, you may have guessed that there would be slightly less than a mole in 34.0 grams.

EXAMPLE 9.2

How many grams of sodium are there in a sample containing 0.226 moles of Na?

Solution Map: | moles of Na | $\longrightarrow$ | grams of Na |

Strategy and Solution:

Step 1. The atomic weight of sodium is 22.9898.

Step 2. Use the molar mass as a conversion factor.

Step 3. Set up and solve.

$$0.226 \text{ mol Na} \times \frac{22.9898 \text{ g Na}}{\text{mol Na}} = 5.20 \text{ g Na}$$

Exercise 9.1

Calculate the number of moles of helium atoms in 10.5 grams of helium.

Exercise 9.2

How many grams of silver are in 0.0562 moles of silver?

Similar calculations can be made for compounds if the chemical formulas of the compounds are known.

EXAMPLE 9.3

How many moles of hydrogen sulfide, H_2S, are there in a sample weighing 92.5 grams?

Solution Map: $\boxed{\text{grams of } H_2S} \longrightarrow \boxed{\text{moles of } H_2S}$

Strategy and Solution:

Step 1. Find the molecular weight of H_2S.

Molecular weight of H_2S :

$2 \times$ atomic weight of $H = 2 \times 1.0079 = $ 2.0158

$1 \times$ atomic weight of $S = 1 \times 32.066 = $ <u>32.066</u>

Molecular weight $= 34.082$

Molar mass $= 34.082$ g H_2S/mol H_2S

Step 2. Use the molar mass as a conversion factor.

Step 3. Set up and solve.

$$92.5 \text{ g } H_2S \times \frac{\text{mol } H_2S}{34.082 \text{ g } H_2S} = 2.71 \text{ mol } H_2S$$

Carbon monoxide created by automobile exhaust is a component of smog.

Exercise 9.3

How many moles of potassium sulfate, K_2SO_4, are in a sample weighing 176.2 grams?

Exercise 9.4

Carbon monoxide, CO, is a pollutant created by automobile exhaust. How many grams of CO are there in 1.55 moles of CO?

9.2 Formulas of Compounds and the Mole

After studying this section, you should be able to:

- Convert from numbers of moles of compound to numbers of moles of elements (and vice versa).
- Convert from numbers of grams of compound to numbers of grams of elements (and vice versa).

In Chapter 2, we learned that molecular formulas can be used to show how many atoms of each element there are in an individual molecule if the compound happens to be molecular. An example is water. The formula H_2O indicates that in one molecule there are two H atoms and one O atom.

Likewise, we can read formulas in terms of *moles* of atoms. For example, the formula for nitric acid, HNO_3, can be read to mean that one mole of nitric acid contains one mole of H atoms, one mole of N atoms, and three moles of O atoms. Similarly, one mole of H_2O contains two moles of H atoms and one mole of O atoms. We can write this kind of information in the form of conversion factors to make chemical calculations. Let's look at some examples.

EXAMPLE 9.4

In 1 mole of phosphoric acid, H_3PO_4, how many moles of H are there? Of P and O?

Solution Map: $\boxed{\text{moles of } H_3PO_4} \longrightarrow \boxed{\text{moles of H, P, and O}}$

Solution: According to the formula, there are three moles of H, one mole of P, and four moles of O for every mole of H_3PO_4. Although it is not necessary in this case, we could have used conversion factors.

$$1 \text{ mol } H_3PO_4 \times \frac{3 \text{ mol H}}{\text{mol } H_3PO_4} = 3 \text{ mol H}$$

$$1 \text{ mol } H_3PO_4 \times \frac{1 \text{ mol P}}{\text{mol } H_3PO_4} = 1 \text{ mol P}$$

$$1 \text{ mol } H_3PO_4 \times \frac{4 \text{ mol O}}{\text{mol } H_3PO_4} = 4 \text{ mol O}$$

EXAMPLE 9.5

Sucrose is ordinary table sugar. In 0.271 moles of sucrose, $C_{12}H_{22}O_{11}$, how many moles of C are there?

Solution Map: $\boxed{\text{moles of } C_{12}H_{22}O_{11}} \longrightarrow \boxed{\text{moles of C}}$

Strategy and Solution:

Step 1. Read the formula in terms of moles and write a conversion factor for the relationship between sucrose and C. The conversion factor is

$$\frac{12 \text{ mol C}}{\text{mol } C_{12}H_{22}O_{11}}$$

Step 2. Set up and calculate.

$$0.271 \text{ mol } C_{12}H_{22}O_{11} \times \frac{12 \text{ mol C}}{\text{mol } C_{12}H_{22}O_{11}} = 3.25 \text{ mol C}$$

Exercise 9.5

How many moles of O are there in 1.31 moles of dihydroxy nitrobenzene, $C_6H_5O_4N$?

As we know, some compounds form substances that have no individual molecules. In its simplest version, a chemical formula tells us the relative numbers of atoms of the different elements in a compound. The formula $KAl(SO_4)_2$ says that in this compound (which in one of its forms is commonly called alum), there is one potassium atom and one aluminum atom for every two sulfur atoms and every eight oxygen atoms. In a similar way, the formula can also be read in terms of moles. There is one mole of potassium atoms and one mole of aluminum atoms for every two moles of sulfur atoms and every eight moles of oxygen atoms in $KAl(SO_4)_2$.

Because a chemical formula indicates the relative numbers of atoms of the different elements in a compound, chemical formulas can be read in terms of moles.

Laboratory measurements (usually in grams) are made with large numbers of particles, not with individual particles. Therefore, when we perform calculations based on mass data, we must read formulas in terms of moles. For many calculations, it may be necessary to convert mass data to moles, make a conversion from moles of compounds to moles of elements, and perhaps convert back to mass data.

EXAMPLE 9.6

How many moles of H are there in 11.0 grams of water?

Solution Map: | grams of H_2O | ⟶ | moles of H_2O | ⟶ | moles of H |

Strategy and Solution:

Step 1.   Find the molecular weight of H_2O by adding the atomic weights of H and O, $2(1.0079) + 1(15.9994) = 18.0152$.

Step 2.   Make conversion factors using the molar mass and the formula.

$$\frac{18.0152 \text{ g } H_2O}{\text{mol } H_2O}$$

$$\frac{2 \text{ mol H}}{\text{mol } H_2O}$$

Step 3.   Set up and solve.

$$11.0 \text{ g } H_2O \times \frac{\text{mol } H_2O}{18.0152 \text{ g } H_2O} \times \frac{2 \text{ mol H}}{\text{mol } H_2O} = 1.22 \text{ mol H}$$

Exercise 9.8

Carbon dioxide, CO_2, is released into our atmosphere when fossil fuels are burned. Calculate the number of moles of O in 156 grams of CO_2.

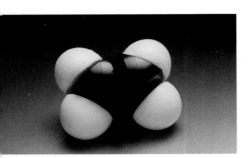

Space-filling model of ethylene, C_2H_4.

EXAMPLE 9.7

Ethylene, C_2H_4, is one of the simplest organic compounds. How many grams of carbon are there in a sample of C_2H_4 that weighs 35.8 grams? How many grams of hydrogen?

Solution Map:

grams of C_2H_4 $\longrightarrow$ moles of C_2H_4 $\longrightarrow$ moles of C $\longrightarrow$

grams of C $\longrightarrow$ grams of H

Strategy and Solution:

Step 1. Find the atomic weight of C and the molecular weight of C_2H_4.

Step 2. Make conversion factors using the molar mass of C, the molar mass of C_2H_4, and the formula. Every molecule of C_2H_4 contains two atoms of carbon; thus every mole of C_2H_4 contains two moles of carbon.

Step 3. Set up and solve.

$$35.8 \text{ g } C_2H_4 \times \frac{\text{mol } C_2H_4}{28.054 \text{ g } C_2H_4} \times \frac{2 \text{ mol C}}{\text{mol } C_2H_4} \times \frac{12.011 \text{ g C}}{\text{mol C}} = 30.7 \text{ g C}$$

The total mass of any compound must be the sum of the masses of the individual elements. In other words, the mass of C_2H_4 must equal the mass of the carbon plus the mass of the hydrogen.

$$30.7 \text{ g C} + \text{mass of H} = 35.8 \text{ g } C_2H_4$$

Subtracting the mass of C from the total mass gives the mass of the H present.

$$35.8 \text{ g } C_2H_4 - 30.7 \text{ g C} = 5.1 \text{ g H}$$

In a 35.8-gram sample of ethylene, there are 30.7 grams of carbon and 5.1 grams of hydrogen.

$Cr_2(SO_4)_3$ crystals.

Exercise 9.9

How many grams of O are in 82 grams of $Cr_2(SO_4)_3$?

Exercise 9.10

Calculate the number of grams of H in 46.3 grams of H_2S.

9.3 **Percent Composition of Compounds from Mass Data**

After studying this section, you should be able to:

- Calculate the percent compositions of compounds from mass data given in grams.

In the previous example, we found that 35.8 grams of C_2H_4 contains 30.7 grams of C and 5.1 grams of H. We could have reported these results in several ways: by giving the actual masses of the two elements in the sample; by giving only the ratio of the masses; or by showing our results as percent values, the method usually employed. In sections to come, we will make many calculations in terms of percent, and we will often be given data in terms of percent. The *percent* of something is 100% multiplied by the ratio of the amount of that something to the total amount. If there are 24 red marbles in 100 marbles, the percent of red marbles is 24.

The word "percent" is derived from Latin per, meaning "for each," and cent, indicating 100. Hence percent, "for each hundred."

$$\frac{24 \text{ red marbles}}{100 \text{ marbles}} \times 100\% = 24\%$$

Similarly, if there are 5 green apples in 25 apples, the percent of green apples is 20.

$$\frac{5 \text{ green apples}}{25 \text{ apples}} \times 100\% = 20\%$$

To have 20% green apples means that for every hundred apples there would be 20 green apples. Let's see how to calculate percentages in chemistry.

Guidelines for Calculating Percent Composition from Mass Data

1. *Find the actual mass of each element in the total mass of the sample.*
2. *For each element, write a fraction in which the mass of the element is the numerator and the mass of the entire sample is the denominator.*

$$\frac{\text{mass of the element}}{\text{mass of the entire sample}}$$

3. *For each element, multiply that fraction by 100%.* The result is the percent of that element in the sample. Steps 2 and 3 are often done together.

$$\frac{\text{mass of the element}}{\text{mass of the entire sample}} \times 100\%$$

The **percent composition** of a compound is a list of the percentages of the elements in the compound. Giving the percent composition of a compound is equivalent to stating the mass of each element in 100 grams of compound.

EXAMPLE 9.8

Calculate the percent of C and of H present in C_2H_4 from the information in Example 9.7.

Solution Map:

$$\boxed{\text{grams of C}} \longrightarrow \boxed{\text{fraction of C}} \longrightarrow \boxed{\text{percent C}} \longrightarrow$$

$$\boxed{\text{grams of H}} \longrightarrow \boxed{\text{fraction of H}} \longrightarrow \boxed{\text{percent H}}$$

Strategy and Solution:

Step 1. The mass of C is 30.7 grams, that of H is 5.1 grams, and the total mass is 35.8 grams.

Step 2. The fraction of C in the sample is 30.7/35.8, and that of H is 5.1/35.8.

Step 3. For each element, multiply the mass fraction by 100%.

$$\text{percent of C is } \frac{30.7 \text{ g C}}{35.8 \text{ g C}_2\text{H}_4} \times 100\% = 85.8\%$$

$$\text{percent of H is } \frac{5.1 \text{ g H}}{35.8 \text{ g C}_2\text{H}_4} \times 100\% = 14\%$$

The percent composition of C_2H_4 is 85.8% C and 14% H. Notice that within our two significant figure precision, the percentages add to 100%.

Exercise 9.11

In the laboratory you determine that a sample of a compound weighing 4.40 grams contains 3.20 grams of oxygen. What is the percent of oxygen in the compound?

Exercise 9.12

The strength of fertilizer is often measured by the percent of nitrogen it contains. A 2.00-kilogram sample of fertilizer contains 0.70 kilograms of N. What is the percent of nitrogen in the sample?

EXAMPLE 9.9

Chemical analysis reveals that a sample of a pure compound contains 51.10 grams of Na, 35.55 grams of S, and 71.10 grams of O. What is its percent composition?

Solution Map: $\boxed{\text{grams}} \longrightarrow \boxed{\text{fraction}} \longrightarrow \boxed{\text{percent}}$

Strategy and Solution:

Step 1. Given the masses of the individual elements, the total mass is

$$51.10 \text{ g} + 35.55 \text{ g} + 71.10 \text{ g} = 157.75 \text{ g}$$

Step 2. The fraction of each element in the compound is

$$\frac{51.10 \text{ g Na}}{157.75 \text{ g}} \qquad \frac{35.55 \text{ g S}}{157.75 \text{ g}} \qquad \frac{71.10 \text{ g O}}{157.75 \text{ g}}$$

Step 3. The percent composition is

$$\frac{51.10 \text{ g Na}}{157.75 \text{ g}} \times 100\% = 32.39\% \text{ Na}$$

$$\frac{35.55 \text{ g S}}{157.75 \text{ g}} \times 100\% = 22.54\% \text{ S}$$

$$\frac{71.10 \text{ g O}}{157.75 \text{ g}} \times 100\% = 45.07\% \text{ O}$$

The sum of the percent values must always equal 100%, except for small differences that sometimes occur because of rounding.

Exercise 9.13

A sample of compound contains 1.20 grams of C, 0.10 grams of H, and 0.70 grams of N. What is its percent composition?

Exercise 9.14

Calculate the percent composition of a sample of an acid (acids are discussed in Chapter 14) that contains 0.403 grams of H, 6.412 grams of S, and 12.800 grams of O.

EXAMPLE 9.10

A compound contains only C, H, and Cl. A sample of this compound weighing 112.55 grams is found to contain 72.06 grams of C and 5.04 grams of H; the rest is Cl. What is its percent composition?

Solution Map: $\boxed{\text{grams}} \longrightarrow \boxed{\text{fraction}} \longrightarrow \boxed{\text{percent}}$

Strategy and Solution:

Step 1. We have the masses of H and C, but we must calculate the mass of Cl. If the compound contains only C, H, and Cl, the total mass of the sample is the sum of the masses of those three elements.

mass of C + mass of H + mass of Cl = mass of sample

72.06 g C + 5.04 g H + mass of Cl = 112.55 g

mass of Cl = 112.55 g − 72.06 g − 5.04 g = 35.45 g

Steps 2 and 3.

$$\frac{72.06 \text{ g C}}{112.55 \text{ g}} \times 100\% = 64.02\% \text{ C}$$

$$\frac{5.04 \text{ g H}}{112.55 \text{ g}} \times 100\% = 4.48\% \text{ H}$$

$$\frac{35.45 \text{ g Cl}}{112.55 \text{ g}} \times 100\% = 31.50\% \text{ Cl}$$

> **Exercise 9.15**
>
> A sample of a compound containing only Na, Br, and O weighs 118.9 grams. What is its percent composition if there are 23.0 grams of Na and 16.0 grams of O in the sample?

9.4 Calculating Percent Composition from Chemical Formulas

After studying this section, you should be able to:

• Calculate the percent compositions of compounds from their formulas.

It is often desirable to calculate the percent composition of a compound from its chemical formula alone. For example, as discussed in Exercise 9.12, fertilizers are often evaluated on the basis of the percentage of nitrogen they contain. If we know the formula of a nitrogen-containing compound, we can determine the percentage of nitrogen. Here are the steps to determine the percent composition from a chemical formula:

Guidelines for Calculating Percent Composition from Chemical Formulas

1. *From the formula of the compound, write ratios for the number of moles of each element in one mole of the compound.*

$$\frac{\text{mol element}}{\text{mol compound}}$$

2. *Convert the number of moles of each element in the list to the number of grams of that element, using the molar mass of the element.* This step yields the number of grams of each element in a sample consisting of one mole of compound.

$$\frac{\text{mol element}}{\text{mol compound}} \times \frac{\text{g element}}{\text{mol element}} = \frac{\text{g element}}{\text{mol compound}}$$

3. *To convert moles of compound to grams of compound, use the molar mass of the compound as a conversion factor. Then multiply by 100%.*

$$\frac{\text{g element}}{\text{mol compound}} \times \frac{\text{mol compound}}{\text{g compound}} \times 100\% = \% \text{ of element}$$

EXAMPLE 9.11

What is the percent composition of nitric acid, HNO_3?

Solution Map: $\boxed{\text{moles}}$ $\longrightarrow$ $\boxed{\text{grams}}$ $\longrightarrow$ $\boxed{\text{percent}}$

Strategy and Solution:

Step 1. One mole of compound contains 1 mole of H, 1 mole of N, and 3 moles of O.

Step 2. The number of grams of each element is

$$\frac{1 \text{ mol H}}{\text{mol compound}} \times \frac{1.0079 \text{ g H}}{\text{mol H}} = \frac{1.0079 \text{ g H}}{\text{mol compound}}$$

$$\frac{1 \text{ mol N}}{\text{mol compound}} \times \frac{14.0067 \text{ g N}}{\text{mol N}} = \frac{14.0067 \text{ g N}}{\text{mol compound}}$$

$$\frac{3 \text{ mol O}}{\text{mol compound}} \times \frac{15.9994 \text{ g O}}{\text{mol O}} = \frac{47.9982 \text{ g O}}{\text{mol compound}}$$

Step 3. Find the molar mass of compound. The molar mass of the compound is 63.0128. Using this as a conversion factor, find the number of grams of each element in one gram of compound. Multiply by 100%.

Notice that the units of percent composition are

$$\frac{\text{grams of element}}{\text{grams of compound}}$$

$$\frac{1.0079 \text{ g H}}{\text{mol compound}} \times \frac{\text{mol compound}}{63.0128 \text{ g compound}} \times 100\% = 1.5995\% \text{ H}$$

$$\frac{14.0067 \text{ g N}}{\text{mol compound}} \times \frac{\text{mol compound}}{63.0128 \text{ g compound}} \times 100\% = 22.2283\% \text{ N}$$

$$\frac{47.9982 \text{ g O}}{\text{mol compound}} \times \frac{\text{mol compound}}{63.0128 \text{ g compound}} \times 100\% = 76.1721\% \text{ O}$$

The percent composition of HNO_3 is 1.5995% H, 22.2283% N, and 76.1721% O. You should always check that the sum of the percentages is 100%.

After you become accustomed to the method, you may find it more convenient to combine steps 2 and 3 into a single equation for each of the elements.

$$\frac{1 \text{ mol H}}{\text{mol HNO}_3} \times \frac{1.0079 \text{ g H}}{\text{mol H}} \times \frac{\text{mol HNO}_3}{63.0128 \text{ g HNO}_3} \times 100\% = 1.5995\% \text{ H}$$

Similar calculations can be written for the other two elements.

Exercise 9.16

Sodium hydroxide, NaOH, is a common base, a kind of compound discussed in Chapter 14. What is the percent composition of NaOH?

Exercise 9.17

What is the percent composition of calcium chlorite, $Ca(ClO_2)_2$?

9.5 Empirical Formulas Calculated from Mass Data or Percent Composition

After studying this section, you should be able to:

- Calculate empirical formulas of compounds, given either laboratory mass data or percent composition.

In Chapter 2, we were introduced to empirical formulas. An *empirical formula* is the simplest formula for a compound. It represents the smallest whole-number ratio of atoms in that compound. On the other hand, *a molecular formula* gives the *actual number* of each kind of atom in a molecule of the compound. For example, HO is the empirical formula for hydrogen peroxide. It tells us that in hydrogen peroxide there is one hydrogen atom for each oxygen atom. The molecular formula for hydrogen peroxide is H_2O_2. Hydrogen peroxide molecules each contain two hydrogen and two oxygen atoms. The molecular formula for hydrogen peroxide is a multiple of its empirical formula. Every molecular formula is a multiple of an empirical formula, even if the multiplier is only 1, as in carbon monoxide, CO, where the empirical and molecular formulas are the same. Ionic compounds do not have molecular formulas.

We can calculate the empirical formula for a compound from the masses of the elements present in the compound. To find an empirical formula from mass data, we need to convert the amount of mass of each element in any sample of the compound to the number of moles of that element. Then we transform this information into a whole-number ratio. Here are the steps:

Guidelines for Calculating Empirical Formula from Mass Data

1. *From laboratory data, write down the number of grams of each element in a sample of compound.*

2. *Convert the number of grams of each element to the number of moles, using the molar mass as a conversion factor.* This will give the actual number of moles of each element in the sample. Except for the fact that these are not whole numbers, they could be used in a chemical formula.

$$\text{g element} \times \frac{\text{mol element}}{\text{g element}} = \text{mol element}$$

3. *To find the empirical formula, divide all the numbers of moles by the smallest number of moles.* If the numbers you get are all integers, then they are the subscripts in the formula. If any of them is not, then go to step 4.

4. *Multiply the numbers from step 3 by the smallest integer (often 2 or 3, rarely bigger than 5 or 6) that will make them all into whole numbers.* The resulting set represents the subscripts in the empirical formula.

EXAMPLE 9.12

A sample of sulfuric acid contains 1.4 grams of H, 22.7 grams of S, and 45.3 grams of O. What is the empirical formula for sulfuric acid?

Solution Map: grams $\longrightarrow$ moles $\longrightarrow$ smallest ratio

Strategy and Solution:

Step 1. We have the number of grams of each element in the sample, so we can go to step 2.

Step 2. Convert the number of grams of each element to the number of moles. Use the molar mass as a conversion factor.

$$1.4 \text{ g H} \times \frac{\text{mol H}}{1.0079 \text{ g H}} = 1.4 \text{ mol H}$$

$$45.3 \text{ g O} \times \frac{\text{mol O}}{15.9994 \text{ g O}} = 2.83 \text{ mol O}$$

$$22.7 \text{ g S} \times \frac{\text{mol S}}{32.066 \text{ g S}} = 0.708 \text{ mol S}$$

Step 3. To reduce them to a set of whole numbers, divide each of the numbers of moles by the smallest number, in this case, 0.708.

$$\frac{1.4 \text{ mol H}}{0.708} = 2.0 \text{ mol H}$$

$$\frac{2.83 \text{ mol O}}{0.708} = 4.00 \text{ mol O}$$

$$\frac{0.708 \text{ mol S}}{0.708} = 1.00 \text{ mol S}$$

The empirical formula, then, is H_2SO_4. (The elements are written in this order because of the chemical properties of the compound. As far as the calculation is concerned, the formula could be O_4SH or HO_4S, etc.)

Exercise 9.18

A sample of a compound contains 3.0 grams of C, 1.0 gram of H, and 4.0 grams of O. What is its empirical formula?

Exercise 9.19

Sodium bicarbonate is used in almost every household. What is its empirical formula? A sample contains 5.06 grams of Na, 2.64 grams of C, 10.56 grams of O, and 0.222 grams of H.

Depending on the precision of the data, the numbers for the formula may not always come out exactly as whole numbers, even after step 4. As in the next example, some rounding will occasionally be necessary. In the example that follows, notice that the laboratory data are given as percent composition, necessitating a slightly different approach to the calculation.

The chemical name for baking soda is sodium bicarbonate.

EXAMPLE 9.13

Potassium carbonate is 56.4% K, 8.70% C, and 34.9% O. What is the empirical formula for potassium carbonate?

Solution Map: $\boxed{\text{percent}} \longrightarrow \boxed{\text{grams}} \longrightarrow \boxed{\text{moles}} \longrightarrow \boxed{\text{smallest ratio}}$

Strategy and Solution:

Step 1.   To obtain the actual mass in grams of each element in a sample, we must imagine that we have an actual sample. It is most convenient to imagine a sample weighing 100.0 grams, because the percent numbers for each element then convert directly into grams. If our imaginary sample of potassium carbonate weighed 100.0 grams, then it would contain the following:

"56.4% K" means there are 56.4 grams of K in a 100.0-gram sample.

$$100.0 \text{ g of sample} \times \frac{56.4 \text{ g K}}{100.0 \text{ g sample}} = 56.4 \text{ g K}$$

$$100.0 \text{ g of sample} \times \frac{8.70 \text{ g C}}{100.0 \text{ g sample}} = 8.70 \text{ g C}$$

$$100.0 \text{ g of sample} \times \frac{34.9 \text{ g O}}{100.0 \text{ g sample}} = 34.9 \text{ g O}$$

Step 2.   Convert each mass to the number of moles.

$$56.4 \text{ g K} \times \frac{\text{mol K}}{39.0983 \text{ g K}} = 1.44 \text{ mol K}$$

$$8.70 \text{ g C} \times \frac{\text{mol C}}{12.011 \text{ g C}} = 0.724 \text{ mol C}$$

$$34.9 \text{ g O} \times \frac{\text{mol O}}{15.9994 \text{ g O}} = 2.18 \text{ mol O}$$

Step 3.   As it stands, the formula is $K_{1.44}C_{0.724}O_{2.18}$. To reduce these values to whole numbers, divide each by the smallest.

$$\frac{1.44}{0.724} = 1.99 \qquad \frac{0.724}{0.724} = 1.00 \qquad \frac{2.18}{0.724} = 3.01$$

This operation yields $K_{1.99}C_{1.00}O_{3.02}$, still not a whole-number formula, but one that we can round to whole numbers. We look at the original data, and find that the data were given to us in only three significant figures, meaning that the last digit is uncertain. We round 1.99 to 2.0 and 3.01 to 3.0, giving the final empirical formula as K_2CO_3. Before you round while calculating formulas, be sure that the data justify it. Rounding incorrectly could lead to an incorrect formula.

Exercise 9.20

A compound contains 52.0% Zn, 9.61% C, and 38.4% O. What is its empirical formula?

In the last two examples, the formulas for the compounds each contained an element that appeared with an unwritten subscript 1, namely, S in $H_2S_1O_4$ and C in $K_2C_1O_3$.

If a subscript is 1, the 1 is not shown.
Example: H_2SO_4.

EXAMPLE 9.14

A compound of phosphorus and oxygen contains 43.7% P and 56.3% O. What is its simplest formula?

Solution Map: | percent | $\longrightarrow$ | grams | $\longrightarrow$ | moles | $\longrightarrow$ | smallest ratio |

Strategy and Solution:

Step 1. A 100.0-gram sample of the compound contains 43.7 grams of P and 56.3 grams of O.

Step 2. Convert each mass to the number of moles of the element.

$$43.7 \text{ g P} \times \frac{\text{mol P}}{30.9738 \text{ g P}} = 1.41 \text{ mol P}$$

$$56.3 \text{ g O} \times \frac{\text{mol O}}{15.9994 \text{ g O}} = 3.52 \text{ mol O}$$

Step 3. Reduce the results to small whole numbers by dividing by the smallest number.

$$\frac{1.41 \text{ mol P}}{1.41 \text{ mol}} = 1.00$$

$$\frac{3.52 \text{ mol O}}{1.41 \text{ mol}} = 2.50$$

Step 4. To obtain subscripts that are whole numbers, we try multiplying the subscripts by different small whole numbers until both subscripts are whole numbers. In this case, we find that 2 will do the trick.

$$2 \times 1.00 = 2.00 \qquad 2 \times 2.50 = 5.00$$

Our formula is P_2O_5.

Exercise 9.21

Hydrocarbons belong to a class of compounds containing only hydrogen and carbon. We are asked to find the empirical formula of a hydrocarbon and are given these data: A sample of the compound contains 11.25 grams of C and 2.52 grams of H. What is its empirical formula?

Exercise 9.22

Iron can form three different oxides. An oxide of iron contains 72.36% Fe and 27.64% O. What is its simplest formula?

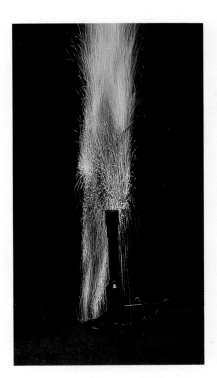

Iron reacts vigorously with pure oxygen to form oxides.

9.6 Calculating the Molecular Formula from the Empirical Formula

After studying this section, you should be able to:

- Determine actual molecular formulas from empirical formulas and approximate molecular weights.

For ionic compounds, there are no molecular formulas because there are no molecules; therefore all formulas for ionic compounds are empirical.

In the previous section we used mass data and molar masses to calculate empirical formulas. In the case of compounds that form molecules, the empirical formula does not necessarily indicate how many atoms of each kind are in a molecule. The molecular formula, however, does show the actual numbers of atoms in the molecule. The molecular formula is always a multiple of the empirical formula, as we saw in the example of H_2O_2.

It is often the case that two entirely different compounds have the same empirical formula. Consider the formula CH_2, which applies to several compounds, two of which are ethylene and cyclohexane.

$$ethylene = (CH_2)_2 = C_2H_4$$

$$cyclohexane = (CH_2)_6 = C_6H_{12}$$

Ethylene is a gas from which the familiar plastic, polyethylene, is made. Cyclohexane is a liquid with entirely different properties, yet both have the same empirical formula. Notice that when two compounds have the same empirical formula, they also have the same percent composition: Ethylene and cyclohexane both contain 85.6% C and 14.4% H.

Whenever possible, it is useful to know the actual molecular formula of a compound rather than just the empirical formula. You can find the molecular formula from the empirical formula if you have an approximate molecular weight—perhaps obtained from a laboratory experiment.

Guidelines for Finding Molecular Formula from Empirical Formula and Molecular Weight

1. *Find the "empirical weight" for the empirical formula.* We will pretend that the formula represents a molecule and proceed as though we were finding an ordinary molecular weight.
2. *Divide the approximate molecular weight by the calculated "empirical weight" and round off the result to the nearest whole number.*
3. *Multiply the empirical formula by the result of step 2.*

EXAMPLE 9.15

The empirical formula for hydrazine is NH_2, and the molecular weight is approximately 30. What is the molecular formula?

Solution Map:

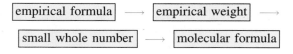

Strategy and Solution:

Step 1. Calculate the "empirical weight" for the empirical formula.

Empirical weight of NH_2:

$1 \times$ atomic weight of $N = 1 \times 14.0067 = 14.0067$

$2 \times$ atomic weight of $H = 2 \times 1.0079 \ = \ \underline{2.0158}$
Empirical weight $\qquad\qquad\qquad\quad = 16.0225$

Step 2. Divide the "empirical weight" into the approximate molecular weight.

$$\frac{30}{16.0225} = 2$$

The answer, 2, is correct because 30 has only one significant figure.

The molecular formula for hydrazine, then, is

$$(NH_2)_2 = N_2H_4$$

An actual molecular weight can then be calculated from the molecular formula. The molecular weight of N_2H_4 is

$$(2 \times 14.0067) + (4 \times 1.0079) = 32.0450$$

Exercise 9.23

The empirical formula of automobile radiator coolant, ethylene glycol, is CH_3O. Its estimated molecular weight is 60. What are its molecular formula and molecular weight?

Exercise 9.24

The empirical formula for a compound of C, H, O, and Cl is C_4H_3OCl, and its molecular weight is approximately 200. What are the molecular formula and the molecular weight of this compound?

In some cases, we may be given only the percent composition of a compound. To find its molecular formula, we first have to find its empirical formula, as shown in the next example.

EXAMPLE 9.16

The compound xylene consists of only carbon and hydrogen. It is 90.49% C, and the rest is hydrogen. Its molecular weight is 106. What are the molecular formula and the molecular weight of xylene?

Ethylene glycol, $C_2H_6O_2$, is the main ingredient in automobile radiator coolant.

Solution Map:

$$\boxed{\text{grams}} \longrightarrow \boxed{\text{moles}} \longrightarrow \boxed{\text{empirical formula}} \longrightarrow$$
$$\boxed{\text{empirical weight}} \longrightarrow \boxed{\text{small whole number}} \longrightarrow \boxed{\text{molecular formula}}$$

Strategy and Solution:

Step 1. Determine the empirical formula. Find the number of grams of each element in a sample of the compound. A 100.00-gram sample contains 90.49 grams of C, and the rest is H.

$$100.00 \text{ g xylene} - 90.49 \text{ g C} = 9.51 \text{ g H}$$

$$90.49 \text{ g C} \times \frac{\text{mol C}}{12.011 \text{ g C}} = 7.534 \text{ mol C} \qquad \frac{7.534 \text{ mol C}}{7.534} = 1.000 \text{ mol C}$$

$$9.51 \text{ g H} \times \frac{\text{mol H}}{1.0079 \text{ g H}} = 9.44 \text{ mol H} \qquad \frac{9.44 \text{ mol H}}{7.534} = 1.25 \text{ mol H}$$

We find the smallest whole number that will change both the results to whole numbers. The number 4 will do this, so we multiply by 4.

$$4 \times 1.000 = 4.000 \quad 4 \times 1.25 = 5.00$$

The empirical formula is C_4H_5. We must now find the molecular formula.

Step 2. Find the "empirical weight." The "empirical weight" for this formula is 53.0835.

Step 3. Divide the "empirical weight" into the approximate molecular weight.

$$\frac{106}{53.0835} = 2.00$$

So we take the actual molecular formula to be twice the empirical formula and write C_8H_{10}, for which the molecular weight is $(8 \times 12.011) + (10 \times 1.0079) = 106.167$.

Exercise 9.25

Citric acid, found in citrus fruits, contains C, H, and O. It is 37.6% C, 4.20% H, and 58.37% O and has a molecular weight of about 190. What are its molecular formula and its molecular weight?

Summary

By interpreting chemical formulas in units of moles, we are able to make useful chemical calculations. Using atomic weights and molecular weights, we can determine the percent composition of a compound either from mass data or from a chemical formula. We can make calculations in which mass data and atomic weights will yield the empirical formula of a compound. Actual molecular formulas can then be determined if other data, such as approximate molecular weights, are available.

Key Term

percent composition   *(p. 239)*

Questions and Problems

Section 9.1 Review of Mass Calculations Using the Mole

Question

1. Name the conversion factor or factors used in a conversion from moles to mass.

Section 9.2 Formulas of Compounds and the Mole

Questions

2. Explain why a chemical formula can be read both in terms of individual atoms and in terms of moles.

3. Is there such a thing as a molecular formula for compounds that do not form molecules?

4. Name the conversion factor or factors used in the conversion from moles of compound to moles of atoms.

5. Name the conversion factor or factors are used in the conversion from grams of compound to grams of atoms.

6. Give solution maps for Problems 15 and 17.

Problems

7. Calculate the number of moles of N atoms in 7.32 moles of N_2H_4.

8. How many moles of H atoms are in 4.8 moles of CH_4?

9. How many moles of H_2CO_3 contain 2.25 moles of O atoms?

10. How many moles of H_3PO_4 contain 4.37 moles of H atoms?

11. How many moles of $C_6H_{12}F_2$ contain as much hydrogen as 6 moles of H_2O?

12. The formula for pentane is C_5H_{12}. How many atoms of carbon are in five molecules of pentane? How many moles of H are in 3.7 moles of pentane? How many atoms of carbon are in 2.2 moles of pentane?

13. How many moles of oxygen atoms are in each of the following samples?
    (a) 66.6 g of NaOH
    (b) 18.0 g of $C_{12}H_{16}O_4$
    (c) 122 g of SO_2
    (d) 9.53 g $\times$ 10^{28} molecules of N_2O_4
    (e) 1.21 g of H_2O_2

14. How many moles of atoms are in each of the samples a, b, d, and e of Problem 13?

15. How many grams of S are in 0.532 moles of $Na_2S_2O_3$?

16. A sample of pentane, C_5H_{12}, weighs 22.7 grams. How many grams of carbon are in that sample?

17. A sample of barium hydroxide, $Ba(OH)_2$, has a mass of 52.11 grams. What is the mass of the Ba in the sample?

18. A sample of potassium chromate, K_2CrO_4, has a mass of 541 grams. How many grams of potassium are in the sample? How many grams of oxygen?

19. How many grams of oxygen are contained in each of the following samples?
    (a) 8.46 g of NO_3F
    (b) 202 g of $NaClO_3$
    (c) 3.03 mol of H_2O_2
    (d) 26.2 mol of $C_6H_{12}O_6$

20. How many grams of lithium are contained in each of the following samples?
    (a) 3.62 g of $Li_2B_4O_7$
    (b) 5.87 g of $LiSO_3F$
    (c) 4.01 mol of Li_3PO_4
    (d) 2.23 mol of $LiAlH_4$
    (e) 4.01 $\times$ 10^{18} molecules of Li_3N
    (f) 1.74 $\times$ 10^3 mol of Li_4SiO_4

Section 9.3 Percent Composition of Compounds from Mass Data

Questions

21. Give a brief definition of percent composition.

22. Outline the procedure for calculating percent composition from laboratory mass data.

Problems

23. A 24.4-gram sample of sodium carbonate contains 10.6 grams of Na, 2.8 grams of C, and 11.5 grams of O. What is its percent composition?

24. Calculate the percent composition of a sample of compound that contains 9.20 grams of Na, 0.40 grams of H, and 6.40 grams of O.

25. What is the percent composition of a sample of compound containing 1.56 grams of K, 2.20 grams of Mn, and 2.56 grams of O?

26. A 32.7-gram sample of calcium carbonate, $CaCO_3$, contains 3.92 grams of C, 13.09 grams of Ca, and the rest is O. What is the percent composition of $CaCO_3$?

27. A sample of the compound $C_7H_6O_4$ weighs 17.6 grams and contains 9.60 grams of C, 7.31 grams of O, and the rest is H. What is its percent composition?

28. A 67.9-gram sample of a compound contains 45.8 grams of C, 8.68 grams of H, and the rest is N. Calculate the percent composition of this compound.

29. A sample of pyridine weighs 9.20 grams and contains 6.98 grams of C, 1.63 grams of N, and the rest is H. What is the percent composition of pyridine?

30. Calculate the percent composition of each of the following samples.
    (a) A 33.29-gram sample of $CaCl_2$ containing 21.27 grams of Cl and 12.02 grams of Ca
    (b) A sample of Na_2S containing 3.68 grams of Na and 2.56 grams of S
    (c) A sample of $MgBr_2$ containing 9.72 grams of Mg and 63.92 grams of Br
    (d) A 23.95-gram sample of LiOH containing 6.94 grams of Li and 1.01 grams of H

31. A sample of a pure acid has a volume of 45.9 mL and a density of 1.50 g/mL. It contains 1.10 grams of H, 15.3 grams of N, and the rest is O. What is its percent composition?

Section 9.4 Calculating Percent Composition from Chemical Formulas

Questions

32. Describe how to calculate the percent composition of a compound from its formula.

33. Suggest a practical use for percent composition.

Problems

34. What is the percent of oxygen in each of the following compounds?
    (a) $AgIO_3$ (b) $KMnO_4$
    (c) C_6H_7O (d) N_2O_5
    (e) $BaSO_4$ (f) $MgMoO_4$
    (g) PtO_2

35. Hydrogen is an excellent fuel. Which of the following compounds has the the highest percent of hydrogen: H_2O, H_2O_2, C_2H_6O (ethanol), or C_8H_{18} (octane)?

36. Calculate the percent of lithium in each of the compounds listed in Problem 20.

37. List the following sulfides in order of increasing percent of S: Co_3S_4, Cs_2S_2, Al_2S_3, K_2S_2, SiS_2.

38. Which has the highest percent of
    (a) sulfur, SeS_2 or SeS?
    (b) sodium, Na_2O or Na_2O_2?
    (c) chlorine, S_2Cl_2 or SCl_4?
    (d) carbon, C_6H_7O or C_6H_5NO?

39. Answer the following with respect to chlorohydroxyl-biphenyl, $C_{12}H_9OCl$:
    (a) What is the percent of Cl?
    (b) What fraction of all the atoms in this compound are carbon atoms?
    (c) What is the weight in grams of one molecule of $C_{12}H_9OCl$?
    (d) At 25°C, chlorohydroxylbiphenyl is a liquid with a density of 1.25 g/mL. How many moles of the compound are contained in a volume of 250 mL?

40. Calculate the percent composition of each of the following: MgB_6, $NaHSO_4$, C_6H_5Cl, ClO_4F.

41. Calculate the percent composition of each of the following: $Ca(NO_3)_2$, NH_3, Fe_2O_3, MgC_2O_4.

Section 9.5 Empirical Formulas Calculated from Mass Data or Percent Composition

Questions

42. What is the relation between an empirical formula and a molecular formula?

43. For compounds that do not form molecules, what is the relation between the empirical formula for the compound and the actual formula?

44. Describe how you can obtain the empirical formula of a compound from the numbers of grams of the individual elements in a sample of the compound.

45. If you are given the percent composition of a compound (rather than the number of grams of each element), what assumption can you make as a starting point for calculating the empirical formula of the compound?

Problems

46. A 13.21-gram sample of magnesium will react with 40.71 grams of arsenic to form a compound. What is the empirical formula of the compound?

47. Ethanol is found in alcoholic beverages. If a 27.64-gram sample of ethanol contains 3.63 grams of H, 14.41 grams of C, and the rest is O, what is the empirical formula of ethanol?

48. A sample of ascorbic acid contains 5.76 grams of C, 0.6464 grams of H, and 7.68 grams of O. Calculate the empirical formula of ascorbic acid.

49. Codeine was formerly used in cough syrups; now it is used primarily as a painkiller. What is its empirical formula if a sample weighing 23.950 grams contains 17.294 grams of C, 1.697 grams of H, 1.121 grams of N, and the rest is O?

50. Find the empirical formula of each of the compounds, using the following percent compositions.
    (a) 31.8% N, 13.6% H, 54.6% C
    (b) 71.04% Ag, 7.90% C, 21.07% O

(c) 83.01% K, 16.99% O

(d) 39.69% K, 27.92% Mn, 32.49% O

(e) 42.86% C, 7.14% H, 50.01% N

51. A certain oxide of chromium contains 23.5% oxygen. What is the empirical formula of this compound?

52. Nitrogen forms six different oxides. Find the empirical formulas (which also happen to be the molecular formulas) for each, given the following.

(a) 46.67% N   (b) 63.64% N
(c) 22.58% N   (d) 30.43% N
(e) 36.84% N   (f) 25.93% N

Section 9.6 Calculating the Molecular Formula from the Empirical Formula

Questions

53. Explain how it is possible for two compounds to have the same empirical formula and completely different physical properties.

54. To determine a molecular formula, you need the empirical formula and an estimated molecular weight. Can you think of some other kind of information that could substitute for the estimated molecular weight and that would still enable you to find the molecular formula? How would such information be used in the calculation?

Problems

55. Vinegar is essentially a solution of water and acetic acid, the empirical formula of which is CH_2O. If the estimated molecular weight of acetic acid is 60, what is its molecular formula?

56. The simplest formula of mercurous iodide is HgI. What is its molecular formula if the estimated molecular weight is 650?

57. The empirical formula for a yellow dye used for wool is C_3H_2O. A mole of the dye is known to have 9.03×10^{24} atoms of C. What is the molecular formula of the dye?

58. The empirical formula for octane (a component of gasoline) is C_8H_{18}. If 3.5×10^{24} atoms of hydrogen are in 0.32 moles of compound, what is the molecular formula of octane?

59. A compound is known to contain C, H, N, and O. A sample of the compound weighing 27.4 grams is 73.69% C, 8.03% H, 8.58% N, and the rest is O. Calculate the weights of each element present in the sample and find the empirical formula of the compound. The molecular weight of the compound is less than 400. What is the actual molecular weight of the compound?

60. An 8.0-gram sample of a hydrocarbon contains enough hydrogen and carbon to form 12.0 grams of H_2O. What is the molecular formula of the hydrocarbon if its molecular weight is 144?

More Difficult Problems

61. Aspirin contains 60.0% carbon and twice as many hydrogen atoms as oxygen atoms. If these are the only elements present, what is the empirical formula of aspirin? Find the molecular formula if 1.00 mole of aspirin molecules contains 4.82×10^{24} atoms of H.

62. The compound PbZ_5 contains 42.77% Z. Calculate the molar mass of Z. What is Z?

63. Let X and Z represent two unknown elements that form a compound with oxygen. A 49.00-gram sample of this compound contains 18.8 grams of X and 9.3×10^{23} atoms of oxygen. If the formula of the compound is $X_3Z_2O_6$, what are the molar masses of X and Z? What elements are they?

Solutions to Exercises

9.1 $\qquad 10.5 \text{ g He} \times \dfrac{\text{mol He}}{4.0026 \text{ g He}} = 2.62 \text{ mol He}$

9.2 $\qquad 0.0562 \text{ mol Ag} \times \dfrac{107.8682 \text{ g Ag}}{\text{mol Ag}} = 6.06 \text{ g Ag}$

9.3 $\qquad 176.2 \text{ g K}_2\text{SO}_4 \times \dfrac{\text{mol K}_2\text{SO}_4}{174.260 \text{ g K}_2\text{SO}_4} = 1.011 \text{ mol K}_2\text{SO}_4$

9.4 $\qquad 1.55 \text{ mol CO} \times \dfrac{28.010 \text{ g CO}}{\text{mol CO}} = 43.4 \text{ g CO}$

9.5 $$1.31 \text{ mol C}_6\text{H}_5\text{O}_4\text{N} \times \frac{4 \text{ mol O}}{\text{mol C}_6\text{H}_5\text{O}_4\text{N}} = 5.24 \text{ mol O}$$

9.6 $$4.6 \text{ mol H} \times \frac{\text{mol H}_2\text{O}}{2 \text{ mol H}} = 2.3 \text{ mol H}_2\text{O}$$

9.7 $$320.5 \text{ g Al}_2\text{O}_3 \times \frac{\text{mol Al}_2\text{O}_3}{101.9612 \text{ g Al}_2\text{O}_3} \times \frac{3 \text{ mol O}}{\text{mol Al}_2\text{O}_3} = 9.430 \text{ mol O}$$

9.8 $$156 \text{ g CO}_2 \times \frac{\text{mol CO}_2}{44.010 \text{ g CO}_2} \times \frac{2 \text{ mol O}}{\text{mol CO}_2} = 7.09 \text{ mol O}$$

9.9 $$82 \text{ g Cr}_2(\text{SO}_4)_3 \times \frac{\text{mol Cr}_2(\text{SO}_4)_3}{392.183 \text{ g Cr}_2(\text{SO}_4)_3} \times \frac{12 \text{ mol O}}{\text{mol Cr}_2(\text{SO}_4)_3} \times \frac{15.9994 \text{ g O}}{\text{mol O}} =$$
$$40 \text{ g O}$$

9.10 $$46.3 \text{ g H}_2\text{S} \times \frac{\text{mol H}_2\text{S}}{34.082 \text{ g H}_2\text{S}} \times \frac{2 \text{ mol H}}{\text{mol H}_2\text{S}} \times \frac{1.0079 \text{ g H}}{\text{mol H}} = 2.74 \text{ g H}$$

9.11 $$\frac{3.20 \text{ g O}}{4.40 \text{ g compound}} \times 100\% = 72.7\% \text{ O}$$

9.12 $$\frac{0.70 \text{ kg N}}{2.00 \text{ kg compound}} \times 100\% = 35\% \text{ N}$$

9.13 $$1.20 \text{ g} + 0.10 \text{ g} + 0.70 \text{ g} = 2.00 \text{ g}$$

$$\frac{1.20 \text{ g C}}{2.00 \text{ g}} \times 100\% = 60.0\% \text{ C}; \quad \frac{0.10 \text{ g H}}{2.00 \text{ g}} \times 100\% = 5.0\% \text{ H};$$

$$\frac{0.70 \text{ g N}}{2.00 \text{ g}} \times 100\% = 35\% \text{ N}$$

9.14 $$0.403 \text{ g} + 6.412 \text{ g} + 12.800 \text{ g} = 19.615 \text{ g}$$

$$\frac{0.403 \text{ g H}}{19.615 \text{ g}} \times 100\% = 2.05\% \text{ H}; \quad \frac{6.412 \text{ g S}}{19.615 \text{ g}} \times 100\% = 32.69\% \text{ S};$$

$$\frac{12.800 \text{ g O}}{19.615 \text{ g}} \times 100\% = 62.256\% \text{ O}$$

9.15 $$118.9 \text{ g} - (23.0 \text{ g} + 16.0 \text{ g}) = 79.9 \text{ g Br}$$

$$\frac{23.0 \text{ g Na}}{118.9 \text{ g}} \times 100\% = 19.3\% \text{ Na}; \quad \frac{16.0 \text{ g O}}{118.9 \text{ g}} \times 100\% = 13.5\% \text{ O};$$

$$\frac{79.9 \text{ g Br}}{118.9 \text{ g}} \times 100\% = 67.2\% \text{ Br}$$

9.16 $\dfrac{1 \text{ mol Na}}{\text{mol NaOH}} \times \dfrac{22.9898 \text{ g Na}}{\text{mol Na}} \times \dfrac{\text{mol NaOH}}{39.9971 \text{ g NaOH}} \times 100\% = 57.4787\%$ Na

$\dfrac{1 \text{ mol O}}{\text{mol NaOH}} \times \dfrac{15.9994 \text{ g O}}{\text{mol O}} \times \dfrac{\text{mol NaOH}}{39.9971 \text{ g NaOH}} \times 100\% = 40.0014\%$ O

$\dfrac{1 \text{ mol H}}{\text{mol NaOH}} \times \dfrac{1.0079 \text{ g H}}{\text{mol H}} \times \dfrac{\text{mol NaOH}}{39.9971 \text{ g NaOH}} \times 100\% = 2.5199\%$ H

9.17 $\dfrac{1 \text{ mol Ca}}{\text{mol Ca(ClO}_2)_2} \times \dfrac{40.078 \text{ g Ca}}{\text{mol Ca}} \times \dfrac{\text{mol Ca(ClO}_2)_2}{174.981 \text{ g Ca(ClO}_2)_2} \times 100\% = 22.904\%$ Ca

$\dfrac{2 \text{ mol Cl}}{\text{mol Ca(ClO}_2)_2} \times \dfrac{35.4527 \text{ g Cl}}{\text{mol Cl}} \times \dfrac{\text{mol Ca(ClO}_2)_2}{174.981 \text{ g Ca(ClO}_2)_2} \times 100\% = 40.5221\%$ Cl

$\dfrac{4 \text{ mol O}}{\text{mol Ca(ClO}_2)_2} \times \dfrac{15.9994 \text{ g O}}{\text{mol O}} \times \dfrac{\text{mol Ca(ClO}_2)_2}{174.981 \text{ g Ca(ClO}_2)_2} \times 100\% = 36.5740\%$ O

9.18 $3.0 \text{ g C} \times \dfrac{\text{mol C}}{12.011 \text{ g C}} = 0.25 \text{ mol C};$ $1.0 \text{ g H} \times \dfrac{\text{mol H}}{1.0079 \text{ g H}} = 0.99 \text{ mol H};$

$4.0 \text{ g O} \times \dfrac{\text{mol O}}{15.9994 \text{ g O}} = 0.25 \text{ mol O};$

$\dfrac{0.25 \text{ mol C}}{0.25} = 1.0 \text{ mol C};$ $\dfrac{0.99 \text{ mol H}}{0.25} = 4.0 \text{ mol H};$

$\dfrac{0.25 \text{ mol O}}{0.25} = 1.0 \text{ mol O}$

CH_4O

9.19 $5.06 \text{ g Na} \times \dfrac{\text{mol Na}}{22.9898 \text{ g Na}} = 0.220 \text{ mol Na}$

$2.64 \text{ g C} \times \dfrac{\text{mol C}}{12.011 \text{ g C}} = 0.220 \text{ mol C}$

$10.56 \text{ g O} \times \dfrac{\text{mol O}}{15.9994 \text{ g O}} = 0.6600 \text{ mol O}$

$0.222 \text{ g H} \times \dfrac{\text{mol H}}{1.0079 \text{ g H}} = 0.220 \text{ mol H}$

$\dfrac{0.220 \text{ mol Na}}{0.220} = 1.00 \text{ mol Na};$ $\dfrac{0.220 \text{ mol C}}{0.220} = 1.00 \text{ mol C};$

$\dfrac{0.6600 \text{ mol O}}{0.220} = 3.00 \text{ mol O};$ $\dfrac{0.220 \text{ mol H}}{0.220} = 1.00 \text{ mol H};$

$NaHCO_3$

9.20
$$52.0 \text{ g Zn} \times \frac{\text{mol Zn}}{65.39 \text{ g Zn}} = 0.795 \text{ mol Zn}$$

$$9.61 \text{ g C} \times \frac{\text{mol C}}{12.011 \text{ g C}} = 0.800 \text{ mol C}$$

$$38.4 \text{ g O} \times \frac{\text{mol O}}{15.9994 \text{ g O}} = 2.40 \text{ mol O}$$

$$\frac{0.795 \text{ mol Zn}}{0.795} = 1.00 \text{ mol Zn};$$

$$\frac{0.800 \text{ mol C}}{0.795} = 1.01 \text{ mol C};$$

$$\frac{2.40 \text{ mol O}}{0.795} = 3.02 \text{ mol O};$$

$$ZnCO_3$$

9.21 $11.25 \text{ g C} \times \dfrac{\text{mol C}}{12.011 \text{ g C}} = 0.9366 \text{ mol C};$ $2.52 \text{ g H} \times \dfrac{\text{mol H}}{1.0079 \text{ g H}} = 2.50 \text{ mol H};$

$$\frac{0.9366 \text{ mol C}}{0.9366} = 1.000 \text{ mol C};   \frac{2.50 \text{ mol H}}{0.9366} = 2.67 \text{ mol H};$$

$$3 \times 1.000 \text{ mol C} = 3.000 \text{ mol C};   3 \times 2.67 \text{ mol H} = 8.01 \text{ mol H};$$

$$C_3H_8$$

9.22 $72.36 \text{ g Fe} \times \dfrac{\text{mol Fe}}{55.847 \text{ g Fe}} = 1.296 \text{ mol Fe};$ $27.64 \text{ g O} \times \dfrac{\text{mol O}}{15.9994 \text{ g O}} = 1.728 \text{ mol O};$

$$\frac{1.296 \text{ mol Fe}}{1.296} = 1.000 \text{ mol Fe};   \frac{1.728 \text{ mol O}}{1.296} = 1.333 \text{ mol O};$$

$$3 \times 1.000 \text{ mol Fe} = 3.000 \text{ mol Fe};   3 \times 1.333 \text{ mol O} = 3.999 \text{ mol O};   Fe_3O_4$$

9.23
$$\frac{60}{31} \approx 2   \text{The molecular formula is } C_2H_6O_2.$$

The molecular weight $= (2 \times 12.011) + (6 \times 1.0079) + (2 \times 15.9994) = 62.068.$

9.24
$$\frac{200}{102} \approx 2   \text{The molecular formula is } C_8H_6O_2Cl_2.$$

The molecular weight $= (8 \times 12.011) + (6 \times 1.0079) + (2 \times 15.9994) + (2 \times 35.4527)$

$$= 205.040$$

9.25 $37.6 \text{ g C} \times \dfrac{\text{mol C}}{12.011 \text{ g C}} = 3.13 \text{ mol C};$ $4.20 \text{ g H} \times \dfrac{\text{mol H}}{1.0079 \text{ g H}} = 4.17 \text{ mol H};$

$$58.37 \text{ g O} \times \dfrac{\text{mol O}}{15.9994 \text{ g O};} = 3.648 \text{ mol O};$$

$$\dfrac{3.13 \text{ mol C}}{3.13} = 1.00 \text{ mol C};\qquad \dfrac{4.17 \text{ mol H}}{3.13} = 1.33 \text{ mol H};$$

$$\dfrac{3.648 \text{ mol O}}{3.13} = 1.17 \text{ mol O};$$

$6 \times 1.00 \text{ mol C} = 6.00 \text{ mol C};$ $6 \times 1.33 \text{ mol H} = 7.98 \text{ mol H};$

$6 \times 1.17 \text{ mol O} = 7.02 \text{ mol O};$ The empirical formula is $C_6H_8O_7$.

The empirical weight $= (6 \times 12.011) + (8 \times 1.0079) + (7 \times 15.994) =$

$$192.125;\qquad \dfrac{190}{192} \approx 1.$$

The molecular formula is $C_6H_8O_7$.
The molecular weight equals the empirical weight.

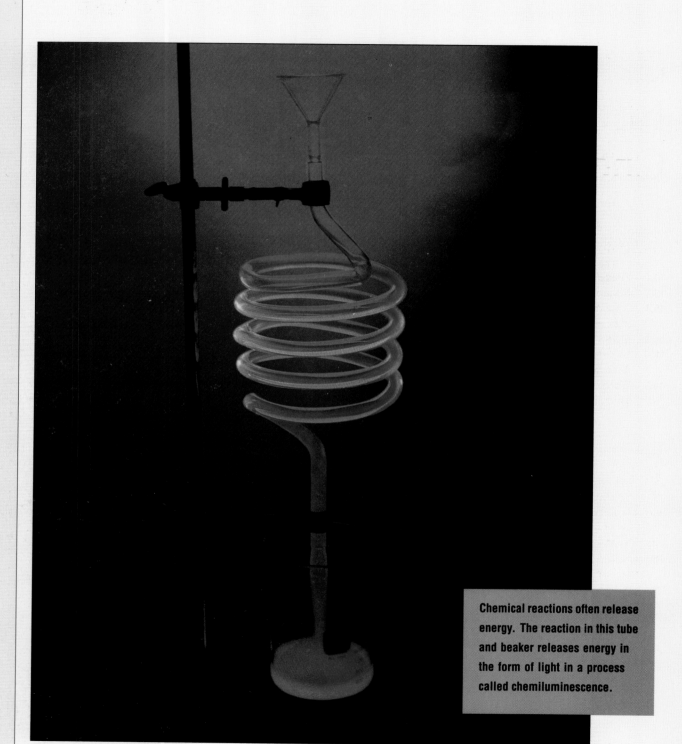

Chemical reactions often release energy. The reaction in this tube and beaker releases energy in the form of light in a process called chemiluminescence.

CHAPTER 10

Calculations Using Information from Chemical Equations

Suppose that you have a technical job with a company that plans to manufacture a small welding set. The design is based on having one kilogram of acetylene, C_2H_2, available for fuel. You are asked to calculate how much oxygen should be included in the welding set to react with the acetylene. You are told that the chemical reaction of acetylene and oxygen produces carbon dioxide and water. Now what do you do? After studying Chapter 10 you will be able to perform chemical calculations that answer questions like this.

First, we learn that chemical reactions can be described clearly and economically by using what is called a chemical equation. We learn that chemical equations must be balanced and how to balance them.

With a full understanding of how chemists interpret chemical equations in terms of moles, rather than individual atoms and molecules, we will be ready to tackle chemical calculations using chemical equations.

10.1 What Is a Chemical Equation?

After studying this section, you should be able to:

- Read and write chemical equations.

When new substances are formed as other substances disappear, a chemical reaction is taking place. It is possible to describe a chemical reaction by writing a simple statement in words, for example, "hydrogen and oxygen combine chemically to form water." Such statements, however, are clumsy to write and leave out much information. A **chemical equation** does the job more completely and simply. The starting substances are called *reactants*. They are usually written on the left side of the equation. The substances formed, the *products*, are usually written on the right side. The reactants are followed by an arrow pointing toward the products ($\longrightarrow$). The reactants are separated from one another, as are the products, by plus signs ($+$).

$$\underset{\text{reactants}}{\text{hydrogen} + \text{oxygen}} \qquad \longrightarrow \qquad \underset{\text{products}}{\text{water}}$$

To simplify the chemical equation further and to give yet more information, formulas are used to represent the reactants and products. Because hydrogen and oxygen do not occur naturally as single atoms but instead as molecules, each containing two atoms, we must use the appropriate formulas, H_2 and O_2. Similarly, we also use the formula for the product, water, H_2O.

$$H_2 + O_2 \longrightarrow H_2O$$

To make certain kinds of calculations or to predict how some kinds of reactions will occur, we need to know the states of the reactants and products. We use the letters (g) for gas, (s) for solid, (l) for liquid, and (aq) for substances dissolved in water. When needed, these abbreviations are written immediately to the right of each formula in the chemical equation.

$$H_2(g) + O_2(g) \longrightarrow H_2O(l)$$

The equations in many of the exercises in this chapter specify the states of the reactants and products.

All chemical equations must obey the law of conservation of matter.

Something important is still missing in our chemical equation. Recall the law of conservation of mass, which says that matter can neither be created nor be destroyed. But according to the equation above, two atoms of hydrogen have combined with two atoms of oxygen to produce a single molecule of water that contains only one oxygen atom. Somewhere we seem to have lost an oxygen atom. According to the law of conservation of mass, this is impossible. What do we do? The solution to this problem is known as *balancing* the chemical equation, a process described in the next section.

10.2 Balancing Chemical Equations

After studying this section, you should be able to:

- Balance chemical equations.

Consider the chemical equation developed in the preceding section. Two atoms of hydrogen have combined with two atoms of oxygen to produce a molecule of water.

$$H_2 + O_2 \longrightarrow H_2O$$

A water molecule, however, contains only one oxygen atom, and we started with two. To write the chemical equation in such a way that the number of each kind of atom will be the same on both sides, we *balance* it. To begin balancing the equation, we change the number of water molecules. We do this by writing a 2, called a **coefficient,** to the left of the symbol for water. The 2 in front of the H_2O then indicates that two water molecules, $H_2O + H_2O$ containing a total of 4 H atoms and 2 O atoms, have been formed.

$$H_2 + O_2 \longrightarrow 2\ H_2O$$

Now, because we have four hydrogen atoms on the right side of the arrow, we must have four hydrogen atoms on the left side. We fix that by writing a 2 in front of H_2 on the left side.

$$2\ H_2 + O_2 \longrightarrow 2\ H_2O$$

Because there are four H atoms and two O atoms on each side (Figure 10.1), this is now a **balanced equation.** No atoms have been created or destroyed in the process we describe.

It is essential to realize that when we balance chemical equations, we must not change the subscripts of the atoms in any of the reactants or products. To do so would turn them into other substances. We balance equations by using only whole number coefficients in front of the symbols for the elements or compounds.

Except for certain kinds of equations, for which specific balancing procedures have been developed, equations usually can be balanced best by inspection and trial and error. Inspection means that you study the equation to find out which substances are in excess and which will have to be increased. Trial and error means that you experiment with some coefficients and see what the results are. If your first attempt doesn't work, try again. It becomes easier with practice.

If no coefficient is written in front of a symbol, the coefficient is assumed to be 1.

A balanced equation has equal numbers of each kind of atom on both sides of the arrow.

When balancing chemical equations, **NEVER** *change the subscripts of chemical formulas.*

For now, balance chemical equations using inspection and trial and error. In Chapter 16, another method is introduced for balancing more complex equations.

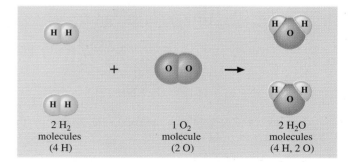

2 H_2 molecules (4 H)	1 O_2 molecule (2 O)	2 H_2O molecules (4 H, 2 O)

Figure 10.1   The reaction of hydrogen and oxygen molecules to produce water. The numbers of hydrogen and oxygen atoms on the left side of the arrow equal the numbers on the right side.

EXAMPLE 10.1

When melted, the compound NaCl can be decomposed into its elements by electricity. The metallic sodium appears in liquid form. Write a balanced equation for this process.

Strategy and Solution:

Step 1. Write the formulas for the reactant and the products in equation form. Remember that chlorine is a diatomic gas, as shown in Table 2.9.

$$NaCl(l) \longrightarrow Na(l) + Cl_2(g)$$

Step 2. Notice that there are two chlorine atoms on the product side; therefore there must be two on the reactant side. If we place the coefficient 2 in front of the NaCl, we get two each of Na and Cl.

$$2\ NaCl(l) \longrightarrow Na(l) + Cl_2(g)$$

Step 3. But the equation is still not balanced. There are two Na's on the left but only one on the right. We fix this by placing a 2 in front of the Na on the product side.

$$2\ NaCl(l) \longrightarrow 2\ Na(l) + Cl_2(g)$$

Step 4. Now we check the entire equation, element by element. Left, two Na's; right, two Na's. OK. Left, two Cl's; right, two Cl's. OK. The equation is now balanced.

EXAMPLE 10.2

Write a balanced equation for the production of ammonia from its elements.

Strategy and Solution: Ammonia is one of the compounds with which you should now be familiar. It is a gas with the formula NH_3. Nitrogen and hydrogen are diatomic gases, N_2 and H_2.

Step 1. Write the reactants and products in a chemical equation.

$$N_2(g) + H_2(g) \longrightarrow NH_3(g)$$

Step 2. Notice that there are two N's on the left side. Let's try putting a 2 in front of NH_3.

$$N_2(g) + H_2(g) \longrightarrow 2\ NH_3(g)$$

Step 3. That gives us six (2×3) H's on the right side, so we must put six H's on the left. We place a 3 in front of H_2.

$$N_2(g) + 3\ H_2(g) \longrightarrow 2\ NH_3(g)$$

Step 4. Checking, we find two N's and six H's on each side.

$$N_2(g) + 3\ H_2(g) \longrightarrow 2\ NH_3(g)$$

EXAMPLE 10.3

Magnesium reacts readily with oxygen, giving solid magnesium oxide, MgO. Write a balanced chemical equation for the reaction.

Strategy and Solution:

Step 1.
$$Mg(s) + O_2(g) \longrightarrow MgO(s)$$

Step 2. There are two O's on the left side. We place a 2 in front of MgO.

$$Mg(s) + O_2(g) \longrightarrow 2\ MgO(s)$$

Step 3. Now balance the magnesium. There are two Mg's on the right. We place a 2 in front of the Mg on the left.

$$2\ Mg(s) + O_2(g) \longrightarrow 2\ MgO(s)$$

Step 4. Now we check. Two Mg's and two O's on each side.

Magnesium reacts brilliantly with oxygen to produce MgO.

EXAMPLE 10.4

Combustion, or burning in oxygen, is a familiar process that we use to heat houses, run cars, produce electricity, and perform many other useful tasks. If a substance containing only C and H, or perhaps C, H, and O, is burned, the products are CO_2 and H_2O, unless the supply of oxygen is restricted. Write a balanced chemical equation for the complete combustion of liquid heptane, C_7H_{16}.

All combustion is chemical reaction. A campfire, for instance, is a series of chemical reactions.

Strategy and Solution:

Step 1. When the words "complete combustion" are used, assume that the products are CO_2 and H_2O. The H_2O usually appears as a gas, steam.

$$C_7H_{16}(l) + O_2(g) \longrightarrow CO_2(g) + H_2O(g)$$

Step 2. Begin with the carbon. Place a 7 in front of CO_2.

$$C_7H_{16}(l) + O_2(g) \longrightarrow 7\ CO_2(g) + H_2O(g)$$

Step 3. Now balance the hydrogen. Placing 8 before H_2O gives the necessary 16 H atoms on the right.

$$C_7H_{16}(l) + O_2(g) \longrightarrow 7\ CO_2(g) + 8\ H_2O(g)$$

Step 4. Balance the oxygen. The total number of O atoms on the right is

$$(7 \times 2) + (8 \times 1) = 22$$

Because 22 O atoms can be obtained from 11 O_2 molecules, write 11 in front of the O_2.

Again, always remember to balance chemical equations by changing coefficients, never subscripts.

$$C_7H_{16}(l) + 11\ O_2(g) \longrightarrow 7\ CO_2(g) + 8\ H_2O(g)$$

Step 5. Check the equation: Seven carbons on each side, 16 hydrogens on each side, and 22 oxygens on each side.

EXAMPLE 10.5

Gaseous methane, CH_4, is burned in a restricted oxygen supply. The products are CO and H_2O. Write the balanced equation.

Strategy and Solution:

Step 1. $CH_4(g) + O_2(g) \longrightarrow CO(g) + H_2O(g)$

Step 2.   First, balance carbon and hydrogen. There is one carbon on each side, so we don't need coefficients for the C's. There are four hydrogens on the left side. We place a 2 in front of H_2O.

$$CH_4(g) + O_2(g) \longrightarrow CO(g) + 2\ H_2O(g)$$

Step 3.   Now, oxygen:

$$CH_4(g) + O_2(g) \longrightarrow CO(g) + 2\ H_2O(g)$$

Three oxygens on the right side, but only even numbers on the left side. Oops! What to do? Answer: Multiply everything except oxygen by two, which makes all the numbers even. Then go on with oxygen.

$$2\ CH_4(g) + O_2(g) \longrightarrow 2\ CO(g) + 4\ H_2O(g)$$

Now we have six oxygen atoms on the right. We put a coefficient of 3 in front of O_2, and the equation balances.

$$2\ CH_4(g) + 3\ O_2(g) \longrightarrow 2\ CO(g) + 4\ H_2O(g)$$

Step 4.   Check: two carbons, eight hydrogens, and six oxygens on each side.

EXAMPLE 10.6

Metallic aluminum reacts with oxygen to give solid aluminum oxide, Al_2O_3. Write a balanced equation for the reaction.

Strategy and Solution:

Step 1. $Al(s) + O_2(g) \longrightarrow Al_2O_3(s)$

Step 2.   This equation is similar to that in Example 10.5. There are two Al's on the right. But multiplying Al on the left by two is not enough, because it leaves an odd number of oxygens on the right. Since there must be an even number of O's on the right, try putting a 2 in front of Al_2O_3.

$$Al(s) + O_2(g) \longrightarrow 2\ Al_2O_3(s)$$

Step 3.   Now we can make the left side balance by putting a 3 in front of O_2 and a 4 in front of Al. A final check shows us that the equation is balanced.

$$4\ Al(s) + 3\ O_2(g) \longrightarrow 2\ Al_2O_3(s)$$

Often, chemical equations are written without the labels (g), (l), (s), that designate the states of reactants and products. Now that we have had practice in using the labels, we will begin to use them only when their information is helpful in interpreting the equation.

EXAMPLE 10.7

Balance $H_2SO_4 + AlCl_3 \longrightarrow Al_2(SO_4)_3 + HCl$.

Strategy and Solution:

Step 1. To begin, notice that S and O appear as SO_4 on both sides of the equation. Because they do, we will treat SO_4 as we would a single symbol, keeping it always together. The equation then becomes easier to balance. We first arrange to get enough SO_4 and Al on the left side to balance the right side. We place a 3 in front of H_2SO_4 and a 2 in front of $AlCl_3$.

$$3\ H_2SO_4 + 2\ AlCl_3 \longrightarrow Al_2(SO_4)_3 + HCl$$

Step 2. That also gives six H's and six Cl's on the left. We balance those by putting a 6 in front of HCl.

$$3\ H_2SO_4 + 2\ AlCl_3 \longrightarrow Al_2(SO_4)_3 + 6\ HCl$$

Step 3. We then check to be sure everything is balanced.

Exercise 10.1

Balance $NaCl + NaBrO \longrightarrow NaClO_3 + NaBr$.

Exercise 10.2

Balance $Zn(s) + NH_4NO_3(s) \longrightarrow N_2(g) + ZnO(s) + H_2O(g)$

Given the right conditions, finely divided Zn metal will react vigorously with ammonium nitrate, NH_4NO_3.

10.3 Reading Chemical Equations in Units of Moles

Here is the equation for the formation of water from its elements. Let's wring all the information we can from it.

$$2\ H_2 + O_2 \longrightarrow 2\ H_2O$$

Two molecules of hydrogen, each containing two hydrogen atoms, combine with one molecule of oxygen, which contains two oxygen atoms, to produce two molecules of water, each of which contains two hydrogen atoms and one oxygen atom.

Chemists in laboratories work with extremely large numbers of molecules, not just a few. Consequently, interpreting chemical equations in terms of individual atoms and molecules is ordinarily not very useful. Recall that chemists use the mole as a unit for handling very large numbers of molecules (refresh your memory by rereading the discussion of the mole in Chapter 8).

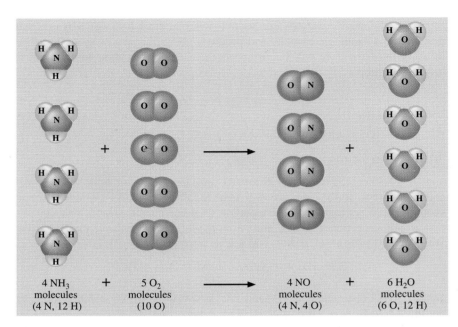

4 NH₃ molecules (4 N, 12 H) + 5 O₂ molecules (10 O) → 4 NO molecules (4 N, 4 O) + 6 H₂O molecules (6 O, 12 H)

Figure 10.2 Four molecules of NH_3 react with five molecules of O_2 to form four molecules of NO and six molecules of H_2O.

Chemical equations can be read in terms of moles equally as well as in terms of individual molecules. Let's read our H_2O equation that way.

$$2 \text{ mol } H_2 + 1 \text{ mol } O_2 \longrightarrow 2 \text{ mol } H_2O$$

Two moles of hydrogen molecules combine with one mole of oxygen molecules to produce two moles of water molecules.

For another example, take the reaction by which ammonia, NH_3, is made into nitrogen monoxide, NO, in a well-known process called the Ostwald oxidation. This reaction is an essential part of the preparation of the industrially important compound nitric acid, HNO_3.

$$4 NH_3 + 5 O_2 \longrightarrow 4 NO + 6 H_2O$$

Interpreted in one way, the equation says that four molecules of NH_3 react with five molecules of O_2 to produce four molecules of NO and six molecules of H_2O. Figure 10.2 illustrates the reaction.

The same chemical equation can be read equally well for sets of dozens of molecules, as in Figure 10.3.

And it can also be read in terms of moles (Figure 10.4).

We have learned that many compounds do not form individual molecules (Section 2.10). In such cases, the concept of the mole—even though there are no individual molecules to be counted—still works quite well. The following equation can be read in mole terms, although NaCl is not molecular.

$$2 Na + Cl_2 \longrightarrow 2 NaCl$$

Two moles of sodium react with one mole of chlorine to produce two moles of sodium chloride.

In the United States, about seven billion kilograms of nitric acid are made every year, much by means of this reaction.

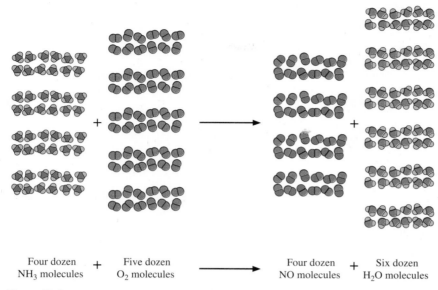

Four dozen + Five dozen   Four dozen + Six dozen
NH₃ molecules   O₂ molecules   NO molecules   H₂O molecules

Figure 10.3 Four dozen molecules of NH_3 react with five dozen molecules of O_2 to form four dozen molecules of NO and six dozen molecules of H_2O.

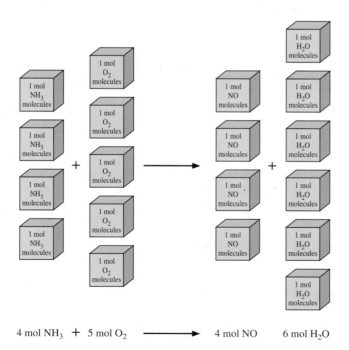

4 mol NH_3 + 5 mol O_2 ⟶   4 mol NO   6 mol H_2O

Figure 10.4 Four moles of NH_3 react with five moles of O_2 to form four moles of NO and six moles of H_2O.

10.4 Mole Ratios

After studying this section, you should be able to:

- Write mole ratios using the coefficients in chemical equations.

We will study this reaction and the man for whom it is named in Chapter 16.

We look again at the first step in the ammonia synthesis, better known as the Ostwald oxidation.

$$4\ NH_3 + 5\ O_2 \longrightarrow 4\ NO + 6\ H_2O$$

Let's read it just once more in terms of moles. After this, we will imagine the mole label to be there when we need it.

$$4\ mol\ NH_3 + 5\ mol\ O_2 \longrightarrow 4\ mol\ NO + 6\ mol\ H_2O$$

The information in this equation, or in any chemical equation, can be written in a series of mathematical ratios. The relation between the number of moles of NH_3 reacting and the number of moles of NO produced can be written in ratio form as

$$\frac{4\ mol\ NH_3}{4\ mol\ NO}$$

which says that there are four moles of NH_3 reacting for every four moles of NO produced. Any ratio between the numbers of moles of substances in an equation is called a **mole ratio.** Several other mole ratios can be written for the Ostwald reaction:

$$\frac{4\ mol\ NH_3}{5\ mol\ O_2} \qquad \frac{4\ mol\ NH_3}{6\ mol\ H_2O} \qquad \frac{5\ mol\ O_2}{6\ mol\ H_2O}$$

$$\frac{6\ mol\ H_2O}{4\ mol\ NO} \qquad \frac{5\ mol\ O_2}{4\ mol\ NO}$$

Any of these six mole ratios can be inverted. For example, the ratio of the numbers of moles of O_2 and NO can be written as

$$\frac{4\ mol\ NO}{5\ mol\ O_2}$$

If we count the inverted forms, there are 12 possible mole ratios that can be written from this one chemical equation, and any of the 12 can be useful in a calculation. Depending on the number of reactants and products, different chemical equations can have more or fewer mole ratios. As we will see in the next section, a mole ratio from a chemical equation can be used as a conversion factor in calculations.

Mole ratios can be used as conversion factors in chemical calculations.

10.5 Calculations Using Mole Ratios

After studying this section, you should be able to:

- Calculate numbers of moles of products, given a chemical equation and the numbers of moles of reactants or of other products.
- Calculate numbers of moles of reactants, given a chemical equation and the numbers of moles of products or of other reactants.

Let's go back to your technical job with a company that plans to manufacture a small welding set. You are asked to calculate how much oxygen will react with one kilogram of acetylene, C_2H_2. You are told that the products are carbon dioxide and water. In chemistry we often need to find the amount of reactant needed to react with a given amount of another reactant. We might also be asked to calculate how much product is produced by a chemical reaction when we know the amount of reactant or reactants with which we started. At other times we may need to determine how much reactant is required to produce a certain amount of product. Although problems of this kind can appear at first to be quite complex, all of them have a single central step in common: the interpretation of a balanced chemical equation in terms of moles, rather than in terms of individual molecules. Calculations involving the relation between amounts of reactants and products are called **stoichiometric calculations.**

To learn how mole ratios from chemical equations can be used in calculations, we can look at a simple reaction.

"Stoichiometric" is a big, impressive term having to do with the exact quantities of substances involved in chemical reactions.

EXAMPLE 10.8

How many moles of N_2O_4 would result from the reaction of 3.27 moles of NO_2? The equation for the reaction is

$$2\ NO_2 \longrightarrow N_2O_4$$

Solution Map: $\boxed{\text{moles of } NO_2}$ $\longrightarrow$ $\boxed{\text{moles of } N_2O_4}$

Strategy and Solution:

Step 1. Be sure that you have a balanced chemical equation for the process involved. The equation is balanced, so we go on.

Step 2. Write the mole ratio for the substances in which you are interested. In this case, there is only one reactant and one product. Write the ratio with N_2O_4 on top; the units in the answer will then be correct.

$$\frac{1\ \text{mol } N_2O_4}{2\ \text{mol } NO_2}$$

Step 3. Multiply the number of moles of reactant, NO_2, by your mole ratio.

$$3.72\ \text{mol } NO_2 \times \frac{1\ \text{mol } N_2O_4}{2\ \text{mol } NO_2} = 1.86\ \text{mol } N_2O_4$$

The answer is in the correct units and is reasonable. If we had seen 186 mol of N_2O_4 from only about 3 mol of NO_2, for example, we would have known something was wrong. The answer has three significant figures, which is the the proper number. (Since there are three significant figures in the value of the number of moles of NO_2 given, and the coefficients in the chemical equation can be taken as exact, there should be three significant figures in the answer.)

In many cities, NO_2 and N_2O_4 contribute substantially to air pollution. The NO_2 is responsible for the reddish-brown color of the haze. We will study the reaction of NO_2 and N_2O_4 in more detail in Chapter 15.

When you perform stoichiometric calculations (as well as all other chemical calculations), be sure to check your answers for significant figures, units, and reasonableness.

Exercise 10.3

The water–gas reaction, the balanced equation for which appears below, is an indispensable industrial process that reduces CO and produces H_2, which is used in the manufacture of chemicals such as ammonia. In this reaction, how many moles of H_2 can be formed from 1.60 moles of CO?

$$H_2O + CO \longrightarrow H_2 + CO_2$$

Exercise 10.4

Using the following reaction, how many moles of H_2 can be formed from 5.45 moles of NaOH?

$$2\ NaOH + Zn \longrightarrow Na_2ZnO_2 + H_2$$

EXAMPLE 10.9

How many moles of O_2 are required to prepare 9.11 moles of NO in the Ostwald oxidation?

Solution Map: | moles of NO | $\longrightarrow$ | moles of O_2 |

Strategy and Solution:

Step 1.   Be sure that you have a balanced chemical equation. (Use the balanced equation given in Section 10.4.)

$$4\ NH_3 + 5\ O_2 \longrightarrow 4\ NO + 6\ H_2O$$

Step 2.   From the mole ratios given by the chemical equation, we select the one that relates the number of moles of O_2 and of NO. To make the answer have the correct units, the ratio is written so that the number of moles of O_2 appears in the numerator.

$$\frac{5\ mol\ O_2}{4\ mol\ NO}$$

Step 3.   Set up and solve.

$$9.11\ mol\ NO \times \frac{5\ mol\ O_2}{4\ mol\ NO} = 11.4\ mol\ O_2$$

Exercise 10.5

How many moles of Na_2O are required to produce 8.72 moles of NaCl using the following reaction?

$$2\ HCl + Na_2O \longrightarrow 2\ NaCl + H_2O$$

Exercise 10.6

In the equation shown in Exercise 10.5, how many moles of HCl would be required to form 3.40 moles of H_2O?

EXAMPLE 10.10

The Haber process for making ammonia, NH_3, is said to be the most important industrial chemical reaction. There are about 350 major ammonia plants in the

Profiles in Science

Fritz Haber (1868–1934)

Born in Breslau, Germany, Fritz Haber became interested in technical matters at an early age. While at school, he specialized in the chemistry of carbon compounds and planned to become an organic chemist. Somewhat later, however, his interests turned to physical chemistry. At age 26, he joined the faculty at Karlsruhe, where his great discoveries were made.

In the years before World War I, Germany suffered from an undersupply of nitrogen compounds, a raw material needed for fertilizers and for manufacturing munitions. In such a situation, a country would not be able to conduct a major war. However, in 1913, after more than three years of work during which he endured constant discouragement from colleagues who thought he was pursuing an impossible goal, Haber announced a way to make hydrogen and nitrogen react efficiently under the urging of a catalyst. Immediately Germany was assured of an ample supply of ammonia with which to make other nitrogen compounds. In one stroke, Haber had made it possible for his country to wage war. World War I started the following spring.

During the war, Haber contributed his talents to the initiation of gas warfare, begun by Germany at Ypres, France, in 1915. Later, after Germany lost the war in 1918, Haber helped rebuild his economically ravaged country.

Even though his invention of the Haber process accelerated World War I, it was nonetheless a major contribution for good; for example, agriculture could never have reached today's production levels without it. For his discoveries, Haber received the Nobel Prize in 1918.

Haber was Jewish. In 1933, he was forced by the Nazis to leave his prestigious research post, abandon his students, and depart Germany. He died, a broken man, in 1934 in Switzerland.

---

world today. The process begins with H_2 and N_2.

$$N_2 + 3\ H_2 \longrightarrow 2\ NH_3$$

According to this equation, how many moles of N_2 are required to react completely with 2.33 moles of H_2?

Solution Map: $\boxed{\text{moles of } H_2} \longrightarrow \boxed{\text{moles of } N_2}$

Strategy and Solution:

Step 1. From the mole ratios given by the chemical equation, we select the one that relates the number of moles of N_2 to the number of moles of H_2.

$$\frac{1 \text{ mol } N_2}{3 \text{ mol } H_2}$$

Step 2. Set up and solve.

$$2.33 \text{ mol } H_2 \times \frac{1 \text{ mol } N_2}{3 \text{ mol } H_2} = 0.777 \text{ mol } N_2$$

> **Exercise 10.7**
>
> How many moles of C_2H_6 are required to react completely with 3.42 moles of O_2, according to the following chemical equation?
>
> $$2\ C_2H_6 + 7\ O_2 \longrightarrow 4\ CO_2 + 6\ H_2O$$

10.6 Stoichiometric Calculations Using Mass Data

After studying this section, you should be able to:

- Calculate numbers of grams of products, given a chemical equation and the amounts of reactants or of other products.
- Calculate numbers of grams of reactants, given a chemical equation and the amounts of products or of other reactants.

Although you will occasionally encounter the kind of problem that we saw in the preceding section, more common are problems in which the data come from laboratory measurements given in such units as grams or liters. After all, chemistry is a laboratory science, and problems are derived from measurements made in the laboratory. The general method of working such problems is to convert the laboratory units into moles, make the mole ratio conversion, and then convert back into laboratory units.

The following example shows you how to find the number of moles of a product made, starting with some number of grams of reactant.

EXAMPLE 10.11

The reaction of methane, CH_4, and oxygen is used in industry, as well as for heating and cooking in homes. The products of the reaction are water and carbon dioxide.

$$CH_4 + O_2 \longrightarrow CO_2 + H_2O$$

How many moles of water can be made from 6.8 grams of CH_4 and excess O_2?

 Solution Map: $\boxed{\text{grams of } CH_4} \longrightarrow \boxed{\text{moles of } CH_4} \longrightarrow \boxed{\text{moles of } H_2O}$

Strategy and Solution:

Step 1. First, as always, be sure that the chemical equation is balanced. Since this one is not, we must balance it.

$$CH_4 + 2\ O_2 \longrightarrow CO_2 + 2\ H_2O$$

We have been told that O_2 is present in excess and that there is only a given amount of CH_4. We base our calculation on the amount of CH_4.

Step 2. Calculate the number of moles of CH_4 present.

$$6.8 \text{ g } CH_4 \times \frac{\text{mol } CH_4}{16.043 \text{ g } CH_4} = 0.424 \text{ mol } CH_4$$

We will carry an extra digit to avoid round-off error.

Step 3. Do the mole ratio calculation using the proper mole ratio obtained from the chemical equation.

$$0.424 \text{ mol CH}_4 \times \frac{2 \text{ mol H}_2\text{O}}{1 \text{ mol CH}_4} = 0.85 \text{ mol H}_2\text{O}$$

After you have had some practice, you may want to combine steps 2 and 3 in a single calculation.

$$6.8 \text{ g CH}_4 \times \frac{\text{mol CH}_4}{16.043 \text{ g CH}_4} \times \frac{2 \text{ mol H}_2\text{O}}{1 \text{ mol CH}_4} = 0.85 \text{ mol H}_2\text{O}$$

From 6.8 grams of CH_4, we can obtain 0.85 moles of H_2O by this combustion reaction.

Exercise 10.8

How many moles of O_2 can be formed from 16.4 grams of $KClO_3$?

$$KClO_3 \longrightarrow KCl + O_2$$

Exercise 10.9

If 24.3 grams of HgO reacted to give Hg and O_2, how many moles of Hg resulted?

$$2 \text{ HgO} \longrightarrow 2 \text{ Hg} + O_2$$

Another similar kind of problem requires that you calculate the number of grams of a reactant necessary to make a required amount of a product.

EXAMPLE 10.12

You wish to prepare 6.6 moles of $Ca(OH)_2$, slaked lime, by mixing CaO with water. If enough H_2O is available to run the reaction, how many grams of CaO will be needed to make the $Ca(OH)_2$?

Solution Map: $\boxed{\text{moles of Ca(OH)}_2} \longrightarrow \boxed{\text{moles of CaO}} \longrightarrow \boxed{\text{grams of CaO}}$

Strategy and Solution:

Step 1. Write a balanced chemical equation.

$$CaO + H_2O \longrightarrow Ca(OH)_2$$

Step 2. Using the mole ratio method, find out how many moles of CaO will be needed.

$$6.6 \text{ mol Ca(OH)}_2 \times \frac{1 \text{ mol CaO}}{1 \text{ mol Ca(OH)}_2} = 6.6 \text{ mol CaO}$$

Science in Action

Although an ordinary person rarely sees any of it, lime, in its various forms, is the fourth most widely used chemical in the United States. "Quicklime," CaO, produces "slaked lime," Ca(OH)$_2$, when mixed with water.

$$CaO + H_2O \longrightarrow Ca(OH)_2$$

CaO is presently made, as it has been for more than a hundred years, by roasting calcium carbonate, CaCO$_3$. A substance containing CaCO$_3$, such as seashells or limestone quarried from the earth, is fed into the top of a kiln. Coal is used to provide the heat, the flames from the coal being passed through the limestone. The reaction is

$$CaCO_3 \longrightarrow CaO + CO_2$$

The process uses large amounts of energy; about a third of a pound of coal is needed for every pound of CaO produced. Nonetheless, about 15 *billion* kilograms of lime are produced in the United States every year. About half of the CaO produced is used in the steel industry for removing impurities from iron. Much of the rest goes into slaked lime, which has many uses such as those in agriculture, in papermaking, as a starting material in making bleach, and for treating city water supplies.

Lime

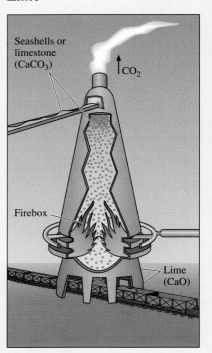

A lime kiln showing how calcium carbonate is roasted.

Step 3. Convert the number of moles of CaO to grams of CaO.

$$6.6 \text{ mol CaO} \times \frac{56.077 \text{ g CaO}}{\text{mol CaO}} = 3.7 \times 10^2 \text{ g CaO}$$

It will require 3.7×10^2 grams of CaO to yield 6.6 moles of Ca(OH)$_2$.

Exercise 10.10

How many grams of Zn are required to produce 10.3 moles of Zn(NO$_3$)$_2$?

$$Zn + 2 \text{ HNO}_3 \longrightarrow Zn(NO_3)_2 + H_2$$

Exercise 10.11

How many grams of HgO are needed to produce 8.50 moles of pure mercury, Hg? (Write the chemical equation first. Check it against Exercise 10.9.)

You will often see problems in which both reactants and products are given in grams. To work such a problem requires that the number of grams of the reactant be converted into numbers of moles of reactant, the mole ratio conversion be made, then the numbers of moles of product be converted into grams of product. Several examples demonstrate this procedure.

EXAMPLE 10.13

Although it is too expensive to produce quicklime this way, metallic calcium, Ca, (like many other metals) can be burned in oxygen to yield its oxide.

$$2 \text{ Ca} + O_2 \longrightarrow 2 \text{ CaO}$$

How many grams of CaO can be made from 31.02 grams of Ca and as much O_2 as is needed for the reaction?

Solution Map:

| grams of Ca | ⟶ | moles of Ca | ⟶ | moles of CaO | ⟶ |

| grams of CaO |

Since weighing instruments cannot directly measure numbers of moles, many calculations must be started and finished by using mass units.

Strategy and Solution:

Step 1. Make sure that the chemical equation is balanced. It is.

Step 2. Use the molar mass to find the number of moles of Ca metal that we have.

$$31.02 \text{ g Ca} \times \frac{\text{mol Ca}}{40.078 \text{ g Ca}} = 0.7740 \text{ mol Ca}$$

Step 3. Use the appropriate mole ratio from the balanced equation to find the number of moles of CaO produced.

$$0.7740 \text{ mol Ca} \times \frac{2 \text{ mol CaO}}{2 \text{ mol Ca}} = 0.7740 \text{ mol CaO}$$

Step 4. Convert the number of moles of CaO into grams. Use the molar mass of CaO.

$$0.7740 \text{ mol CaO} \times \frac{56.077 \text{ g CaO}}{\text{mol CaO}} = 43.40 \text{ g CaO}$$

From 31.02 grams of Ca, we can produce 43.40 grams of CaO.

You will save time and be less likely to make errors if you do problems in a single calculation that combines the three steps.

$$31.02 \text{ g Ca} \times \frac{\text{mol Ca}}{40.078 \text{ g Ca}} \times \frac{2 \text{ mol CaO}}{2 \text{ mol Ca}} \times \frac{56.077 \text{ g CaO}}{\text{mol CaO}} = 43.40 \text{ g CaO}$$

EXAMPLE 10.14

In the human body, glucose, $C_6H_{12}O_6$, reacts with O_2 and releases the energy needed for living.

The balanced equation for the overall reaction is

$$C_6H_{12}O_6 + 6 O_2 \longrightarrow 6 CO_2 + 6 H_2O$$

Plants use the energy of the sun to create glucose and oxygen from CO_2 and H_2O. Other living things obtain energy by reacting glucose with oxygen.

How many grams of H_2O result from the reaction of 19.3 grams of glucose with excess oxygen?

Solution Map:

$$\boxed{\text{grams of } C_6H_{12}O_6} \longrightarrow \boxed{\text{moles of } C_6H_{12}O_6} \longrightarrow$$

$$\boxed{\text{moles of } H_2O} \longrightarrow \boxed{\text{grams of } H_2O}$$

Solution: We can solve the problem in a single calculation.

$$19.3 \text{ g } C_6H_{12}O_6 \times \frac{\text{mol } C_6H_{12}O_6}{180.157 \text{ g } C_6H_{12}O_6} \times \frac{6 \text{ mol } H_2O}{1 \text{ mol } C_6H_{12}O_6} \times \frac{18.0152 \text{ g } H_2O}{\text{mol } H_2O} =$$

$$11.6 \text{ g } H_2O$$

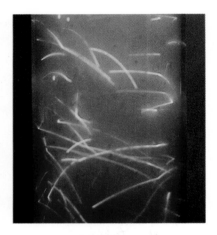

The streaks of light indicate a vigorous reaction between antimony, Sb, and chlorine, Cl_2.

Exercise 10.12

How many grams of Na_2ZnO_2 can be produced if 76.8 grams of NaOH react with an excess of zinc? Write the chemical equation for the reaction. (Check it against the one in Exercise 10.4.)

Exercise 10.13

How many grams of $SbCl_3$ will be produced if 18.6 grams of Sb react with an excess of Cl_2?

$$2 \text{ Sb} + 3 \text{ Cl}_2 \longrightarrow 2 \text{ SbCl}_3$$

When you see a problem in chemistry, especially one that looks something like those we have just seen, stop and ask yourself if the solution depends in some way on the information in a chemical formula or a chemical equation. If it does, and it often will, ask yourself how you can convert the given data into mole language, so that you can use the formula or the equation. You can ordinarily use the steps you have just learned.

Guidelines for Calculating Masses of Reactants and Products in Chemical Reactions

1. *Balance the chemical equation.*
2. *If amounts are given as masses (grams), convert the known mass of reactant or product to the equivalent number of moles.*
3. *Use the balanced chemical equation to set up the proper mole ratio.*
4. *Multiply the mole ratio(s) by the number of moles of the known substance (reactant or product) to find the number of moles in question.*
5. *Convert back from number of moles to mass.*

10.7 Percentage Yield Calculations

After studying this section, you should be able to:

• Calculate percentage yield, given a chemical equation and the amounts of reactants.

Chemical reactions are ordinarily run to yield products, and, of course, chemists are greatly interested in how much product is made. The amount of product actually obtained is known as the **actual yield.** Often the actual yield is less than the amount of product determined by calculation, the **theoretical yield.**

In industrial chemistry, expensive and complex efforts are taken to increase actual yields.

    paper & pencil

The actual yield can be less than the theoretical yield for many reasons. Going from reactants to final products often requires several steps, and during any one of these steps some of the product can be lost through handling, purification, or side reactions. (Side reactions lead to substances other than the desired product.)

The relation of the actual yield and the theoretical yield is reported as the **percentage yield.** To calculate the percentage yield, just multiply the ratio of the actual yield to the theoretical yield by 100%.

$$\frac{\text{actual yield}}{\text{theoretical yield}} \times 100\% = \text{percentage yield}$$

Here are two examples.

EXAMPLE 10.15

Ammonia, NH_3, is produced industrially by the reaction of N_2 and H_2. (See Example 10.2 if you don't remember the chemical equation.) Theoretically, 49.1 kilograms of NH_3 should have been produced during a small production run. However, only 25.9 kilograms were actually obtained. Calculate the percentage yield.

Solution: Find the ratio of the actual yield to the theoretical yield and multiply by 100%.

$$\frac{25.9 \text{ kg}}{49.1 \text{ kg}} \times 100\% = 52.7\%$$

EXAMPLE 10.16

One of the principal reactions in making iron, Fe, from its ore, Fe_2O_3, is the reaction of the ore with carbon monoxide. The carbon monoxide is obtained from coke, a form of coal.

$$Fe_2O_3 + 3 \text{ CO} \longrightarrow 2 \text{ Fe} + 3 \text{ CO}_2$$

In a laboratory experiment, 312 grams of Fe_2O_3 reacted with CO, yielding 185 grams of Fe metal. What was the percentage yield of Fe?

Strategy and Solution:

Step 1. Balance the chemical equation if it is not already balanced.

Step 2. Determine the theoretical yield.

$$312 \text{ g Fe}_2O_3 \times \frac{\text{mol Fe}_2O_3}{159.692 \text{ g Fe}_2O_3} \times \frac{2 \text{ mol Fe}}{1 \text{ mol Fe}_2O_3} \times \frac{55.847 \text{ g Fe}}{\text{mol Fe}} = 218 \text{ g Fe}$$

    theoretical yield

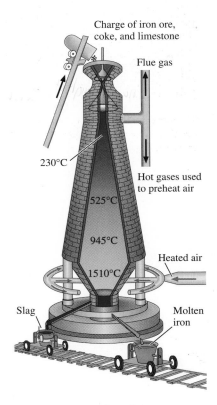

Charge of iron ore, coke, and limestone

Flue gas

230°C

Hot gases used to preheat air

525°C

945°C

Heated air

1510°C

Slag

Molten iron

Iron (Fe) is made from its ore (Fe_2O_3) in a blast furnace. Smelting of iron (turning the ore into metal) is explained in detail in Section 16.3.

Step 3.   Find the ratio of the actual yield to the theoretical yield and multiply by 100%.

$$\frac{185 \text{ g}}{218 \text{ g}} \times 100\% = 84.9\%$$

Exercise 10.14

What was the percentage yield if 5.7 grams of NaCl were obtained when 4.6 grams of HCl reacted with NaOH?

$$HCl + NaOH \longrightarrow NaCl + H_2O$$

10.8 Limiting Reactant Calculations

After studying this section, you should be able to:

- Determine which reactant is the limiting reactant in a reaction.
- Calculate the number of moles of product formed in a limited reaction, given the number of moles of the reactants.

Several kinds of chemistry problems involve the relation between amounts of reactants and products. One such problem requires that we find out in advance whether any of the reactants will be used up, stopping the reaction. The reactant that is completely consumed first is called the **limiting reactant** or **limiting reagent.** The limiting reactant runs out, and thereby limits the amount of product that can be formed.

no enough of element for more R [handwritten margin note]

A look at an everyday problem will help you understand the concept of limiting reactants. You want to determine how many ham sandwiches you can make from the amount of bread and ham you have on hand. (You know you have plenty of cheese, mustard, mayonnaise, pickles, and so on.) Each sandwich will consist of 2 slices of bread and 1 slice of ham. You have 12 slices of bread and 8 slices of ham. How many sandwiches can you make? Here's the sandwich equation.

$$2 \text{ slices bread} + 1 \text{ slice ham} \longrightarrow 1 \text{ sandwich}$$

Most likely you have already figured out that you could make six sandwiches. How did you calculate this answer? You figured that you can make six sandwiches from the amount of bread or eight sandwiches from the amount of ham. However, the smaller number of sandwiches, six, is really all you can make because after making six, you have used all the bread. In the sandwich reaction, the limiting reactant is the bread.

Limiting reactant problems in chemistry are quite similar to this example. When solving such a problem, determine which reactant will produce the least amount of product and then base your answer on the amount of product produced by that reactant.

Again, the ammonia reaction:

$$N_2 + 3 \text{ H}_2 \longrightarrow 2 \text{ NH}_3$$

According to this chemical equation, the nitrogen and hydrogen react in the mole ratio of one mole of nitrogen to three moles of hydrogen. Suppose we mix three moles of N_2 and three moles of H_2. Considering only the amount of N_2, we would say that six moles of NH_3 would be produced. However, from the three moles of H_2, the reaction can produce only two moles of NH_3. Therefore, when two moles of NH_3 have been produced, the H_2 will be consumed, and the reaction will stop. At this point, there will be two moles of N_2 left. Let's look at several examples.

EXAMPLE 10.17

How many moles of H_2O are formed in the Ostwald oxidation reaction when 4.0 moles of NH_3 and 4.0 moles of O_2 are mixed and caused to react until one of the reactants is used up?

$$4\ NH_3 + 5\ O_2 \longrightarrow 4\ NO + 6\ H_2O$$

Solution Map:

$$\boxed{\text{moles of } NH_3} \longrightarrow \boxed{\text{moles of } H_2O}$$

$$\boxed{\text{moles of } O_2} \longrightarrow \boxed{\text{moles of } H_2O}$$

Strategy and Solution: Determine which reactant will run out first (the limiting reactant) by calculating how many moles of H_2O will be formed by the given amount of each reactant. The reactant that forms the smallest number of moles of H_2O is clearly the one that will run out first and is therefore the limiting reactant.

Step 1.   Make sure your chemical equation is balanced.

Step 2.   Determine which reactant is limiting by calculating how much product will be formed by each reactant in the amount given.

$$4.0 \text{ mol } NH_3 \times \frac{6 \text{ mol } H_2O}{4 \text{ mol } NH_3} = 6.0 \text{ mol } H_2O$$

$$4.0 \text{ mol } O_2 \times \frac{6 \text{ mol } H_2O}{5 \text{ mol } O_2} = 4.8 \text{ mol } H_2O$$

There is enough NH_3 to make 6.0 moles of H_2O, but only enough O_2 to produce 4.8 moles of H_2O. When 4.8 moles of H_2O have been formed, the reaction will cease because there is no more O_2. Some NH_3 will be left over. By comparing the number of moles of product that can be made by each of the reactants, we have found that O_2 is the limiting reactant.

Step 3.   In this problem, the smaller amount of product was 4.8 moles of H_2O. That is your answer: 4.8 moles of H_2O are formed.

If the problem had required any further calculations, such as converting the moles to grams, you would have used 4.8 moles of H_2O to continue.

EXAMPLE 10.18

Given the reaction between sulfuric acid, H_2SO_4, and aluminum, Al,

$$3\ H_2SO_4 + 2\ Al \longrightarrow Al_2(SO_4)_3 + 3\ H_2$$

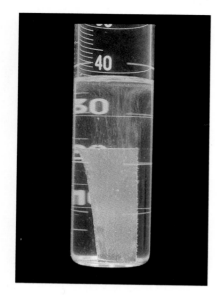

Aluminum as well as many other metals will react with H_2SO_4.

how many moles of H_2 will be produced if 1.7 moles of sulfuric acid are mixed with 2.2 moles of aluminum and allowed to react? Which is the limiting reactant?

Solution Map:

$$\boxed{\text{moles of } H_2SO_4} \longrightarrow \boxed{\text{moles of } H_2}$$

$$\boxed{\text{moles of Al}} \longrightarrow \boxed{\text{moles of } H_2}$$

Strategy and Solution:

Step 1. Make sure your chemical equation is balanced.

Step 2. Calculate separately how many moles of H_2 will be produced by 1.7 moles of H_2SO_4 and by 2.2 moles of Al.

$$1.7 \text{ mol } H_2SO_4 \times \frac{3 \text{ mol } H_2}{3 \text{ mol } H_2SO_4} = 1.7 \text{ mol } H_2$$

$$2.2 \text{ mol Al} \times \frac{3 \text{ mol } H_2}{2 \text{ mol Al}} = 3.3 \text{ mol } H_2$$

Step 3. The reaction will stop when the smaller number of moles (1.7 moles of H_2) has been produced because the H_2SO_4 will be gone. The H_2SO_4 is, therefore, the limiting reactant. Al will be left over, and 1.7 moles of H_2 will be produced.

EXAMPLE 10.19

In the preparation of nitric acid, the NO that comes from the Ostwald oxidation is reacted with oxygen to give NO_2; then the NO_2 is reacted with water to give the acid. The balanced equation for the second reaction is

$$3 \text{ } NO_2 + H_2O \longrightarrow 2 \text{ } HNO_3 + NO$$

How many moles of HNO_3 can be made from 2.2 moles of NO_2 and 0.56 moles of H_2O, if these react together until one is used up? Which is the limiting reactant?

Solution Map:

$$\boxed{\text{moles of } NO_2} \longrightarrow \boxed{\text{moles of } HNO_3}$$

$$\boxed{\text{moles of } H_2O} \longrightarrow \boxed{\text{moles of } HNO_3}$$

Strategy and Solution:

Step 1. Make sure you have a balanced equation.

Step 2. Calculate answers for the amounts of both reactants.

$$2.2 \text{ mol } NO_2 \times \frac{2 \text{ mol } HNO_3}{3 \text{ mol } NO_2} = 1.5 \text{ mol } HNO_3$$

$$0.56 \text{ mol } H_2O \times \frac{2 \text{ mol } HNO_3}{1 \text{ mol } H_2O} = 1.1 \text{ mol } HNO_3$$

Step 3. Choose the answer that is smaller. The answer is 1.1 moles of HNO_3 can be made. H_2O is the limiting reactant.

Nitric acid, HNO_3, reacts readily with all but a few metals such as gold or platinum.

Exercise 10.15

In the equation below, Ag reacts with HNO_3 to form $AgNO_3$. If 2.5 moles of Ag are mixed with 2.0 moles of HNO_3, how many moles of $AgNO_3$ can be produced?

$$3 \text{ Ag} + 4 \text{ HNO}_3 \longrightarrow 3 \text{ AgNO}_3 + \text{NO} + 2 \text{ H}_2\text{O}$$

Exercise 10.16

Copper reacts readily with concentrated HNO_3 according to the following equation.

$$\text{Cu} + 4 \text{ HNO}_3 \longrightarrow \text{Cu(NO}_3)_2 + 2 \text{ NO}_2 + 2 \text{ H}_2\text{O}$$

How many moles of NO_2 can be produced from 0.245 moles of copper and 0.320 moles of HNO_3?

A copper penny reacts readily with HNO_3, forming a green solution of copper(II) nitrate and brown nitrogen dioxide gas.

EXAMPLE 10.20

The following reaction was run in the laboratory, starting with 3.3 moles of HNO_3, 2.9 moles of As, and 1.5 moles of H_2O. How many moles of H_3AsO_4 were produced? What reactants were left over?

$$5 \text{ HNO}_3 + 3 \text{ As} + 2 \text{ H}_2\text{O} \longrightarrow 5 \text{ NO} + 3 \text{ H}_3\text{AsO}_4$$

Solution Map: moles of HNO_3 $\longrightarrow$ moles of H_3AsO_4

moles of As $\longrightarrow$ moles of H_3AsO_4

moles of H_2O $\longrightarrow$ moles of H_3AsO_4

Strategy and Solution: This problem is solved in exactly the same way as the preceding ones except that three reagents are involved instead of two.

Step 1.   Make sure you have a balanced equation.

Step 2.
$$3.3 \text{ mol HNO}_3 \times \frac{3 \text{ mol H}_3\text{AsO}_4}{5 \text{ mol HNO}_3} = 2.0 \text{ mol H}_3\text{AsO}_4$$

$$2.9 \text{ mol As} \times \frac{3 \text{ mol H}_3\text{AsO}_4}{3 \text{ mol As}} = 2.9 \text{ mol H}_3\text{ASO}_4$$

$$1.5 \text{ mol H}_2\text{O} \times \frac{3 \text{ mol H}_3\text{AsO}_4}{2 \text{ mol H}_2\text{O}} = 2.2 \text{ mol H}_3\text{AsO}_4$$

Step 3.   The smallest amount, 2.0 moles of H_3AsO_4, is produced. HNO_3 is the limiting reactant. As and H_2O are left over.

Exercise 10.17

Phosphoric acid, one of the oxyacids, is H_3PO_4. It is used in many products we see in our everyday lives—detergents, baking powder, fertilizers, fireproof textiles, soft drinks, and so on. It can be made by reacting tetraphosphorus decoxide, P_4O_{10}, with water.

$$P_4O_{10} + 6\ H_2O \longrightarrow 4\ H_3PO_4$$

How many moles of phosphoric acid can be prepared from 0.55 moles of P_4O_{10} and 1.20 moles of water?

10.9 Limiting Reactant Calculations Using Mass Data

After studying this section, you should be able to:

- Calculate the number of grams of product formed in a limited reaction, given the numbers of grams of the reactants.

Just as mole ratio problems often begin with mass units, so do problems involving a limiting reactant. Such problems require one more step than do ordinary mass–mass problems, but they are not harder to understand. A limiting reactant problem given in mass units can be worked in three steps, all of which we have seen before.

Guidelines for Limiting Reactant Calculations Using Mass Data

1. *Convert the values given in mass units to numbers of moles.*
2. *Using the methods of Section 10.8, find out which of the reactants limits the reaction, that is, runs out.* By doing this step, you automatically find out how many moles of product are produced.
3. *When you have the number of moles of product produced, convert that value back to mass units.*

EXAMPLE 10.21

In an experiment, 4.0 grams of Na metal were placed in 7.7 grams of water. If the reaction ran according to the balanced equation given below, how many grams of NaOH were produced?

$$2\ Na + 2\ H_2O \longrightarrow H_2 + 2\ NaOH$$

Solution Map:

grams of Na $\longrightarrow$ moles of Na $\longrightarrow$ moles of NaOH

grams of H_2O $\longrightarrow$ moles of H_2O $\longrightarrow$ moles of NaOH

Strategy and Solution:

Step 1.   The numbers of moles of Na and of H_2O present are

$$4.0\ \text{g Na} \times \frac{\text{mol Na}}{22.9898\ \text{g Na}} = 0.17\ \text{mol Na}$$

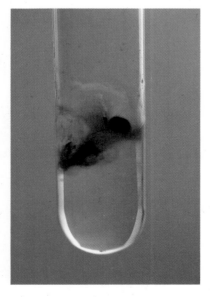

Sodium metal reacts violently with water to produce hydrogen gas, H_2, and NaOH.

$$7.7 \text{ g H}_2\text{O} \times \frac{\text{mol H}_2\text{O}}{18.0152 \text{ g H}_2\text{O}} = 0.43 \text{ mol H}_2\text{O}$$

Step 2. Find the number of moles of NaOH produced by 0.17 moles of Na and the number produced by 0.43 moles of H_2O.

$$0.17 \text{ mol Na} \times \frac{2 \text{ mol NaOH}}{2 \text{ mol Na}} = 0.17 \text{ mol NaOH}$$

$$0.43 \text{ mol H}_2\text{O} \times \frac{2 \text{ mol NaOH}}{2 \text{ mol H}_2\text{O}} = 0.43 \text{ mol NaOH}$$

From this we find that Na is the limiting reactant and that 0.17 moles of NaOH are produced in the reaction.

Step 3. Convert the number of moles of NaOH to grams, using the molar mass of NaOH.

$$0.17 \text{ mol NaOH} \times \frac{39.9971 \text{ g NaOH}}{\text{mol NaOH}} = 6.8 \text{ g NaOH}$$

EXAMPLE 10.22

Sulfuric acid reacts with ammonia according to the following chemical equation:

$$2 \text{ NH}_3 + \text{H}_2\text{SO}_4 \longrightarrow (\text{NH}_4)_2\text{SO}_4$$

For this reaction, 22.4 grams of ammonia were added to 22.4 grams of sulfuric acid. How many grams of $(\text{NH}_4)_2\text{SO}_4$ were produced?

Solution Map:

| grams of H_2SO_4 | $\longrightarrow$ | moles of H_2SO_4 | $\longrightarrow$ | moles of $(\text{NH}_4)_2\text{SO}_4$ |

| grams of NH_3 | $\longrightarrow$ | moles of NH_3 | $\longrightarrow$ | moles of $(\text{NH}_4)_2\text{SO}_4$ |

Strategy and Solution:

Step 1. Make sure you have a balanced equation.

Step 2. Convert to numbers of moles of the reactants.

$$22.4 \text{ g H}_2\text{SO}_4 \times \frac{\text{mol H}_2\text{SO}_4}{98.079 \text{ g H}_2\text{SO}_4} = 0.228 \text{ mol H}_2\text{SO}_4$$

$$22.4 \text{ g NH}_3 \times \frac{\text{mol NH}_3}{17.0304 \text{ g NH}_3} = 1.32 \text{ mol NH}_3$$

Step 3. Find the limiting reactant and the number of moles of product.

$$0.228 \text{ mol H}_2\text{SO}_4 \times \frac{1 \text{ mol } (\text{NH}_4)_2\text{SO}_4}{1 \text{ mol H}_2\text{SO}_4} = 0.228 \text{ mol } (\text{NH}_4)_2\text{SO}_4$$

$$1.32 \text{ mol NH}_3 \times \frac{1 \text{ mol } (\text{NH}_4)_2\text{SO}_4}{2 \text{ mol NH}_3} = 0.660 \text{ mol } (\text{NH}_4)_2\text{SO}_4$$

H_2SO_4 is limiting.

Step 4. Calculate the number of grams of $(\text{NH}_4)_2\text{SO}_4$ formed.

$$0.228 \text{ mol } (\text{NH}_4)_2\text{SO}_4 \times \frac{132.140 \text{ g } (\text{NH}_4)_2\text{SO}_4}{\text{mol } (\text{NH}_4)_2\text{SO}_4} = 30.1 \text{ g } (\text{NH}_4)_2\text{SO}_4$$

Exercise 10.18

Calculate how many grams of $AlBr_3$ would result from the reaction of 22.6 grams of Al and 38.5 grams of Br_2.

$$2\ Al + 3\ Br_2 \longrightarrow 2\ AlBr_3$$

Exercise 10.19

How many grams of $Ca_3(PO_4)_2$ will be produced if 76.5 grams of $Ca(OH)_2$ react with 100.0 grams of H_3PO_4?

$$Ca(OH)_2 + H_3PO_4 \longrightarrow Ca_3(PO_4)_2 + H_2O$$

10.10 Stoichiometric Calculations Involving Density Data

After studying this section, you should be able to:

- Combine density measurements with the mole ratio method to make calculations from chemical equations.

Laboratory data based on the reactions of liquids or gases are often expressed in units of volume because measuring the volume of a liquid or gas is usually easier than measuring its mass. Volumes can be converted to masses if the densities of the liquid are known. As you may recall, density is defined as the mass of a sample divided by its volume.

Often it is more convenient to measure volume than mass; in such cases, you must make use of density data in chemical calculations.

$$density = \frac{mass}{volume}$$

Stoichiometric calculations in which some of the data are given in volumes are similar to the mass data problems described in the previous section, but we need to add another conversion step. Here are the steps we will follow when we solve problems in which the reactant data are given in units of volume.

Guidelines for Stoichiometric Calculations Involving Density Data

1. *Convert reactant volume data into numbers of grams by using the density. (If you have to find the limiting reactant, you will need to do this step and the next two steps for more than one reactant.)*
2. *Convert the gram values into numbers of moles.*
3. *Use the mole ratio calculation to find out how much product is formed.*
4. *Convert the numbers of moles of product into other units, as necessary.*

EXAMPLE 10.23

The density of pure H_2SO_4 is 1.85 g/mL. To an excess of Na_2CO_3 is added 43.2 mL of pure H_2SO_4, and the reaction runs until all the H_2SO_4 is used up.

The balanced chemical equation is

In running a reaction, painstaking measurement is required to assemble all the reactants in exact stoichiometric amounts. Usually reactions are run with an excess of one or more reactants.

$$H_2SO_4 + Na_2CO_3 \longrightarrow Na_2SO_4 + CO_2 + H_2O$$

How many grams of CO_2 are formed in this process?

Solution Map:

$\boxed{\text{milliliters of } H_2SO_4} \longrightarrow \boxed{\text{grams of } H_2SO_4} \longrightarrow \boxed{\text{moles of } H_2SO_4} \longrightarrow$

$\boxed{\text{moles of } CO_2} \longrightarrow \boxed{\text{grams of } CO_2}$

Strategy and Solution:

Step 1. Make sure you have a balanced equation.

Step 2. We are told that Na_2CO_3 is present in excess, so we will make our calculation using H_2SO_4. We calculate the number of grams of H_2SO_4 present by multiplying the volume times the density.

$$43.2 \text{ mL } H_2SO_4 \times \frac{1.85 \text{ g } H_2SO_4}{\text{mL } H_2SO_4} = 79.9 \text{ g } H_2SO_4$$

Step 3. We find the number of moles of H_2SO_4.

$$79.9 \text{ g } H_2SO_4 \times \frac{\text{mol } H_2SO_4}{98.079 \text{ g } H_2SO_4} = 0.815 \text{ mol } H_2SO_4$$

Step 4. We perform the mole ratio calculation for the product.

$$0.815 \text{ mol } H_2SO_4 \times \frac{1 \text{ mol } CO_2}{1 \text{ mol } H_2SO_4} = 0.815 \text{ mol } CO_2$$

Step 5. We can now convert the number of moles of CO_2 into grams.

$$0.815 \text{ mol } CO_2 \times \frac{44.010 \text{ g } CO_2}{\text{mol } CO_2} = 35.9 \text{ g } CO_2$$

EXAMPLE 10.24

Octane, C_8H_{18}, is one of the ingredients in gasoline. It burns in air by reacting with O_2. The density of octane is 0.702 g/mL. How many grams of CO_2 are produced if 1.00 L of octane is burned according to the following balanced chemical equation? Assume that there is sufficient O_2 to use all the octane.

$$2 \, C_8H_{18} + 25 \, O_2 \longrightarrow 16 \, CO_2 + 18 \, H_2O$$

Solution Map:

$\boxed{\text{liters of } C_8H_{18}} \longrightarrow \boxed{\text{milliliters of } C_8H_{18}} \longrightarrow \boxed{\text{grams of } C_8H_{18}} \longrightarrow$

$\boxed{\text{moles of } C_8H_{18}} \longrightarrow \boxed{\text{moles of } CO_2} \longrightarrow \boxed{\text{grams of } CO_2}$

Strategy and Solution:

Step 1. Make sure you have a balanced equation.

Step 2. Find the number of grams of octane.

$$1.00 \text{ L } C_8H_{18} \times \frac{1000 \text{ mL}}{\text{L}} \times \frac{0.702 \text{ g } C_8H_{18}}{\text{mL } C_8H_{18}} = 702 \text{ g } C_8H_{18}$$

Step 3. Find the number of moles of C_8H_{18}.

$$702 \text{ g } C_8H_{18} \times \frac{\text{mol } C_8H_{18}}{114.230 \text{ g } C_8H_{18}} = 6.15 \text{ mol } C_8H_{18}$$

Step 4.   Find the number of moles of CO_2.

$$6.15 \text{ mol } C_8H_{18} \times \frac{16 \text{ mol } CO_2}{2 \text{ mol } C_8H_{18}} = 49.2 \text{ mol } CO_2$$

Step 5.   Find the number of grams of CO_2.

$$49.2 \text{ mol } CO_2 \times \frac{44.010 \text{ g } CO_2}{\text{mol } CO_2} = 2.17 \times 10^3 \text{ g } CO_2$$

Let's now do a problem in which we make all the conversions in a single calculation.

EXAMPLE 10.25

Ethyl acetate is one of a family of compounds called esters. The pleasant smells of many perfumes come from esters.

Acetic acid is what makes vinegar sour; its formula is $HC_2H_3O_2$. As a pure compound, acetic acid has a density of 1.05 g/mL. It can be made to react with ethyl alcohol, C_2H_6O, to produce a sweet-smelling compound, ethyl acetate, $C_4H_8O_2$, which is used in fingernail polish remover, among other things.

$$HC_2H_3O_2 + C_2H_6O \longrightarrow C_4H_8O_2 + H_2O$$

A sample of acetic acid with a volume of 0.733 mL is reacted with excess ethyl alcohol. How many grams of ethyl acetate are produced?

Solution Map:

$$\boxed{\text{milliliters of } HC_2H_3O_2} \longrightarrow \boxed{\text{grams of } HC_2H_3O_2} \longrightarrow$$

$$\boxed{\text{moles of } HC_2H_3O_2} \longrightarrow \boxed{\text{moles of } C_4H_8O_2} \longrightarrow \boxed{\text{grams of } C_4H_8O_2}$$

Solution:

$$0.733 \text{ mL } HC_2H_3O_2 \times \frac{1.05 \text{ g } HC_2H_3O_2}{\text{mL } HC_2H_3O_2} \times \frac{\text{mol } HC_2H_3O_2}{60.052 \text{ g } HC_2H_3O_2} \times$$

$$\frac{1 \text{ mol } C_4H_8O_2}{1 \text{ mol } HC_2H_3O_2} \times \frac{88.106 \text{ g } C_4H_8O_2}{\text{mol } C_4H_8O_2} = 1.13 \text{ g } C_4H_8O_2$$

Exercise 10.20

Excess Al was mixed with 12.8 mL of Br_2. How many grams of $AlBr_3$ were formed? The density of Br_2 is 3.12 g/mL. Use the reaction in Exercise 10.18.

Exercise 10.21

HCl gas has a density of 1.47 g/L at 25°C. If 4.32 grams of Fe_2O_3 are mixed with 25.0 L of HCl, how many grams of $FeCl_3$ can be obtained?

$$Fe_2O_3 + 6 \text{ HCl} \longrightarrow 2 \text{ FeCl}_3 + 3 \text{ H}_2O$$

10.11 **Energy Changes in Chemical Reaction**

One of the factors that characterizes our twentieth-century industrialized society is the lavish, widespread use of energy. And much of the energy that powers our modern world is chemical energy, the energy produced when chemical reaction occurs.

Many chemical reactions spontaneously release energy, often in the form of heat. The combustion of hydrocarbons is an example. In the laboratory, we can measure the amount of heat produced in a combustion reaction by using a bomb calorimeter, as illustrated in Figure 10.5. Chemical reactions that produce heat are said to be **exothermic.** Other reactions consume energy, and can run only as long as they are supplied with energy. Such reactions are said to be **endothermic.** The decomposition of H_2O into H_2 and O_2 is an example of an endothermic reaction.

Because energy can neither be created nor destroyed, the products of an exothermic reaction contain less energy than did the reactants; the products of an endothermic reaction contain more energy than did the reactants. Scientists use the term **enthalpy** to refer to the energy contained by a substance. In an exothermic reaction, the enthalpy of the products is less than the enthalpy of the reactants; the reverse is the case in endothermic reactions.

The symbol H represents the amount of enthalpy of a system, and $\triangle H$ represents the change of enthalpy during a reaction or other process ($\triangle$ is used to mean *a change of*). If a reaction is exothermic, energy is lost as the reactants change to products, therefore $\triangle H$ is negative. If the reaction is endothermic, energy is gained. $\triangle H$ is positive.

$$-\triangle H \text{ means the reaction is exothermic}$$

$$+\triangle H \text{ means the reaction is endothermic}$$

If we want to show the energetics of a reaction, we follow the chemical equation with an indication of the enthalpy change. Here are two examples.

The combustion of butane, C_4H_{10} (sometimes called liquefied petroleum gas or LPG), is exothermic.

$$2\ C_4H_{10}(g) + 13\ O_2(g) \longrightarrow 8\ CO_2(g) + 10\ H_2O(g) \qquad \triangle H = -330 \text{ kJ}$$

Electricity can be used to decompose water. The reaction is endothermic.

$$2\ H_2O(l) \longrightarrow 2\ H_2(g) + O_2(g) \qquad \triangle H = +572 \text{ kJ}$$

The amount of enthalpy shown for a reaction is the amount of energy lost or gained by the amounts, measured in moles, of reactants and products shown in the chemical equation.

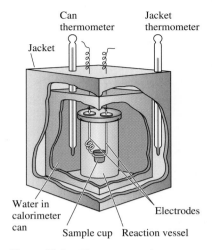

Figure 10.5 The amount of energy released by a combustion reaction can be measured using a bomb calorimeter. The volume of the reaction vessel is constant.

Butane is used as the fuel in lighters.

Summary

Chemical reactions are described by chemical equations that give the formulas of reactants and products. A chemical equation must be balanced to be complete; that is, the same total number of atoms of each element must be shown on each side of the equation.

Balanced chemical equations tell us the numbers of atoms and molecules of reactants and products. Such equations can also be read in terms of moles, and it is in this language that they are most useful to us. It is easy to read directly from a balanced chemical equation the ratios of the numbers of moles of the substances involved. We can then use a mole ratio to calculate the amount of a substance produced or consumed in a reaction with a known amount of another substance.

If quantities of reacting substances are given in laboratory units such as grams, those quantities must be converted to numbers of moles before we can make mole ratio calculations. We use the molar mass to make the conversion to moles. It is often necessary to convert the result back into mass units after we have made a mole ratio calculation. If the quantities of the substances are initially given in terms of volume, and if the densities are known, the first conversion is into mass units, followed by a conversion to moles and a mole ratio calculation. The same sequence can be followed in reverse if answers are required to be in units of volume.

Reactions often stop because one reactant runs out before the others do. The reactant used up first is called the limiting reactant. Unless we know in advance which reactant is limiting (that is, know that the other reactants are present in excess), we need to make mole ratio calculations for all the reactants present before we can determine how much product will result. Once we know the limiting reactant, we can base an ordinary mole ratio calculation on the quantity of that limiting reactant.

The change in energy during a reaction is called the enthalpy change, and an indication of this change is often used to show how much energy is gained or lost. An exothermic reaction is one in which the enthalpy change is negative; that is, exothermic reactions lose enthalpy. Endothermic reactions, on the other hand, gain enthalpy.

Key Terms

chemical equation	*(p. 260)*	percentage yield	*(p. 277)*
coefficient	*(p. 261)*	limiting reagent	*(p. 278)*
balanced equation	*(p. 262)*	limiting reactant	*(p. 278)*
mole ratio	*(p. 268)*	exothermic reaction	*(p. 287)*
stoichiometric calculation	*(p. 269)*	endothermic reaction	*(p. 287)*
actual yield	*(p. 277)*	enthalpy	*(p. 287)*
theoretical yield	*(p. 277)*		

Questions and Problems

Section 10.2 Balancing Chemical Equations

Questions

1. When chemical equations are balanced, only the coefficients may be rewritten. Why not change the subscripts?

2. Describe how to balance a chemical equation by inspection and trial and error.

Problems

3. Balance the following equations.
   (a) $H_2 + O_2 \longrightarrow H_2O$
   (b) $N_2 + I_2 \longrightarrow NI_3$
   (c) $HCl + NaOH \longrightarrow NaCl + H_2O$
   (d) $KClO_3 \longrightarrow KCl + O_2$
   (e) $CuCl_2 + H_2S \longrightarrow CuS + HCl$

4. Balance the following equations.
   (a) $Ca(NO_3)_2 + Pb \longrightarrow Ca(NO_2)_2 + PbO$
   (b) $H_2SO_4 + KOH \longrightarrow K_2SO_4 + H_2O$
   (c) $C_6H_{14}N + O_2 \longrightarrow CO_2 + NO_2 + H_2O$
   (d) $BaBr_2 + Cs_2CO_3 \longrightarrow BaCO_3 + CsBr$
   (e) $CuO + NH_3 \longrightarrow Cu + N_2 + H_2O$

5. Balance the following equations.
   (a) $Na + Cl_2 \longrightarrow NaCl$
   (b) $SiO + C \longrightarrow SiC + CO$

(c) $CaCO_3 \longrightarrow CaO + CO_2$
(d) $NH_3 + O_2 \longrightarrow NO + H_2O$
(e) $MnO_2 + HCl \longrightarrow MnCl_2 + Cl_2 + H_2O$

6. Balance the following equations.
   (a) $Al + Fe_3O_4 \longrightarrow Al_2O_3 + Fe$
   (b) $H_3PO_4 + NaOH \longrightarrow Na_3PO_4 + H_2O$
   (c) $Cu + H_2SO_4 \longrightarrow CuSO_4 + SO_2 + H_2O$
   (d) $I_2 + HNO_3 \longrightarrow HIO_3 + NO_2 + H_2O$
   (e) $Ag + HNO_3 \longrightarrow AgNO_3 + NO + 2\ H_2O$

Section 10.3 Reading Chemical Equations in Units of Moles

Question

7. Why do chemists often interpret chemical equation coefficients in terms of numbers of moles rather than numbers of molecules?

Section 10.4 Mole Ratios

Problem

8. List all possible mole ratios that can be written for the chemical equation in Problem 45.

Section 10.5 Calculations Using Mole Ratios

Question

9. Why must you have a balanced chemical equation before you can use the mole ratio method?

Problems

10. How many moles of I_2 can be produced from 2.31×10^{-2} moles of HI?

$$2\ HI \longrightarrow H_2 + I_2$$

11. How many moles of product can be formed from 2.80 moles of the first reactant in each of the equations given below?
    (a) $4\ Al + 3\ O_2 \longrightarrow 2\ Al_2O_3$
    (b) $2\ Cu + S \longrightarrow Cu_2S$
    (c) $2\ NH_3 + H_2SO_4 \longrightarrow (NH_4)_2SO_4$
    (d) $Cl_2O_7 + H_2O \longrightarrow 2\ HClO_4$

12. How many moles of product can be formed according to each equation in Problem 11 if you have 2.54×10^{-2} moles of the second reactant?

13. Iron (Fe) can be produced according to the following equation:

$$Fe_3O_4 + 4\ H_2 \longrightarrow 3\ Fe + 4\ H_2O$$

Beginning with 2.74 moles of H_2 and excess Fe_3O_4, how many moles of Fe can be produced by this reaction?

14. How many moles of H_2 would you need to form 8.69 mol of H_2O?

$$2\ H_2 + O_2 \longrightarrow 2\ H_2O$$

15. How many moles of O_2 would be required to produce 3.85 moles of CO_2 by the following equation?

$$2\ CO + O_2 \longrightarrow 2\ CO_2$$

16. In each of the following equations, how many moles of the first reactant are required to produce 7.22 moles of the second product?
    (a) $FeO + C \longrightarrow Fe + CO$
    (b) $2\ Al + 6\ HCl \longrightarrow 2\ AlCl_3 + 3\ H_2$
    (c) $Al(NO_3)_3 + 4\ NaOH \longrightarrow NaAlO_2 + 3\ NaNO_3 + 2\ H_2O$
    (d) $C_3H_8 + 5\ O_2 \longrightarrow 3\ CO_2 + 4\ H_2O$

17. Given the equations in Problem 16, how many moles of the second reactant will you need to form 6.74 moles of the first product?

Section 10.6 Stoichiometric Calculations Using Mass Data

Problems

18. How many moles of NaCl can be produced from 24.3 grams of Cl_2 and excess Na?

$$2\ Na + Cl_2 \longrightarrow 2\ NaCl$$

19. How many moles of NO can be prepared from 10.7 grams of NH_3 and excess O_2? (Use the chemical equation from Problem 5d.)

20. The combustion of butane, C_4H_{10}, yields carbon dioxide and water.

$$C_4H_{10} + O_2 \longrightarrow CO_2 + H_2O$$

How many moles of CO_2 and of H_2O can be obtained from the reaction of 420 grams of butane and excess oxygen?

21. How many moles of H_2 are required to produce 9.00 grams of Fe? (Use the chemical equation in Problem 13.)

22. How many moles of HCl are required to produce 23.5 grams of $CaCl_2$? How many moles of $Ca(OH)_2$ are required?

$$Ca(OH)_2 + HCl \longrightarrow CaCl_2 + H_2O$$

23. How many moles of Fe_2O_3 and of CO are needed to produce 10.5 grams of Fe?

$$Fe_2O_3 + 3\ CO \longrightarrow 2\ Fe + 3\ CO_2$$

24. How many grams of H_2SO_4 can be produced from 27.0 g of SO_3 and excess H_2O?

$$SO_3 + H_2O \longrightarrow H_2SO_4$$

25. If you have 310.5 grams of NO and excess O_2, how many grams of NO_2 can you produce?

$$2\ NO + O_2 \longrightarrow 2\ NO_2$$

26. According to the following equation, how many grams of $FeCl_3$ and of H_2 will be formed from 205 grams of Fe and excess HCl?

$$Fe + HCl \longrightarrow FeCl_3 + H_2$$

27. Using the equation in Problem 41, calculate the number of grams of PbO that could be produced starting with 37.0 grams of PbS and excess O_2.

28. How many grams of H_2 are required to produce 15.8 grams of HCl?

$$H_2 + Cl_2 \longrightarrow 2\ HCl$$

29. Calculate the number of grams of $KClO_3$ that would be needed to produce 11.0 grams of KCl. (See Exercise 10.8 for the chemical equation.)

30. How many grams of $MgCO_3$ would you need to produce 8.30 grams of MgO? To form 50.0 grams of CO_2?

$$MgCO_3 \longrightarrow MgO + CO_2$$

31. Silver chloride, AgCl, is quite insoluble and can easily be recovered as a solid from a solution. $CaCl_2$, calcium chloride, reacts with silver nitrate, $AgNO_3$.

$$CaCl_2(aq) + 2\ AgNO_3(aq) \longrightarrow$$

$$2\ AgCl(s) + Ca(NO_3)_2(aq)$$

If 15.4 grams of solid AgCl were recovered in an experiment using this reaction, how many grams of $CaCl_2$ were used?

Section 10.7 Percentage Yield Calculations

Problems

32. A reaction run in the laboratory produces 43.75 grams of product. Theoretically, 55.32 grams of product should be obtained. What is the percentage yield?

33. Calculate the percentage yield if the actual yield is 3.560 grams and the theoretical yield is 5.210 grams.

34. Theoretically, you should obtain 32.0 grams of product. You actually collect 29.1 grams after reaction. What is the percentage yield?

35. In a laboratory experiment, 23.81 grams of $BaSO_4$ were recovered after reacting 20.00 grams of Na_2SO_4 with excess $BaCl_2$. Calculate the percentage yield.

$$Na_2SO_4 + BaCl_2 \longrightarrow BaSO_4 + 2\ NaCl$$

36. What is the percentage yield if 16.92 grams of $CaCN_2$ react to produce 4.92 grams of NH_3?

$$CaCN_2 + 3\ H_2O \longrightarrow CaCO_3 + 2\ NH_3$$

37. After reacting 2.45 grams of H_2S, 12.38 grams of PbS are obtained. What is the percentage yield?

$$H_2S + Pb(NO_3)_2 \longrightarrow PbS + 2\ HNO_3$$

Section 10.8 Limiting Reactant Calculations

Questions

38. Define the term "limiting reactant."

39. If we run a reaction of X and Z starting with 6 moles of X and 4 moles of Z, is it possible for X to be the limiting reactant? Explain.

Problems

40. If 9.50 moles of O_2 are reacted with 7.65 moles of Fe, how many moles of Fe_2O_3 can result?

$$4\ Fe + 3\ O_2 \longrightarrow 2\ Fe_2O_3$$

41. If 4.30 moles of O_2 are reacted with 6.40 moles of PbS, how many moles of PbO can be produced?

$$2\ PbS + 3\ O_2 \longrightarrow 2\ PbO + 2\ SO_2$$

42. If 1.4 moles of NH_3 react with 1.0 moles of O_2, how many moles of N_2 will be produced?

$$4\ NH_3 + 3\ O_2 \longrightarrow 2\ N_2 + 6\ H_2O$$

43. How many moles of iron can be produced from 2.05 moles of Al and 0.57 moles of Fe_3O_4?

$$8\ Al + 3\ Fe_3O_4 \longrightarrow 4\ Al_2O_3 + 9\ Fe$$

44. Beginning with 4.50 moles of Al, 4.00 moles of NaOH, and 3.00 moles of H_2O, how many moles of H_2 can be formed?

$$Al + NaOH + H_2O \longrightarrow NaAlO_2 + H_2$$

45. How many moles of $MnCl_2$ can be produced from 3.65 moles of $H_2C_2O_4$, 4.40 moles of HCl, and 1.50 moles of $KMnO_4$?

$$2\ KMnO_4 + 5\ H_2C_2O_4 + 6\ HCl \longrightarrow$$

$$2\ MnCl_2 + 10\ CO_2 + 8\ H_2O + 2\ KCl$$

46. According to the chemical equation below, how many moles of $CaCO_3$ can be formed from 1.40 moles of $Fe_2(CO_3)_3$, 4.00 moles of $Ca(OH)_2$, and 5.75 moles of HCl? How many moles of $FeCl_3$ will also be produced?

$$6\ HCl + Fe_2(CO_3)_3 + 3\ Ca(OH)_2 \longrightarrow$$

$$2\ FeCl_3 + 3\ CaCO_3 + 6\ H_2O$$

Section 10.9 Limiting Reactant Calculations Using Mass Data

Problems

47. Find the limiting reactant in each of these reaction systems.

    (a) $CaO + 3 C \longrightarrow CaC_2 + CO$

    50.0 g CaO, 30.5 g C

    (b) $4 FeS_2 + 11 O_2 \longrightarrow 2 Fe_2O_3 + 8 SO_2$

    40.0 g FeS_2, 100.0 g O_2

    (c) $2 C_3H_6O_2 + 7 O_2 \longrightarrow 6 CO_2 + 6 H_2O$

    15.0 g $C_3H_6O_2$, 200.0 g O_2

    (d) $2 Al + 3 H_2SO_4 \longrightarrow Al_2(SO_4)_3 + 3 H_2$

    25.0 g Al, 90.0 g H_2SO_4

48. If 75.6 grams of $KMnO_4$ react with 0.68 grams of NO, how many moles of MnO_2 can be produced? How many grams of KNO_3?

$$KMnO_4 + NO \longrightarrow MnO_2 + KNO_3$$

49. Methane, CH_4, can be produced from Al_4C_3 and water. Calculate the number of grams of CH_4 and of $Al(OH)_3$ that can be formed from 41.0 grams of Al_4C_3 and 65.0 grams of H_2O.

$$Al_4C_3 + 12 H_2O \longrightarrow 4 Al(OH)_3 + 3 CH_4$$

50. NaOH is often prepared commercially from calcium hydroxide, $Ca(OH)_2$, and Na_2CO_3.

$$Na_2CO_3 + Ca(OH)_2 \longrightarrow NaOH + CaCO_3$$

How many grams of NaOH can be produced if 50.0 grams of $Ca(OH)_2$ react with 50.0 grams of Na_2CO_3?

51. The chemical equation for the preparation of phosphorus in an electric furnace is

$$2 Ca_3(PO_4)_2 + 6 SiO_2 + 10 C \longrightarrow 6 CaSiO_3 + 10 CO + P_4$$

Calculate the number of grams of P_4 that can be produced from 50.0 grams of $Ca_3(PO_4)_2$, 10.0 grams of C, and 20.0 grams of SiO_2.

52. Ferrous chloride, $FeCl_2$, reacts with hydrochloric acid and potassium dichromate, $K_2Cr_2O_7$, to form ferric chloride, $FeCl_3$, and chromic chloride, $CrCl_3$.

$$6 FeCl_2 + 14 HCl + K_2Cr_2O_7 \longrightarrow$$

$$6 FeCl_3 + 2 KCl + 2 CrCl_3 + 7 H_2O$$

If 12 grams of each reactant are mixed together, how many grams of KCl will result?

Section 10.10 Stoichiometric Calculations Involving Density Data

Problems

53. Ethanol, C_2H_5OH, is a product of the fermentation of sugar, $C_6H_{12}O_6$.

$$C_6H_{12}O_6 \longrightarrow 2 C_2H_5OH + 2 CO_2$$

How many grams of alcohol and of carbon dioxide can be produced from 750 grams of sugar? How many milliliters of alcohol result if the density of the alcohol is 0.789 g/mL?

54. When heated, limestone, $CaCO_3$, produces "quicklime," CaO, and carbon dioxide.

$$CaCO_3(s) \longrightarrow CaO(s) + CO_2(g)$$

If carbon dioxide has a density of 1.98 g/L, how many grams of limestone were originally heated when 6.50 L of carbon dioxide resulted?

55. Calcium carbide, CaC_2, reacts with water to produce acetylene gas, C_2H_2.

$$CaC_2(s) + 2 H_2O(l) \longrightarrow C_2H_2(g) + Ca(OH)_2(aq)$$

This reaction is used in carbide lamps, in which the acetylene is burned to provide light. How many grams of water are required to react with 6.4 grams of calcium carbide? How many liters of acetylene are produced if the density of the gas is 0.618 g/L?

56. Hematite iron ore, Fe_2O_3, can be reacted with carbon monoxide to form iron.

$$Fe_2O_3 + 3 CO \longrightarrow 2 Fe + 3 CO_2$$

How many grams of hematite are required to produce 1.0 kilogram of iron? How many liters of carbon monoxide are required if its density is 1.25 g/L? Calculate the number of liters of carbon dioxide produced at a density of 1.98 g/L. How many grams of water are produced?

57. Calculate the number of liters of O_2, measured at a density of 1.429 g/L, that can be produced when you begin with 5.50 L of gaseous H_2O. Under the conditions of this particular reaction, which was run at high temperature, the H_2O has a density of 0.4980 g/mL.

$$2 H_2O(g) \longrightarrow 2 H_2(g) + O_2(g)$$

58. How many grams of CO_2 can be produced from the complete combustion of 1.00 L of CH_4, measured at a density of 0.992 g/L, and 10.0 L of O_2, measured at a density of 0.409 g/L? Write the equation.

59. If 24 grams of mercurous chloride, Hg_2Cl_2, and 2.00 grams of H_2O react according to the following equation, how many grams of oxygen will be produced?

$$2 Hg_2Cl_2 + 2 H_2O \longrightarrow 4 HCl + O_2 + 4 Hg$$

How many milliliters of mercury will be formed in this reaction? The density of mercury is 13.6 g/cm^3.

Solutions to Exercises

10.1 $NaCl + 3\ NaBrO \longrightarrow NaClO_3 + 3\ NaBr$

10.2 $Zn(s) + NH_4NO_3(s) \longrightarrow N_2(g) + ZnO(s) + 2\ H_2O(g)$

10.3 $$1.60\ mol\ CO \times \frac{1\ mol\ H_2}{1\ mol\ CO} = 1.60\ mol\ H_2$$

10.4 $$5.45\ mol\ NaOH \times \frac{1\ mol\ H_2}{2\ mol\ NaOH} = 2.72\ mol\ H_2$$

10.5 $$8.72\ mol\ NaCl \times \frac{1\ mol\ Na_2O}{2\ mol\ NaCl} = 4.36\ mol\ Na_2O$$

10.6 $$3.40\ mol\ H_2O \times \frac{2\ mol\ HCl}{1\ mol\ H_2O} = 6.80\ mol\ HCl$$

10.7 $$3.42\ mol\ O_2 \times \frac{2\ mol\ C_2H_6}{7\ mol\ O_2} = 0.977\ mol\ C_2H_6$$

10.8 $$16.4\ g\ KClO_3 \times \frac{mol\ KClO_3}{122.5492\ g\ KClO_3} \times \frac{3\ mol\ O_2}{2\ mol\ KClO_3} = 0.201\ mol\ O_2$$

   Did you remember to balance the chemical equation?

10.9 $$24.3\ g\ HgO \times \frac{mol\ HgO}{216.59\ g\ HgO} \times \frac{2\ mol\ Hg}{2\ mol\ HgO} = 0.112\ mol\ Hg$$

10.10 $$10.3\ mol\ Zn(NO_3)_2 \times \frac{mol\ Zn}{mol\ Zn(NO_3)_2} \times \frac{65.39\ g\ Zn}{mol\ Zn} = 674\ g\ Zn$$

10.11 $$8.50\ mol\ Hg \times \frac{2\ mol\ HgO}{2\ mol\ Hg} \times \frac{216.59\ g\ HgO}{mol\ HgO} = 1.84 \times 10^3\ g\ HgO$$

10.12 $$76.8\ g\ NaOH \times \frac{mol\ NaOH}{39.9971\ g\ NaOH} \times \frac{1\ mol\ Na_2ZnO_2}{2\ mol\ NaOH} \times \frac{143.37\ g\ Na_2ZnO_2}{mol\ Na_2ZnO_2} =$$

$$138\ g\ Na_2ZnO_2$$

10.13 $$18.6\ g\ Sb \times \frac{mol\ Sb}{121.75\ g\ Sb} \times \frac{2\ mol\ SbCl_3}{2\ mol\ Sb} \times \frac{228.11\ g\ SbCl_3}{mol\ SbCl_3} = 34.8\ g\ SbCl_3$$

10.14 $$4.6\ g\ HCl \times \frac{mol\ HCl}{36.4606\ g\ HCl} \times \frac{1\ mol\ NaCl}{1\ mol\ HCl} \times \frac{58.4425\ g\ NaCl}{mol\ NaCl} =$$

$$7.4\ g\ NaCl\ theoretically$$

$$\frac{5.7\ g}{7.4\ g} \times 100\% = 77\%$$

10.15 $$2.5\ mol\ Ag \times \frac{3\ mol\ AgNO_3}{3\ mol\ Ag} = 2.5\ mol\ AgNO_3$$

$$2.0\ mol\ HNO_3 \times \frac{3\ mol\ AgNO_3}{4\ mol\ HNO_3} = 1.5\ mol\ AgNO_3$$

   HNO_3 is limiting. 1.5 moles of $AgNO_3$ will be produced.

10.16 $$0.245 \text{ mol Cu} \times \frac{2 \text{ mol NO}_2}{1 \text{ mol Cu}} = 0.490 \text{ mol NO}_2$$

$$0.320 \text{ mol HNO}_3 \times \frac{2 \text{ mol NO}_2}{4 \text{ mol HNO}_3} = 0.160 \text{ mol NO}_2$$

HNO_3 is limiting. 0.160 moles of NO_2 will be produced.

10.17 $$0.55 \text{ mol P}_4\text{O}_{10} \times \frac{4 \text{ mol H}_3\text{PO}_4}{1 \text{ mol P}_4\text{O}_{10}} = 2.20 \text{ mol H}_3\text{PO}_4$$

$$1.20 \text{ mol H}_2\text{O} \times \frac{4 \text{ mol H}_3\text{PO}_4}{6 \text{ mol H}_2\text{O}} = 0.800 \text{ mol H}_3\text{PO}_4$$

Water is limiting; therefore 0.800 moles of H_3PO_4 will be produced.

10.18 $$22.6 \text{ g Al} \times \frac{\text{mol Al}}{26.9815 \text{ g Al}} \times \frac{2 \text{ mol AlBr}_3}{2 \text{ mol Al}} = 0.838 \text{ mol AlBr}_3$$

$$38.5 \text{ g Br}_2 \times \frac{\text{mol Br}_2}{159.81 \text{ g Br}_2} \times \frac{2 \text{ mol AlBr}_3}{3 \text{ mol Br}_2} = 0.161 \text{ mol AlBr}_3; \text{Br}_2 \text{ is limiting}$$

$$0.161 \text{ mol AlBr}_3 \times \frac{266.69 \text{ g AlBr}_3}{\text{mol AlBr}_3} = 42.9 \text{ g AlBr}_3$$

10.19 $$76.5 \text{ g Ca(OH)}_2 \times \frac{\text{mol Ca(OH)}_2}{74.093 \text{ g Ca(OH)}_2} \times \frac{1 \text{ mol Ca}_3(\text{PO}_4)_2}{3 \text{ mol Ca(OH)}_2} =$$

$$0.344 \text{ mol Ca}_3(\text{PO}_4)_2$$

$$100.0 \text{ g H}_3\text{PO}_4 \times \frac{\text{mol H}_3\text{PO}_4}{97.9951 \text{ g H}_3\text{PO}_4} \times \frac{1 \text{ mol Ca}_3(\text{PO}_4)_2}{2 \text{ mol H}_3\text{PO}_4} =$$

$$0.5102 \text{ mol Ca}_3(\text{PO}_4)_2$$

$Ca(OH)_2$ is limiting;

$$0.344 \text{ mol Ca}_3(\text{PO}_4)_2 \times \frac{310.177 \text{ g Ca}_3(\text{PO}_4)_2}{\text{mol Ca}_3(\text{PO}_4)_2} = 107 \text{ g Ca}_3(\text{PO}_4)_2$$

Did you remember to balance the chemical equation?

10.20 $$12.8 \text{ mL Br}_2 \times \frac{3.12 \text{ g Br}_2}{\text{mL Br}_2} \times \frac{\text{mol Br}_2}{159.81 \text{ g Br}_2} \times \frac{2 \text{ mol AlBr}_3}{3 \text{ mol Br}_2} \times \frac{266.69 \text{ g AlBr}_3}{\text{mol AlBr}_3} =$$

$$44.4 \text{ g AlBr}_3$$

10.21 $$25.0 \text{ L HCl} \times \frac{1.47 \text{ g HCl}}{\text{L}} \times \frac{\text{mol HCl}}{36.4606 \text{ g HCl}} \times \frac{2 \text{ mol FeCl}_3}{6 \text{ mol HCl}} =$$

$$0.336 \text{ mol FeCl}_3$$

$$4.32 \text{ g Fe}_2\text{O}_3 \times \frac{\text{mol Fe}_2\text{O}_3}{159.69 \text{ g Fe}_2\text{O}_3} \times \frac{2 \text{ mol FeCl}_3}{1 \text{ mol Fe}_2\text{O}_3} = 0.0541 \text{ mol FeCl}_3;$$

$$\text{Fe}_2\text{O}_3 \text{ is limiting}$$

$$0.0541 \text{ mol FeCl}_3 \times \frac{162.205 \text{ g FeCl}_3}{\text{mol FeCl}_3} = 8.78 \text{ g FeCl}_3$$

The balloons rise because the warm air in the balloons is less dense than the cooler air of the atmosphere.

Properties and Chemical Reactions of Gases

I t may seem strange that something like a gas could have chemical and physical properties important enough to rate an entire chapter in a textbook, but studies of the behavior of gases have, in fact, given us much valuable scientific knowledge. Gases are one of the three states of matter; many of the substances we see every day as solids and liquids can exist also as gases. We might think that because we can't pick up a gas with our hands, or ordinarily even see it, we can't measure its properties as we can those of solids and liquids—but this is not the case. A sample of gas enclosed in a container has definite properties that characterize it completely and that can be used to understand its behavior. Once we have examined gases and their properties, they will seem much less mysterious.

One of the most important reasons for studying the behavior of gases is that it is easy to determine the number of molecules in a gas sample without knowing the molar mass or even the kind of gas present. Knowing the number of molecules present in reacting samples of substances helps us to understand

We take the existence of gases for granted unless we are in a situation in which we can't breathe.

Bromine, chlorine, nitrogen dioxide, and a few other gases are colored and thus can be seen. Several gases such as ammonia and hydrogen sulfide have very strong and distinctive odors.

the reactions themselves. For this reason, studies of gas reactions have been important historically and remain important.

We begin our study by looking at some of the properties of gases and by reviewing the theory that explains those properties. We then learn the laws that describe the behavior of gases. Finally, we find out how to do calculations involving gas properties and chemical reactions of gases. The chapter ends with a discussion of some of the limitations of the gas laws.

11.1 Properties of Gases

After studying this section, you should be able to:

- List the properties that characterize samples of gases.
- Explain how a barometer measures pressure.

Because most known gases are invisible and odorless, and because they can't be picked up and handled like other samples, people often forget that gases even exist (unless they get into a situation in which they can't breathe for a moment or two). Gases, however, are matter, and, like other matter, can be characterized by physical and chemical properties. The properties of gases in which we will be interested are volume, mass, temperature, and pressure.

Volume.   A sample of gas fills its entire container. Therefore, the volume of any sample of gas is equal to the volume of its container.

Mass.   Like samples of liquids and solids, samples of gas have mass. For example, a glass bulb filled with air has a greater mass than the same bulb containing nothing (Figure 11.1).

Temperature.   Like other samples of matter, samples of gases have temperatures. Because temperature variation causes a much greater change in the properties of a gas than it does in a liquid or a solid, it is particularly important to study the effects of temperature on the behavior of gases.

Pressure.   Gases have a property that solids and liquids don't have. A sample of gas exerts force on *all* the walls of the container that holds it. A liquid, on the other hand, exerts force only on the part of the container it contacts. Although it is usually inconvenient to measure the total force exerted by a gas on all the walls of a container, it is ordinarily sufficient to measure the force on

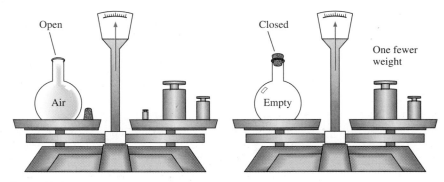

Figure 11.1   A flask containing air weighs more than an empty flask.

Figure 11.2 The left and middle tubes are more than 760 mm tall. Some of the mercury has run out, leaving 760 mm of mercury in each. The tube on the right is less than 760 mm tall. No mercury has run out.

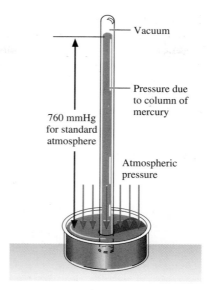

Vacuum

Pressure due to column of mercury

760 mmHg for standard atmosphere

Atmospheric pressure

Figure 11.3 The pressure of the atmosphere equals the pressure exerted by the column of mercury.

Figure 11.4 A barometer.

any square centimeter or other convenient unit of area. This works because the force is the same on all parts of the walls. ~~The force exerted on a unit of area~~ is called the **pressure.**

Evangelista Torricelli, an Italian scientist and one-time assistant to Galileo, discovered more than two centuries ago, that if a tube closed at one end is filled with mercury and inverted in a dish of mercury, only part of the mercury will run out of the tube (Figure 11.2). No matter how long a tube Torricelli used, the mercury would run out, leaving a column of a particular height, the height of the column always being the same for all the tubes longer than that height. If Torricelli used a tube shorter than that height, no mercury ran out.

Many people tried to explain what held up the mercury in Torricelli's tubes. Some even thought that there was an invisible string somehow attached to the liquid surface. Today we know that the air molecules of the atmosphere exert pressure on the dish of mercury outside the tube. This pushes the liquid mercury up inside the tube. The pressure exerted by the air outside is then balanced by pressure exerted by the mercury in the tube (Figure 11.3).

Atmospheric pressure is reported by weather stations in inches or millimeters of mercury and is measured by a barometer (Figure 11.4). The barometer is a Torricelli tube with scales and adjustments attached. On an average day at sea level, the pressure of the atmosphere will cause a column of mercury to stand 760 mm high in a barometer (29.9 in.). At higher altitudes where there is less atmosphere, the pressure is much lower. At the top of Mount Everest, the atmospheric pressure is only about 270 mm of mercury.

In honor of Torricelli, the millimeter of mercury has been given the name **torr.** For example, 760 mm of mercury equals 760 torr. Another pressure

If a tire is inflated to a pressure of 32 psi (pounds per square inch), it means that the air inside is exerting a force equivalent to 32 pounds of weight on every square inch of the inside surface of the tire.

The total force exerted by air on the top of a card table is approximately the same as the weight of five automobiles. Fortunately, that same force is exerted upward on the bottom side of the table.

keeping mercury from all leaking out

The height of the mercury in a Torricelli tube varies somewhat from day to day as well as from place to place, depending on the weather and altitude.

higher up lower pressure

Torr is both singular and plural.

1 mm Hg = 1 torr

1 atm = 760 torr

unit, the **atmosphere** (atm), is defined as 760 torr; it is the pressure of the atmosphere at sea level at 0°C. The SI unit of pressure is the **pascal,** abbreviated Pa; 101,325 Pa = 760 torr. Table 11.1 shows the relations between one atmosphere and other pressure units.

Table 11.1   Pressure Conversions

One Atmosphere (atm) Equals
760 torr
760 mm Hg
101,325 Pa
29.9 in. Hg

EXAMPLE 11.1

Convert 746 torr to atmospheres.

Solution: Use the unit conversion factor between atm and torr.

$$746 \text{ torr} \times \frac{\text{atm}}{760 \text{ torr}} = 0.982 \text{ atm}$$

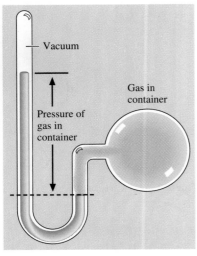

Exercise 11.1

Convert 0.455 atm to torr.

Exercise 11.2

How many pascals equal 642 mm Hg?

Gas pressures are usually measured in the laboratory with an instrument called a **manometer.** Such pressures are usually stated as the height of a mercury column. As you can see in Figure 11.5, a manometer works on the same principle as a barometer.

Figure 11.5 Closed tube manometer. This instrument measures pressure by taking the difference in height of the mercury in the u-shaped tube. In the left side, the only pressure on the mercury below the dotted line is from the mercury above the line. That pressure matches the pressure exerted at the same level by the gas acting on the mercury surface in the right side. Manometers in which the left side is open to the air are also in use. Pressure readings in such instruments (like those taken with ordinary pressure gauges) must be corrected for atmospheric pressure.

11.2 Kinetic Molecular Theory

After studying this section, you should be able to:

• List the assumptions in the kinetic molecular theory.

Why do gases behave as they do? To explain the behavior of gases, scientists have devised what is called the **kinetic molecular (KM) theory.** The theory is molecular because it assumes that samples of gases are made up of individual molecules; kinetic because the molecules are in constant random motion. Let's look at the individual features of the KM theory and justify each by a simple physical observation.

1. *A sample of gas consists of individual molecules, all of which are in rapid, random, independent motion* (Figure 11.6). This is reasonable. We know that gases expand rapidly to fill their containers. They could not do this if their molecules were not moving.*

2. *The molecules of a gas are far apart in comparison to their own sizes.* Placed in a large, closed container, a small sample of liquid water occupies

*The kinetic theory proposes that the individual particles of solids and liquids are also in constant motion. The particles of a solid are quite close together and held in place. They move only by vibrating. Molecules in a liquid are nearly as close together as those in a solid. They too vibrate, but they also crowd one another from place to place, allowing the liquid to flow.

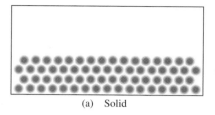

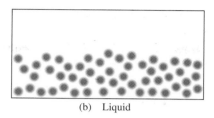

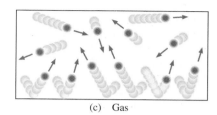

| (a)   Solid | (b)   Liquid | (c)   Gas |

Figure 11.6   (a) In a solid, the particles are closely packed. They vibrate around their places in the pattern. (b) In a liquid, the pattern is gone, and holes appear. The particles vibrate and exchange places. (c) In a gas, the molecules are widely separated and are in rapid, independent motion.

little space. The molecules of the liquid are essentially touching each other. However, if the water is heated and converted to its gaseous form, steam, the steam will occupy the entire volume of the container (see Figure 11.7). The water molecules (which themselves have gotten no larger) must now be separated by relatively long distances. Gases are mostly empty space.

3. *The pressures of gases are produced by collisions of the gas molecules with the walls of the container.* This is illustrated in Figure 11.8.

4. *The molecules of a gas collide with one another and against the container walls without an overall gain or loss of energy, although energy can be exchanged in collisions.* If the gas were to lose energy because of collisions, the molecules would soon slow down and stop. There would be no pressure. We know, however, that at constant temperature a sample of gas will exert a constant pressure indefinitely. We conclude, then, that the gas as a whole does not gain or lose energy as long as the temperature is maintained. elastic

The KM theory, then, explains how gases fill their containers and exert pressure. The molecules are far apart and are moving independently in all directions, bombarding the walls. Unless the temperature drops, the overall motion never ceases or slows. We must not think, however, that all the molecules in a gas

Under ordinary conditions, air molecules are moving at an average speed of about 10^5 cm/s, or about 2200 mi/hr. They each travel an average distance of several hundred times their own size in about 1/10,000,000,000 of a second.

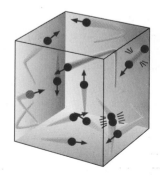

Figure 11.7   (a) A flask contains a small amount of liquid water. (b) When the flask is heated, the liquid water turns into steam, which fills the entire flask.

Figure 11.8   Gas molecules collide with one another and the walls of their container. Collisions with the walls produce the pressure of the gas.

are moving at the same velocity; that is, that they all have the same energy. At any particular moment, a few of the molecules are moving at high velocities, others are moving slowly, and most have velocities in between.

11.3 Temperature and the Energies of Gas Molecules

After studying this section, you should be able to:

- Describe the effects of increasing the temperature on the kinetic energy of a gas.
- Explain why some gases effuse more rapidly than others.

We learned in Section 2.4 that adding energy to a sample of matter increases its temperature. In fact, the temperature of a sample of matter is a direct measure of the average kinetic energy of its molecules. This is true of gases as well as liquids and solids.

As we saw in Section 2.2, velocity and kinetic energy are related, $\frac{1}{2}mv^2 = $ KE. Increasing the temperature of a gas increases the kinetic energy of its molecules and, therefore, increases the average molecular velocity. Since the kinetic molecular theory tells us that pressure is caused by collisions of moving molecules with the walls of the container, we might predict that there is a relation between the temperature of a sample and its pressure. We will explore this thought further as we go along.

Because velocity and kinetic energy are related, we can conclude that if gas molecules of different mass have the same average kinetic energy, the heavy molecules must be moving more slowly than the lighter molecules. Constant-temperature experiments with gas molecules of different masses moving out of their containers through small holes bear out this conclusion. At any given temperature, light gas molecules, such as H_2, will escape through a small hole in a container far more rapidly than a gas with more massive molecules, simply because the H_2 molecules are moving faster. The process by which a gas escapes through a small hole is known as *effusion*. Let's look at this in more detail.

Assume that the molecules in a sample of H_2 have the same average kinetic energy, KE, as those in a sample of O_2 (that is, that both have the same temperature). Then, on the average,

$$KE_{H_2} = KE_{O_2}$$

But, as we know, $KE = \frac{1}{2}mv^2$, so that for the average H_2 and O_2 molecules,

$$\frac{1}{2}m_{H_2}v_{H_2}^2 = \frac{1}{2}m_{O_2}v_{O_2}^2,$$

or

$$\frac{v_{H_2}}{v_{O_2}} = \sqrt{\frac{m_{O_2}}{m_{H_2}}}$$

Because $m_{O_2} = 32$ and $m_{H_2} = 2$,

$$\sqrt{\frac{m_{O_2}}{m_{H_2}}} = \sqrt{\frac{32}{2}} = \sqrt{16} = 4$$

Rubber party balloons are not very satisfactory because the fast-moving helium atoms can escape fairly rapidly through microscopic pores in the rubber. Because they have smaller pores, balloons made of aluminum-coated mylar plastic are better.

Science in Action

I n late 1938, a small band of German scientists discovered that uranium atoms, when bombarded with slow neutrons, would split into fragments (a process called fission), releasing enormous amounts of energy.

Shortly after World War II began, scientists of several countries—England and the United States in particular—started to investigate whether the energy of nuclear fission could be used for military purposes. Very soon, it was discovered that only one isotope of uranium, ^{235}U, could be used to produce useful fission. Natural uranium is mostly the isotope ^{238}U; it contains less than one part in a hundred of the lighter ^{235}U. The fact that there was no simple way to separate out the ^{235}U, however, presented a serious problem.

Without being sure that it would even work, in 1945 U.S. scientists gambled billions of dollars building a plant in Oak Ridge, Tennessee, to separate ^{235}U

Gaseous Effusion and Nuclear Weapons

Effusion drums in an isotope separation plant.

by gaseous effusion, using the gas, UF_6. In a sample of gas, all the molecules have an average kinetic energy, but to have that energy, the lighter molecules must be moving faster than molecules of greater mass. (After all, kinetic energy $= \frac{1}{2}$ mass $\times$ velocity^2.) If a gas is allowed to seep (effuse) through a series of tiny holes, the lighter molecules will get through somewhat faster than the heavier ones. In the Oak Ridge plant, tubes that were hand-made of the element nickel allowed ^{235}U to be separated from natural uranium via a several-thousand-step process of gas effusion.

Gas effusion is still employed to make ^{235}U for nuclear power plants as well as nuclear weapons.

The ratio $\frac{v_{H_2}}{v_{O_2}} = 4$ shows that the hydrogen molecules must be moving four times faster than the oxygen molecules at the same temperature.

because smaller & lighter

The equation

$$\frac{v_{H_2}}{v_{O_2}} = \sqrt{\frac{m_{O_2}}{m_{H_2}}}$$

is a mathematical statement of the relation between the velocities of gas molecules and their masses. It is called Graham's law of effusion, after the English scientist who discovered it in 1846. Graham's law is used to calculate relative rates of effusion for gases.

Experiments made at the same temperature with different gases show that a mole of gas enclosed in a given volume will exert the same pressure, no matter what kind of gas it is; that is, the mass of the gas molecules makes no difference. This means, for example, that *one-mole samples of gases can have identical pressure, volume, and temperature even if their molecules have very different masses.* This important property of gases is discussed in Section 11.5.

11.4 Boyle's Law: The Dependence of Pressure on Volume

After studying this section, you should be able to:

- Describe an experiment that verifies Boyle's law.
- State Boyle's law verbally and mathematically.
- Perform calculations using Boyle's law.

Imagine a cylindrical container of gas that has a movable piston similar to that of an old-fashioned bicycle pump (Figure 11.9). We will make sure that there is a constant amount of gas in the cylinder by not allowing any gas to leak in or out. We will also be careful to keep the temperature constant. Now we perform a simple experiment. By placing weights on the piston, we push it down with varying amounts of force and look at its position. As we might expect, we find that when we push harder, the plunger goes farther down. As the plunger goes down, the volume of the gas in the cylinder gets smaller. Since our downward force on the piston is directly related to the upward pressure the gas is exerting on the inside, we can say that the volume of the gas decreases as its pressure increases. We can further imagine how our experiment could be refined so that we could measure the volume and the pressure of the gas as we moved the piston.

While keeping the gas sample at a constant temperature (an important factor as we shall see), a student used an apparatus similar to the piston and cylinder for measuring several pressure–volume experiments. The results of those measurements are shown in Table 11.2. The pressures (P) in atmospheres are shown in the first column and the volumes (V) in milliliters in the second column. The products of the pressure and the volume ($P \times V$) are in the third column. Notice that the values in the third column are all the same.

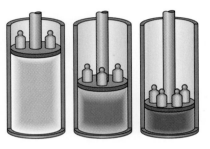

Figure 11.9   An experimental apparatus for determining the volume of a gas at different pressures. As the external pressure increases, the volume of the gas decreases.

Table 11.2	Pressure–Volume Data Taken by a Student	
Pressure (atm)	Volume (mL)	Pressure × Volume (atm mL)
1.623	9.25	15.0
1.344	11.18	15.0
1.133	13.22	15.0
0.997	15.07	15.0
0.758	19.75	15.0
0.718	20.92	15.0
0.671	22.36	15.0
0.612	24.55	15.0

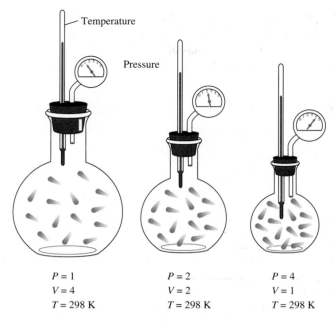

$P = 1$
$V = 4$
$T = 298$ K

$P = 2$
$V = 2$
$T = 298$ K

$P = 4$
$V = 1$
$T = 298$ K

Figure 11.10   Illustration of Boyle's law. These three flasks contain the same number of molecules. At 298 K, $P \times V = 4$ in all three flasks.

Robert Boyle performed similar experiments more than 200 years ago. His experiments demonstrated that *at constant temperature, the product of the pressure and the volume of a fixed mass of gas is constant* (see Figure 11.10). This is an expression of **Boyle's law,** which can be written mathematically as

$$PV = \text{constant} \qquad \text{or} \qquad PV = k$$

where k represents a constant number whose value we may or may not know. When the product of two variables (in this case, P and V) is equal to a constant number, the two variables are said to be **inversely proportional** to one another.

In Table 11.2, $k = 15.0$ atm mL, and, for the first PV pair,

$$P_1 V_1 = 15.0 \text{ atm mL}$$

For the second pair also,

$$P_2 V_2 = 15.0 \text{ atm mL}$$

Therefore,

$$P_1 V_1 = P_2 V_2$$

In general, for any gas sample of fixed mass at constant temperature,

$$P_1 V_1 = P_2 V_2 = P_3 V_3$$

and so on, for any number of measurements in which P and V are changed. This is another mathematical statement of Boyle's law, one that is convenient for making calculations.

in other words
decrease one
increase other

We use Boyle's law constantly. Air-cushioned rides in trains, landing gear struts on airliners, lift devices on automobile hatchbacks, and pressure tanks for rural water systems are just a few examples.

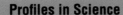

Profiles in Science

Robert Boyle (1627–1691)

Born to royalty as the fourteenth child of the Earl of Cork in Ireland, Robert Boyle became a perfect example of the wealthy, amateur, gentleman scientist of the seventeenth century. Like most, he was self-taught, learning science as he experimented in his laboratory. In his prime, Boyle was regarded as the chief English scientist of his time, an "ornament of English science." Distinguished foreign visitors, in science and other fields as well, considered it an honor to be received by him.

Boyle felt chemistry to be the ideal experimental science. Decades before Dalton's atomic theory, and even before the experiments that led to it, Boyle published a book called *The Skeptical Chymist,* in which he attacked the old theories of matter (air, earth, fire, and water, and so on) and proposed that all matter must be made of "physical corpuscles"—a hint of the atomic theory to come. Although his experiments on gases overshadow his other work, it is fair to state that his contributions to chemistry and his experimental and philosophical work contributing to the atomic theory were just as important.

Boyle is best known now for his invention of the air pump and for his discovery in 1660 of the law that bears his name.

Calculations Using Boyle's Law

Here we will learn how to work problems using Boyle's law. In all of our Boyle's law calculations, we must remember that the temperature and the number of gas molecules (the number of moles) must be kept constant.

First, let's use Boyle's law to calculate the change in pressure of a sample of gas caused by a change in its volume. We will use the first two sets of values (rounded) in Table 11.2 as an example.

$$P_1 = 1.62 \text{ atm} \qquad V_1 = 9.25 \text{ mL}$$
$$P_2 = 1.34 \text{ atm} \qquad V_2 = 11.2 \text{ mL}$$

Suppose that we have finished with the first set of measurements, $P_1 V_1$, and have changed the volume of the gas to $V_2 = 11.2$ mL, but that we haven't yet measured the new pressure, P_2. Can we calculate what the new pressure will be? Yes. We algebraically rearrange Boyle's law,

$$P_1 V_1 = P_2 V_2$$

to give the new pressure, P_2.

$$P_2 = P_1 \times \frac{V_1}{V_2}$$

Then we substitute the known values in the equation and solve.

$$P_2 = 1.62 \text{ atm} \times \frac{9.25 \text{ mL}}{11.2 \text{ mL}} = 1.34 \text{ atm}$$

When we make the measurement, we find that our calculation is correct, $P_2 = 1.34$ atm.

Notice that the original pressure was multiplied by a volume ratio numerically less than one.

$$\frac{V_1}{V_2} = \frac{9.25 \text{ mL}}{11.2 \text{ mL}} < 1$$

Multiplying any number by a number *less* than one reduces the original number. From Boyle's law, we know that if the volume increases, the pressure must decrease. Therefore, we must multiply the original pressure by a ratio smaller than one, that is, a ratio in which the smaller volume is in the numerator and the larger volume is in the denominator. For a decreased volume, we would use a ratio *larger* than one, a ratio with the larger volume on top. We can approach Boyle's law problems in two ways, either by rearranging the equation to fit the circumstances or by thinking what each change means physically and setting up conversion factors accordingly.

Now let's try a similar calculation to predict a new volume. We can begin with our same sample at the same conditions as before, $P_1 = 1.62$ atm and $V_1 = 9.25$ mL, but this time we will change the *pressure*, P_2, to 0.758 atm. We use Boyle's law in a slightly different arrangement to predict the new volume.

$$V_2 = V_1 \times \frac{P_1}{P_2}$$

$$V_2 = 9.25 \text{ mL} \times \frac{1.62 \text{ atm}}{0.758 \text{ atm}} = 19.8 \text{ mL}$$

When we do the actual measurement (the fifth experiment down in Table 11.2) we find that the new volume indeed has increased to 19.8 mL. We have *increased* the volume by multiplying the orginal volume by a ratio of pressures *greater* than one. To obtain a ratio greater than one, we put the larger pressure in the numerator and the smaller pressure in the denominator. But how did we know to increase the volume? Boyle's law states that if the pressure decreases, the volume increases.

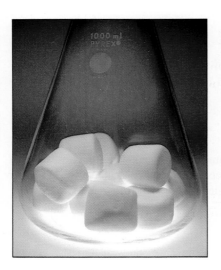

An illustration of Boyle's law. Marshmallows are spongy because of thousands of tiny bubbles of trapped air. When the pressure is reduced by pumping air out of the flask, the volume of air in the bubbles increases, and the marshmallows expand.

EXAMPLE 11.2

A fixed mass of gas having a volume of 3.20 L is at a pressure of 721 torr. While the temperature remains constant, the gas is brought to a new pressure of 618 torr. What is the new volume of the gas?

Strategy and Solution: First, we identify the variables. In this problem, $V_1 = 3.20$ L, $P_1 = 721$ torr, $P_2 = 618$ torr. We are looking for V_2, the new volume.

Step 1. Rearrange $P_1V_1 = P_2V_2$, solving for V_2. The new volume will equal the original volume multiplied by a conversion factor containing the pressure values.

$$V_2 = V_1 \times \frac{P_1}{P_2}$$

Step 2. We can now substitute the numerical values. We know from Boyle's law that if the pressure is reduced, the volume increases. We want to multiply the original volume by a ratio of the pressures numerically greater than 1. The larger pressure should be in the numerator and the smaller pressure in the denominator.

$$V_2 = 3.20 \text{ L} \times \frac{721 \text{ torr}}{618 \text{ torr}} = 3.73 \text{ L}$$

Step 3. Check the answer. Does it obey Boyle's law? As the pressure is decreased from 721 torr to 618 torr, the volume is increased from 3.20 L to 3.73 L. The answer obeys Boyle's law.

Notice that the mass of the gas did not enter the calculation in Example 11.2. In calculations involving only pressure and volume, we need only to be sure that the mass and temperature have not changed during the experiment.

EXAMPLE 11.3

A fixed mass of gas at constant temperature of 293 K has a volume of 447 mL at a pressure of 1.033 atm. If the volume is reduced to 212 mL, what is the new pressure?

Strategy and Solution: Let $P_1 = 1.033$ atm, $V_1 = 447$ mL, $V_2 = 212$ mL.

Step 1. Rearrange $P_1V_1 = P_2V_2$, to solve for P_2. The new pressure will be given by the original pressure multiplied by a conversion factor containing the volume values.

$$P_2 = P_1 \times \frac{V_1}{V_2}$$

Step 2. We can now substitute the numerical values. Multiply the original pressure by a ratio of the volumes numerically greater than one. The larger volume is in the numerator and the smaller volume is in the denominator.

$$P_2 = 1.033 \text{ atm} \times \frac{447 \text{ mL}}{212 \text{ mL}} = 2.18 \text{ atm}$$

Step 3. Check the answer. From Boyle's law, we know that if the volume decreases, the pressure increases. Our answer agrees with Boyle's law.

EXAMPLE 11.4

A cylinder with a volume of 1.00 L contains a fixed mass of gas at 22°C and 760. torr. To what new volume must the gas be brought for the new pressure to be 1100. torr?

Strategy and Solution: Let $V_1 = 1.00$ L, $P_1 = 760.$ torr, $P_2 = 1100.$ torr.

Step 1. Rearrange Boyle's law to find V_2. The new volume will be given by the original volume multiplied by a conversion factor containing the pressure values.

$$V_2 = V_1 \times \frac{P_1}{P_2}$$

Step 2. We can now substitute the numerical values. We want to multiply the original volume by a ratio of the pressures that is numerically less than 1.

$$V_2 = 1.00 \text{ L} \times \frac{760. \text{ torr}}{1100. \text{ torr}} = 0.691 \text{ L}$$

Step 3. Check the answer. Since 0.691 L is smaller than 1.00 L, we know that we did indeed decrease the volume to increase the pressure.

Remember that in Chapter 1 we learned always to check the answer to see if it's reasonable. In this pressure-volume problem, imagine what is actually happening in the cylinder.

Exercise 11.3

A 2.00-L sample of an ideal gas has a pressure of 1.00 atm. What would be the volume of the gas at the same temperature but at a pressure of 0.750 atm?

Exercise 11.4

A sample of gas has a volume of 6.50 L and a pressure of 900. torr. While the temperature remains constant, the volume of the gas is increased to 10.0 L. What is the new pressure of the gas?

EXAMPLE 11.5

A fixed mass of gas at constant temperature occupies a volume of 554 mL and has a pressure of 1.13 atm. What will be its new pressure in torr if the volume is decreased to 198 mL?

Strategy and Solution: Let $V_1 = 554$ mL, $P_1 = 1.13$ atm, $V_2 = 198$ mL.

Even though this problem has new units and contains unit changes at the end, it is basically the same problem.

Step 1. We are looking for a new pressure. Using Boyle's law, set up an equation for P_2.

$$P_2 = P_1 \times \frac{V_1}{V_2}$$

Step 2. Insert the numbers and do the arithmetic. Multiply the original pressure by a ratio of the volumes numerically greater than one.

$$P_2 = 1.13 \text{ atm} \times \frac{554 \text{ mL}}{198 \text{ mL}} = 3.16 \text{ atm}$$

Step 3. Check the answer. P_2 is larger than P_1, which is to be expected from Boyle's law.

Step 4. Convert the answer to torr, using the conversion factor one atm = 760 torr.

$$3.16 \text{ atm} \times \frac{760 \text{ torr}}{\text{atm}} = 2.40 \times 10^3 \text{ torr}$$

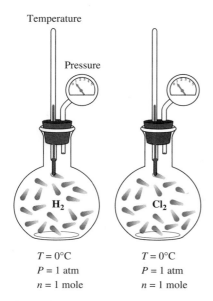

Figure 11.11 Illustration of Avogadro's hypothesis. The flasks contain the same number of molecules at equal temperature, pressure, and volume, although one flask contains hydrogen and the other contains chlorine.

Caution: Avogadro's law is true only for gases, not for liquids or solids.

Exercise 11.5

The volume of a gas sample is increased from 3.1 L to 5.7 L at constant temperature. The original pressure of the gas was 695 torr. In atmospheres, what is the new pressure of the sample?

11.5 Relation of Volume to Amount of Sample: Avogadro's Law

After studying this section, you should be able to:

- State Avogadro's hypothesis.
- State Avogadro's law in words and mathematically.
- Perform calculations using Avogadro's law.

In the early 1800's, a French chemist, Joseph Gay-Lussac, experimented with chemical reactions of gases and found that, at the same temperature and pressure, gases of different kinds reacted chemically in volumes that were always in ratios of whole numbers. For example, exactly two liters of H_2 gas react with exactly one liter of O_2 gas to form H_2O. Although there was an important fundamental idea to be learned from these facts, they were unexplained until Amadeo Avogadro proposed the theory <u>that *equal volumes of gases at equal temperature and pressure contain equal numbers of molecules, no matter what kinds of molecules make up the gases*</u>. This statement is known as **Avogadro's hypothesis,** illustrated in Figure 11.11.

Avogadro's hypothesis leads to another useful relation called *Avogadro's law*. Suppose we have one liter of gas. At the same temperature and pressure, it will contain one half as many molecules as do two liters of gas and one-third as many molecules as three liters. In fact, *at constant temperature and pressure, the volume of a gas sample is directly proportional to the number of moles of gas in the sample*. That is, as the number of moles increases, the volume increases; as the number of moles decreases, the volume decreases. More gas, more volume. Less gas, less volume. Figure 11.12 illustrates Avogadro's law.

Shown mathematically, Avogadro's law looks like this:

$$V = kn \qquad \text{(at constant pressure and temperature)}$$

The number of moles in this equation is given by n. As before, k is a constant whose value we may or may not know, depending on the kind of experiment. When a variable (in this case, V) is equal to the product of a constant number and another variable (in this case, n) the two variables are said to be **directly** **proportional** to one another.

increase one increase other etc.

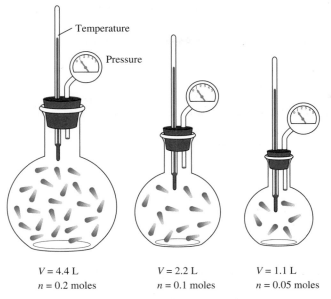

$V = 4.4$ L
$n = 0.2$ moles

$V = 2.2$ L
$n = 0.1$ moles

$V = 1.1$ L
$n = 0.05$ moles

Figure 11.12 Illustration of Avogadro's law. For all three flasks to be at $0°C$ and 1 atm pressure, the smaller flasks must contain fewer molecules. In all these cases, $V = 22 \times n$.

We can write the $V = kn$ equation for an experiment in which the number of moles and the volume are changed.

$$V_1 = kn_1 \qquad V_2 = kn_2$$

or

$$\frac{V_1}{n_1} = k \qquad \text{and} \qquad \frac{V_2}{n_2} = k$$

In general, for any gas sample at constant temperature and pressure,

$$\frac{V_1}{n_1} = \frac{V_2}{n_2}$$

This equation is convenient for working problems.

EXAMPLE 11.6

A sample of gas containing 4.66 moles has a volume 2.25 L. Under conditions of constant pressure and temperature, the number of moles of gas is increased to 5.22. What is the new volume of the sample?

Strategy and Solution: Let $V_1 = 2.25$ L, $n_1 = 4.66$ mol, $n_2 = 5.22$ mol.

Step 1. We set up an equation to give V_2. The new volume will be given by the original volume multiplied by a conversion factor containing the numbers of moles.

$$V_2 = V_1 \times \frac{n_2}{n_1}$$

Step 2. Multiply the original volume by a mole ratio numerically greater than one.

$$V_2 = 2.25 \text{ L} \times \frac{5.22 \text{ mol}}{4.66 \text{ mol}} = 2.52 \text{ L}$$

Step 3. Check the answer. Since 2.52 L is larger than 2.25 L, we know that we have matched the problem to the physical change that occurred. Don't forget to do this kind of check each time.

Exercise 11.6

A 2.50-mole sample of a certain gas has a volume of 8.74 L. At constant temperature and pressure, the number of moles is decreased to 1.25. Calculate the new volume of the gas.

Exercise 11.7

A sample of gas decreases in volume from 30.4 L to 18.6 L when some of the gas escapes. The final number of moles is 5.32. How many moles of gas were originally in the sample? (Temperature and pressure are constant.)

This is the same Gay-Lussac who later, in 1808, discovered the combining volumes relation that led Avogadro to make his hypothesis.

11.6 Relation Between Pressure and Temperature: Gay-Lussac's Law

After studying this section, you should be able to:

- State Gay-Lussac's law in words and mathematically.
- Perform calculations using Gay-Lussac's law.

We know that temperature is a measure of molecular motion, that is, of kinetic energy. So when the temperature of a gas sample increases, its molecules move faster, colliding with the container walls more often and harder. We remember from the kinetic molecular theory that the pressure of a gas is caused by collisions of its molecules with the walls of its container. Therefore, it should come as no surprise that the pressure of a gas sample increases when the temperature increases. In fact, it is found that *the pressure of a fixed mass of gas at constant volume is directly proportional to its absolute temperature*. This law was reported in 1802 by Joseph Gay-Lussac, who became famous for this and several other important discoveries. The algebraic statement of **Gay-Lussac's law** is

$$P = kT$$

or, for a sample of gas at two sets of temperature and pressure,

$$\frac{P_1}{T_1} = k \qquad \text{and also} \qquad \frac{P_2}{T_2} = k$$

therefore

$$\frac{P_1}{T_1} = \frac{P_2}{T_2}$$

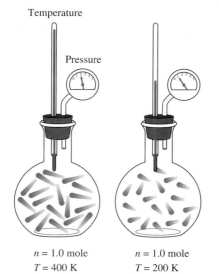

Temperature

Pressure

$n = 1.0$ mole	$n = 1.0$ mole
$T = 400$ K	$T = 200$ K
$P = 2$ atm	$P = 1$ atm

Figure 11.13 Illustration of Gay-Lussac's law. As the Kelvin temperature decreases, the pressure decreases proportionately.

$273°K + C°$

where T is the Kelvin temperature. Figure 11.13 is an illustration of Gay-Lussac's law.

Notice that Gay-Lussac's law is stated in terms of the absolute temperature scale rather than the Celsius scale. Graphs of the pressure of a sample of gas at different temperatures show us why we must use the Kelvin temperature for gas calculations. Figure 11.14 graphs the pressure of a fixed amount of gas at constant volume between the temperatures of 0°C and 100°C. Notice that when the temperature reaches 0°C, the gas still has most of its pressure.

If we look at the equation $P = kT$, where we let T be the temperature on the Celsius scale, we find that the equation requires the pressure of the gas to be zero when the Celsius temperature falls to zero ($P = k \times 0 = 0$). But physically, the gas simply does not do that. The temperature in your food freezer is well below 0°C, but the air in the freezer still has pressure (otherwise, the freezer might implode). The Celsius scale clearly can't be used in Gay-Lussac's law.

Since we can easily reach temperatures below 0°C (the temperature of dry ice, for example, is −78.5°C), let's measure the pressure of our gas sample at temperatures as low as possible, then extend the straight line on the graph even beyond that point by drawing it in (a process called extrapolation). Figure 11.15 shows what such a graph looks like. The line goes through the point representing our last measurement and continues (dashed) until it reaches the baseline, where the value of the pressure goes to zero. The temperature at this point is −273.15°C. This is the temperature at which the pressure of the sample would, according to our graph, become zero, that is, the point at which the motion of the molecules would cease. Absolute zero (−273.15°C), which is the defined zero on the Kelvin scale, was discovered in just this way by the Englishman Lord Kelvin while studying the laws of Gay-Lussac and of Charles (see the next section).

Technically, absolute zero remains theoretical because that temperature has never been achieved in the laboratory (although some experiments have come

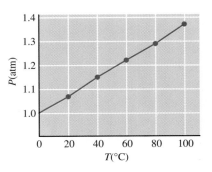

Figure 11.14 Pressure of a sample of N_2 gas in a 1-L flask as it is cooled from 100°C to 0°C. Notice that when the temperature reaches 0°C, the gas still has a pressure greater than zero.

With considerable effort, substances have been brought close to 0 K, but theoretically, no substance can have a temperature this low. However, some processes are carried out at 4 K, the boiling temperature of helium.

To cool something is to decrease the energies of its molecules. Near 0 K (−273.15°C), this becomes difficult, for there is no less-energetic place to transfer the energy.

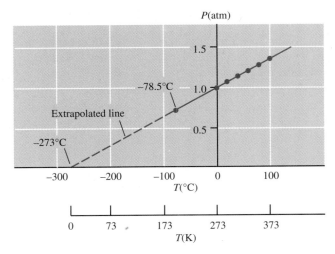

Figure 11.15 The data from Figure 11.14 are again plotted here, and a measurement at −78.5°C has been added. The line has been extrapolated to the left until it reaches a pressure of zero. This line intersects at a temperature of −273.15°C.

within a small fraction of a degree). Nor has any sample of gas ever been brought to zero pressure. Absolute zero does, however, have physical meaning when interpreted by the kinetic molecular theory and has a practical use in gas law calculations.

The practical effect of the relationship we have learned is that we must remember to convert all temperatures to the Kelvin scale in solving gas law problems. If we forget to convert to the Kelvin scale and use the Celsius temperature instead, we will get a wrong answer.

Here is a typical Gay-Lussac's law problem solved by the same kind of approach that we used with Boyle's law. Since Gay-Lussac's law involves only temperature and pressure, we use it when volume and number of moles are held constant.

EXAMPLE 11.7

A sample of gas with a fixed mass of 2.4 grams is contained in a flask with a constant volume. At a temperature of 21°C, this gas has a pressure of 745 torr. What is the new pressure if the temperature of the sample is raised to 81°C?

Strategy and Solution:

Step 1. Convert temperatures to the Kelvin scale. The initial temperature is 21°C + 273 = 294 K and the final temperature is 81°C + 273 = 354 K. The initial pressure, P_1, is 745 torr.

Step 2. Rearrange Gay-Lussac's law, solving for P_2. P_1 is multiplied by a conversion factor containing the temperatures.

$$P_2 = P_1 \times \frac{T_2}{T_1}$$

Step 3. Substitute the values and calculate. Multiply the original pressure by a ratio of the temperatures numerically greater than one.

$$P_2 = 745 \text{ torr} \times \frac{354 \text{ K}}{294 \text{ K}} = 897 \text{ torr}$$

Step 4. Check your result by visualizing a physical experiment. Temperature and pressure are directly proportional; therefore, when the temperature increases, as in this example, the pressure also increases; P_2 must be higher than P_1.

Exercise 11.8

At 298 K, a sample of gas occupies a container of constant volume. The measured pressure is 1.50 atm. When the gas is heated, the pressure is found to have increased to 5.00 atm. What is the new Kelvin temperature?

Exercise 11.9

What will be the new pressure of a sample of gas if its original pressure was 795 torr at 185°C and the temperature is lowered to 2°C?

11.7 Relation of the Volume of a Gas to Its Temperature: Charles' Law

After studying this section, you should be able to:

- State Charles' law in words and mathematically.
- Perform calculations using Charles' law.

Picture again the piston and cylinder apparatus similar to the one we used for our pressure–volume measurements in Section 11.4. No gas can leak in or out. We imagine that the pump has a frictionless piston that can move freely, so that the pressure inside will always be the same as that outside, that is, constant. Now we heat the pump and the gas inside. What happens to the piston? Look at Figure 11.16. We observe that it moves up. As the temperature of the gas inside the cylinder increases, its volume also increases, pushing the piston up.

The law describing the effect of temperature on the *volume* of a gas sample was discovered by the Frenchman Jacques Charles and announced in the early 1800's at about the same time as Gay-Lussac's law. **Charles' law** states that *the volume of a fixed quantity of gas at constant pressure is directly proportional to its absolute temperature*. At constant pressure, Charles' law is stated mathematically as

$$V = kT \qquad \text{or} \qquad \frac{V}{T} = k$$

and

$$\frac{V_1}{T_1} = k \qquad \frac{V_2}{T_2} = k$$

$$\frac{V_1}{T_1} = \frac{V_2}{T_2}$$

An illustration of Charles' law is shown in Figure 11.17.

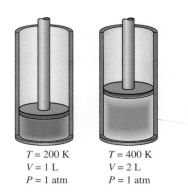

$T = 200$ K $T = 400$ K
$V = 1$ L $V = 2$ L
$P = 1$ atm $P = 1$ atm

Figure 11.16 Illustration of Charles' law. As the temperature rises, the volume increases proportionately. The cylinders contain the same amount of gas at the same pressure.

When immersed in liquid nitrogen ($-196°C$), the volume of air in a balloon decreases so much that the balloon is flattened.

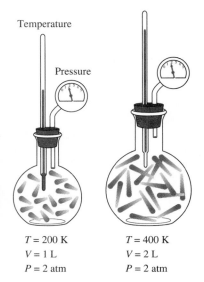

$T = 200$ K $T = 400$ K
$V = 1$ L $V = 2$ L
$P = 2$ atm $P = 2$ atm

Figure 11.17 An illustration of Charles' law. At the same pressure and with the same amount of gas, the volume increases proportionally as the temperature increases.

Profiles in Science

Charles and Gay-Lussac

December 1, 1783 Charles made his historic ascent in a hydrogen-filled balloon.

Both Charles and Gay-Lussac were Frenchmen who lived during the late eighteenth and early nineteenth centuries. Although their names are linked through the gas laws named after them and although they had some strikingly similar interests, they did not collaborate.

Jacques Charles (1746–1825) was a minor clerk in a government office. Dismissed because of budget cuts, he took up the study of gases more or less as a hobby. He became well-known in France because he was the second person ever to ascend in a balloon and the first to pioneer the use of hydrogen as a lifting gas. Charles' balloon was made of silk covered with a rubber solution. Inflating the balloon with hydrogen took several days. The balloon stayed aloft for almost 45 minutes and traveled about 15 miles. When it landed in a village, however, the people were so frightened that they tore it to shreds.

Charles' advances in ballooning led the king to set him up in a laboratory of his own, where Charles continued his investigations of gas behavior. In 1787, he discovered what we now know as Charles' law, namely that the volume of a gas at constant pressure is directly proportional to its temperature.

Joseph-Louis Gay-Lussac (1775–1850), on the other hand, was a science student from his earliest years. He was long considered to be the best student that the École Polytechnique had ever had and in his later years became one of its most famous professors. Gay-Lussac lived to become France's premier chemist. He is remembered not only for the law that bears his name, but for discovering the law by which gases combine, for pioneer work with alkali metals (at that time only a laboratory curiosity), for methods of analyzing metals, and for numerous other substantial contributions to science.

Charles did not even publish the law that carries his name. That action was taken 15 years later by Gay-Lussac, who had heard of Charles' experiments and arranged to get them into print. Given that he knew Boyle's law and had Charles' results, it was not hard for Gay-Lussac to perform and publish experiments of his own relating *pressures* and temperatures of gases, just as those of Charles related *volumes* and temperatures.

Interestingly enough, Gay-Lussac was also fascinated by ballooning and made several ascents to conduct experiments in the thin atmosphere. In 1804, he reached a record altitude of more than four miles. That record stood for decades.

To solve problems involving the effects of temperature on the volumes of gases at constant pressure, we can use exactly the same method that we used to work problems using Boyle's law and Gay-Lussac's law.

EXAMPLE 11.8

A fixed amount of gas at constant pressure of 755 torr has a volume of 16.0 L at a temperature of 298 K. What will be its volume if the temperature is raised to 500. K?

Strategy and Solution: Let $V_1 = 16.0$ L, $T_1 = 298$ K, $T_2 = 500.$ K.

Step 1. Convert temperatures to the absolute scale if they are not already given that way. In this problem, the temperatures are already in kelvins.

Step 2. Identify the variable whose new value is required, in this case volume. Rearrange Charles' law, solving for V_2. The initial volume must be multiplied by a conversion factor containing the temperatures.

$$V_2 = V_1 \times \frac{T_2}{T_1}$$

Step 3. Multiply the original volume by a temperature ratio greater than one. The higher temperature is in the numerator and the lower temperature is in the denominator.

$$V_2 = 16.0 \text{ L} \times \frac{500. \text{ K}}{298 \text{ K}} = 26.8 \text{ L}$$

Step 4. Check your result. Because the temperature increased in this problem, the volume also increased.

EXAMPLE 11.9

A fixed mass of gas occupies a volume of 1.00 L at a temperature of 100.°C. At what Celsius temperature will the volume be 2.00 L?

Strategy and Solution:

Step 1. Convert the given temperature to kelvins.

$$100.°\text{C} + 273 = 373 \text{ K}$$

Step 2. Identify the variable whose new value is required. Rearrange Charles' law, solving for T_2.

$$T_2 = T_1 \times \frac{V_2}{V_1}$$

Step 3. Because the volume increases, the temperature increases. Multiply the initial Kelvin temperature by a ratio of volumes numerically greater than one.

$$T_2 = 373 \text{ K} \times \frac{2.00 \text{ L}}{1.00 \text{ L}} = 746 \text{ K}$$

Step 4. Check your result and convert the new absolute temperature to degrees Celsius.

$$746 \text{ K} - 273 = 473°\text{C}$$

Exercise 11.10

A sample of an ideal gas at 298 K occupies a volume of 5.25 L. Calculate the volume of the sample at 455 K.

Exercise 11.11

A 4.28-L sample of an ideal gas at 0°C is cooled at constant pressure. What is the final Celsius temperature of the sample if the final volume is 2.74 L?

11.8 The Combined Gas Law

After studying this section, you should be able to:

- Use the combined gas law to calculate the effects of changing the conditions under which samples of gas exist.

The laws of Boyle, Gay-Lussac, and Charles can be mathematically combined into a single relation. For a fixed amount of any gas,

$$\frac{PV}{T} = k$$

$$\frac{P_1V_1}{T_1} = k \qquad \frac{P_2V_2}{T_2} = k$$

$$\frac{P_1V_1}{T_1} = \frac{P_2V_2}{T_2}$$

This is called the **combined gas law.** In working with the three individual laws, we were required to hold the amount of gas and one other variable constant, but the combined law requires only that the amount of gas be constant. As with the other gas laws, we must use Kelvin temperature. Following the same approach we used with the individual laws, we can solve the combined law for the change in any one of the variables in terms of the other two.

EXAMPLE 11.10

Rearrange the combined gas law equation so that the new temperature, T_2, appears alone on the left side of the equation.

Strategy and Solution:

Step 1. Begin with the combined equation. Multiply both sides by T_2.

$$T_2 \times \frac{P_1V_1}{T_1} = \frac{P_2V_2}{T_2} \times T_2$$

Step 2. Then by T_1.

$$T_1 \times T_2 \times \frac{P_1V_1}{T_1} = P_2V_2 \times T_1$$

Step 3. Divide both sides by P_1V_1.

$$T_2 \times \frac{P_1V_1}{P_1V_1} = \frac{P_2V_2}{P_1V_1} \times T_1$$

$$T_2 = T_1 \times \frac{P_2}{P_1} \times \frac{V_2}{V_1}$$

Exercise 11.12

Rearrange the combined equation so that the new volume, V_2, appears alone on the left side.

EXAMPLE 11.11

A sample of gas occupies a volume of 2.13 L at 50.°C and 720. torr. What will its new pressure be if its volume is reduced to 1.50 L and its temperature is reduced to 25°C?

Strategy and Solution: Let $P_1 = 720.$ torr, $V_1 = 2.13$ L, $T_1 = 50.°C$ $V_2 = 1.50$ L, $T_2 = 25°C$.

Step 1. Convert the Celsius degrees to kelvins.

$$50.°C + 273 = 323 \text{ K} \qquad 25°C + 273 = 298 \text{ K}$$

Step 2. Set up the equation to solve for P_2.

$$P_2 = P_1 \times \frac{V_1}{V_2} \times \frac{T_2}{T_1}$$

As we might expect, the new pressure is given by the product of the old pressure and, this time, two conversion factors, one for volume change and one for temperature change.

Step 3. Because the volume of the gas decreases, we know that the pressure will increase. Multiply the initial pressure by a ratio of volumes numerically greater than one. Because the temperature decreases, multiply the original pressure by a ratio of temperatures numerically less than one.

$$P_2 = 720. \text{ torr} \times \frac{2.13 \text{ L}}{1.50 \text{ L}} \times \frac{298 \text{ K}}{323 \text{ K}} = 943 \text{ torr}$$

Step 4. Check your answer.

Exercise 11.13

A sample of gas has a pressure of 2.15 atm and a volume of 719 mL at 21°C. What will its new Celsius temperature be when it occupies a volume of 600. mL at a pressure of 3.50 atm?

11.9 The Ideal Gas Law

After studying this section, you should be able to:

- Calculate values for the ideal gas constant, using various sets of units.
- Perform calculations using the ideal gas law.

In these equations, k is used only to represent a constant number, but the k's in the three equations do not represent the same number.

Here are three of the gas laws we have studied.

$$PV = k \qquad \frac{V}{T} = k \qquad \frac{V}{n} = k$$

Boyle's law   Charles' law   Avogadro's law

These three equations can be mathematically combined into a single equation.

$$\frac{PV}{nT} = R$$

This is usually written as

$$PV = nRT$$

where R is a constant number (replacing all of the k's above) called the **ideal gas constant,** and the equation itself is a statement of the **ideal gas law.** The value and the units of R depend on the units used for the variables. Kelvins are always used for temperature. Ordinarily, liters are used for volume and moles are used for the amount of gas. Pressure is usually given either in atmospheres or torr. Let's calculate R when pressure is in atmospheres, volume in liters, and the amount of gas in moles. If we measure the actual volume of one mole of gas at a pressure of one atmosphere and 273 K (0°C), we find it to be 22.4 L. We put these values in the ideal gas equation.

Although many consider the law itself to be an ideal one, the name of the ideal gas law comes from the fact that it works perfectly only for an "ideal" gas (see Section 11.14).

$$PV = nRT$$

therefore

$$R = \frac{PV}{nT}$$

at STP all gases occupy 22.4L / mol

$$R = \frac{22.4 \text{ L} \times 1.00 \text{ atm}}{1.00 \text{ mol} \times 273 \text{ K}} = 0.0821 \frac{\text{L atm}}{\text{mol K}}$$

Similarly, if pressure is in torr instead of atmospheres, then

$$R = \frac{62.4 \text{ L torr}}{\text{mol K}}$$

The ideal gas equation can be used in many different kinds of problems. For example, it allows us to calculate the value of any of the variables by measuring all the others.

EXAMPLE 11.12

From this point onward, we will assume that gases obey the ideal gas law unless informed otherwise.

A sample of unknown gas has a pressure of 651 torr, a temperature of 295 K, and a volume of 3.66 L. How many moles of gas are in the sample?

Strategy and Solution:

Step 1. If necessary, convert the given temperature to kelvins. We note that the temperature is already given in kelvins, so we need not convert it.

Step 2. Rearrange the ideal gas equation to give the number of moles.

$$n = \frac{PV}{RT}$$

Step 3.   Insert the values and do the arithmetic.

$$n = \frac{(651 \text{ torr})(3.66 \text{ L})}{(62.4 \text{ L torr/mol K})(295 \text{ K})} = 0.129 \text{ mol}$$

Exercise 11.14

A sample of gas occupies a 4.00-L flask at 52°C and has a pressure of 1.75 atm. How many moles of gas are present?

EXAMPLE 11.13

What is the pressure in torr exerted by 0.331 moles of the gas Ar, when the sample is contained in a flask with a volume of 400. mL and a temperature of 22°C?

Strategy and Solution:

Step 1.   Convert temperatures to the absolute scale.

$$22°C + 273 = 295 \text{ K}$$

Step 2.   Notice that we are given a problem that contains units not found in any of the forms of R that we already know. We will have to convert the units of the problem to fit an R that we know or else find a value of R in the appropriate units. We choose to use $R = 62.4 \text{ L torr/mol K}$. Now we must convert the 400. mL in the problem to 0.400 L.

Step 3.   Rearrange the ideal gas equation to give the required variable.

$$P = \frac{nRT}{V}$$

Step 4.   Insert the values of the variables and of R and do the arithmetic.

$$P = \frac{(0.331 \text{ mol})(62.4 \text{ L torr/mol K})(295 \text{ K})}{0.400 \text{ L}} = 1.52 \times 10^4 \text{ torr}$$

Exercise 11.15

A flask with a volume of 250. mL contains 0.0130 moles of gas at a pressure of 1.25 atm. What is the Celsius temperature of the gas?

Exercise 11.16

A 55.5-gram sample of N_2 gas has a pressure of 850. torr at 273 K. What is the volume of the sample? (Hint: You will need to use the molecular weight of N_2 to find the number of moles in the sample.)

11.10 Standard Temperature and Pressure; Standard Molar Volume

After studying this section, you should be able to:

• Use STP values in working gas-law problems.

A set of conditions called **standard temperature and pressure, STP,** has been established for convenience in reporting the results of experiments. The standard pressure is exactly one atmosphere, and the standard temperature is 0°C. In scientific writing, gas volumes are usually given at STP.

As we saw in Section 11.9, the volume of one mole of gas measured at one atm and 0°C, that is, at STP, is 22.4 L. Because it is measured at standard conditions, this volume is given the special name, the **standard molar volume.** Because data concerning gases are often given at STP, the standard molar volume is a useful value to memorize so that it can be used in solving gas law problems.

When you work problems starting with the standard molar volume, you must always remember two things:

1. *The standard molar volume can be used only for gases, not for liquids or solids.* (For example, 22.4 L of liquid water contains not just one mole but 1,244 moles of water.)

2. *As defined, the standard molar volume is used only at STP.*

Within these restrictions, the standard molar volume is handy for several kinds of calculations, one of which is to find the number of moles in a sample of gas whose volume at STP is known. The standard molar volume is a conversion factor between volume at STP and the number of moles of gas.

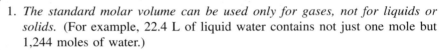

$$\frac{22.4\ \text{L}}{\text{mol}} \qquad \text{or} \qquad \frac{\text{mol}}{22.4\ \text{L}}$$

28.2 cm

28.2 cm

28.2 cm

$(28.2\ \text{cm})^3 = 22,400\ \text{cm}^3 = 2.24\ \text{L}$

The volume of one mole of gas at STP is 22.4 L. This 22.4-L box, which is 28.2 cm on a side, is large enough to hold a basketball, leaving room to spare.

EXAMPLE 11.14

A sample of gas occupies a volume of 8.32 L at STP. How many moles of gas are in the sample?

Solution Map: ⎡volume⎤ ⟶ ⎡moles⎤

Solution: Use the molar gas volume. Set up and calculate.

$$8.32\ \text{L} \times \frac{\text{mol}}{22.4\ \text{L}} = 0.371\ \text{mol}$$

At STP there must be less than one mole of gas in a volume smaller than 22.4 L.

Exercise 11.17

How many moles of gas will occupy 17.2 liters at STP?

Exercise 11.18

What volume will 5.2 moles of gas occupy at STP?

11.11 Gas Densities and Molar Mass

After studying this section, you should be able to:

- Calculate the molar mass of a gas sample from the values of its other properties.
- Calculate the density of a gas from its molar mass.

Given the pressure, volume, and temperature of a sample of gas, we can calculate its molar mass. The following example illustrates how.

EXAMPLE 11.15

A sample of gas weighing 1.55 grams has a volume of 0.250 L at a temperature of 21°C and a pressure of 0.966 atm. What is its molar mass?

Strategy and Solution:

Step 1.   Convert the temperature to kelvins.

$$21°C + 273 = 294 \text{ K}$$

Step 2.   Rearrange the ideal gas law, solving for n.

$$n = \frac{PV}{RT}$$

Step 3.   Insert the values and do the arithmetic.

$$n = \frac{(0.966 \text{ atm})(0.250 \text{ L})}{(0.0821 \text{ L atm/mol K})(294 \text{ K})} = 0.0100 \text{ mol}$$

Step 4.   Find the molar mass. Divide the grams by the number of moles.

$$\frac{1.55 \text{ g}}{0.0100 \text{ mol}} = 155 \text{ g/mol}$$

Exercise 11.19

A sample of gas weighing 0.700 grams has a volume of 1.23 L at 300. K and 0.500 atm. What is the molar mass of the gas?

We know that we can calculate the number of moles of a gas from its mass and its molar mass. For example, the number of moles of 4.0 grams of H_2 gas is

$$4.0 \text{ g } H_2 \times \frac{\text{mol } H_2}{2.0158 \text{ g } H_2} = 2.0 \text{ mol } H_2$$

In general,

$$n = \frac{\text{mass}}{\text{molar mass}}$$

We can substitute for n in $PV = nRT$.

$$PV = \frac{\text{mass}}{\text{molar mass}} RT$$

Rearranging, we get

$$P \times \text{molar mass} = \frac{\text{mass}}{V} RT$$

Recall that the ratio of mass to volume is density.

$$\frac{\text{mass}}{V} = d$$

So, if we substitute d, we have a new form of the ideal gas law.

$$P \times \text{molar mass} = dRT$$

Like other matter, gases can be characterized by their densities, d, usually given as the number of grams of gas per liter at STP. Table 11.3 lists the densities of several gases. Although the densities of solids and liquids are little affected by pressure and temperature changes, the density of a gas depends entirely on its temperature and pressure. Gases at STP are usually only about one-thousandth as dense as common liquids or solids.

Table 11.3 Densities of Gases Compared to Those of Solids and Liquids

Substance	State	Density g/cm^3 ($0°C$, 1 atm)
H_2	Gas	0.000089
O_2	Gas	0.0014
Freon 113 ($C_2Cl_3F_3$)	Gas	0.0084
Water	Liquid	1.000
Aluminum	Solid	2.70
Gold	Solid	19.3

If we know the density of a gas at a particular pressure and temperature, we can calculate its molar mass using the new form of the ideal gas law. Let's look at an example.

EXAMPLE 11.16

Measuring the density of a gas is a widely used method for determining molar mass.

The density of a gas at 0.934 atm and 255 K is 0.714 g/L. What is the molar mass of the gas?

Strategy and Solution:

Step 1. Rearrange the ideal gas law, solving for molar mass.

$$\text{molar mass} = \frac{dRT}{P}$$

323 Gas Densities and Molar Mass

Step 2. Substitute and calculate.

$$\text{molar mass} = \frac{(0.714 \text{ g/L })(0.0821 \text{ L atm/mol K})(255 \text{ K})}{0.934 \text{ atm}} = 16.0 \text{ g/mol}$$

Exercise 11.20

The STP density of a gas is 5.28 g/L. What is the molar mass of the gas?

To find the density of a sample of gas whose formula is known, you need to know only the conditions under which the gas exists.

EXAMPLE 11.17

What is the density of gaseous oxygen, O_2, at STP?

Strategy and Solution:

Step 1. Find the molar mass. For O_2 it is 31.9988 g/mol.

Step 2. Rearrange the ideal gas law, solving for density.

$$d = \frac{P \times \text{molar mass}}{RT}$$

Step 3. Substitute and calculate. Notice how the units cancel to give grams per liter.

$$d = \frac{(1.00 \text{ atm})(31.9988 \text{ g/mol})}{(0.0821 \text{ L atm/mol K})(273 \text{ K})} = 1.43 \text{ g/L}$$

EXAMPLE 11.18

What is the density of Cl_2 gas at 298 K and 750. torr?

Strategy and Solution:

Step 1. Find the molar mass. It is 70.9054 g/mol.

Step 2. Rearrange the ideal gas law, solving for density.

$$d = \frac{P \times \text{molar mass}}{RT}$$

Step 3. Substitute and calculate.

$$d = \frac{(750. \text{ torr})(70.9054 \text{ g/mol})}{(62.4 \text{ L torr/mol K})(298 \text{ K})} = 2.86 \text{ g/L}$$

Exercise 11.21

What is the density of carbon dioxide, CO_2, at 31°C and 801 torr?

Exercise 11.22

What is the density of gaseous ammonia, NH_3, at 0.824 atm and 25°C?

11.12 Dalton's Law of Partial Pressures

After studying this section, you should be able to:

- Calculate the total pressure of a mixture of gases from their partial pressures.
- Calculate the pressure of a gas collected over water.

Let's consider an imaginary experiment. At 0°C we place one mole of He gas in a box that has a volume of 22.4 liters. The pressure that the He exerts will be one atmosphere. (Why?) Suppose, instead, that we place one mole of O_2 in the same box. The pressure of the O_2 would also be one atmosphere. The individual pressures of the two gases depend only on the number of moles in the box, the volume, and the temperature—not on the kind of molecules or their masses. Nothing new so far. But suppose we put both the mole of He and the mole of O_2 into the same box. There will be two moles of gas in the box, and the pressure will be two atmospheres. Look at Figure 11.18. The O_2 molecules will exert one atmosphere of pressure and the He molecules will exert one atmosphere of pressure, making the total pressure the sum of the individual pressures of the two gases. This is an example of **Dalton's law of partial pressures,** which states that *when two or more gases are present in the same container, the total pressure is the sum of the pressures exerted by the individual gases, each behaving as if it were in the container alone*. The pressure exerted by a single kind of gas in a mixture of gases is called the **partial pressure** of that gas. Dalton's law can be expressed mathematically as the sum of the partial pressures.

$$P_{total} = P_1 + P_2 + P_3 + \cdots$$

In this equation, the combined pressure of all the gases in the mixture is P_{total}, and the individual partial pressures of the gases (1, 2, 3) are P_1, P_2, P_3. For the case discussed above, we might write the equation this way:

$$P_{total} = P_{He} + P_{O_2}$$

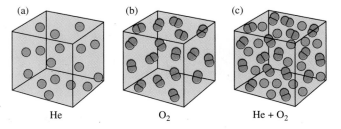

(a) He   (b) O_2 (c) He + O_2

Figure 11.18 Each box has a volume of 22.4 L and a temperature of 0°C. (a) One mole of He. Pressure is 1 atm. (b) One mole of O_2. Pressure is 1 atm. (c) One mole of He and one mole of O_2. Pressure of He is 1 atm; pressure of O_2 is 1 atm. Total pressure is 2 atm.

EXAMPLE 11.19

What is the total pressure of a mixture of 0.972 atm of O_2, 0.012 atm of CO, and 0.133 atm of NO?

Solution: From Dalton's law, we know that the total pressure is the sum of the partial pressures.

$$0.972 \text{ atm} + 0.012 \text{ atm} + 0.133 \text{ atm} = 1.117 \text{ atm}$$

EXAMPLE 11.20

A mixture of gas contains 3.8 gram of N_2 and 0.80 gram of O_2 in a 10.00-L container at 305 K (Figure 11.19). What is the total pressure of the mixture?

Strategy and Solution:

Step 1. Find the number of moles of each gas.

$$3.8 \text{ g } N_2 \times \frac{\text{mol } N_2}{28.0134 \text{ g } N_2} = 0.14 \text{ mol } N_2$$

$$0.80 \text{ g } O_2 \times \frac{\text{mol } O_2}{31.9988 \text{ g } O_2} = 0.025 \text{ mol } O_2$$

Step 2. Find the partial pressure of each gas using the ideal gas law.

$$PV = nRT \longrightarrow P = \frac{nRT}{V}$$

$$P = \frac{(0.14 \text{ mol})(0.0821 \text{ L atm/mol K})(305 \text{ K})}{10.00 \text{ L}} = 0.35 \text{ atm}$$

$$P = \frac{(0.025 \text{ mol})(0.0821 \text{ L atm/mol K})(305 \text{ K})}{10.00 \text{ L}} = 0.063 \text{ atm}$$

Step 3. Calculate the total pressure using Dalton's law.

$$0.35 \text{ atm} + 0.063 \text{ atm} = 0.41 \text{ atm}$$

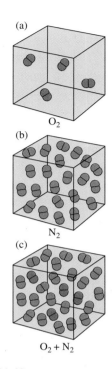

Figure 11.19 (a) Oxygen, 0.025 moles, $P = 0.063$ atm. (b) Nitrogen, 0.14 moles, $P = 0.35$ atm. (c) 0.025 moles of oxygen and 0.14 moles of nitrogen together, $P = 0.063$ atm + 0.35 atm = 0.41 atm.

Exercise 11.23

What is the total pressure of a mixture of 15.5 grams of N_2 and 0.918 atm of SO_2 occupying a 7.00-L flask at 0°C?

Mixtures of gases are used in many experiments and chemical reactions. In several common laboratory experiments, for example, various gases are collected in bottles of water (Figure 11.20). If gases are in contact with liquid water, there will always be some gaseous water mixed with the other gas. In the experiment shown in Figure 11.20, the H_2 gas trapped in the bottle also contains some

What we call humidity at any given temperature is the ratio of the actual partial pressure of gaseous water in the atmosphere to the partial pressure of gaseous water if the atmosphere were saturated with water.

Figure 11.20 H_2 gas is collected by replacement of liquid water. The collected gas is a mixture of H_2 and gaseous H_2O.

gaseous H_2O. The pressure of the gas above the liquid is caused by both the gaseous H_2O molecules and the H_2 molecules.

$$P_{total} = P_{H_2} + P_{H_2O}$$

To find the pressure of the H_2 alone, we need to measure the total pressure of the gas mixture and subtract from that value the partial pressure of the gaseous H_2O.

$$P_{H_2} = P_{total} - P_{H_2O}$$

Vapor pressure is discussed in detail in Section 12.5

The pressure exerted by gaseous water in the presence of liquid water is called the **vapor pressure** of the water. Convenient tables like the one in Appendix B allow you to look up the value of water vapor pressure at various temperatures.

EXAMPLE 11.21

The total pressure of a sample of H_2 collected over water at 25°C is 749 torr. The vapor pressure of water at 25°C is 24 torr. What is the partial pressure of the H_2 in the sample?

Solution: We know from Dalton's law that

$$P_{H_2O} + P_{H_2} = P_{total}$$

Then

$$P_{H_2} = P_{total} - P_{H_2O} = 749\,\text{torr} - 24\,\text{torr} = 725\,\text{torr}$$

> **Exercise 11.24**
>
> A sample of nitrogen is collected in a flask over water at 30°C, at which temperature the vapor pressure of water is 32 torr. The total pressure in the flask is 0.750 atm. What is the pressure of the nitrogen?

11.13 Stoichiometric Calculations Involving Gases

After studying this section, you should be able to:

- Use the gas laws to make stoichiometric calculations involving gases.

In any calculation that uses a chemical equation and includes both gases and substances that are not gases, work the problem by converting the starting data into numbers of moles. Use the mole ratios from the chemical equation, then convert back to the desired units. The only difference between this kind of problem and an ordinary mole ratio problem is that, with these problems, the gas laws must sometimes be used to find the numbers of moles.

EXAMPLE 11.22

Metallic sodium will react vigorously with liquid water to give hydrogen gas. The balanced equation for this reaction is

$$2\ Na(s) + 2\ H_2O(l) \longrightarrow 2\ NaOH(aq) + H_2(g)$$

In an experiment, 3.31 grams of Na were reacted with excess H_2O to give H_2 gas. How many liters of H_2 at STP were produced?

Solution Map: grams of Na $\longrightarrow$ moles of Na $\longrightarrow$

moles of H_2 $\longrightarrow$ volume of H_2 at STP

Strategy and Solution:

Step 1. Calculate the numbers of moles of the reactants. Because Na is limiting, we don't have to worry about the number of moles of H_2O.

$$3.31\ g\ Na \times \frac{mol\ Na}{22.9898\ g\ Na} = 0.144\ mol\ Na$$

Step 2. Use the mole ratio method to convert moles of Na to moles of H_2.

$$0.144\ mol\ Na \times \frac{1\ mol\ H_2}{2\ mol\ Na} = 7.20 \times 10^{-2}\ mol\ H_2$$

Step 3. Convert the number of moles of H_2 to liters at STP. Use the standard molar volume as a conversion factor.

$$7.20 \times 10^{-2}\ mol\ H_2 \times \frac{22.4\ L\ H_2}{mol\ H_2} = 1.61\ L\ H_2$$

If you wish you can put all the steps together in a single calculation.

$$3.31\ g\ Na \times \frac{mol\ Na}{22.9898\ g\ Na} \times \frac{1\ mol\ H_2}{2\ mol\ Na} \times \frac{22.4\ L\ H_2}{mol\ H_2} = 1.61\ L\ H_2$$

EXAMPLE 11.23

A sample of O_2 with a volume of 15.8 L at 20.°C and 755 torr is reacted with excess H_2 to form water. How many grams of water are produced?

$$2 H_2(g) + O_2(g) \longrightarrow 2 H_2O(l)$$

Solution Map: volume of O_2 $\longrightarrow$ moles of O_2 $\longrightarrow$

moles of H_2O $\longrightarrow$ grams of H_2O

Strategy and Solution:

Step 1. Find the number of moles of O_2, using $PV = nRT$.

$$n = \frac{(755 \text{ torr})(15.8 \text{ L})}{(62.4 \text{ L torr/mol K})(293 \text{ K})} = 0.652 \text{ mol } O_2$$

Step 2. Calculate the number of moles of H_2O using the mole ratio method.

$$0.652 \text{ mol } O_2 \times \frac{2 \text{ mol } H_2O}{1 \text{ mol } O_2} = 1.30 \text{ mol } H_2O$$

Step 3. Convert to grams of H_2O.

$$1.30 \text{ mol } H_2O \times \frac{18.0152 \text{ g } H_2O}{\text{mol } H_2O} = 23.4 \text{ g } H_2O$$

Exercise 11.25

Limestone, $CaCO_3$, reacts when heated to form CaO and CO_2. A sample of limestone weighing 5.20 grams was heated to produce reaction. How many liters of CO_2, measured at 300. K and 730. torr, were produced?

EXAMPLE 11.24

A sample of $KClO_3$ mixed with inert material is heated strongly until all of the O_2 is driven off. The O_2 is collected over water, and the volume of the gas is found to be 0.981 L at 20.°C and a total pressure of 720. torr. What was the mass of the $KClO_3$ alone? The vapor pressure of H_2O at 20.°C is 17.5 torr.

$$2 KClO_3(s) \longrightarrow 3 O_2(g) + 2 KCl(s)$$

Solution Map:

total pressure $\longrightarrow$ partial pressure of O_2 $\longrightarrow$

moles of O_2 $\longrightarrow$ moles of $KClO_3$ $\longrightarrow$ grams of $KClO_3$

Strategy and Solution:

Step 1. Find the partial pressure of O_2. The partial pressure of O_2 is the total gas pressure less the vapor pressure of the water.

$$(720. \text{ torr total}) - (17.5 \text{ torr } H_2O) = 702 \text{ torr } O_2$$

Step 2. Use the ideal gas law to find the number of moles of O_2 produced. We must convert the temperature to kelvins: $20.°C + 273 = 293$ K. Then

$$\text{mol } O_2 = \frac{PV}{RT} = \frac{(702 \text{ torr})(0.981 \text{ L})}{(62.4 \text{ L torr/mol K})(293 \text{ K})} = 3.77 \times 10^{-2} \text{ mol } O_2$$

Step 3. Use the mole ratio method to find the number of moles of $KClO_3$.

$$3.77 \times 10^{-2} \text{ mol } O_2 \times \frac{2 \text{ mol } KClO_3}{3 \text{ mol } O_2} = 2.51 \times 10^{-2} \text{ mol } KClO_3$$

Step 4. Use the molar mass to calculate the number of grams of $KClO_3$.

$$2.51 \times 10^{-2} \text{ mol } KClO_3 \times \frac{122.549 \text{ g } KClO_3}{\text{mol } KClO_3} = 3.08 \text{ g } KClO_3$$

Exercise 11.26

Liquid pentane, C_5H_{12}, has a density of 0.626 g/mL at 25°C. If 0.265 L of pentane is burned in an excess of oxygen, how many liters of gaseous CO_2, measured at 25°C and 0.950 atm, will be produced? (The products are CO_2 and H_2O. You will need to write a balanced equation before starting your calculation.)

11.14 Nonideality in Gases

After studying this section, you should be able to:

- List the characteristics of an ideal gas.

Careful thought about what we already know tells us that, under some circumstances, the gas laws can't operate at all. For example, we have learned that the volume of a fixed mass of gas at constant pressure is proportional to the absolute temperature, $V = kT$. Suppose we let the temperature go to zero? True, absolute zero has never been experimentally achieved, but temperatures only a few thousandths of a kelvin away have been reached. Does the volume of a gas really disappear when the temperature reaches zero? Obviously not. Matter can't have zero volume. Gas molecules take up space just as other molecules do.

Because the molecules themselves have volume that will not change appreciably with pressure, conditions in which those molecules are forced to be close together, either at high pressures or low temperatures, cause gases to deviate from the behavior described by the ideal gas law. No gas behaves ideally at temperatures very close to 0 K. Many gases are not ideal even at much higher temperatures. Consider gaseous H_2O at 600 K and 1 atm. At this temperature, the gas obeys the ideal gas law, but it begins to deviate as the temperature is lowered to about 400 K. Because the molecules of the gaseous water no longer have enough kinetic energy to resist the forces that attract them to one another at 373 K (100°C), the volume suddenly decreases to about a thousandth of what it was, and the sample becomes a liquid.

The ideal gas law gets its name not because the law is ideal, but because it requires ideal, perfect behavior of gases. An imaginary gas that would obey the ideal gas law under all conditions is called an **ideal gas,** or a perfect gas. The molecules of an ideal gas have no volume and experience no attraction to or repulsion from other molecules. No such gases exist, but many gases are nearly ideal, and all gases obey the law quite closely under certain conditions. Within their generous limits, the gas laws are useful tools that we will employ again and again for real gases.

no perfect gases but close

Summary

A sample of gas can be characterized by the values of four properties: pressure, volume, temperature, and mass (or number of moles). These four properties are interdependent. It is not possible to change the value of one property without causing a corresponding change in at least one of the other properties. Individual pairs of properties are related by various gas laws, such as Boyle's law (PV = a constant when T and n are fixed). The individual gas laws are often combined in a form known as the ideal gas law.

$$PV = nRT$$

The ideal gas law is ordinarily used to calculate the value of one property of a gas when the values of the other three properties are known.

The properties of gases are explained by the kinetic molecular theory, which describes gases as collections of widely separated molecules in rapid, random motion. According to the KM theory, pressure is caused by the collisions of molecules with the walls of their container.

Because they are commonly measured by barometers or similar devices, pressures of gases are often stated in units of torr (millimeters of mercury). A second useful pressure unit is the atmosphere; one atmosphere equals 760 torr. The SI unit of pressure is the pascal; 101,325 Pa equals 760 torr.

If we know the conditions of temperature and pressure under which a sample of gas exists and we know its density, we can calculate its molar mass. Likewise if we know the molar mass or the chemical formula of a gas and know its temperature and pressure, we can calculate its density.

Dalton's law (the total pressure of a sample of mixed gases is the sum of the partial pressures of the individual gases, each acting as if it were alone) allows us to make calculations for individual gases that are part of a mixture, such as gases collected along with water vapor.

Key Terms

pressure	*(p. 297)*	Gay-Lussac's law	*(p. 310)*
torr	*(p. 297)*	Charles' law	*(p. 313)*
atmosphere	*(p. 298)*	combined gas law	*(p. 316)*
pascal	*(p. 298)*	ideal gas constant	*(p. 318)*
manometer	*(p. 298)*	ideal gas law	*(p. 318)*
kinetic molecular theory	*(p. 298)*	standard temperature and pressure (STP)	*(p. 320)*
Boyle's law	*(p. 303)*	standard molar volume law	*(p. 320)*
inversely proportional	*(p. 303)*	Dalton's law of partial pressures	*(p. 324)*
Avogadro's hypotheses	*(p. 308)*	partial pressure	*(p. 324)*
Avogadro's law	*(p. 308)*	vapor pressure	*(p. 326)*
directly proportional	*(p. 308)*	ideal gas	*(p. 330)*

Questions and Problems

Section 11.1 Properties of Gases

Questions

1. List four properties common to all gases.
2. Describe how gases, liquids, and solids differ in terms of volume.
3. How does the pressure exerted by a gas differ from that exerted by a liquid or a solid?
4. Explain the origin of the term "torr."

Problems

Note: in the problems that follow, if a variable such as P, V, n, or T is not mentioned, its value should be assumed to be constant. In Problem 15, for example, the temperature and number of moles are constant. (In these problems, assume that all gases are ideal.)

5. Make the following pressure conversions.
   (a) 700 torr $\longrightarrow$ mm Hg
   (b) 958 mm Hg $\longrightarrow$ atm
   (c) 0.035 atm $\longrightarrow$ torr
6. Make the following pressure conversions.
   (a) 1020 torr $\longrightarrow$ atm
   (b) 156 mm Hg $\longrightarrow$ torr
   (c) 342 torr $\longrightarrow$ atm

Section 11.2 Kinetic Molecular Theory

Questions

7. List the four assumptions of the kinetic molecular theory.
8. What does the kinetic molecular theory of gases explain?

Section 11.3 Temperature and Energies of Gas Molecules

Questions

9. Explain what is meant by the average kinetic energy of the molecules of a gas.
10. How is it possible that samples of different gases at the same temperature and volume can have the same pressure?
11. Gases effuse easily. Can liquids or solids effuse easily? Explain.

Section 11.4 Boyle's Law: The Dependence of Pressure on Volume

Questions

12. State Boyle's law in your own words.
13. Design and explain an experiment that can verify Boyle's law.
14. Write three different mathematical equations, all of which are statements of Boyle's law.

Problems

15. If a sample of gas occupies 450. mL at 850. torr, what volume will it occupy
    (a) at 524 torr?
    (b) at 638 mm Hg?
    (c) at 3.69 atm?
16. A 9.25-L sample of gas has a pressure of 22.86 atm. What will be the pressure if the volume is changed
    (a) to 5.42 L?
    (b) to 2043 mL?

17. In a laboratory experiment, a gas had a volume of 2.50 L. When the volume was changed to 3.55 L, the pressure measured 514 torr. What was the original pressure in atmospheres?

18. A sample of gas has a volume of 7.20 L and pressure of 1.22 atm. After the volume is changed, the pressure measures 0.542 atm. What is the new volume?

19. If a sample of gas has a pressure of 134 torr and occupies 0.297 L, what will be its pressure when the volume is increased to 3.79 L?

Section 11.5 Relation of Volume to Amount of Sample: Avogadro's Law

Problems

20. A sample containing 3.40 moles of a certain gas occupies 148 L. What would be the volume of the gas if the sample size were increased to 5.00 moles?

21. A 3.55-L sample of gas contains 2.00 moles. What will be its volume if the number of moles is doubled?

22. Exactly 5 moles of a particular gas occupy 2.60 L. How many moles of gas are there in a 175-mL sample under the same conditions?

23. Assuming constant temperature and pressure, how many moles of gas must be added to 1.50 L of gas to increase its volume to 3.10 L? (If you cannot answer this question as it is written, explain why and state what further data are needed, if any.)

Section 11.6 Relation Between Pressure and Temperature: Gay-Lussac's Law

Questions

24. Why must you use absolute temperature when using Gay-Lussac's law?

25. What is the average kinetic energy of the molecules of a gas at 0 K?

26. How does Gay-Lussac's law explain what happens to automobile tires when they are driven for a long time on a hot day?

27. Write two mathematical statements for Gay-Lussac's law.

Problems

28. Make the following temperature conversions.
    (a) 225 K $\longrightarrow$ °C
    (b) 33 °C $\longrightarrow$ K
    (c) 432 °C $\longrightarrow$ K
    (d) 145 K $\longrightarrow$ °C

29. At 122°C, a sample of gas has a pressure of 2.25 atm. What will be its pressure at 25°C?

30. A sample of gas is heated from 317 K to 428 K. If its initial pressure was 983 torr, what will be its final pressure?

31. A sample of gas at 4.00 atm has a temperature of 211°C. What will be its temperature at 733 torr?

32. By how many degrees Kelvin would you have to raise the temperature of a gas at 175 torr to obtain a pressure of 760. torr? The initial temperature is 155°C.

33. Before lift-off, the pressure of the gas inside a space capsule is 760. torr at sea level, where the temperature is 26°C. The sealed capsule experiences an increase in temperature to 534°C during lift-off. What is the pressure inside the satellite during that time? In space, the same satellite comes to a temperature of −126°C. What is the pressure inside at that time?

Section 11.7 Relation of the Volume of a Gas to Its Temperature: Charles' Law

Questions

34. What is Charles' law? How does it explain the behavior of a balloon taken out of a deep freeze on a hot day?

35. Write two mathematical equations that are statements of Charles' law.

36. Why must you use absolute temperature when using Charles' law?

37. Describe an experiment that can verify Charles' law. (You may want to include principles from Boyle's and Gay-Lussac's laws.)

Problems

38. A sample of gas occupies 6.50 L at 298 K. What will be its volume at 555 K?

39. A 4.75-L sample of gas has a temperature of 22°C. What would be the Celsius temperature of the same number of moles of gas in a 2.55-L flask at the same pressure?

40. If a 0.500-L sample of gas increases in temperature from 155 K to 555 K, what will be its final volume?

41. By how many kelvins would you have to lower the temperature of a sample of gas to change its volume from 1.22 L to 50.0 mL if the initial temperature is 37 K? If the initial temperature is 475 K?

42. A balloon filled with gas has a volume of 2.00 L at 22°C, about room temperature. The balloon is placed in a refrigerator at a temperature of 0°C. What will be the new volume of the balloon after it has cooled to the new temperature?

Section 11.8 The Combined Gas Law

Problems

43. A sample of gas occupies 4.5 L at 25°C and 746 torr. What volume will the same gas occupy at 45°C and 760. torr?

44. A gas exerts a pressure of 1.50 atm at 100.°C. The same gas is transferred to a 5.00-L flask at 0°C and 1.00 atm. What was the original volume of the gas?

45. A 2.50-L flask contains a gas with a pressure of 435 torr at 28°C. What pressure will the gas exert in a 1.00-L flask at 0°C?

46. At 308 K, a 10.0-L sample of gas has a pressure of 0.948 atm. The gas is allowed to expand into a 25.0-L container at 318 K. What will be the new pressure?

47. A 50.0-L cylinder contains a gas at 25°C and 256 torr. At what temperature will the gas occupy 10.0 L at 455 torr?

48. At 22°C, a sample of gas occupies 200. mL. At what temperature will the gas occupy 500. mL if the pressure is doubled?

49. A sample of gas occupies a volume of 155 mL. The pressure is halved, and the temperature is tripled. What volume will the gas now occupy?

50. A sample of gas is collected in a 250.-mL flask at 305 K. The gas is allowed to expand into a 2.00-L flask at 100.°C. Will the new pressure be higher or lower than the original pressure?

51. A sample of gas occupies 20.6 L. Half the gas escapes. The final pressure is 468 torr and the final volume is 11.3 L. (The temperature has remained constant.) What was the initial pressure?

Section 11.9 The Ideal Gas Law

Question

52. What is the universal gas constant? Where does it come from? What symbol is ordinarily used for it? Give two different values of the constant, along with their corresponding units.

Problems

53. Calculate the value of R in units of L atm/mol K, starting with 62.4 L torr/mol K.

54. A certain sample of gas occupies 250. mL at 38°C and 765 torr. How many moles of gas are present?

55. If a sample of gas occupies 2.20 L, has a pressure of 1.56 atm, and contains 0.0058 mol, what is its Celsius temperature?

56. How many moles of gas occupy 555 mL at 725 torr and 15°C?

57. What is the temperature of a 55.7-gram sample of CH_4 gas occupying 10.0 L at 2.75 atm?

58. A 1,275-mL flask contains 11.3 grams of oxygen at 22°C. What pressure does the gas exert on the walls of the flask?

59. At 298 K, what volume would 13.4 grams of CO occupy at 1.16 atm? At 695 torr?

60. If a 2.16-gram sample of HCl has a pressure of 5.50 atm at 23°C, how much HCl will be present if the volume is halved and the pressure and temperature are changed to 2.00 atm and 57°C?

61. A sample containing 6.65 moles of gas occupies 7.33 L at 48°C. What is the pressure?

62. A 4.58-gram sample of hydrogen gas occupies 6.75 L at 25°C. What is the pressure?

63. A sample of gas occupies 3.00 L at 25°C and 756 torr. How many moles of gas are present?

64. At constant temperature and volume, a container of helium contains 45.0 grams of He at a pressure of 2.00 atm. The container develops a leak, allowing helium to escape for several hours. When the leak is discovered, the pressure is 952 torr. How many grams of helium are left in the container?

Section 11.10 Standard Temperature and Pressure; Standard Molar Volume

Questions

65. Define STP.

66. What is the standard molar volume? Under what conditions is it defined?

Problems

67. How many moles of gas at STP are contained in a 500.-mL container?

68. Calculate the number of moles of gas that occupy 40.3 liters at STP.

69. What is the volume in liters of 0.222 moles of gas at STP?

70. Find the volume of gas in milliliters of 0.0152 moles of gas at STP.

71. How many liters will 3.6 moles of gas occupy at STP?

Section 11.11 Gas Densities and Molar Mass

Problems

72. What is the molar mass of a gas weighing 5.40 grams and occupying 2.24 L at STP?

73. A sample of gas occupies 5.26 L at STP and weighs 12.23 grams. What is its molar mass?

74. Chloroazide is an explosive gas. At 25°C and a pressure of 752 torr, a sample of this gas has a volume of 3.89 L and a mass of 12.16 grams. What is the molar mass of chloroazide?

75. What is the molar mass of a gas weighing 0.0841 grams at 15.6°C and 688 torr if its volume is 50.0 mL?

76. Find the density of carbon dioxide, CO_2, at 555 torr and 35°C.

77. What is the density of natural gas, CH_4, at 0.826 atm and 335 K?

78. Carbon monoxide, CO, is produced by automobiles. What is the density of CO at 0.952 atm and 29°C?

79. Hydrogen sulfide, H_2S, is a bad-smelling, highly poisonous gas. At 734 torr and 295 K, what is its density?

80. What is the density of Ne at STP?

81. What is the density of N_2 at STP?

82. What is the density of CO_2 at -12°C and 725 torr?

83. Calculate the density of O_2 at -25°C and 0.444 atm.

84. What is the molar mass of a gas having a density of 3.92 g/L at STP?

85. The density of a gas is 1.3402 g/L at STP. What is its molar mass?

86. Calculate the molar mass of a gas that has a density of 2.144 g/L at STP.

Section 11.12 Dalton's Law of Partial Pressures

Questions

87. In your own words, state Dalton's law of partial pressures.

88. Define the terms "partial pressure" and "vapor pressure."

89. What steps must you take to find the pressure of a gas that has been collected over water?

Problems

90. Calculate the total pressure exerted by a gas sample containing 1.02 atm of NO, 2.50 atm of N_2O, and 0.505 atm of NO_2.

91. If a mixture of gases contains 833 torr CO_2, 557 torr O_2, and the rest is CO, what is the partial pressure of CO? The total pressure is 3.05 atm.

92. Calculate the total pressure in torr of a mixture of oxygen, hydrogen, and nitrogen. The oxygen has a partial pressure of 0.550 atm. The composition of the gas mixture is 35.4% hydrogen, 50.8% nitrogen, and the rest is oxygen.

93. In a 5.50-L container, what is the total pressure of a mixture containing 43.7 grams of O_2 and 1.00 atm of N_2 at 0°C?

94. Calculate the total pressure exerted by 1.57 grams of CH_4 and 1.57 grams of O_2 at 25°C in a 1.00-L flask.

95. What will be the total pressure of 16.3 grams of krypton, 14.8 grams of CO, and 2.37 grams of helium, all contained in a 20.0-L flask at 293 K?

96. A 10.0-L flask contains NO_2F and NOCl. The total pressure is 4.00 atm at 309 K. If the flask contains 15.0 grams of NO_2F, what is the partial pressure of NOCl?

97. What is the partial pressure of CO_2 in a mixture of CO_2 and CO with a total pressure of 2.00 atm? The 1.00-L flask contains 0.400 grams of CO at 22°C.

98. A sample of SO_2 at 304 K in a container with a volume of 2.75 L exerts a pressure of 225 torr. Enough NO is added to increase the pressure to 367 torr. How many grams of NO were added?

99. The total pressure of a sample of CO_2 collected over water at 25°C is 0.748 atm. The vapor pressure of water at 25°C is 23.8 torr. What is the partial pressure of CO_2 in the sample?

100. In an experiment, a 2.00-L sample of O_2 is collected over water at 25°C. The vapor pressure of water at this temperature is 23.8 torr. How many grams of O_2 are collected if the total pressure is 0.680 atm?

11.13 Stoichiometric Calculations Involving Gases

Problems

101. How many liters of CO_2 and H_2O, measured together at 701 K, can be produced from the combustion of 5.75 L of CH_4 and as much O_2 as needed?

$$CH_4(g) + 2\,O_2(g) \longrightarrow CO_2(g) + 2\,H_2O(g)$$

102. Given the chemical equation

$$2\ H_2S(g) + O_2(g) \longrightarrow 2\,S(g) + 2\ H_2O(g)$$

how many liters of H_2S are necessary to produce 6.80 L of H_2O, with both gases measured at 349°C and at a pressure of 1.1 atm? How many grams of sulfur will be produced?

103. Nitrogen gas and oxygen gas can produce nitric monoxide.

$$N_2 + O_2 \longrightarrow 2\ NO$$

How many liters of NO can be produced from 2.25 L of N_2 if reactants and products are all measured at the same T and P?

104. Trichloromethane, $CHCl_3$, can be formed by the reaction of Cl_2 and CH_4.

$$3 \ Cl_2(g) + CH_4(g) \longrightarrow CHCl_3(l) + 3 \ HCl(g)$$

If 2.40 L of Cl_2 reacted with 5.75 L of CH_4, how many liters of HCl were produced (all measured at 25°C and 1.0 atm)? How many grams of $CHCl_3$ were formed? Which reactant was present in excess?

105. How many liters of N_2 at STP can be produced from 63.2 grams of KNO_3?

$$2 \ KNO_3(s) + 4 \ C(s) \longrightarrow K_2CO_3(s) + 3 \ CO(g) + N_2(g)$$

106. A 23.0-gram sample of Ca metal reacts with 450.0 mL of NH_3 at STP. How many grams of Ca_3N_2 are produced?

$$6 \ Ca(s) + 2 \ NH_3(g) \longrightarrow 3 \ CaH_2(s) + Ca_3N_2(s)$$

107. A 2.96-gram sample of an unknown gas containing only carbon and hydrogen reacts with an excess of oxygen to form 4.05 L of CO_2 at STP. What is the molar mass of the gas? (The equation is not balanced.)

$$X + O_2 \longrightarrow CO_2 + H_2O$$

108. At a constant temperature and pressure, 2.67 L of H_2 are mixed with 1.75 L of SO_2. According to the following equation, how many liters of H_2O can be produced, still at the same T and P?

$$2 \ H_2(g) + SO_2(g) \longrightarrow S(s) + 2 \ H_2O(g)$$

109. A sample of phosphine, PH_3, occupying 1.50 L at 25°C and 0.080 atm, reacts with 2.28 L of O_2 at 20.° C and 0.100 atm according to this equation.

$$4 \ PH_3(g) + 8 \ O_2(g) \longrightarrow P_4O_{10}(s) + 6 \ H_2O(g)$$

How many grams of P_4O_{10} are produced?

110. If 4.85 L of N_2 at STP are mixed with 3.00 grams of O_2, how many liters of nitrogen dioxide, N_2O, at 21°C and 748 torr can be obtained?

$$2 \ N_2 + O_2 \longrightarrow 2 \ N_2O$$

111. A sample containing 54.0 grams of SO_2 reacts with 28.0 grams of O_2 in a 5.00-L flask. When the reaction reaches completion at 500.°C, what is the total pressure in the flask?

$$2 \ SO_2(g) + O_2(g) \longrightarrow 2 \ SO_3(g)$$

112. The following reaction takes place.

$$Zn(s) + 2 \ HCl(aq) \longrightarrow ZnCl_2(aq) + H_2(g)$$

The H_2 is collected over water in a 1.00-L flask at 22°C and a pressure of 1.60 atm. The vapor pressure of water at 22°C is 19.8 torr. How many grams of Zn are used?

11.14 Nonideality of Gases

Questions

113. State two characteristics of an ideal gas.

114. Under what conditions do gases deviate the most from ideal behavior?

More Difficult Problems

115. A balloon containing 0.26 moles of gas has a volume of 1.55 L. Some of the gas is allowed to escape. How many moles escaped if the final volume of the balloon is 0.78 L? Temperature and pressure are constant.

116. Calculate the partial pressure of hydrogen if a 1500.0-mL flask contains hydrogen and oxygen at STP. There is an equal mass of each gas.

117. A 10.0-L flask contains oxygen and hydrogen. The total pressure is 725 torr at 25°C. If there are twice as many hydrogen molecules as oxygen molecules in the flask, what is the partial pressure of the hydrogen?

118. Gaseous NO_2 forms N_2O_4 (also a gas) under certain conditions. If 46% of the molecules in a mixture of NO_2 and N_2O_4 are NO_2 molecules, how many grams of N_2O_4 are in a 250.-mL flask of the mixture at 25°C and 769 torr?

119. At 105°C and 1.0 atm in a 5.00-L flask, 45% of the molecules are NH_3 and the rest are O_2. The following (familiar?) reaction takes place.

$$4 \ NH_3(g) + 5 \ O_2(g) \longrightarrow 4 \ NO(g) + 6 \ H_2O(g)$$

What is the pressure in the flask at 250.°C after the reaction has finished?

Solutions to Exercises

11.1 $$0.455 \text{ atm} \times \frac{760 \text{ torr}}{\text{atm}} = 346 \text{ torr}$$

11.2 $$642 \text{ mm Hg} \times \frac{101,325 \text{ Pa}}{760 \text{ mm Hg}} = 8.56 \times 10^4 \text{ Pa}$$

11.3 $$2.00 \text{ L} \times \frac{1.00 \text{ atm}}{0.750 \text{ atm}} = 2.67 \text{ L}$$

11.4 $$900. \text{ torr} \times \frac{6.50 \text{ L}}{10.0 \text{ L}} = 585 \text{ torr}$$

11.5 $$695 \text{ torr} \times \frac{3.1 \text{ L}}{5.7 \text{ L}} \times \frac{\text{atm}}{760 \text{ torr}} = 0.50 \text{ atm}$$

11.6 $$8.74 \text{ L} \times \frac{1.25 \text{ mol}}{2.50 \text{ mol}} = 4.37 \text{ L}$$

11.7 $$5.32 \text{ mol} \times \frac{30.4 \text{ L}}{18.6 \text{ L}} = 8.70 \text{ mol}$$

11.8 $$298 \text{ K} \times \frac{5.00 \text{ atm}}{1.50 \text{ atm}} = 993 \text{ K}$$

11.9 $$795 \text{ torr} \times \frac{275 \text{ K}}{458 \text{ K}} = 477 \text{ torr}$$

11.10 $$5.25 \text{ L} \times \frac{455 \text{ K}}{298 \text{ K}} = 8.02 \text{ L}$$

11.11 $$273 \text{ K} \times \frac{2.74 \text{ L}}{4.28 \text{ L}} = 175 \text{ K}; \quad 175 \text{ K} - 273 = -98°\text{C}$$

11.12 $$\frac{P_2 V_2}{T_2} = \frac{P_1 V_1}{T_1};$$

$$\frac{T_2}{P_2} \times \frac{P_2 V_2}{T_2} = \frac{P_1 V_1}{T_1} \times \frac{T_2}{P_2};$$

$$V_2 = V_1 \times \frac{P_1}{P_2} \times \frac{T_2}{T_1}$$

11.13 $21°\text{C} + 273 = 294 \text{ K}$ $$294 \text{ K} \times \frac{600. \text{ mL}}{719 \text{ mL}} \times \frac{3.50 \text{ atm}}{2.15 \text{ atm}} = 399 \text{ K};$$

$$399 \text{ K} - 273 = 126°\text{C}$$

11.14 $$\frac{1.75 \text{ atm} \times 4.00 \text{ L}}{(0.0821 \text{ L atm/mol K}) \times 325 \text{ K}} = 0.262 \text{ mol}$$

11.15 $$\frac{1.25 \text{ atm} \times 0.250 \text{ L}}{0.0130 \text{ mol} \times (0.0821 \text{ L atm/mol K})} = 293 \text{ K}$$

$$293 \text{ K} - 273 = 20.°\text{C}$$

11.16
$$55.5 \text{ g N}_2 \times \frac{\text{mol N}_2}{28.0134 \text{ g N}_2} = 1.98 \text{ mol N}_2$$

$$\frac{1.98 \text{ mol} \times (62.4 \text{ L torr/mol K}) \times 273 \text{ K}}{850. \text{ torr}} = 39.7 \text{ L}$$

11.17
$$17.2 \text{ L} \times \frac{\text{mol}}{22.4 \text{ L}} = 0.768 \text{ mol}$$

11.18
$$5.2 \text{ mol} \times \frac{22.4 \text{ L}}{\text{mol}} = 1.2 \times 10^2 \text{ L}$$

11.19
$$\frac{(0.500 \text{ atm})(1.23 \text{ L})}{(0.0821 \text{ L atm/mol K})(300. \text{ K})} = 0.0250 \text{ mol}; \quad \frac{0.700 \text{ g}}{0.0250 \text{ mol}} = 28.0 \text{ g/mol}$$

11.20
$$\frac{(5.28 \text{ g/L})(0.0821 \text{ L atm/mol K})(273 \text{ K})}{1.00 \text{ atm}} = 118 \text{ g/mol}$$

11.21
$$\frac{(801 \text{ torr})(44.010 \text{ g/mol})}{(62.4 \text{ L torr/mol K})(304 \text{ K})} = 1.86 \text{ g/L}$$

11.22
$$\frac{(0.824 \text{ atm})(17.0304 \text{ g mol})}{(0.0821 \text{ L atm/mol K})(298 \text{ K})} = 0.574 \text{ g/L}$$

11.23
$$15.5 \text{ g N}_2 \times \frac{\text{mol N}_2}{28.0134 \text{ g N}_2} = 0.553 \text{ mol N}_2$$

$$\frac{(0.553 \text{ mol})(0.0821 \text{ L atm/mol K})(273 \text{ K})}{7.00 \text{ L}} = 1.77 \text{ atm}$$

$$1.77 \text{ atm} + 0.918 \text{ atm} = 2.69 \text{ atm}$$

11.24
$$32 \text{ torr} \times \frac{1 \text{ atm}}{760 \text{ torr}} = 0.042 \text{ atm}$$

$$0.750 \text{ atm} - 0.042 \text{ atm} = 0.708 \text{ atm}$$

11.25
$$CaCO_3 \longrightarrow CaO + CO_2$$

$$5.20 \text{ g CaCO}_3 \times \frac{\text{mol CaCO}_3}{100.087 \text{ g CaCO}_3} \times \frac{1 \text{ mol CO}_2}{1 \text{ mol CaCO}_3} = 0.0520 \text{ mol CO}_2$$

$$\frac{(0.0520 \text{ mol})(62.4 \text{ L torr/mol K})(300. \text{ K})}{730. \text{ torr}} = 1.33 \text{ L}$$

11.26
$$C_5H_{12} + 8 \text{ O}_2 \longrightarrow 5 \text{ CO}_2 + 6 \text{ H}_2O$$

$$0.265 \text{ L C}_5\text{H}_{12} \times \frac{1000 \text{ mL}}{\text{L}} \times \frac{0.626 \text{ g C}_5\text{H}_{12}}{\text{mL C}_5\text{H}_{12}} \times \frac{\text{mol C}_5\text{H}_{12}}{72.150 \text{ g C}_5\text{H}_{12}} \times \frac{5 \text{ mol CO}_2}{1 \text{ mol C}_5\text{H}_{12}} =$$

$$11.5 \text{ mol CO}_2$$

$$\frac{(11.5 \text{ mol})(0.0821 \text{ L atm/mol K})(298 \text{ K})}{0.950 \text{ atm}} = 296 \text{ L}$$

Much of the enormous power of a hurricane comes from the energy released when water vapor in the air condenses to liquid water.

The Condensed States: Liquids and Solids

In Chapter 2, we learned that most kinds of matter can exist in any of three states: gas, liquid, or solid. In Chapter 11, we studied the properties of gases. We took a quantitative approach to our study because there is a clear connection between the number of molecules in a sample of gas and the physical properties of the gas. No such relationship exists for liquids and solids. The structures of liquids and solids vary so widely that the theories and calculations concerning those structures are themselves complex and varied. For this reason, we take a more descriptive approach to the properties of liquids and solids.

We will study the effects of energy changes. When liquids or solids absorb or release energy, they change either their temperature or their phase. Therefore, we will be particularly interested in temperature and pressure. For the first time, we will investigate dynamic equilibrium, a situation that is important for phase changes.

We begin with a study of the kinds of intermolecular forces and conclude with a discussion of several types of solids.

12.1 Condensation

After studying this section, you should be able to:

- Describe the process of condensation.
- Explain why condensation of a gas occurs when temperature is lowered.

Let's cool some argon gas and see what happens. At ordinary temperatures, argon behaves like an ideal gas, but if we lower its temperature enough, ideality disappears. In the process of cooling, the molecules of gas lose kinetic energy (remember from Section 11.3 that temperature is a measure of average molecular kinetic energy). Eventually, some of the argon atoms are no longer moving fast enough to overcome any attractive forces that exist between them. Consequently, some of the argon atoms stick together when they collide, and, at temperatures far below 0°C, the argon becomes a liquid. The process of a gas turning into a liquid is called **condensation.**

A liquid or solid is held together by attractive forces between its molecules.

Under certain conditions, a gas can turn into a solid without ever being a liquid. This process is also condensation. Liquids and solids, therefore, are known as **condensed states.**

How do we know that argon atoms attract one another? The very fact that argon *can* become a liquid means that something must be able to hold the atoms together. In this chapter, we'll look at the kinds of forces that can exist between individual particles—atoms, molecules, or ions.

12.2 Forces Between Individual Particles

After studying this section, you should be able to:

- Explain the differences between dipole–dipole forces, temporary dipole forces, and hydrogen bonding.
- Predict, on the basis of chemical bonding, the type of intermolecular force in a sample.

As we know, molecules are held together *internally* by chemical bonds. Chemical bonds are responsible for the existence of compounds—substances that can be identified by formulas. The attractive forces acting *between* one molecule and another are *not* chemical bonds. It is important to understand this difference: Forces *within* molecules are chemical bonds, but *between* molecules there are other forces that are not chemical bonds. In general, the forces that act between molecules are not as strong as chemical bonds. In most cases, the distinction between chemical bonds and nonchemical forces is clear, but in others it is indistinct. We will see examples of the second case when we discuss hydrogen bonds.

Somewhat more than a hundred years ago, a Dutch physicist named Johannes Diderik van der Waals began an investigation of the nature of the forces acting between molecules (intermolecular forces). These forces, now known collectively as **van der Waals forces,** have been classified into two categories: **dipole–dipole forces** and **temporary dipole forces.**

Dipole Forces

To understand how dipole forces arise, let's first review briefly what we learned about polar covalent bonds in Sections 6.5 and 6.13. If a covalent bond is

Profiles in Science

Johannes Diderik van der Waals (1837–1923)

Many discoveries in science were made by people who began looking for one thing and, in the process, opened up whole new fields of understanding. The work for which van der Waals is best known is an example.

Van der Waals was what is sometimes called a "late bloomer." A self-made man, he was late getting the resources to go to college and was almost middle-aged (36) before finishing his education. The progress of his work from that point, however, made up for lost time. Within four years, he had become a professor at his university in Amsterdam and was well on the way to becoming famous in scientific circles.

Van der Waals was interested in the relations between liquids and gases, believing that molecules did not undergo any changes in molecular properties when changing their state, from liquid to gas or vice versa. This led him to investigate why real gases do not always obey the ideal gas law, such as when they condense into liquid. Van der Waals was the first to show that the forces that molecules exert on one another in liquids exist also between the molecules when they are gases. When gas molecules are close enough together (high pressure) and slow enough (low temperature), these forces cause them to stick. Intermolecular forces are called van der Waals forces, not only because they were discovered by van der Waals but also because much of our understanding of the liquid and gas states is due to his pioneer work. In his lifetime, van der Waals was honored by the greatest scientists, many of whom visited Amsterdam to learn from him. He received the Nobel Prize in 1910.

formed by two atoms of unequal electronegativity (attraction for a shared pair of electrons), the electrons in the bond will spend their time on the average closer to one atom than the other. In other words, the molecule has a partial negative charge toward one end of the bond and a partial positive charge toward the other. This results in the formation of a polar covalent bond, one that has an electric dipole. The greater the difference in the electronegativities, the larger the dipole. In a molecule with more than one polar covalent bond, the dipoles may reinforce one another, or they may cancel so that the molecule itself has no dipole.

Consider the compound hydrogen chloride, HCl, which is a gas at room temperature. A hydrogen chloride molecule is held together internally by a single, strong polar covalent bond. Because of the difference in electronegativity between H and Cl ($3.0 - 2.1 = 0.9$), an HCl molecule has a rather strong electric dipole (Figure 12.1).

Two or more molecules with electric dipoles exert attractive forces on one another because the positive end of one attracts the negative end of the other and vice versa (Figure 12.2). The attraction of two HCl molecules to each other is called a dipole–dipole force or just dipole force. Other HCl molecules are attracted in the same way (Figure 12.3). At low temperatures, HCl molecules do not have enough energy to overcome the dipole forces between them, and a liquid forms. At one atmosphere pressure, HCl liquefies at $-85°C$ and freezes at $-115°C$.

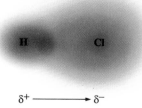

$\delta^+ \longrightarrow \delta^-$

Figure 12.1 An HCl molecule has an electric dipole, making it a polar molecule. When HCl molecules are placed in an electric field, their positive ends are attracted to the negative plate and their negative ends to the positive plate.

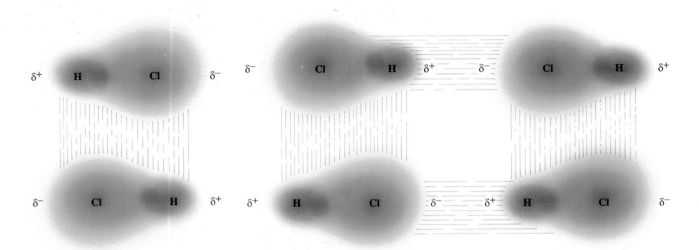

Figure 12.2 The positive end of an HCl molecule is attracted to the negative end of another HCl molecule, and the negative end is attracted to the positive end of another HCl molecule.

Figure 12.3 Dipole attractions can extend beyond two molecules.

Temporary Dipole Forces

Iodine, I_2, is a solid at room temperature. Its molecules are obviously held together by substantial forces, but we know that I_2 molecules have no permanent dipoles because the two atoms in each molecule are exactly alike. Evidently, another kind of intermolecular force exists.

The intermolecular force in substances like I_2 is created by short-lived dipoles. Over time, electrons in a molecule are uniformly distributed; however, at any particular instant, the electron density can be greater at one end of the molecule than at the other. This difference creates an electric dipole for a tiny fraction of a second. Then the electrons of any nearby molecule respond by shifting in the reverse direction to create another dipole. The two dipoles attract each other. This process repeated over and over results in an intermolecular force, which is usually not strong, but which nonetheless exists. It is the only explanation for the properties of I_2 and many other substances. Study the drawings in Figure 12.4.

Because they have more electrons and because their electrons are held more loosely, large molecules and large atoms have larger temporary dipole forces than do small atoms and molecules. The temperatures at which the halogens liquefy demonstrate the effect of molecular size. Since fluorine molecules are small and hold their electrons tightly, the forces between F_2 molecules are weak,

Figure 12.4 A temporary dipole develops when the electron density shifts to one end of a molecule. Once a temporary dipole has formed, other molecules respond by forming temporary dipoles of their own. The resulting dipoles attract one another.

At room temperature, chlorine is a gas, bromine is a liquid, and iodine is a solid.

and fluorine must be cooled to $-199°C$ before it will liquefy at one atmosphere pressure. Chlorine, Cl_2, liquefies at $-34.6°C$. Br_2 is one of the few elements that is a liquid at room temperature; it liquefies at $+59°C$. As we have seen, I_2 is a solid at room temperature. The forces between I_2 molecules are stronger than those between bromine molecules.

For all but the largest molecules, temporary dipole forces are weaker than permanent dipole forces. Temporary dipole forces occur between all molecules, even in the noble gas "molecules" that contain only one atom each. Intermolecular forces are often a result of permanent dipole forces and temporary dipole forces acting together.

Hydrogen Bonds

Hydrogen bonding is a kind of intermolecular force that is stronger than van der Waals forces but weaker than covalent bonds. Hydrogen bonds occur only among certain kinds of molecules, whereas van der Waals forces occur among molecules of all kinds. Although they are limited in the kinds of molecules they affect, hydrogen bonds are nonetheless extremely important. The properties of water, on which biological organisms depend, are caused to a large extent by hydrogen bonding. The forces that hold proteins together are hydrogen bonds. The genetic code is written with hydrogen-bonded code units. A tree can stand because of hydrogen bonds. Life, as we know it, depends on hydrogen bonds. What are hydrogen bonds?

To understand how hydrogen bonds occur, let's consider water, H_2O. Oxygen is a strongly electronegative element. In an O—H bond, the oxygen atom exerts a powerful attraction on the pair of electrons shared with H. Because hydrogen atoms have no inner electrons, the shift of the electron pair in the O—H bond leaves the nucleus of the H atom somewhat exposed. This concentrates the positive end of the O—H dipole in a tiny area that can get very close to unshared

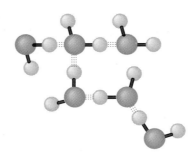

Figure 12.5   Hydrogen bonding in water.

Hydrogen bonds are stronger than ordinary dipole–dipole forces, which are generally stronger than temporary dipole forces.

The exact balance of forces that causes H bonds to work exists only between H and a small number of other atoms: N, O, and F.

electron pairs on other molecules. The attraction between a partly bare hydrogen nucleus and an available electron pair at the negative end of another O—H dipole constitutes a hydrogen bond. A hydrogen bond has some of the characteristics of an ionic bond and therefore is stronger than ordinary dipole forces.

Because the two O—H polar covalent bonds in a water molecule lie at an angle to one another, they do not cancel as do the bonds in CO_2; thus water is a polar molecule. One water molecule can form hydrogen bonds to two other molecules (Figure 12.5).

Hydrogen bonds occur in molecules in which hydrogen atoms are bonded to oxygen, nitrogen, or fluorine—the only elements whose atoms are small enough and electronegative enough to have the necessary effect on hydrogen.

12.3 Properties of Condensed Phases

After studying this section, you should be able to:

- Explain the differences between solids, liquids, and gases.

While a gas takes the shape and volume of its container, a liquid assumes only the shape of its container. It has its own volume. An individual piece of a solid has its own volume and shape, without any dependence on a container.

Gases consist mostly of empty space, but in liquids and solids, the particles are close together. In liquids and solids, the individual particles can be said to be "touching" one another, that is, there is little or no excess space between them (Figure 12.6). That is why liquids and solids cannot be easily compressed as gases can, and the densities of liquids and solids are far greater than the densities of gases.

Liquids can flow; solids can't. The particles in a liquid stay close together but can exchange places, playing musical chairs with the tiny amount of excess space available, the "holes." The particles in solids, however, are tightly held in fixed positions relative to one another. That is why solids are rigid.

In both liquids and solids, molecules still have kinetic energy; they constantly vibrate. In a liquid, however, molecules move from place to place, jostling each other out of the way. As we will see in Section 15.5, not all the molecules in any sample have the *same* energy. At any instant, some are moving extremely

Gas   Liquid   Solid

Figure 12.6   The particles of a gas are far apart. In liquids and solids the particles are closer together.

Recall the discussion of glassy solids in Section 2.1. In some cases, it is difficult to tell whether something is a solid or just a stiff liquid.

In every bit of matter—solid, liquid, or gas—the atoms are always in constant motion.

rapidly, while others may be almost at rest. Most will have about the average energy. This distribution of energies has several consequences, both in changes of state and in chemical reaction.

12.4 Vaporization

After studying this section, you should be able to:

- Explain the process of evaporation.
- Explain the difference between sublimation and vaporization.
- Show how energy changes accompany vaporization and condensation.
- Explain the connection between the values of heats of vaporization and intermolecular forces.
- Calculate the amount of energy required to vaporize a given sample.

We all know that if we leave a glass of water uncovered long enough, it all disappears in a process called **evaporation.** The process by which a liquid becomes a gas is called **vaporization.** If the gas molecules escape, the process is *evaporation*.

The molecules at the surface of water are held close together by hydrogen bonds. Although all of the water molecules have energy, not all have the same amount. The most energetic molecules, however, have enough kinetic energy to overcome the forces holding them, and they leave the surface. Unless these are contained, they will drift away, more will leave, and the liquid will eventually disappear (Figure 12.7).

When molecules leave the surface of a liquid, the departing molecules carry their energy away from the liquid. The remaining molecules then have less average kinetic energy, and the liquid is cooler. (You could think of this as the hot molecules being the ones that leave.) This is why evaporating perspiration cools your skin. Unless more energy is supplied from outside, any sample of liquid will get colder and colder and finally cease to evaporate. Usually, however, energy is absorbed from the surroundings, and the liquid continues to evaporate.

Both liquids and solids evaporate. An example of the evaporation of a solid is the disappearance of dry ice (solid CO_2) or the vanishing of snow at high altitudes on a sunny day. The change of a solid directly into a gas is called **sublimation.**

The energy that one gram of escaping molecules carries away from a liquid is called the **heat of vaporization** of that liquid, and from one gram of a solid, the **heat of sublimation.**

Heats of vaporization are different for different substances. If two molecules are strongly attracted to one another, the energy required to separate them will be substantially larger than if their attractions are weak. Correspondingly, the heat of vaporization is also larger. The heat of vaporization of argon is much less than that of water because the temporary dipole forces between argon molecules are much weaker than the hydrogen bonds between water molecules.

Heats of vaporization have been measured for many substances, a few of which are listed in Table 12.1. If you know the heat of vaporization, you can calculate the amount of energy required to vaporize a sample of any size. Since energy must be added, vaporization is an endothermic process.

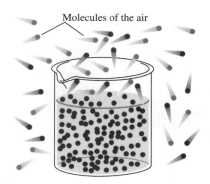

Figure 12.7 Molecules of a liquid continually evaporate from an open container.

Actually, evaporation *implies the process in which molecules leave the surface and are lost.* Vaporization *includes all situations in which liquid becomes gas.*

Evaporation of perspiration cools you because the escaping water molecules take energy away from your skin.

Camphor sublimes and condenses to a solid elsewhere in its container. Some solid nasal inhalers use camphor.

Table 12.1 Values of Some Heats of Vaporization

Substance	Heat of Vaporization (kJ/g)
Argon	0.163
Ethanol	0.85
Ammonia	1.38
Zinc	1.76
Water	2.26
Sodium chloride	2.92
Sodium	4.27

For any given substance, the heat of condensation numerically equals the heat of vaporization.

EXAMPLE 12.1

How many kilojoules of energy are required to vaporize 15.1 grams of water?

Solution: Use the heat of vaporization (Table 12.1) as a conversion factor.

$$15.1 \text{ g} \times \frac{2.26 \text{ kJ}}{\text{g}} = 34.1 \text{ kJ}$$

Exercise 12.1

(a) How much energy is required to vaporize 21.0 grams of ammonia? (b) Of 1.35 kilograms of Na?

If a gas condenses into its liquid form, the energy that was required to vaporize it will be released. It is then known as the **heat of condensation.** Since energy is released, condensation is an exothermic process. For the compound ethanol, the heat of condensation (which is numerically equal to the heat of vaporization) is 0.85 kJ/g. When a gram of ethanol condenses, 0.85 kilojoules of energy are released. Flesh burns from steam are especially bad, not only because the steam is hot, but primarily because the steam condenses into liquid on the skin. Additional damage results as the heat of condensation is released.

12.5 Vapor Pressure

After studying this section, you should be able to:

- Show how the rates of opposing processes can become equal, resulting in a state of dynamic equilibrium.
- Define and explain vapor pressure.
- Discuss the relation between vapor pressure and intermolecular forces.

Vapor is a gas that can be condensed readily into either liquid or solid with which it is in contact.

We have seen what happens if we leave an open glass of water for a long time, but what if the glass is tightly covered so that nothing can escape? What then?

To do a simple experiment, begin with a closed container that has nothing whatever in it, not even air. We inject some water and keep track of what happens. As they would if the container were open, water molecules leave the surface of the liquid and become water vapor. The presence of the vapor has no effect on the molecules leaving the surface of the liquid; they continue to depart at a constant rate. Molecules of the vapor, however, begin colliding with the surface of the liquid, rejoining it. As the number of vapor molecules increases, the number of molecules returning to the liquid also increases. Finally, the number rejoining the liquid becomes the same as the number leaving, and the number of molecules in the vapor no longer changes (see Figure 12.8). At this point, the liquid and vapor are said to be in a state of *dynamic phase equilibrium*, *dynamic* meaning that the processes of vaporization and condensation are still going on and *equilibrium* meaning that the ongoing processes have balanced each other so that no further changes occur.

Equilibrium is established when the rate of condensation equals the rate of vaporization. Chemical equilibrium, in which the rate of a forward reaction equals the rate of its reverse reaction, is discussed in Chapter 15.

$$\text{liquid} \underset{\text{condensation}}{\overset{\text{vaporization}}{\rightleftharpoons}} \text{vapor}$$

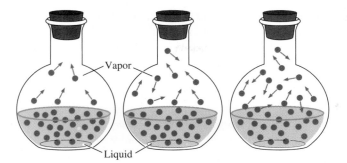

Figure 12.8 In the left figure, liquid has just been added to the container. Molecules are beginning to leave the surface of the liquid to become vapor. After a short while, some vapor molecules begin to join the surface of the liquid. Later still, the number of vapor molecules rejoining the liquid surface equals the number of liquid molecules becoming vapor. Equilibrium has been established.

We know from the gas laws that if the volume, temperature, and number of molecules of a gas are constant, the pressure also is constant. Because the number of vapor molecules in our simple experiment has become constant, the water vapor has reached a constant pressure. In a closed container at equilibrium, the pressure exerted by a vapor on its liquid is called the **vapor pressure** of that liquid. The vapor pressure is a property of the liquid and is dependent only on the type of liquid and the temperature. Vapor pressures of many substances have been measured (Figure 12.9); some are given in Table 12.2. Notice the enormous differences, all due to the nature of the liquid.

If a substance has small intermolecular forces, the molecules can easily escape the liquid, and the vapor pressure will be high. Substances that evaporate easily (have high vapor pressures) are said to be **volatile.** Gasoline is a volatile substance; butane, the clear liquid seen in inexpensive cigarette lighters, is extremely volatile. On the other hand, substances that have strong interparticle forces have correspondingly low vapor pressures.

Vapor pressure can be established only if there is always some liquid remaining.

Table 12.2 Vapor Pressures of a Few Liquids

Substance	Vapor Pressure (torr at 20°C)
Carbon dioxide	4.3×10^4
Ammonia	2.5×10^3
Butane	1.6×10^3
Water	17.5
Mercury	1.2×10^{-3}

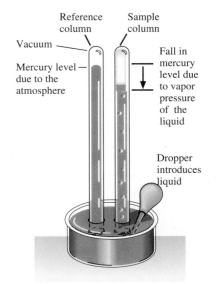

Figure 12.9 To measure the vapor pressure of a liquid, two Torricelli tubes are set up, one being used for a comparison. Liquid is introduced into the bottom of the other tube; it floats to the surface of the mercury and some of it evaporates into the vacuum at the top of the tube, where it establishes its vapor pressure. The difference in the two mercury levels gives the vapor pressure. This method is used frequently as a classroom demonstration.

Liquid butane is a volatile liquid. As a vapor it burns readily in the presence of air.

At constant temperature, any liquid confined in a closed container will occupy the volume above it with vapor, the pressure of which soon will become constant. The process is independent of the presence of another gas; the liquid develops its vapor anyway, and its pressure adds to the total pressure according to Dalton's law (Section 11.12). We have already learned that when gases are collected in the presence of water, the total pressure includes the water vapor pressure.

The liquid $\rightleftharpoons$ vapor equilibrium condition is a balance between the rates of opposing processes. It is independent of the *amounts* of the phases as long as some of each phase is present. If 50 grams of liquid water are placed in a closed 500-milliliter flask at 20°C, with or without other gases, a characteristic H_2O vapor pressure of 17.5 torr will be established. If another 50 grams of water are placed in a closed 5-liter flask, the water will also vaporize until the pressure of the water vapor is 17.5 torr. There will be more liquid and less vapor in the first flask than in the second, but in both the pressure of the water vapor will equal the H_2O vapor pressure of 17.5 torr. If one gram of water is placed in a 100-liter flask, all the water will vaporize before the vapor pressure can be established. There will be no equilibrium because there is no liquid water.

Solids also develop vapor pressures, but because intermolecular forces are quite large in solids, their vapor pressures often are too small to measure. A joke among chemists is that tungsten has a vapor pressure of one atom per universe.

Chemists find that funny.

12.6 Boiling and Boiling Temperatures

After studying this section, you should be able to:

- Explain the process of boiling.
- Define and explain the normal boiling point for a liquid.
- Explain the reason for the change of boiling point with elevation.
- Explain the connection between intermolecular forces and boiling temperature.

If liquid water is placed in contact with dry air at one atmosphere pressure and room temperature, the pressure of the water vapor (17.5 torr) is much less than the pressure the air exerts on the surface of the liquid. Even so, individual water molecules leave the surface and find their way among the air molecules. In this way, the liquid slowly establishes its vapor pressure, which then adds to the pressure of the air (Figure 12.10).

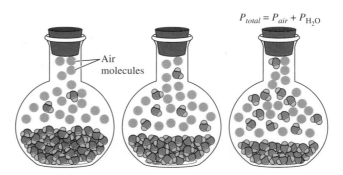

$$P_{total} = P_{air} + P_{H_2O}$$

Air molecules

Figure 12.10   The vapor pressure of the water adds to the air pressure already present.

As liquid water is heated in an open pan, its vapor pressure increases until it becomes equal to the pressure of the air. To see how that happens, look at the graph of the vapor pressure of water as the temperature is raised (Figure 12.11). Toward the bottom of the curve, the vapor pressure of water at 20°C is only 17.5 torr, but as we follow the graph to higher temperatures, we see that the vapor pressure rises. When the vapor pressure of a liquid equals or exceeds the pressure of the atmosphere, molecules leave the liquid in such numbers that they no longer merely diffuse between air molecules, they simply collide with the air molecules rapidly enough to push the air away. Bubbles of vapor form in the liquid, and the water is said to be **boiling.**

The **boiling temperature** or *boiling point* is the temperature at which the vapor pressure of a liquid equals the pressure on its surface. The boiling temperature of a liquid in the presence of air at one atmosphere pressure is called the **normal boiling point.** Because it is directly related to the vapor pressure of a substance, the normal boiling temperature is a distinctive property of that substance. Table 12.3 gives normal boiling temperatures for several liquids.

Molecules with stronger intermolecular forces ordinarily have higher normal boiling temperatures. For example, HCl boils at 188.3 K at one atmosphere pressure, whereas the normal boiling point for Ar is only 87.5 K, even though the two molecules have the same number of electrons and about the same mass. The two liquids with the lowest boiling points of all, He and H_2 (boiling temperatures of 4 K and 20 K, respectively), are nonpolar with extremely weak temporary dipole forces.

At high altitudes, such as on a mountain top, the pressure of air is less than at sea level. Consequently, less vapor pressure is required to push back the air. At high altitudes, therefore, the temperature at which water boils is lower than its normal boiling temperature, and at altitudes below sea level, it is higher (see Table 12.4). Pressure cookers cook food rapidly because the water in the cooker boils at temperatures higher than 100°C. Typically, a pressure cooker operates at about an atmosphere above the outside pressure, yielding boiling points up to 120°C.

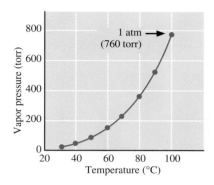

Figure 12.11   Vapor pressure increases strongly with a rise in temperature.

A four-minute egg boiled at La Paz, Bolivia (altitude 3,660 m) is practically raw.

Table 12.3   Normal Boiling Temperatures of Some Liquids

Liquid	Normal Boiling Point (°C)
Sodium	892
Sulfur	444
Octane (similar to gasoline)	125.6
Water	100
Ethyl alcohol	78.5
Ethyl ether	34.6
Ammonia	−33
Chlorine	−34.6
Propane (lpg)	−44.5
Methane (natural gas)	−161
Oxygen	−183

Table 12.4   Boiling Point of Water at Various Locations

Location	Elevation (feet)	Boiling Point (°C)
Top of Mount Everest	29,028	76.5
La Paz, Bolivia	12,795	91.4
Denver, CO	5,280	95.0
Newport Beach, CA	3	100.0
Death Valley, CA	−282	100.3

When energy is supplied to liquid and vapor in equilibrium at the boiling temperature, all the liquid vaporizes before either the liquid or the vapor increases its temperature. This means that *if we continue to heat boiling water, the temperature of the water will not increase, but the water will continue to boil away.* If energy is removed from a vapor at its boiling temperature, all the vapor will condense before the temperature decreases.

Table 12.5	Normal Melting Temperatures of Some Solids	
Solid		Normal Melting Temperature (°C)
Hydrogen		−259
Oxygen		−219
Argon		−189
Ammonia		−78
Bromine		−7
Water		0
Sodium		98
Lithium bromide		550

12.7 Melting and Freezing

After studying this section, you should be able to:

- Explain why a liquid freezes when the temperature is lowered sufficiently.
- Explain the connection between intermolecular forces and freezing temperatures.
- Calculate the energy required to melt a given sample.

Melting/Freezing Temperatures

We all know that when the temperature of a liquid decreases, eventually the liquid **freezes.** During the cooling process, molecules of the liquid lose kinetic energy and can no longer move relative to one another; they are literally frozen in position. Consequently, the liquid turns into a solid.

When heated to a high enough temperature, molecules of a solid gain enough kinetic energy to overcome intermolecular forces and begin to change positions. The solid **melts,** forming a liquid. The temperature at which a solid melts is called either the **melting temperature** or the **freezing temperature** (or sometimes the *melting* or *freezing point*), depending on which phase is disappearing and which is growing. The melting temperature is a distinctive property of each substance and depends only on the nature of the substance and the pressure. Melting temperatures taken at one atmosphere pressure are called **normal melting temperatures.** Table 12.5 lists the normal melting temperatures of several substances.

Heat of Fusion

Just as it requires energy to change a liquid into a vapor, so energy is needed to change a solid into a liquid. At 0°C, 0.335 kJ of energy are required to melt one gram of ice. The amount of energy required to melt one gram of a solid substance into a liquid is called the **heat of fusion.** That same amount of energy is released when a liquid freezes into a solid. At 0°C, 0.335 kJ of energy are *released* when one gram of liquid water freezes. Table 12.6 lists the heats of fusion of several substances. Again, notice the wide variation in values due to the different types of intermolecular forces in the substances.

EXAMPLE 12.2

How many kilojoules of energy are required to melt 15.1 grams of ice at 0°C to liquid water at the same temperature?

Solution: Multiply the mass of the sample by the heat of fusion (Table 12.6).

$$15.1 \text{ g} \times \frac{0.335 \text{ kJ}}{\text{g}} = 5.06 \text{ kJ}$$

Exercise 12.2

How many kilojoules of energy are required to melt 1.35 grams of NaCl at a constant temperature?

As in the case of vaporization, if we add energy to a solid at its freezing temperature, *all the solid will melt before the temperature of the liquid or solid increases.* Likewise, if we take energy away from a liquid at the freezing temperature, all the liquid will freeze before the temperature decreases.

Freezing/Melting Equilibrium

Just as a liquid and its vapor or a solid and its vapor establish an equilibrium, so can a solid and a liquid be in equilibrium. If you put some ice in a glass of water at room temperature, the ice will melt, taking heat from the liquid water and cooling it. In this process, while the surface of the ice is warmed by the warm liquid, molecules leave the ice much faster than they return to it from the liquid. But when the temperature of the entire mixture reaches 0°C, the rate of escape of molecules from ice to liquid is matched by the rate of return of molecules from liquid to ice. No more ice will melt unless energy is added from the surroundings. The solid water, ice, is in equilibrium with the liquid. The freezing/melting temperature of a substance is the temperature at which the solid/liquid equilibrium exists.

12.8 **Crystalline Solids**

After studying this section, you should be able to:

- Make rough predictions of the properties of liquids and solids based on the kinds of intermolecular forces likely to exist in them.

In Section 2.1, we were introduced to the nature of solid matter and the differences between the organized solids called *crystals* and the unorganized solids such as plastics, glasses, and gels. Now we examine crystalline solids further and learn about how the structure of a crystal affects its properties.

Crystalline solids are highly ordered. The atoms of a crystal are arranged in a regular, repeating, three-dimensional pattern. This is illustrated by the structure of potassium chloride in Figure 12.13 below. These organized crystalline solids produce beautiful crystals like those shown in Figure 12.12.

There are three classes of crystalline solids: ionic solids, molecular solids, and covalent solids. An ionic solid is composed of ions, a molecular solid of molecules, and a covalent solid of individual atoms.

Table 12.6	Values of Some Heats of Fusion
Substance	Heat of Fusion (kJ/g)
Argon	0.029
Zinc	0.100
Sodium	0.113
Water	0.335
Sodium chloride	0.520

If you are cooling soft drinks with ice in a cooler, all the ice will melt into liquid before the temperature of the cooler rises (assuming that everything in the cooler is at equilibrium).

Ice is in equilibrium with the liquid water at 0°C.

(a)

(b)

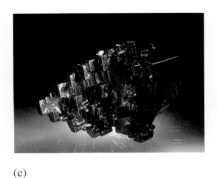

(c)

Figure 12.12 (a) Amethyst is a crystalline solid composed of silicon dioxide. The purple color comes from traces of the element manganese. (b) Quartz is silicon dioxide without visible impurities. (c) Elemental bismuth forms beautifully organized crystals.

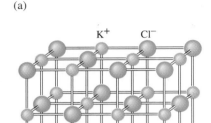

Figure 12.13 The structure of KCl crystals.

Ionic Solids

We learned in Section 6.2 that when potassium and chlorine react, electron transfer occurs, creating positive and negative ions. Under ordinary conditions, potassium and chloride ions combine to make an ionic crystal of potassium chloride, KCl, which is quite similar to everyday table salt. In a potassium chloride crystal, each K^+ is surrounded by six Cl^- ions, and each Cl^- is similarly surrounded by six K^+ as its closest neighbors (Figure 12.13).

The KCl crystal is held together by the strong attractions between the positive charges on the K^+ ions and the negative charges on their nearest–neighbor Cl^- ions. These attractive forces far outweigh the repulsive forces exerted on each ion by its next nearest neighbors, which are ions of the same charge.

The properties of ionic crystals are clearly related to their structures. One such property is their typically high melting temperatures. For example, KCl crystals melt at 776°C. High melting temperatures, like large heats of fusion, are caused by the strong attractions between the ions, in this case K^+ and Cl^-. Disrupting these strong forces requires large amounts of energy. During melting, and particularly during vaporization, when the distances between the ions are being greatly increased, large amounts of energy are absorbed.

Not all ionic crystals are like KCl. Many ionic crystals have more complicated arrangements caused by the sizes, shapes, and charges of their ions. All, however, stick together because of the attractive forces between the ions. Many ionic compounds form crystals that are harder and tougher than those of KCl; for example, various forms of $CaCO_3$ (limestone and marble) are used for building. These owe their strength primarily to the strong attractive forces between the doubly charged ions of Ca^{2+} and CO_3^{2-}.

The ionic solid, calcite is a mineral composed of calcium carbonate.

Covalent Crystals

Just as the atoms in a molecule are held together by covalent bonds, so can covalent bonds join the atoms in a much larger structure called a **covalent crystal** or **covalent solid.** In a sense, every atom in a covalent solid is a part of one giant crystalline molecule. Covalent solids have high melting points because of the strong attractions between the covalently bonded atoms. Such crystals are hard and most are tough.

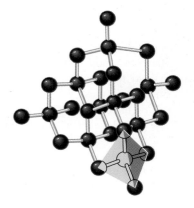

Diamonds.

Figure 12.14 An illustration of the bonding in diamond.

A diamond is an example of a covalent crystal. Every diamond is a crystal made of carbon atoms. Each carbon atom shares electron pairs with four equally close neighboring carbons (see Figure 12.14). The bonds in diamond are all extremely strong; diamond is one of the hardest substances known. Diamond also has the highest melting temperature known, about 3,800 K.

Graphite, another form of carbon, is also a covalent solid. It is interesting because it is soft but has a melting point close to that of diamond. High melting temperatures are usually associated with strong bonds, and softness, with weak bonds; these two properties are generally inconsistent and are not ordinarily found in the same substance. Graphite, however, has both strong and weak bonds. The graphite crystal consists of sheets of carbon atoms joined by weak bonds (Figure 12.15). The carbon atoms within each sheet are bonded by extremely strong covalent bonds. Because the sheets are held one to the next by rather weak bonds, a force applied to the crystal easily causes the stacks of sheets to slip out of place. To change solid graphite into a liquid, however, all the bonds must be disrupted, both strong and weak; this gives graphite a high melting temperature.

The melting temperature is not precisely known, because it is difficult to find a container in which to melt the diamond crystals and because diamond loses its structure and sublimes.

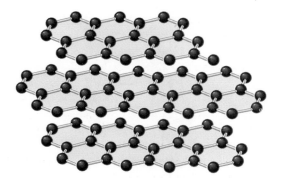

Figure 12.15 The structure of graphite. Notice that the bonds within the layers are quite short; they are also strong. The bonding between the layers is much weaker.

Sugar decomposes as it melts. The resulting substance is caramel, or, with more decomposition, charcoal.

Molecular Crystals

A third kind of solid, called a **molecular crystal,** is composed of individual, internally bonded molecules held in their positions only by van der Waals forces or hydrogen bonds, not by chemical bonds. Many common materials form molecular crystals. Some, like CO_2 and I_2, are held together by temporary dipole forces; others, like HCl, are united by permanent dipoles.

The properties of a molecular crystal are not difficult to predict. Because the forces between molecules are usually weak, we would expect to find the structure of a molecular solid much easier to destroy than that of a covalent crystal. Our prediction is confirmed by the low melting points of molecular substances compared to the much higher melting points of covalent and ionic crystals. Methane, CH_4, a typical molecular crystal, melts at about 90 K. Water melts at 273 K, and Br_2 at 266 K. Compare these temperatures with the melting points of typical, simple ionic crystals (NaCl at 1076 K, CaO at 2850 K, K_2CO_3 at 1275 K) or with the melting point of diamond.

Not all molecular crystals, however, are easily melted. Ordinary table sugar, $C_{12}H_{22}O_{11}$, can be melted only with great care. Because many dipole attractions and hydrogen bonds exist between each pair of sugar molecules, it is nearly as difficult to separate these molecules from one another as it is to decompose the molecules themselves. The melting temperature of sucrose is 460 K, a higher temperature than we might expect for a molecular solid and one that causes the delicate molecules to decompose. The familiar process of caramelization, a chemical reaction in which sugar molecules are destroyed, occurs when sugar is heated under ordinary conditions. Caramel flavor comes from partly decomposed sugar.

As we would expect, molecular crystals are much softer than ionic or covalent crystals—another demonstration of the weakness of the intermolecular forces.

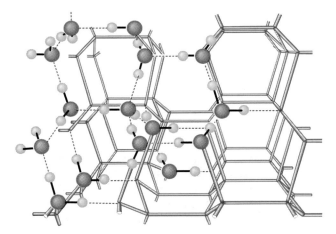

Figure 12.16 The structure of ice. Notice that the ring structure creates empty spaces. These cause the ice to be less dense than liquid water.

Science in Action

As we have seen, in crystalline molecular solids the molecules are locked into a repeating pattern. Usually, when a crystal melts, the molecules simultaneously begin to move about and to rotate, changing both their positions and their orientations. The crystal liquefies. However, certain kinds of solids—those whose molecules form long, stiff rods—melt, keeping some degree of order. That is, the molecules still line up somewhat. Substances in this state are called *liquid crystals* because they have some of the properties of liquids and some of crystals. In liquid crystals, the amount of ordering is extremely sensitive to electric fields.

In 1991, Pierre-Gilles de Gennes of France received the Nobel Prize for explaining how liquid crystals change their level of ordering (the degree to which they line up) under electric fields. This is the effect responsible for the devices called liquid crystal diodes (LCDs). In LCDs, liquid crystals are sandwiched in thin layers between transparent electrified plates. When the charges on parts of the plates are changed, the LCDs either transmit or stop polarized light from going through those areas.

LCDs can serve as displays in electronic devices such as watches. A watch face is designed so that the parts making the characters can be connected selectively to the watch battery at a given time, while other parts are not. The portion connected to the battery appears black. The portion not connected appears white because the light can go through the liquid crystal.

Billions of LCD displays have been made, not only for watches, but for computer screens, auto speedometers, clocks, electric meters, and numerous other applications. LCD screens are inexpensive and use little power. They are a direct application of sophisticated chemistry.

Liquid Crystal Displays

a) Current off

b) Current on

(a) Without an electrical current, liquid crystal molecules are twisted. If you try to pass polarized light through them, the light will twist and can pass through a filter. (b) If a current is turned on, the liquid crystals will line up, block the passage of light, and form dark numbers on a watch face.

Solid water forms molecular crystals that have a hydrogen-bonded structure. If you look at Figure 12.16, an expanded sketch of several water molecules in an ice crystal, you can see how each H atom is closely bonded to one O atom and joined by a hydrogen bond (dotted line) to another O atom; this pattern forms tiny cages of water molecules, each cage having a hollow place in the center. When ice melts, most of this structure is destroyed and most of the hollow places disappear. Liquid water is therefore more dense than ice, although the liquids of most substances are less dense than their solids.

Ice cubes float in liquid water. They do so because ice is less dense than liquid water. If ice were more dense than liquid water, ice would sink. In winter, many lakes and, in places, oceans would freeze from the bottom up, often never thawing completely. World climate would be entirely different.

Metallic Solids

In metals, the valence electrons belong to no particular atom.

Although we will study metallic solids in less detail, we can look briefly at the structures and properties of metals. It is thought that when metal atoms join to form solids, each atom gives up one or more of its valence electrons. These electrons then become the shared property of the entire metallic solid. What results is a "sea" of electrons, acting as a glue to cement the metal ions together. The forces uniting metal atoms together are called **metallic bonds.** Figure 12.17 illustrates the phenomenon. Because the metal ions can easily change their positions, metallic solids bend without breaking and can be hammered into various shapes.

Because the electrons are not bound to individual atoms and are free to be anywhere in the crystal, metals readily conduct electricity. Metals also readily conduct heat because their electrons can rapidly carry kinetic energy from one place to another in the metal crystal. Metals ordinarily feel cold to the touch because they conduct heat away from us, but when they are heated, they feel very hot.

The melting and boiling temperatures of metals depend on how many electrons each atom gives to the electron "sea." When more valence electrons are available, more "electron glue" is formed and greater positive charges exist on the metal ions: Both factors increase the strength of the metallic solid. Al (three valence electrons) has a higher melting temperature than Mg (two valence electrons), which, in turn, has a higher melting temperature than Na (one valence electron). The atoms of some metals, like those in the transition series (Fe and Cr, for example) have some *d*-electron interactions as well as the electron "sea" holding them together. Crystals of such metals are hard and have high melting points. Table 12.7 gives melting temperatures for several metals.

Metal crystals bend without breaking and can be hammered into different shapes. This is possible because the metal structure has no specific bonds, but is merely a collection of positive ions held together by negative "electron glue." The ions can easily change their positions within the crystal.

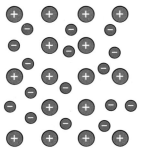

Figure 12.17 Electron sea model of Na. Here the electrons are shown as tiny circles, but in the actual metal, they are smeared out over the entire sample.

Table 12.7 Melting Temperatures of Some Metals

Metal	Melting Temperature (°C)
Cesium	28.5
Sodium	97.8
Magnesium	651
Aluminum	660
Iron	1,535
Chromium	1,890
Tungsten	3,410

Summary

Many substances can exist in any of the three states: solid, liquid, or gas. All substances can exist as solids. Whether a substance will be solid, liquid, or gas depends on the forces between its individual particles and on the temperature and pressure.

Van der Waals forces arise either from the mutual attraction of permanent dipoles in molecules or from the interaction of temporary dipoles. In some cases, both kinds of force operate together. Van der Waals forces are generally much weaker than chemical bonds.

Hydrogen bonds are special cases of dipole interaction that exist between molecules that have hydrogen atoms bonded to oxygen, nitrogen, or fluorine. Although hydrogen bonds lead to stable structural arrangements, they are not usually classed as chemical bonds because they do not form structures with characteristic chemical formulas.

Individual molecules can leave the surface of a liquid or solid to become a gas in a process known as vaporization. The rate at which this process occurs depends on the forces between the particles and on the temperature and pressure. When particles leave a liquid, diffusing away through the atmosphere and becoming lost, the process is called evaporation. The amount of energy gained when a liquid vaporizes to a gas is called the heat of vaporization.

In a closed vessel, gaseous molecules form a vapor in contact with the liquid. The pressure of that vapor rises until the rate of return of vapor molecules to the liquid is the same as the rate of vaporization, at which time a state of equilibrium is reached. From this moment, the pressure of the vapor remains constant unless the system is disturbed. The equilibrium pressure of a vapor in contact with its liquid is called the vapor pressure of that liquid and is a property of the liquid. Solids in closed containers can also develop vapor pressures, but for any given substance, the vapor pressure of a solid is usually much smaller than that of a liquid.

The boiling point of a liquid is the temperature at which the vapor pressure of the liquid equals the pressure being exerted on the liquid surface. The normal boiling point is the temperature at which the vapor pressure equals one atmosphere.

The process by which a solid becomes a liquid is called melting, and the process of a liquid turning into a solid is known as freezing. The amount of energy released during freezing is called the heat of fusion.

Various attractive forces can exist between particles. Solids can be held together by chemical bonds, either ionic or covalent, or by metallic bonds. Under ordinary conditions, the molecules of gases, liquids, and some solids are attracted to one another either by hydrogen bonds or by van der Waals forces.

Water is a highly polar substance. Hydrogen bonding in the liquid and solid forms gives properties to water that distinguish it from nearly all other substances. In addition to abnormally high boiling and melting points, water has a high heat of vaporization.

Table 12.8 lists several examples of each kind of solid and examples of typical properties. Be sure to study this table carefully; it will help you tie together what you have learned.

Table 12.8   Properties of Crystalline Solids

Kind of Crystal	Forces	Examples	Comments
Ionic	Electrostatic attraction	NaCl, CaCO₃, K₂SO₄, Sr(NO₃)₂, KOH	• Often quite hard, depending on the charges of the ions • Occasionally very brittle • Fairly high to very high melting points • All are solids at ordinary temperatures
Covalent	Covalent	Diamond, graphite, SiO₂, SiC (carborundum)	• Usually very hard and tough (exception: graphite) • Very high melting points • Strongest interparticle forces • All are solids
Molecular H-bonded	H bonds (see text)	H₂O, HF, NH₃	• Mostly liquids or gases at room temperature • Melting and boiling points much lower than those of ionic crystals • Soft and easily shattered
Dipolar	Dipole–dipole interaction	ICl, HCl	• Almost all are liquids or gases at room temperature • Melting and boiling points generally lower than in H-bonded substances
Temporary dipole	Interaction of temporary dipoles	I₂, Br₂, N₂, H₂, Ar, CO₂	• Most are gases, with a few liquids and solids (Br₂, I₂, for example) • Weakest interparticle forces of all • Temporary dipole forces often contribute to attractions in other kinds of substances (as in IBr, which has mostly temporary dipole but also some dipole–dipole forces)

Key Terms

condensation	(p. 340)	volatile	(p. 347)
condensed states	(p. 340)	boiling	(p. 349)
van der Waals forces	(p. 340)	boiling temperature	(p. 349)
dipole–dipole forces	(p. 340)	normal boiling point	(p. 349)
temporary dipole forces	(p. 340)	freeze	(p. 350)
hydrogen bonding	(p. 343)	melt	(p. 350)
evaporation	(p. 345)	melting temperature	(p. 350)
sublimation	(p. 345)	freezing temperature	(p. 350)
heat of vaporization	(p. 345)	normal melting temperature	(p. 350)
heat of sublimation	(p. 345)	heat of fusion	(p. 350)
heat of condensation	(p. 346)	covalent crystal	(p. 352)
vapor pressure	(p. 347)	molecular crystal	(p. 354)
		metallic bond	(p. 356)

Questions and Problems

Section 12.1 Condensation

Question

1. Look up the word "condensed" in the dictionary. Why are liquids and solids called condensed states? Why is the process of changing from a gas to a liquid or solid called condensation?

Section 12.2 Forces Between Individual Particles

Questions

2. What are van der Waals forces? Explain at least two ways in which these forces differ from chemical bonds.

3. Explain the causes of electric dipoles.

4. With the help of diagrams, explain how dipole forces operate between molecules.

5. Explain how intermolecular forces can develop even among molecules that have no permanent dipoles.

6. Explain how a molecule can have polar covalent bonds yet not be polar.

7. Name two different kinds of molecules that are polar. Draw their structures, indicating the dipoles.

8. Draw structures for two molecules that have polar bonds but are not polar.

9. What is a hydrogen bond and in what circumstances does it exist?

10. How does a hydrogen bond differ from an ionic or covalent chemical bond? How does it differ from an ordinary dipole force, as found, for example, between H_2S molecules?

Section 12.3 Properties of Condensed Phases

Questions

11. Why can gases be more easily compressed than liquids and solids?

12. Why are liquids and solids more dense than gases?

13. Why can liquids flow while solids can't?

Section 12.4 Vaporization

Questions

14. Explain the process of evaporation. Will evaporation take place if a liquid is in a closed container? Explain.

15. What is of difference between vaporization and evaporation? What is the difference between vaporization and sublimation?

16. How are heat of vaporization and heat of condensation related?

17. The heat of vaporization of water is far greater than that of liquid H_2. Why?

18. Burns caused by steam are usually far more serious than those caused by boiling water. Why?

19. Many animals, including humans, perspire on hot days. Explain why this is advantageous in terms of heat of vaporization.

20. Isopropyl alcohol feels cool on your skin, but salad oil, also a liquid, does not. Explain the difference.

Problems

21. Calculate the amount of energy required to vaporize 35.7 grams of ethanol.

22. How much energy is required to vaporize 1.35 L of water at a constant temperature of $100°C$?

23. Calculate the number of kilojoules that must be added to vaporize 3.00 grams of sodium chloride; 3.00 grams of argon. Why does sodium chloride require more energy?

24. How much energy is required to vaporize 32 grams of zinc? 32 grams of sodium? What do you think causes the difference?

25. How much energy will be released when 19 grams of steam condense? 19 grams of argon? Why does the condensation of water release more energy?

26. Calculate the amount of energy released when a 152-gram sample of ethanol condenses.

27. The normal boiling point of ammonia is $-33.4°C$. A sample of liquid ammonia containing 1.00 mol was vaporized at $-33.4°C$, a process requiring 23.4 kJ. What is the heat of vaporization of NH_3 at its boiling point?

Section 12.5 Vapor Pressure

Questions

28. What is a vapor?

29. What is vapor pressure?

30. Why can't vapor pressure be established unless the substance is in a closed container?

31. Why do some substances have higher vapor pressures than others? Name some substances with high vapor pressures and some with low.

32. If a substance has a high vapor pressure, it is likely to have a low heat of vaporization and vice versa. Explain.

33. What is dynamic phase equilibrium? What processes come to equilibrium when vapor pressure is established?

Section 12.6 Boiling and Boiling Temperatures

Questions

34. What is boiling?

35. Explain the connection between boiling temperature and atmospheric pressure.

36. What is the normal boiling point of a liquid?

37. In liquids, what is the relation between boiling point and the strength of the intermolecular forces?

38. A refrigerator operates by causing a liquid to boil inside coils of tubing. Explain how this causes cooling.

39. Compare the heat of vaporization and the boiling point of ammonia to those of other substances listed in Tables 12.1 and 12.3; then explain why ammonia makes a good refrigerant.

Section 12.7 Melting and Freezing

Questions

40. Why is the freezing temperature the same as the melting temperature for any given liquid? What is the difference between the two processes?

41. How do intermolecular forces affect heats of fusion?

42. What are the two opposing processes when an equilibrium is established between a solid and a liquid?

43. How is it possible for a liquid to boil and freeze simultaneously? Explain fully.

44. By means of an electric heating coil, energy is added to a piece of ice that has an initial temperature of $-10°C$. The temperature of the ice slowly increases to $0°C$, at which time the ice begins to melt. Energy continues to be added, but now the temperature of the ice and water remains constant until all the ice is melted. Why does the temperature not change during melting?

45. Why is the heat of vaporization of a substance usually larger than the heat of melting?

46. Make a rough graph showing how the temperature of a sample of water changes as energy is slowly added to it at a uniform rate. Start with a temperature of $-10°C$, when the water is frozen, and finish at $120°C$, with steam. Here is the way to set up the graph:

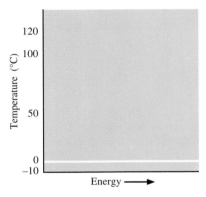

Problems

47. Calculate the amount of energy required to melt 35.7 grams of argon.

48. How much energy is required to melt 92 grams of sodium chloride? 92 grams of ice? Why does sodium chloride require more energy?

49. Calculate the number of kilojoules that will be released when 45.0 grams of zinc solidify; when 45.0 grams of sodium solidify.

50. How much energy is released when 124 grams of argon solidify? 124 grams of water? Why does water release more energy?

Section 12.8 Crystalline Solids

Questions

51. What are the essential differences between ionic crystals, covalent crystals, and molecular crystals? What is the main reason for these differences?

52. What forces unite metal atoms in a metallic solid? How are these forces different from ionic and covalent bonds?

53. Without referring to Table 12.8, construct a table that shows covalent crystals, ionic crystals, and molecular crystals. For each type, give the kinds of particles involved, the interparticle forces, the properties of the crystal, and some examples.

Solutions to Exercises

12.1 (a)
$$21.0 \text{ g} \times \frac{1.38 \text{ kJ}}{\text{g}} = 29.0 \text{ kJ}$$

(b)
$$1.35 \text{ kg} \times \frac{1000 \text{ g}}{\text{kg}} \times \frac{4.27 \text{ kJ}}{\text{g}} = 5.76 \times 10^3 \text{ kJ}$$

12.2
$$1.35 \text{ g} \times \frac{0.520 \text{ kJ}}{\text{g}} = 0.702 \text{ kJ}$$

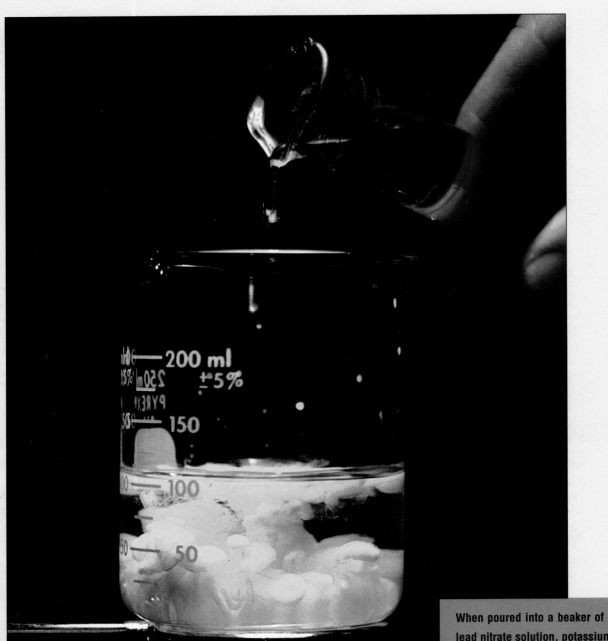

When poured into a beaker of lead nitrate solution, potassium chromate solution reacts to give solid, yellow lead chromate.

Solutions

A bumper sticker that occasionally appears on college and university automobiles says, "Chemists Have Solutions"—which is true. Chemistry could hardly have begun without solutions. Numerous chemical reactions, especially many of those performed in experimental laboratories, are carried out in solution. Chemical measurements are often made in solution; compounds are prepared and sold in solution; many phenomena important to chemistry occur only in solution.

In this chapter, we are introduced to the language of solutions, then we investigate why and how substances mix and dissolve in one another. We learn which substances form solutions and which do not. We find that there are limits to how much of one substance will dissolve in another. We see that many solutions are created by chemical reaction rather than by simple mixing.

We will apply our knowledge to determine the concentrations of solutions. We then use concentration information to determine the freezing and boiling points of solutions and to perform stoichiometric calculations.

13.1 What Is a Solution?

After studying this section, you should be able to:

- Distinguish between solvents and solutes.

We learned in Section 2.7 that *a solution is a homogeneous mixture of two or more substances.* Solutions can exist in any of the three phases: gas, liquid, or solid. Air is a gaseous solution mostly of N_2 and O_2. A brass doorknob is a solid solution of copper and zinc. Gasoline, seawater, tea with sugar, tap water—and most of the other liquids with which we come into contact—are solutions.

Solutions are composed of **solvents** and **solutes.** In most cases, it is easy to determine which is solute and which is solvent. If substances of different phases are involved, the one that has the same phase as the resulting solution is called the solvent. For example, if solid sugar is dissolved in liquid water to form a liquid solution, the water is said to be the solvent and the sugar the solute. If all of the substances in a solution are of the same phase, then the substance present in the greatest quantity is said to be the solvent. In a soft drink, the solvent is water because more than nine tenths of the solution is water. In a few cases, the distinction between solvent and solute is difficult to make; however, we will be concerned only with straightforward situations, usually with solids dissolved in liquids.

In a solution, a solvent is the substance that has the same phase as the solution or that is present in the greatest quantity.

Because water can dissolve thousands of substances, it is sometimes called a *universal solvent.* Solutions in which the solvent is water are called **aqueous solutions.** Unless otherwise stated, the solvent in a solution is usually understood to be water.

In aqueous solutions, water is the solvent.

Solutions are often confused with samples of pure substances because both are homogeneous. You can distinguish between the two if you remember that every compound contains its elements in definite proportions according to a formula. In a solution, however, solvents and solutes can be present in various proportions.

13.2 Solubility; Saturated Solutions

After studying this section, you should be able to:

- Explain how saturated and supersaturated solutions can be prepared.
- Explain how temperature affects solubility.
- Explain how pressure affects the solubility of gases.

The amount of a substance that can dissolve in a given amount of another substance is often limited. A cup of water will dissolve a teaspoon of sugar or even two teaspoons if you stir for awhile, but if you put in ten teaspoons of sugar, most of the sugar will remain as a solid in the bottom of the cup, no matter how long you stir (see Figure 13.1). When a solution contains as much

Figure 13.1 Sugar solutions. (a) When added to an unsaturated solution, sugar dissolves. (b) When added to a saturated solution, sugar remains undissolved.

(a)

(b)

Figure 13.2 When a small crystal of sodium acetate, $NaC_2H_3O_2$, is added to a supersaturated solution of sodium acetate, solid sodium acetate immediately forms.

solute as will dissolve at a given temperature, it is said to be **saturated** with that solute at that temperature. Saturated solutions are usually prepared by adding to the solvent more solute than can be dissolved, mixing until no more solute disappears, then leaving some excess solute in contact with the solution. The maximum amount of solute that can be dissolved in a given amount of solvent to give a saturated solution at a particular temperature is called the **solubility** of that solute in that solvent at that temperature.

A saturated solution contains as much dissolved solute as possible at a given temperature, and it is usually in contact with some solid solute.

max amount

Under extraordinary circumstances, **supersaturated** solutions can be created. Such solutions contain more dissolved solute than they would in an ordinary saturated solution. Supersaturated solutions are unstable and can exist only if none of the solid form of the solute is present (Figure 13.1). A popular lecture demonstration (Figure 13.2) begins with a cool, supersaturated solution of sodium acetate into which a tiny crystal, a "seed," of solid sodium acetate is dropped. Crystals of the solid compound grow rapidly on the seed, and within seconds, the container is full of solid sodium acetate.

Sometimes simply shaking a supersaturated solution will cause it to lose its excess solute.

Effect of Temperature on Solubility

$T \propto S$ Gases $T \propto \frac{1}{S}$

In most cases, the amount of solid that will dissolve in a liquid solvent to form a saturated solution increases with increasing temperature. More sugar will dissolve in a cup of hot water than in a cup of cold water. At 20°C, 65.2 grams of KBr will dissolve in 100 grams of water; the solubility is 65.2 g/100 g of water. At 80°C, the solubility of KBr is 95.0 g/100 g of water.

Increasing the temperature will increase the solubility of solids in liquid solvents.

On the other hand, the solubility of most gases in liquid solvents decreases with increasing temperature. At 25°C, the solubility of CO_2 is 0.145 g/100 g of water. At 60°C, however, the solubility of CO_2 is only 0.058 g/100 g of water.

Increasing the temperature will decrease the solubility of most gases in liquid solvents.

Effect of Pressure on Solubility

Changes in pressure have little effect on the solubility of solids and liquids in liquid solvents. The solubility of gases in liquid solvents, however, is greatly affected by changes in pressure. An increase in pressure of a gas over a liquid solvent increases the solubility of the gas. If the pressure of a gas over a liquid solvent is reduced, the solubility of the gas will decrease. In champagne bottles,

$P \propto S$
Gases
Increasing the pressure will increase the solubility of most gases in liquid solvents.

CO_2 gas is dissolved at greater than atmospheric pressure. In locker rooms of winning World Series teams, the dissolved CO_2 gas is released with results seen on TV screens everywhere. (Agitation accelerates removal of the CO_2 bubbles from the champagne!)

13.3 Kinds of Substances That Dissolve in One Another

After studying this section, you should be able to:

- Determine which solutes will dissolve in polar solvents and which in nonpolar solvents.
- Determine whether a compound will dissolve in water.

It is not easy to predict exactly which substances will dissolve which, but a rule of thumb helps in most cases: Chemists are fond of saying that *like dissolves like,* meaning that substances of similar polarity will form solutions. Let's see why.

Remember from the discussion in Section 6.13 that some molecules are polar. Polar molecules are attracted to other polar molecules by fairly strong dipole–dipole forces. If two polar substances are mixed, all the molecules attract one another, and the substances mix well. Recall that water is polar and forms hydrogen bonds. Other polar substances, especially those that can form hydrogen bonds, dissolve well in water. Ethanol, C_2H_5OH, for instance, dissolves easily because each molecule has a polar OH group with which a water molecule can form a hydrogen bond (Figure 13.3).

On the other hand, the molecules of nonpolar substances have less attraction for each other. So if two nonpolar substances are mixed, neither excludes the

Dipole–dipole forces are discussed in Section 12.2.

Figure 13.3 Hydrogen bonding between water and ethanol.

Figure 13.4 The attractions that the water molecules have for one another keep the oil droplet from dissolving in the water.

Figure 13.5 The attractions that the water molecules have for one another keep the droplets together. Water will not dissolve in oil.

Table 13.1	Solubilities Substances	
Substances	Example	Dissolves?
Polar and polar	Water and sugar	Yes
Nonpolar and nonpolar	Kerosene and grease	Yes
Nonpolar and polar	Oil and water	No

other. Grease and gasoline are both nonpolar; therefore, grease will dissolve in gasoline.

However, if a polar substance is mixed with a nonpolar substance, the polar molecules attract one another and exclude the nonpolar molecules, for which they have little attraction. Water is polar and oil is not. Try to dissolve a drop of oil in water (Figure 13.4). The water molecules attract one another strongly and refuse to separate so that the oil molecules can mix with them. Try to dissolve a drop of water in oil. The water molecules stay in their drop because they are attracted to one another and are reluctant to separate (Figure 13.5). The oil molecules have essentially no attraction for water molecules and therefore do not dissolve the water.

Knowing the structures of two substances will allow you to make some good guesses about whether they will dissolve in each other. Look at Table 13.1.

Substances of like polarity more easily dissolve one another than do substances of different polarities.

13.4 Electrolytes

After studying this section, you should be able to:

- Explain how ionic solids dissolve and dissociate in water.
- Explain the reason for the name "electrolyte."
- Tell why solutions of strong electrolytes are not all necessarily strong conductors of electricity.
- Explain the difference between strong and weak electrolytes.

Water readily dissolves NaCl, but NaCl is insoluble in nonpolar liquids. We might have predicted this behavior because we know that an NaCl pair is extremely polar; the Cl is present as a negative ion and the Na as a positive one. Actually, there is more to it than that. Ionic substances dissolve in water by separating completely into their individual ions. We were first introduced to this process, called *dissociation*, in Section 6.2.

$$NaCl(s) \longrightarrow Na^+ + Cl^-$$

NaCl can dissolve and separate into ions because the polar water molecules attract the positive Na^+ and negative Cl^- ions of the solid salt. These attractions replace the strong ionic forces holding the ions together in the solid NaCl and promote the dissolving process.

Study Figure 13.6, which shows how ions are removed from the surface of a crystal. Notice that when the ions are finally taken into solution, they remain as individuals.

It would be proper to indicate that the ions of NaCl, for example, existed in aqueous solution by writing "(aq)" after each ion, but since nearly all reactions of ions occur in aqueous solution, the "(aq)" is usually left out for ions.

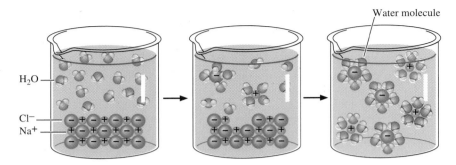

Figure 13.6 Dissociation of NaCl in water. The polar water molecules interact with the positive and negative ions of the salt. These interactions partly replace the strong ionic forces holding the ions together in the undissolved solid and assist in the dissolving process.

A substance that forms ions in solution is called an **electrolyte** because solutions of ions conduct electricity, a subject that is discussed in detail in Sections 16.4 and 16.5. If a substance is to conduct electricity, it must contain charged particles, and those particles must be able to travel through the substance. That NaCl solutions conduct electricity tells us that charged particles, ions, exist in the solution, and we conclude that those ions came from the dissociation of NaCl. Solid NaCl, however, does not conduct; the ions are present, but they are locked in place, unable to travel. Aqueous solutions of sucrose (table sugar) do not conduct because they contain no charged particles. Sucrose does not dissociate when it dissolves; it is not an electrolyte.

Ionic substances that are completely ionized in solution are called **strong electrolytes.** $K_2Cr_2O_7$ is a good example. No $K_2Cr_2O_7$ particles exist in solution; the compound is present only as its ions, K^+ and $Cr_2O_7^{2-}$. To the extent to which they are soluble, most ionic substances are strong electrolytes; however, not all ionic substances are necessarily very soluble. Calcium sulfate, for example, is a strong electrolyte, but only a little of it will dissolve in water. The forces between the Ca^{2+} ions and the SO_4^{2-} ions are so strong that water cannot easily disrupt the structure of the $CaSO_4$ crystals; therefore, relatively few ions are formed. Because there are so few ions in solution, the solution conducts electricity only slightly.

Strong electrolytes, just because they are strong, are not necessarily very soluble.

Many compounds form ionic solutions but are not themselves ionic. These compounds dissolve by means of a chemical reaction, called **ionization,** that creates ions where there were none before. An example is hydrogen chloride, HCl, which is covalent yet forms ions in solution and is a strong electrolyte. The ionization reaction of HCl is

$$HCl(g) + H_2O(l) \longrightarrow H_3O^+ + Cl^-$$

We will learn more about reactions of this kind and about H_3O^+ in Chapters 14 and 15.

Some solutes form solutions in which only a small proportion of the compound in solution is in the form of ions; such substances are called **weak electrolytes.** Acetic acid, $HC_2H_3O_2$, the compound that makes vinegar sour, is a good example. When $HC_2H_3O_2$ dissolves, most of the compound exists as $HC_2H_3O_2$ molecules, and only a small fraction forms the ions, H_3O^+ and

Weak electrolytes, such as NH_3, are sometimes very soluble, but even so only a small portion of the dissolved compound forms ions.

Figure 13.7 The light bulb glows when electricity is conducted through the solution in the beaker. In the left picture, the beaker contains pure water, which supplies a negligible number of ions. There is no visible conduction by the water. In the center picture, some of the weak electrolyte, acetic acid ($HC_2H_3O_2$) has been added. The light has a dim glow, indicating that only a little electricity is conducted. Acetic acid dissociates only a little, furnishing relatively few ions to the solution. In the right picture, the strong electrolyte, potassium dichromate ($K_2Cr_2O_4$) has been added. The bright glow indicates that electricity is strongly conducted, showing that the ($K_2Cr_2O_4$) is highly dissociated into its ions, K^+ and CrO_4^{2-}.

$C_2H_3O_2^-$. We will discuss weak electrolytes in greater detail in Chapters 14 and 15. The conduction abilities of strong electrolytes, weak electrolytes, and nonelectrolytes are shown in Figure 13.7.

13.5 Hydrates (optional)

After studying this section, you should be able to:

- Explain how water of hydration can be trapped in an ionic crystal.
- Explain how water of hydration can be driven from a compound.
- Identify a hydrate from its chemical formula.

When a dissolving ion leaves the surface of its crystal, it carries with it a layer of solvent molecules (look back at Figure 13.6) and is said to be **solvated.** If the solvent happens to be H_2O, the ion is said to be **hydrated.**

Sometimes the water molecules are only loosely attached, as in solutions containing Na^+, but in many cases, they are strongly attracted to the ion and form stable structures that stay with the ion even when the compound becomes a solid. Compounds containing water molecules are called **hydrates.** For example, in many of its compounds Cu^{2+} is present in a structure in which four H_2O molecules surround each ion. In solid $CuSO_4$ as it is ordinarily found, both the Cu^{2+} and the SO_4^{2-} are hydrated; four H_2O molecules are associated with Cu^{2+} and one with the SO_4^{2-}. To represent water of hydration in formulas, we write the formula of the compound, followed by a centered dot and the number of water molecules associated with one unit of the formula. The hydrate of cupric sulfate, for example, would appear as $CuSO_4 \cdot 5\ H_2O$.

Table 13.2 lists the formulas and names of some common laboratory hydrates.

Table 13.2   Some Common Hydrates

Name	Formula
$Ba(OH)_2 \cdot 8\ H_2O$	Barium hydroxide octahydrate
$CaCl_2 \cdot 2\ H_2O$	Calcium chloride dihydrate
$CaSO_4 \cdot \frac{1}{2}\ H_2O$	Calcium sulfate hemihydrate
$CoCl_2 \cdot 2\ H_2O$	Cobalt(II) chloride dihydrate
$CuSO_4 \cdot 5\ H_2O$	Copper(II) sulfate pentahydrate
$MgSO_4 \cdot 7\ H_2O$	Magnesium sulfate heptahydrate
$Na_2CO_3 \cdot 10\ H_2O$	Sodium carbonate decahydrate

When hydrates are heated, the water molecules vaporize and leave the solid. The resulting substances are said to be in the **anhydrous** form. When exposed to water, anhydrous solids become hydrates once again. Upon exposure to water, some hydrates may incorporate even more water molecules. The powder,

When heated, $CuSO_4 \cdot 5\ H_2O$ loses its waters of hydration and turns from a blue crystalline solid to a white powder.

The blue substance (top left) is the pentahydrate of copper sulfate, $CuSO_4 \cdot 5\ H_2O$. The white substance below it is anhydrous copper sulfate, $CuSO_4$. Cobalt(II) chloride hexahydrate, $CoCl_2 \cdot 6\ H_2O$ (top right) is a deep red, while the anhydrous form, $CoCl_2$ (below), is lavender. You may have seen this compound added to drying agents to indicate the presence of water. In the dry form, the pellets are pale blue, but when even a little water is present, the blue becomes pink.

Science in Action

Because sources of energy are becoming expensive, hard to find, and polluting, people are beginning to build houses that waste as little energy as possible and get as much of their energy as possible from the sun. The difficulty is that at night, when heat is most needed in a house, the sun doesn't shine. One solution is to store solar energy, and one promising storage medium is epsom salts, a compound once used as a laxative.

During the day, water is circulated through panels oriented toward the sun. The sun heats the water, which is then sent through coils surrounded by epsom salts. Epsom salts is the popular name for magnesium sulfate heptahydrate, $MgSO_4 \cdot 7 H_2O$. This compound has the property that, when heated to 48.4°C, it loses one of its seven water molecules and absorbs energy in the process.

$$MgSO_4 \cdot 7 H_2O \longrightarrow MgSO_4 \cdot 6 H_2O + H_2O \qquad \Delta H = 58.7 \text{ kJ}$$

The energy to power the reaction is absorbed from the hot water in the coils.

At night, when heat is needed, cool water is circulated through the same coils. Now the dehydration is reversed and energy is released.

$$MgSO_4 \cdot 6 H_2O + H_2O \longrightarrow MgSO_4 \cdot 7 H_2O \qquad \Delta H = -58.7 \text{ kJ}$$

The energy that is released heats the water in the coils. This is then circulated through pipes hidden in the floor, and the house is warmed. One barrel, about 50 gallons, will hold about 1,200 moles of epsom salts. This will supply about as much heat as will a medium-sized, forced-air furnace running for an hour. Thus several barrels of epsom salts, with the necessary plumbing, are enough to warm a house through the night. The only operating expense is for the small pump that moves the water through the coils.

A solar house.

---

plaster of paris, is used for making wallboard, statuary, casts for broken bones, and many other things. It is the hemihydrate of calcium sulfate, $CaSO_4 \cdot \frac{1}{2} H_2O$. When water is added to the powder, it becomes the dihydrate, $CaSO_4 \cdot 2 H_2O$, which forms a hard, rocklike solid.

13.6 Solubility Rules

After studying this section, you should be able to:

- Use the solubility rules and the periodic table to determine the approximate solubility of any given ionic compound.

How can we predict whether or not a solid will dissolve in water and to what extent it will dissolve? Through experimentation, chemists have developed a set of general guidelines for determining the solubility of ionic compounds. These solubility rules are listed in Table 13.3.

Soluble can be taken to mean that several tens of grams of the compound will dissolve in 100 grams of water at room temperature. *Moderately soluble* might mean from 5 to about 15 grams of compound in 100 grams of water. Only a gram or so of a *slightly soluble* compound, and much less still of a *very slightly soluble* compound, will dissolve in 100 grams of water at room temperature.

It is not necessary to memorize all the solubility rules at once, but you should study them now, then go through them again when we discuss the chemical reactions of ionic solutions.

(memorize)

Table 13.3 Ionic Compounds Soluble in Water

Compounds	Exceptions
Most compounds of the alkali metals and of ammonium ions, NH_4^+	A few Li compounds
Nitrates (NO_3^-), acetates $(C_2H_3O_2^-)$, and chlorates (ClO_3^-)	$HgC_2H_3O_2$ and $AgC_2H_3O_2$
Chlorides, bromides, and iodides	Compounds of Ag, Hg(I), Bi(III), and Sb; $PbCl_2$ and $PbBr_2$ are slightly soluble.
Sulfates (SO_4^{2-})	Compounds of Ca, Sr, Ba, Ag, Hg, and Pb, which are slightly to moderately soluble
A few hydroxides and sulfides	Hydroxides and sulfides of the alkali metals and ammonium ion are soluble; hydroxides and sulfides of Ba, Sr, and Ca are slightly to moderately soluble; all other hydroxides and sulfides are insoluble.

Exercise 13.1

Which of the following compounds is soluble in water: Hg_2S, K_2SO_4, $CaSO_4$, Ag_3PO_4?

13.7 Reactions of Ions in Solution

After studying this section, you should be able to:

* Write chemical equations for reactions of ions in solution.
* Use the solubility rules to predict whether mixing solutions of ionic compounds will result in reaction.

Now that we know some solubility rules and something about solutions of ions, we can use that knowledge to predict the results of many reactions.

In some cases, mixtures of ions in solution are unreactive. For example, $BaCl_2$ and $NaNO_3$ can be dissolved in the same solution, or solutions of the two compounds can be mixed. The result in either case will be a solution that contains all four kinds of ions: No chemical reaction occurs. However, if we were to mix solutions of $AgNO_3$ and KCl, we would find that a white solid formed, indicating that a reaction occurred between the two compounds.

Could we predict that $BaCl_2$ and $NaNO_3$ would not react and that KCl and $AgNO_3$ would? Yes, by knowing the three general ways in which ions react in solution.

White solid AgCl instantly forms when solutions of $AgNO_3$ and NaCl are mixed.

1. *Precipitate formation.* When a solid forms after solutions of two ionic compounds are mixed, that solid is called a **precipitate,** and the reaction is called a precipitation reaction.

Mixing solutions of $AgNO_3$ and $NaCl$ produces a white precipitate, $AgCl$.

$$AgNO_3(aq) + NaCl(aq) \longrightarrow NaNO_3(aq) + AgCl(s)$$

2. *Formation of water.* When two ions combine in solution to form water, they cease to exist as ions. Because water is a stable compound, the formation of water is a strong driving force to make reaction occur. In the next chapter, we will study acid-base reactions, a classic case of reaction by the formation of water. The reaction of HCl and NaOH solutions is an example. The final result is an aqueous solution of NaCl.

$$HCl(aq) + NaOH(aq) \longrightarrow NaCl(aq) + H_2O(l)$$

3. *Gas formation.* When ions in solution can react to form a gas, the gas escapes. A reaction has occurred. The loss of gas from the solution is what drives the reaction. For example, when HCl combines with a carbonate, such as Na_2CO_3, carbonic acid, H_2CO_3 is formed.

$$Na_2CO_3(aq) + 2\ HCl(aq) \longrightarrow 2\ NaCl(aq) + H_2CO_3(aq)$$

Carbonic acid is unstable, however, and immediately decomposes to form water and carbon dioxide, a gas.

$$H_2CO_3(aq) \longrightarrow CO_2(g) + H_2O(l)$$

The reaction of HCl and Na_2CO_3, therefore, can be written as

$$Na_2CO_3(aq) + 2\ HCl(aq) \longrightarrow 2\ NaCl(aq) + H_2O(l) + CO_2(g)$$

The formation of a gas from ionic solutions is not common, but even so, several industrial processes are designed to take advantage of this kind of reaction.

There are other ways in which reactions of ions can be made to occur. Examples are electrolysis (which we will study in Chapter 16) and the formation of what are called complex ions (a subject not covered in this textbook).

The following examples, along with the solubility rules of Section 13.6, will help you to predict the outcomes of many reactions.

Hydrochloric acid, HCl, reacts vigorously with sodium carbonate, Na_2CO_3, to produce CO_2 gas.

EXAMPLE 13.1

Will mixing solutions of KI and $Pb(NO_3)_2$ result in reaction? Write a chemical equation for any reaction that occurs.

Strategy and Solution:

Step 1. Mentally separate the soluble compounds into their respective ions. KI becomes K^+ and I^-. $Pb(NO_3)_2$ becomes Pb^{2+} and $2\ NO_3^-$.

Step 2. Combine pairs of ions to form possible products and write their formulas. Combine K^+ with NO_3^- and Pb^{2+} with I^-. Notice that 2 I^- ions must be written with one Pb^{2+} to form PbI_2. The possible products are

$$PbI_2 \qquad KNO_3$$

Step 3. Determine if any of the possible products forms an insoluble solid, water, or a gas. To determine if either product forms a precipitate, look at the solubility rules. PbI_2 is insoluble, while KNO_3 is soluble. Because a precipitate will form, a reaction will occur.

A can of carbonated soft drink contains carbonic acid. When the can is opened, carbonic acid decomposes, releasing CO_2.

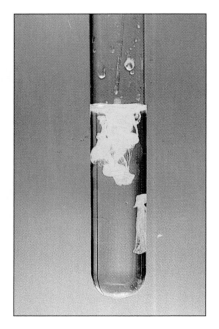

Yellow PbI_2 solid forms immediately upon mixing solutions of KI and $Pb(NO_3)_2$.

Step 4. Write and balance the chemical equation for the reaction.

$$2 \text{ KI(aq)} + \text{Pb(NO}_3)_2(\text{aq}) \longrightarrow 2 \text{ KNO}_3(\text{aq}) + \text{PbI}_2(\text{s})$$

(Note that $KNO_3(aq)$ represents K^+ ions and NO_3^- ions in solution rather than dissolved KNO_3 in molecular form.)

EXAMPLE 13.2

Will mixing a solution of CsI with one of $Ca(NO_3)_2$ result in reaction? Write a chemical equation for any reaction that occurs.

Strategy and Solution:

Step 1. Separate the soluble compounds into their respective ions. CsI becomes Cs^+ and I^-. $Ca(NO_3)_2$ becomes Ca^{2+} and NO_3^-.

Step 2. Combine pairs of ions to form possible products and write their formulas. Cs^+ combines with NO_3^-, while Ca^{2+} combines with I^-.

$$\text{CsNO}_3 \qquad \text{CaI}_2$$

Step 3. Determine if any of the possible products forms an insoluble solid, water, or a gas. To determine if either product forms a precipitate, look at the solubility rules. Both possible products are soluble. Neither forms a gas or a slightly ionized compound. Therefore, no reaction will occur.

> **Exercise 13.2**
>
> Will mixing solutions of the following pairs of compounds result in reaction? Write the formulas of any products.
> (a) $K_3PO_4 + SrCl_2$
> (b) $Ba(OH)_2 + HClO_3$
> (c) $Cu(NO_3)_2 + Na_2S$

13.8 Net Ionic Equations

After studying this section, you should be able to:

• Write net ionic equations for reactions.

The chemistry of ionic solutions occurs as reactions between individual ions, regardless of the source of those ions. For example, a solution of calcium ions, Ca^{2+}, will form solid calcium sulfate when mixed with a solution containing SO_4^{2-}, whether the SO_4^{2-} came from H_2SO_4, Na_2SO_4, or some other soluble sulfate. It is sometimes helpful, therefore, to write a chemical equation as a **net ionic equation**, that is, to include only the ions that actually react.

For the reaction of solutions of $Ca(NO_3)_2$ and Na_2SO_4, we might have written

$$\text{Na}_2\text{SO}_4(\text{aq}) + \text{Ca(NO}_3)_2(\text{aq}) \longrightarrow \text{CaSO}_4(\text{s}) + 2 \text{ NaNO}_3(\text{aq})$$

If our purpose is to indicate only what kind of reaction takes place, we could write our equation as follows.

$$Ca^{2+} + SO_4^{2-} \longrightarrow CaSO_4(s)$$

This net ionic equation, with its shorter form, is accurate because the Na^+ ions and the NO_3^- ions do not react to form a solid, a gas, or a slightly ionized compound. These ions were in the solution before reaction occurred and remain in solution afterward. Because the Na^+ and NO_3^- do not react, we need not show them. We need not write "(aq)" after each ion in these equations because it is understood that the ions are in aqueous solutions.

EXAMPLE 13.3

Write and balance the net ionic equation for the reaction of $AgNO_3$ and $NaCl$ solutions.

Strategy and Solution:

Step 1. Separate the soluble compounds into their respective ions. $AgNO_3$ becomes Ag^+ and NO_3^-. NaCl become Na^+ and Cl^-.

Step 2. Combine pairs of ions to form possible products and write their formulas. Na^+ and NO_3^- become $NaNO_3$. Ag^+ and Cl^- become $AgCl$.

Step 3. Determine if any of the possible products forms an insoluble solid, water, or a gas. $AgCl$ is an insoluble solid.

Step 4. Write the equation using ions in place of soluble compounds.

$$Ag^+ + NO_3^- + Na^+ + Cl^- \longrightarrow AgCl(s) + Na^+ + NO_3^-$$

Step 5. Remove those ions that appear on both sides of the equation. The resulting equation is the net ionic equation. It is balanced as it stands.

$$Ag^+ + Cl^- \longrightarrow AgCl(s)$$

Exercise 13.3

Write and balance the net ionic equation for the reaction of $NaOH$ and $MgCl_2$ solutions.

13.9 **Concentration**

After studying this section, you should be able to:

- Calculate the weight percent of a solution, given mass data.
- Calculate the molarity of a solution.
- Calculate the molality of a solution.
- Calculate the molarity of a solution from the weight percent and the density.
- Calculate the weight percent of a solution from the molarity and the density.
- Calculate the molality of a solution from the weight percent and vice versa.

Many experiments in chemistry are done in solution. In such experiments, it is usually necessary to know how much solute the solution contains. *The amount of solute dissolved in a given amount of solvent or in a given amount of solution* is called the **concentration** of that solute in the solution. We usually characterize solutions by saying what the solute is, what the solvent is, and what the value of the concentration is. We will study several sets of units used to state concentration values.

A solution is said to be **concentrated** if the solution has a high proportion of solute in relation to solvent. If there is only a small proportion of solute, the solution is **dilute.** There is no particular proportion of solute beyond which every solution is concentrated; the use of the two terms depends on the kind of solution and the circumstances of its use. Any solution can be made less concentrated by diluting it with more solvent. Solvent used to make a solution more dilute is often called a diluent.

Weight Percent of Solute

The weight percent of a solution is really the mass percent, grams of solute per 100 grams of solution.

The concentration of solute in a solution is often given by stating the ratio of the mass of solute to the mass of the entire solution, usually in percent terms. Because mass is usually measured by weighing, the concentration is called the **weight percent.** The weight percent of a solution is given by *the mass of solute divided by the mass of solution times 100%*.

$$\text{weight percent} = \frac{\text{weight of solute}}{\text{weight of solution}} \times 100\%$$

Weight percent is most often used when the chemistry of the solution is not under study and it is not important to know how many moles of solute are present in a sample of solution.

EXAMPLE 13.4

A solution is made by dissolving 33 grams of glucose in 66 grams of water. What is the weight percent of glucose in the solution?

Strategy and Solution:

Step 1. Find the total weight of the solution by adding the weight of the solute and the solvent.

$$33 \text{ g glucose} + 66 \text{ g water} = 99 \text{ g solution}$$

Step 2. To find the weight percent of a solution, divide the weight of the solute by the total weight of the solution and multiply by 100%.

$$\frac{33 \text{ g glucose}}{99 \text{ g solution}} \times 100\% = 33\% \text{ glucose}$$

EXAMPLE 13.5

Ethylene glycol, $C_2H_6O_2$, is the main ingredient in many commercial antifreezes. A solution is prepared by dissolving 12.4 grams of ethylene glycol in 78.0 grams of water. What is the weight percent of ethylene glycol in this solution?

Strategy and Solution:

Step 1.   Find the total weight of the solution.

$$12.4 \text{ g C}_2\text{H}_6\text{O}_2 + 78.0 \text{ g water} = 90.4 \text{ g solution}$$

Step 2.   Find the weight percent.

$$\frac{12.4 \text{ g C}_2\text{H}_6\text{O}_2}{90.4 \text{ g solution}} \times 100\% = 13.7\% \text{ C}_2\text{H}_6\text{O}_2$$

Commercial antifreezes contain ethylene glycol, $C_2H_6O_2$.

Exercise 13.4

What is the weight percent of NaH_2PO_4 in a solution prepared by dissolving 21.3 grams of NaH_2PO_4 in 455 grams of H_2O?

The weight percent of a solution is numerically the same as *the number of grams of solute in 100 grams of solution.* We can use this definition as a conversion factor when solving problems.

EXAMPLE 13.6

How many grams of KBr are required to make 35.0 grams of a 23.0% solution?

Solution Map: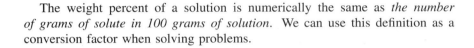

Strategy and Solution:

Step 1.   Use the weight percent as a conversion factor. A solution that is 23.0% contains 23.0 grams of KBr in 100 grams of solution.

$$\frac{23.0 \text{ g KBr}}{100 \text{ g solution}}$$

Step 2.   Set up and calculate.

$$35.0 \text{ g solution} \times \frac{23.0 \text{ g KBr}}{100 \text{ g solution}} = 8.05 \text{ g KBr}$$

Exercise 13.5

Find the number of grams of ethanol, C_2H_5OH, in 355 grams (about the mass of one 12-oz can of beer) of a 3.2% by weight solution.

EXAMPLE 13.7

Sodium carbonate is sometimes obtained as the decahydrate, $Na_2CO_3 \cdot 10 \text{ H}_2O$. How many grams of $Na_2CO_3 \cdot 10 \text{ H}_2O$ must be used to prepare 250.0 grams of a 10.0% solution of Na_2CO_3?

Solution Map:

$$\boxed{\text{grams of solution}} \longrightarrow \boxed{\text{grams of Na}_2\text{CO}_3} \longrightarrow$$

$$\boxed{\text{moles of Na}_2\text{CO}_3} \longrightarrow \boxed{\text{moles of Na}_2\text{CO}_3 \cdot 10 \text{ H}_2\text{O}} \longrightarrow$$

$$\boxed{\text{grams of Na}_2\text{CO}_3 \cdot 10 \text{ H}_2\text{O}}$$

Strategy and Solution:

Step 1.   Find the weight of Na_2CO_3 needed. A solution that is 10.0% by weight contains 10.0 grams of solute in 100 grams of solution.

$$250.0 \text{ g solution} \times \frac{10.0 \text{ g Na}_2\text{CO}_3}{100 \text{ g solution}} = 25.0 \text{ g Na}_2\text{CO}_3$$

Step 2.   Find the weight of $Na_2CO_3 \cdot 10 \text{ H}_2\text{O}$ needed.

$$25.0 \text{ g Na}_2\text{CO}_3 \times \frac{\text{mol Na}_2\text{CO}_3}{105.989 \text{ g Na}_2\text{CO}_3} \times \frac{\text{mol Na}_2\text{CO}_3 \cdot 10 \text{ H}_2\text{O}}{\text{mol Na}_2\text{CO}_3} \times$$

$$\frac{286.141 \text{ g Na}_2\text{CO}_3 \cdot 10 \text{ H}_2\text{O}}{\text{mol Na}_2\text{CO}_3 \cdot 10 \text{ H}_2\text{O}} = 67.5 \text{ g Na}_2\text{CO}_3 \cdot 10 \text{ H}_2\text{O}$$

Exercise 13.6

How many grams of $SnCl_2 \cdot 2 \text{ H}_2\text{O}$ are needed to make 75 grams of an 8.0% solution of $SnCl_2$?

EXAMPLE 13.8

How many grams of solvent are required to make 225 grams of solution that is 15.0% by weight?

Solution Map: $\boxed{\text{grams of solution}} \longrightarrow \boxed{\text{grams of solvent}}$

Strategy and Solution:

Step 1.   Determine how many grams of solvent are in 100 grams of solution.

$$100 \text{ g solution} = 15.0 \text{ g solute} + ? \text{ g solvent}$$

$$100 \text{ g solution} - 15.0 \text{ g solute} = 85.0 \text{ g solvent}$$

Step 2.   Use the weight percent of solvent as a conversion factor.

$$225 \text{ g solution} \times \frac{85.0 \text{ g solvent}}{100 \text{ g solution}} = 191 \text{ g solvent}$$

Exercise 13.7

Calculate the number of grams of solvent required to prepare 650 grams of solution that is 35% by weight NaCl.

Molarity

To describe concentration, most chemists who run chemical reactions in solution use the relation between the number of moles of solute and the number of liters of solution: moles per liter. When the concentration of a solution is expressed in moles per liter, it is called the **molarity.**

The adjective that describes molarity is *molar.* For example, if a solution has a molarity of 1 mol/L, it is said to be 1 molar. The abbreviation for molar is the capital letter M. Another abbreviation for molarity is the use of [ ]. For example, the molarity of a KBr solution can be written as [KBr].

To find the molarity of any given solution, divide the number of moles of solute by the number of liters of solution. If the number of moles of solute is not known, it must be calculated.

Molarity is the number of moles of solute divided by the number of liters of solution, not liters of solvent.

$$\text{molarity} = \frac{\text{moles of solute}}{\text{liters of solution}} = \frac{\text{mol}}{\text{L}}$$

Let's look at several examples.

EXAMPLE 13.9

A solution is made by dissolving 0.88 moles of ethanol, C_2H_5OH, in enough water to make a total of 5.1 L of solution. What is the molarity of C_2H_5OH in the solution?

Solution: Divide the number of moles of solute by the number of liters of solution.

$$\frac{0.88 \text{ mol } C_2H_5OH}{5.1 \text{ L solution}} = 0.17 \text{ M } C_2H_5OH$$

Exercise 13.8

What is the molarity of a solution that is prepared by dissolving 4.5 moles of HCl in enough water to make 35 L of solution?

EXAMPLE 13.10

What is the molarity of 0.250 L of solution containing 18.2 grams of potassium cyanide, KCN?

Solution Map: grams of KCN $\longrightarrow$ moles of KCN $\longrightarrow$ molarity

Strategy and Solution:

Step 1. Find the number of moles of solute.

$$18.2 \text{ g KCN} \times \frac{\text{mol KCN}}{65.116 \text{ g KCN}} = 0.280 \text{ mol KCN}$$

Step 2. Divide the number of moles of solute by the volume of solution in liters.

$$\frac{0.280 \text{ mol KCN}}{0.250 \text{ L solution}} = 1.12 \text{ M KCN}$$

Exercise 13.9

What is the molarity of 500.0 mL of solution containing 7.20 grams of sodium acetate, $NaC_2H_3O_2$?

Exercise 13.10

Calculate the molarity of a solution prepared by dissolving 40.3 grams of $Fe(ClO_4)_2$ in enough water to make 750 mL of solution.

EXAMPLE 13.11

A solution is 0.155 M in sucrose, $C_{12}H_{22}O_{11}$. How many grams of sucrose are there in 45 mL of solution?

Solution Map:

$$\boxed{\text{mL of solution}} \longrightarrow \boxed{\text{L of solution}} \longrightarrow$$
$$\boxed{\text{moles of } C_{12}H_{22}O_{11}} \longrightarrow \boxed{\text{grams of } C_{12}H_{22}O_{11}}$$

Strategy and Solution:

Step 1. Write the molarity as a conversion factor.

$$\frac{0.155 \text{ mol } C_{12}H_{22}O_{11}}{\text{L solution}}$$

Step 2. Set up and solve. Use the molar mass to find the number of grams of sucrose.

$$45 \text{ mL solution} \times \frac{\text{L solution}}{1000 \text{ mL solution}} \times \frac{0.155 \text{ mol } C_{12}H_{22}O_{11}}{\text{L solution}} \times$$
$$\frac{342.30 \text{ g } C_{12}H_{22}O_{11}}{\text{mol } C_{12}H_{22}O_{11}} = 2.4 \text{ g } C_{12}H_{22}O_{11}$$

Exercise 13.11

Calculate the number of grams of NaOH in 250 mL of a 0.45 M NaOH solution.

Although molarity is used for solutions with various kinds of solvents, it ordinarily refers to aqueous solutions. If no solvent is designated in a concentration term, it is usually safe to assume that the solvent is water.

When you use molar concentrations, be sure to remember that the volume refers to the number of liters of *solution,* not solvent. If you dissolve 2.3 moles of C_2H_5OH in one liter of water, you will finish with more than one liter of solution, and the molarity will not be 2.3 M. To make a solution that is 2.3 M in C_2H_5OH, you would place 2.3 moles of C_2H_5OH in a container such as a volumetric flask and add water until the volume of the solution reached one liter. Figure 13.8 illustrates how to make a molar solution beginning with solid $KMnO_4$.

Molarity can be used to refer to substances that are actually present in the solution, as well as substances with which the solution is made. Carelessness about this point can lead to considerable confusion. For example, if you were

The volume of solvent will not equal the volume of solution.

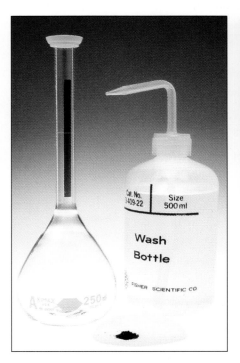
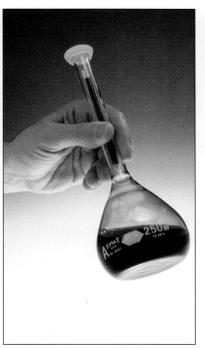
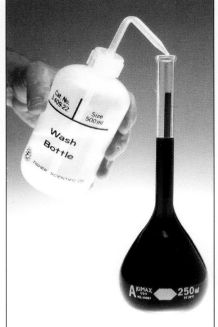

Figure 13.8 Preparation of a potassium permanganate, $KMnO_4$ solution. Place the correct mass of solid $KMnO_4$ in a volumetric flask and add about half the required solvent. Swirl the flask to dissolve the solid. Add enough solvent so that the total volume of the solution reaches the mark on the neck of the flask.

to place one mole of NaCl in a container and add enough water to make one liter of solution, the solution would not actually be 1 M in NaCl because all the NaCl would have dissociated to make Na^+ and Cl^-. The solution would be 1 M in each of these ions, but not in NaCl.

Even so, it is commonly said that an ionic solution is, say, 1 M in NaCl, meaning that *one mole of NaCl was used to prepare one liter of solution.* For the same solution, however, it would be more correct to say that it was 1 M in Na^+ or 1 M in Cl^-. This distinction may seem fussy, but when we discuss chemical equilibrium in Chapter 15, we will need to use molarity to refer only to substances actually present in solution, and when we come to Section 13.11, discussing the physical properties of solutions, we will learn that many properties depend on the total concentration of dissolved ions.

EXAMPLE 13.12

In preparing a KI solution, 9.1 grams of KI were dissolved in water, making 225 mL of solution. What is the molarity of K^+? (The question assumes that you remember that KI is completely dissociated in solution.)

$$KI(s) \longrightarrow K^+ + I^-$$

Solution Map:

$$\boxed{\text{grams of KI}} \longrightarrow \boxed{\text{moles of KI}} \longrightarrow \boxed{\text{moles of } K^+} \longrightarrow$$

$$\boxed{\text{molarity of } K^+}$$

Strategy and Solution:

Step 1.   Calculate the number of moles of KI.

$$9.1 \text{ g KI} \times \frac{\text{mol KI}}{166.0028 \text{ g KI}} = 0.055 \text{ mol KI}$$

Step 2.   Find the number of moles of K^+. There is one mole of K^+ for every mole of KI dissolved. (Unless this relation is obvious, it is best to write the equation for the dissociation.) Therefore, there are 0.055 moles of K^+ in solution.

Step 3.   Divide the number of moles of K^+ by the volume of the solution. Be sure to convert milliliters to liters.

$$\frac{0.055 \text{ mol } K^+}{225 \text{ mL solution}} \times \frac{1000 \text{ mL solution}}{\text{L solution}} = 0.24 \text{ M } K^+$$

EXAMPLE 13.13

The concentration of a solution containing Na_2SO_4 is 0.85 M. How many grams of Na^+ are contained in 350 mL of solution? (Assume that the Na_2SO_4 dissociates completely.)

Solution Map:

$$\boxed{\text{mL of solution}} \longrightarrow \boxed{\text{L of solution}} \longrightarrow$$

$$\boxed{\text{moles of } Na_2SO_4} \longrightarrow \boxed{\text{moles of } Na^+} \longrightarrow \boxed{\text{grams of } Na^+}$$

Strategy and Solution:

Step 1. Write the chemical equation for the dissociation or reaction that occurs when the solution is made.

$$Na_2SO_4(aq) \longrightarrow 2\ Na^+ + SO_4^{2-}$$

Step 2. Write the molarity as a conversion factor.

$$\frac{0.85\ mol\ Na_2SO_4}{L\ solution}$$

Step 3. Set up and solve. Use the molar mass to find the number of grams of Na^+.

$$350\ mL\ solution \times \frac{L\ solution}{1000\ mL\ solution} \times \frac{0.85\ mol\ Na_2SO_4}{L\ solution} \times$$

$$\frac{2\ mol\ Na^+}{mol\ Na_2SO_4} \times \frac{22.9898\ g\ Na^+}{mol\ Na^+} = 14\ g\ Na^+$$

EXAMPLE 13.14

A solution of K_3PO_4 is 0.37 M in K^+.
(a) What is the molarity of PO_4^{3-} in the solution?
(b) How many grams of K_3PO_4 were needed to prepare 800.0 mL of the solution?

Part(a)

Strategy and Solution:

Step 1. Write the equation for the dissociation of K_3PO_4.

$$K_3PO_4(aq) \longrightarrow 3\ K^+ + PO_4^{3-}$$

Step 2. Calculate the molarity of PO_4^{3-}, knowing that there is only one mole of PO_4^{3-} for every three moles of K^+ produced.

$$\frac{0.37\ mol\ K^+}{L\ solution} \times \frac{mol\ PO_4^{3-}}{3\ mol\ K^+} = 0.12\ M\ PO_4^{3-}$$

Part(b)

Solution Map:

$$\boxed{\text{mL of solution}} \longrightarrow \boxed{\text{L of solution}} \longrightarrow \boxed{\text{moles of } K^+} \longrightarrow$$
$$\boxed{\text{moles of } K_3PO_4} \longrightarrow \boxed{\text{grams of } K_3PO_4}$$

Strategy and Solution:

Step 1. Write the molarity of K^+ as a conversion factor.

$$\frac{0.37\ mol\ K^+}{L\ solution}$$

Step 2. Set up and solve. Use the molar mass to find the number of grams of K_3PO_4.

$$800.0 \text{ mL solution} \times \frac{\text{L solution}}{1000 \text{ mL solution}} \times \frac{0.37 \text{ mol K}^+}{\text{L solution}} \times \frac{\text{mol K}_3\text{PO}_4}{3 \text{ mol K}^+} \times$$

$$\frac{212.266 \text{ g K}_3\text{PO}_4}{\text{mol K}_3\text{PO}_4} = 21 \text{ g K}_3\text{PO}_4$$

Exercise 13.12

Find the number of grams of $CoCl_3$ in 137 mL of solution that is 0.25 M Cl^-. Notice, as in Example 13.14, that the value of the molarity of species actually in solution is different from the molarity based on the way the solution was prepared.

Exercise 13.13

What is the molarity of Na^+ in 2.0 M Na_3PO_4?

Exercise 13.14

A solution was made by dissolving 119.3 grams of beryllium phosphate, $Be_3(PO_4)_2$, in enough water to make 1.73 L of solution. In this solution, what were the molarities of Be^{2+} and PO_4^{3-}? (Assume that complete dissociation occurred and that the dissolved species were Be^{2+} and PO_4^{3-}.)

Occasionally you may want to determine the molarity of a solution that has a known weight percent. Knowing the density of a solution allows you to convert from weight percent of a solution to molarity and vice versa. The most important thing to remember about the density of a solution is that it is mass of *solution* per volume of *solution*. Let's look at an example.

EXAMPLE 13.15

A solution of ethanol is 45.0% C_2H_5OH by weight. The density of the solution is 1.17 g/mL. What is the molarity of the solution?

We want to find the number of moles of solute in a liter of solution. Start by imagining that we have a 100-gram sample of solution. We find the number of grams of solute in the sample, convert to number of moles of solute, use the density to find out how many mL are in the sample, then convert this to liters.

Solution Map:

weight percent of C_2H_5OH $\longrightarrow$ moles of C_2H_5OH per 100 grams of solution —

moles of C_2H_5OH per mL of solution $\longrightarrow$ molarity

Strategy and Solution:

Step 1.   Write the weight percent as a conversion factor.

$$\frac{45.0 \text{ g C}_2\text{H}_5\text{OH}}{100 \text{ g solution}}$$

Step 2.   Write the density of the solution as a conversion factor.

$$\frac{1.17 \text{ g solution}}{\text{mL solution}}$$

Step 3.   Set up and calculate. Follow the roadmap, converting to liters as the last step.

$$\frac{45.0 \text{ g C}_2\text{H}_5\text{OH}}{100 \text{ g solution}} \times \frac{\text{mol C}_2\text{H}_5\text{OH}}{46.069 \text{ g C}_2\text{H}_5\text{OH}} \times \frac{1.17 \text{ g solution}}{\text{mL solution}} \times$$

$$\frac{1000 \text{ mL solution}}{\text{L solution}} = 11.4 \text{ M C}_2\text{H}_5\text{OH}$$

Exercise 13.15

Calculate the molarity of a solution that is 35.2% NaCl and has a density of 1.05 g/mL.

Exercise 13.16

What is the molarity of K^+ in a solution that is 87.3% K_2CO_3 and has a density of 1.12 g/mL?

EXAMPLE 13.16

What is the weight percent of a solution that is 6.8 M KBr? The density of the solution is 1.28 g/mL.

Solution Map:

$$\boxed{\text{molarity of KBr}} \longrightarrow \boxed{\text{grams of KBr per L of solution}} \longrightarrow$$

$$\boxed{\text{grams of KBr per mL of solution}} \longrightarrow \boxed{\text{weight fraction}} \longrightarrow \boxed{\text{weight percent}}$$

Strategy and Solution:

Step 1.   Write the molarity and density as conversion factors.

$$\frac{6.8 \text{ mol KBr}}{\text{L solution}} \qquad \frac{1.28 \text{ g solution}}{\text{mL solution}}$$

Step 2.   Find the weight percent beginning with molarity. (Notice that it is necessary to invert the density conversion factor.)

$$\frac{6.8 \text{ mol KBr}}{\text{L solution}} \times \frac{119.002 \text{ g KBr}}{\text{mol KBr}} \times \frac{\text{L solution}}{1000 \text{ mL solution}} \times$$

$$\frac{\text{mL solution}}{1.28 \text{ g solution}} \times 100\% = 63\% \text{ KBr}$$

Exercise 13.17

A solution of ethanol, C_2H_5OH, has a density of 0.894 g/mL and is 12 M C_2H_5OH. What is its weight percent?

Molality (omit)

Concentration can also be reported in **molality,** *the number of moles of solute per kilogram of solvent.* Similar to molarity, the adjective that describes molality is *molal;* for example, a solution with a molality of 1 mol/kg is said to be 1 molal. The abbreviation for molal is the small letter m. It is important to understand the difference between molality and molarity. Molarity is the number of moles in a *liter of solution,* while molality is the number of moles *in a kilogram of solvent.* Molality is not often used for laboratory solutions. However, it is useful for calculations involving colligative properties, which we will discuss in detail in Section 13.11.

To find the molality of any given solution, divide the number of moles of solute by the number of kilograms of solvent.

$$\text{molality} = \frac{\text{moles of solute}}{\text{kilograms of solvent}} = \frac{\text{mol}}{\text{kg}}$$

Here are a few examples.

EXAMPLE 13.17

What is the molality of a solution that is prepared by mixing 0.72 moles of solute with 3.9 kilograms of solvent? Assume that the solute does not dissociate or react in solution.

Solution: Divide the number of moles of solute by the number of kilograms of solvent.

$$\frac{0.72 \text{ mol solute}}{3.9 \text{ kg solvent}} = 0.18 \text{ m}$$

EXAMPLE 13.18

Calculate the molality of a solution that is prepared by adding 56.1 grams of hexane, C_6H_{14}, to 450.0 grams of benzene, C_6H_6.

Solution Map: | grams of C_6H_{14} | $\longrightarrow$ | moles of C_6H_{14} | $\longrightarrow$ | molality |

Strategy and Solution:

Step 1.   Calculate the number of moles of solute. Benzene is present in the greatest quantity, so benzene is the solvent. Change grams of hexane to moles of hexane.

$$56.1 \text{ g C}_6\text{H}_{14} \times \frac{\text{mol C}_6\text{H}_{14}}{86.177 \text{ g C}_6\text{H}_{14}} = 0.651 \text{ mol C}_6\text{H}_{14}$$

Step 2.   Find the molality.

$$\frac{0.651 \text{ mol C}_6\text{H}_{14}}{450.0 \text{ g C}_6\text{H}_6} \times \frac{1000 \text{ g C}_6\text{H}_6}{\text{kg C}_6\text{H}_6} = 1.45 \text{ m C}_6\text{H}_{14}$$

Exercise 13.18

What is the molality of a solution containing 7.9 grams of Cl_2 in 20.0 grams of CCl_4?

Exercise 13.19

Calculate the molality of a solution that was prepared by mixing 0.333 grams of benzoic acid (molecular weight = 122.1) and 1.632 grams of camphor.

You can convert from weight percent of a solution to molality and vice versa. Let's look at a few examples.

EXAMPLE 13.19

What is the molality of a solution that is 9.0% acetone, C_3H_6O.

Solution Map:

| grams of C_3H_6O per 100 grams of solution | $\longrightarrow$ | grams of C_3H_6O per gram of solvent | $\longrightarrow$ |

| moles of C_3H_6O per gram of solvent | $\longrightarrow$ | molality |

Strategy and Solution:

Step 1.   Determine the amount of solvent in 100 grams of solution.

$$100 \text{ g solution} - 9.0 \text{ g C}_3\text{H}_6\text{O} = 91.0 \text{ g solvent}$$

Step 2.   Write the information from step 1 as a conversion factor.

$$\frac{9.0 \text{ g C}_3\text{H}_6\text{O}}{91.0 \text{ g solvent}}$$

Step 3.   Find the molality.

$$\frac{9.0 \text{ g C}_3\text{H}_6\text{O}}{91.0 \text{ g solvent}} \times \frac{\text{mol C}_3\text{H}_6\text{O}}{58.080 \text{ g C}_3\text{H}_6\text{O}} \times \frac{1000 \text{ g solvent}}{\text{kg solvent}} = 1.7 \text{ m C}_3\text{H}_6\text{O}$$

Exercise 13.20

What is the molality of a 13.8% C_6H_{14} solution?

EXAMPLE 13.20

What is the weight percent of a solution that has a molality of 0.428 $C_6H_{12}O_6$?

Solution Map:

$\boxed{\text{moles of } C_6H_{12}O_6} \longrightarrow \boxed{\text{grams of } C_6H_{12}O_6} \longrightarrow \boxed{\text{grams of solution}} \longrightarrow$
$\boxed{\text{weight fraction}} \longrightarrow \boxed{\text{weight percent}}$

Strategy and Solution:

Step 1.   Find the number of grams of solute in one kilogram of solvent.

$$0.428 \text{ mol } C_6H_{12}O_6 \times \frac{180.157 \text{ g } C_6H_{12}O_6}{\text{mol } C_6H_{12}O_6} = 77.1 \text{ g } C_6H_{12}O_6$$

Step 2.   Calculate the total mass of solution. Remember that one kilogram of solvent is equal to 1,000 grams of solvent.

$$1,000 \text{ g solvent} + 77.1 \text{ g } C_6H_{12}O_6 = 1077.1 \text{ g solution}$$

Step 3.   Find the weight percent.

$$\frac{77.1 \text{ g } C_6H_{12}O_6}{1077.1 \text{ g solution}} \times 100\% = 7.16\% \text{ } C_6H_{12}O_6$$

Exercise 13.21

Sucrose, table sugar, has the formula $C_{12}H_{22}O_{11}$. What is the weight percent of a solution that is 0.475 m $C_{12}H_{22}O_{11}$?

13.10 Dilution

After studying this section, you should be able to:

• Calculate concentration changes brought about by diluting or mixing solutions.

Adding solvent to a solution dilutes that solution; that is, *it reduces the concentration of the solute.* Some chemistry experiments involve either adding solvent to solutions or mixing solutions that have different concentrations or contain different solutes.

Often it is necessary to determine the molarity of a solution after dilution. To make our calculations more convenient, let's label our solutions so that a subscript 1 designates the original solution and a subscript 2 designates the diluted solution. For example, the volume of the original solution would be V_1, and the molarity of the diluted solution would be M_2. The equations relating the concentrations of the two solutions can be written as follows.

$$M_1V_1 = \frac{\text{mol solute}}{\text{L solution 1}} \times \text{L solution 1} = \text{mol solute in solution 1}$$

$$M_2V_2 = \frac{\text{mol solute}}{\text{L solution 2}} \times \text{L solution 2} = \text{mol solute in solution 2}$$

The number of moles of solute in a solution is not changed by dilution. That is, the new solution will contain the same amount of solute as did the original solution.

$$\text{moles of solute in solution 1} = \text{moles of solute in solution 2}$$

Therefore,

$$M_1V_1 = M_2V_2$$

We can rearrange this equation to give us the molarity of solution 2, the new solution.

$$M_2 = M_1 \times \frac{V_1}{V_2}$$

This equation says that we can find the molarity of a diluted solution by multiplying the original molarity by a ratio of the volumes. Because we know that the new solution will not be as concentrated as the original, we put the smaller volume in the numerator.

EXAMPLE 13.21

A volume of 100.00 mL of a solution that is 0.100 M potassium dichromate, $K_2Cr_2O_7$, is diluted to 1000.0 mL. What is the molarity of the new solution?

Solution: $M_1 = 0.100$ M, $V_1 = 100.00$ mL, and $V_2 = 1000.0$ mL. Substitute and solve.

$$M_2 = 0.100 \text{ M} \times \frac{100.00 \text{ mL}}{1000.0 \text{ mL}} = 0.0100 \text{ M}$$

Diluting a $K_2Cr_2O_7$ solution. A solution of known volume and molarity is placed in a volumetric flask. Enough solvent is added to fill the flask to the mark on the neck.

Exercise 13.22

What is the final molarity of a solution created by adding enough water to 43 mL of 2.00 M $Mn(SO_3)_2$ to make 155 mL of solution?

Exercise 13.23

A 50.0-mL sample of 0.60 M solution is diluted to 500.0 mL. What is the final molarity?

EXAMPLE 13.22

To 505 mL of a solution that is 0.33 M in ethanol, C_2H_5OH, are added 217 mL of water. What is the molarity of the new solution?

Solution: Because we were not told otherwise, we assume that solvent of the original solution was water and we assume that adding 217 mL of water to the original sample made its new volume 722 mL. Therefore,

$$M_1 = 0.33 \text{ M}, V_1 = 505 \text{ mL}, \text{ and } V_2 = 722 \text{ mL}.$$

Substitute and solve.

$$M_2 = 0.33 \text{ M} \times \frac{505 \text{ mL}}{722 \text{ mL}} = 0.23 \text{ M}$$

In most cases, we can assume that when solvent is added to a dilute solution, the final volume will be the sum of the volumes of the original solution and the solvent.

Exercise 13.24

A solution is 2.50 M. What will be the molarity if 25.0 mL of solvent are added to 155.0 mL of solution?

In situations in which two solutions of the same solute are mixed, the simple dilution equation cannot be used because the beginning molarity is no longer a single value.

EXAMPLE 13.23

In most cases, we can assume that when one solution is added to another, the final volume is the sum of the original volumes.

In the laboratory, a solution of desired molarity is often made by mixing two solutions. What is the molarity of a solution prepared by adding 23.6 mL of 0.45 M KBr to 45.0 mL of 1.0 M KBr?

Solution Map: | total mols and total volume | $\longrightarrow$ | molarity |

Strategy and Solution:

Step 1. Find the number of moles of solute in the two original solutions. Add these together; this will be the number of moles of solute in the final solution. (a) Calculate the number of moles of solute in each solution.

$$23.6 \text{ mL solution} \times \frac{\text{L solution}}{1000 \text{ mL solution}} \times \frac{0.45 \text{ mol KBr}}{\text{L solution}} = 0.011 \text{ mol KBr}$$

$$45.0 \text{ mL solution} \times \frac{\text{L solution}}{1000 \text{ mL solution}} \times \frac{1.0 \text{ mol KBr}}{\text{L solution}} = 0.045 \text{ mol KBr}$$

(b) Add these values together to find the total amount of solute that will be in the final solution.

$$0.011 \text{ mol KBr} + 0.045 \text{ mol KBr} = 0.056 \text{ mol KBr}$$

Step 2. Find the total volume of the final solution.

$$23.6 \text{ mL solution} + 45.0 \text{ mL solution} = 68.6 \text{ mL solution}$$

Step 3. Calculate the molarity by dividing the total moles of solute by the total volume of solution in liters.

$$\frac{0.056 \text{ mol KBr}}{68.6 \text{ mL solution}} \times \frac{1000 \text{ mL solution}}{\text{L solution}} = 0.82 \text{ M KBr}$$

Notice that when solutions are mixed, the value of the final molarity of any given solute always lies between the molarities of that solute in the two solutions.

Exercise 13.25

34.0 mL of 3.5 M HCl are mixed with 78.5 mL of 1.4 M HCl. What is the molarity of the final solution?

Exercise 13.26

A solution is created by mixing 4.0 mL of 10.0 M NaCl and 2.0 mL of 7.5 M NaCl. What is the molarity of the final solution?

13.11 Colligative Properties of Solutions (optional)

After studying this section, you should be able to:

- Explain why the freezing and boiling temperatures of solutions are different from those of pure solvents.
- Determine the freezing point and boiling point of a given solution.
- Use the boiling point or the freezing point of a solution to determine the molar mass of the solute.

Several physical properties of solutions differ from those of pure solvents. Those properties of solutions that depend on the amount of solute dissolved are called **colligative properties.** Vapor pressure, melting point, and boiling point are examples of colligative properties. Let's see why these properties of solutions are different from those of pure solvents.

Consider the surface of an aqueous solution of sugar (see Figure 13.9). The surface consists of many water molecules and relatively few sugar molecules. Water molecules overcome the attractive forces of the liquid and become vapor. The amount of water vapor above the solution, however, is not as great as the amount of water vapor above a sample of pure water. The solute molecules in the surface get in the way of vaporizing water molecules and reduce the rate of vaporization. This means that the vapor pressure of the solution must be less than the vapor pressure of pure water. This effect holds for any solution containing a nonvolatile solute such as sugar.

The more solute in solution, the greater the reduction in the amount of vapor at a particular temperature. A very dilute solution will have a vapor pressure similar to that of the pure solvent. A more concentrated solution will have a vapor pressure lower than that of the pure solvent. Look at Figure 13.10.

Figure 13.9 Because sugar molecules take up space in the surface of the solution, the rate of vaporization of water molecules is reduced. Thus the vapor pressure of the solution is less than the vapor pressure of pure water.

Solutions containing two volatile substances sometimes behave differently. This section concerns only liquid solutions that contain one or more nonvolatile solutes.

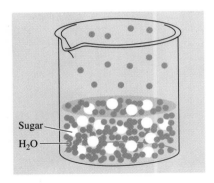

Figure 13.10 At greater concentrations of sugar molecules, there are more sugar molecules at the surface of the solution. This causes a correspondingly lower water vapor pressure.

A nonelectrolyte produces one mole of solute particles for every mole dissolved. Because sugar is a nonelectrolyte, one mole of dissolved sugar yields one mole of sugar molecules in solution. Electrolytes, however, produce more than one mole of solute particles for every mole of solute *dissolved*. Because NaCl dissociates upon dissolving, one mole of dissolved NaCl will give two moles of solute particles, one mole of Na^+ and one mole of Cl^-. A 1 M solution of NaCl, therefore, will have a lower vapor pressure than a 1 M solution of sugar. In fact, dissolving one mole of NaCl has the same effect as dissolving two moles of sugar. In general, because they produce multiple solute particles upon dissolving, electrolytes have a greater effect on vapor pressure (or, for that matter, on all colligative properties) than do nonelectrolytes.

Boiling Point Elevation

From the preceding discussion, you may have already thought of a consequence of reducing the vapor pressure of a solvent. Recall from Chapter 12 that a liquid boils when its vapor pressure equals or exceeds the pressure of the gas above it. At any temperature, a solution has a lower vapor pressure than that of the pure solvent. At the boiling temperature of the pure solvent, a solution will not boil because its vapor pressure is not as great as that of the gas above it. For a solution to boil, its temperature must be higher than that of the pure solvent. That is, *the boiling point of a solution is higher than the boiling point of the pure solvent.* This effect is called **boiling point elevation.** Figure 13.11

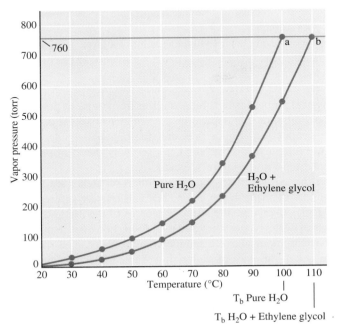

Figure 13.11 At every temperature, the vapor pressure of the solution is less than the vapor pressure of the pure water. Therefore, the boiling temperature of the solution (b) is higher than the boiling temperature of the pure solvent (a).

shows that the boiling temperature of a solution of ethylene glycol (commercial antifreeze) and water is higher than the boiling temperature of pure water.

The difference between the boiling point of a pure solvent and the boiling point of a solution is ΔT_b. Recall from Section 2.6 that $\Delta T = T_{final} - T_{initial}$. For boiling point elevation calculations, $\Delta T_b = T_{b\,solution} - T_{b\,pure\,solvent}$. ΔT_b can be calculated by using the molality of dissolved particles, m, and the **boiling point elevation constant,** K_b, for the solvent. The relation is

$$\Delta T_b = m\,K_b$$

of which the units are

$$degrees = \frac{mol\ solute}{kg\ solvent} \times \frac{deg\ kg\ solvent}{mol\ solute}$$

Table 13.4 lists the normal boiling temperatures and boiling point elevation constants for several solvents.

Table 13.4	Boiling Points and Boiling Point Elevation Constants for Some Solvents	
Solvent	T_b (°C)	K_b ($\frac{deg\ kg}{mol\ solute}$)
Water	100.00	0.512
Benzene	80.0	2.53
Cyclohexane	80.7	2.79
Acetic acid	117.9	2.93
Chloroform	61.2	3.63
Nitrobenzene	210.8	5.24
Camphor	209	5.95

EXAMPLE 13.24

What is the difference between the boiling temperature of pure water and that of a solution prepared by mixing 555 grams of water with 35 grams of ethylene glycol, $C_2H_6O_2$?

Solution Map:

$$\boxed{grams\ of\ solute} \longrightarrow \boxed{moles\ of\ solute} \longrightarrow \boxed{molality} \longrightarrow$$
$$\boxed{temperature\ change}$$

Strategy and Solution:

Step 1. Find the molality of the solution.

$$35\ g\ C_2H_6O_2 \times \frac{mol\ C_2H_6O_2}{62.068\ g\ C_2H_6O_2} = 0.56\ mol\ C_2H_6O_2$$

$$\frac{0.56\ mol\ C_2H_6O_2}{555\ g\ H_2O} \times \frac{1000\ g\ H_2O}{kg\ H_2O} = 1.0\ m\ C_2H_6O_2$$

Adding ethylene glycol to the water in an automobile radiator keeps the coolant from boiling over in the summer.

Ethylene glycol is used in most commercial antifreezes. When mixed with water, the resulting solution has a boiling temperature higher (and a freezing temperature lower—p. 396) than that of pure water.

Step 2. Substitute the molality and the boiling point elevation constant for water into the equation and solve.

$$\Delta T_b = \frac{1.0 \text{ mol solute}}{\text{kg } H_2O} \times \frac{0.512 \text{ deg kg } H_2O}{\text{mol solute}} = 0.52 \text{ deg}$$

The difference in the boiling temperature is 0.52 degrees, or in other words, the boiling temperature of the solution is 0.52 degrees higher than the boiling temperature of pure water. Therefore, a 1.0 m solution of $C_2H_6O_2$ will boil at 100.52°C.

EXAMPLE 13.25

Find the boiling temperature of a solution that contains 46 grams of NaCl and 225 grams of H_2O.

Solution Map:

grams of NaCl $\longrightarrow$ moles of NaCl $\longrightarrow$ molality of NaCl $\longrightarrow$

molality of ions $\longrightarrow$ temperature change $\longrightarrow$ boiling temperature

Strategy and Solution:

Step 1. Find the molality of the solution.

$$\frac{46 \text{ g NaCl}}{225 \text{ g } H_2O} \times \frac{\text{mol NaCl}}{58.4425 \text{ g NaCl}} \times \frac{1,000 \text{ g } H_2O}{\text{kg } H_2O} = 3.5 \text{ m NaCl}$$

Step 2. Substitute the molality and the boiling point elevation constant for water into the equation and solve. Remember that one mole of NaCl dissociates in solution and produces *two moles of solute* particles.

$$\Delta T_b = \frac{3.5 \text{ mol NaCl}}{\text{kg } H_2O} \times \frac{2 \text{ mol solute}}{\text{mol NaCl}} \times \frac{0.512 \text{ deg kg } H_2O}{\text{mol solute}} = 3.6 \text{ deg}$$

Step 3. Find the boiling temperature of the solution.

$$100.00°C + 3.6 \text{ deg} = 103.6°C$$

Exercise 13.27

A solution contains 6.2 grams of KBr and 42 grams of H_2O. What is the boiling temperature of the solution?

Using boiling point elevation data, it is possible to determine the molar mass of a solute. An example shows us how.

EXAMPLE 13.26

A solution containing 92.0 grams of camphor and 3.8 grams of an unknown solute has a boiling temperature 4.6 degrees higher than pure camphor. Assuming that the solute is a nonelectrolyte, what is the molar mass of the solute? The boiling point elevation constant for camphor is 5.95°C kg/mol.

Solution Map:

temperature change $\longrightarrow$ molality $\longrightarrow$

moles of solute $\longrightarrow$ molar mass of solute

Strategy and Solution:

Step 1. Use the boiling point elevation equation to find the molality of the solution.

$$4.6 \text{ deg} = m \times \frac{5.95 \text{ deg kg camphor}}{\text{mol solute}}$$

$$m = \frac{0.77 \text{ mol solute}}{\text{kg camphor}}$$

Step 2. Calculate the number of moles of solute.

$$92.0 \text{ g camphor} \times \frac{\text{kg camphor}}{1{,}000 \text{ g camphor}} \times \frac{0.77 \text{ mol solute}}{\text{kg camphor}} = 0.071 \text{ mol solute}$$

Step 3. Calculate the molar mass of the solute.

$$\frac{3.8 \text{ g solute}}{0.071 \text{ mol solute}} = 54 \text{ g/mol}$$

Exercise 13.28

A solution has a boiling temperature of 83.5°C. It contains 32.0 grams of benzene and 5.2 grams of unknown solute. What is the molar mass of the solute?

Freezing Point Depression

Freezing of a liquid occurs at the temperature at which the motions of the individual molecules can no longer overcome the attractive forces that tend to hold the molecules together. The molecules stick, and solid begins to form. We have seen that solute molecules lower vapor pressure by getting in the way of molecules leaving a liquid to form a vapor. Similarly, solute molecules get in the way of solvent molecules trying to leave the liquid to form a solid, thereby requiring lower temperatures for freezing. That is, *the freezing point of a solution is lower than the freezing point of the pure solvent.* This effect is called **freezing point depression.**

Dissolved substances lower the freezing temperature and raise the boiling temperature.

Calculation of the difference between the freezing point of a solution and the freezing point of pure solvent, ΔT_f, is similar to a boiling point elevation calculation.

$$\Delta T_f = m \, K_f$$

where $\Delta T_f = T_{f\,\text{pure solvent}} - T_{f\,\text{solution}}$, m is the molality of the solution, and K_f is known as the **freezing point depression constant.** Freezing temperatures and freezing point depression constants for several solvents appear in Table 13.5.

Table 13.5	Freezing Points and Freezing Point Depression Constants for Some Solvents	
Solvent	T_f (°C)	K_f ($\frac{\text{deg kg solvent}}{\text{mol solute}}$)
Water	0.00	1.86
Acetic acid	16.6	3.90
Benzene	5.50	5.10
Cyclohexane	6.5	20.2
Camphor	179.8	40.0

EXAMPLE 13.27

A solution contains 35 grams of ethylene glycol, $C_2H_6O_2$, and 555 grams of H_2O. What is the freezing point of this solution?

Solution Map:

$$\boxed{\text{grams of solute}} \longrightarrow \boxed{\text{moles of solute}} \longrightarrow \boxed{\text{molality}} \longrightarrow$$

$$\boxed{\text{temperature change}} \longrightarrow \boxed{\text{freezing temperature}}$$

Strategy and Solution:

Step 1. Find the molality of the solution.

$$\frac{35 \text{ g } C_2H_6O_2}{555 \text{ g } H_2O} \times \frac{\text{mol } C_2H_6O_2}{62.068 \text{ g } C_2H_6O_2} \times \frac{1{,}000 \text{ g } H_2O}{\text{kg } H_2O} = 1.0 \text{ m } C_2H_6O_2$$

Step 2. Substitute the molality and the freezing point depression constant for water into the equation and solve.

$$\Delta T_f = \frac{1.0 \text{ mol solute}}{\text{kg } H_2O} \times \frac{1.86 \text{ deg kg } H_2O}{\text{mol solute}} = 1.9 \text{ deg}$$

Step 3. Calculate the freezing temperature of the solution.

$$0.00°C - 1.9 \text{ deg} = -1.9°C$$

Exercise 13.29

Rock salt, NaCl, is added to ice in home ice cream freezers to lower the freezing temperature of the water. What is the freezing temperature of a mixture containing 2.0 kilograms of NaCl and 15.0 kilograms of H_2O?

Other kinds of calculations involving freezing point depression are also worked in the same way as boiling point elevation problems.

Exercise 13.30

A solution containing 2.3 grams of unknown solute and 16.8 grams of benzene freezes at 6.7°C. What is the molar mass of the unknown solute? (Hint: See Example 13.26.)

13.12 Stoichiometry Using Solutions

After studying this section, you should be able to:

- Make stoichiometric calculations for chemical reactions performed in solutions.

Many important reactions begin with ions or other species in solution. To make stoichiometric calculations for those reactions, it is often necessary to start with solution concentrations. Using the language of molarity usually simplifies the problem because the given volumes and molarities can so easily be translated into numbers of moles.

EXAMPLE 13.28

Solid NaCl will react with $AgNO_3$ solution to produce insoluble AgCl.

$$NaCl(s) + AgNO_3(aq) \longrightarrow AgCl(s) + NaCl(aq)$$

How many grams of NaCl are required to react exactly with 250.0 mL of 0.0900 M $AgNO_3$ solution?

Solution Map:

$$\boxed{\text{mL of AgNO}_3 \text{ solution}} \longrightarrow \boxed{\text{L of AgNO}_3 \text{ solution}} \longrightarrow$$

$$\boxed{\text{moles of AgNO}_3} \longrightarrow \boxed{\text{moles of NaCl}} \longrightarrow \boxed{\text{grams of NaCl}}$$

Strategy and Solution:

Step 1. Find the number of moles $AgNO_3$.

$$250.0 \text{ mL solution} \times \frac{\text{L solution}}{1,000 \text{ mL solution}} \times \frac{0.0900 \text{ mol AgNO}_3}{\text{L solution}} =$$

$$0.0225 \text{ mol AgNO}_3$$

Step 2. Find the number of moles of NaCl needed.

$$0.0225 \text{ mol AgNO}_3 \times \frac{1 \text{ mol NaCl}}{1 \text{ mol AgNO}_3} = 0.0225 \text{ mol NaCl}$$

Step 3. Use the molecular weight of NaCl to find the number of grams of NaCl.

$$0.0225 \text{ mol NaCl} \times \frac{58.4425 \text{ g NaCl}}{\text{mol NaCl}} = 1.32 \text{ g NaCl}$$

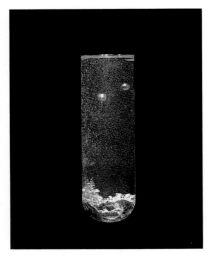

Metallic zinc reacts readily with H_2SO_4 solution, forming aqueous $ZnSO_4$ and releasing H_2 gas.

EXAMPLE 13.29

Metallic zinc reacts readily with H_2SO_4.

$$Zn(s) + H_2SO_4(aq) \longrightarrow ZnSO_4(aq) + H_2(g)$$

How many liters of 0.33 M H_2SO_4 solution are required to react exactly with a sample containing 4.4 g of Zn metal?

Solution Map:

$$\boxed{\text{grams of Zn}} \longrightarrow \boxed{\text{moles of Zn}} \longrightarrow \boxed{\text{moles of } H_2SO_4} \longrightarrow$$

$$\boxed{\text{volume of } H_2SO_4 \text{ solution}}$$

Solution:

$$4.4 \text{ g Zn} \times \frac{\text{mol Zn}}{65.39 \text{ g Zn}} \times \frac{1 \text{ mol } H_2SO_4}{1 \text{ mol Zn}} \times \frac{\text{L solution}}{0.33 \text{ mol } H_2SO_4} = 0.20 \text{ L solution}$$

Exercise 13.31

Using the following chemical equation, calculate how many mL of 3.50 M HNO_3 solution are required to react with 2.83 grams of Cu.

$$3 \text{ Cu} + 8 \text{ HNO}_3 \longrightarrow 3 \text{ Cu(NO}_3)_2 + 2 \text{ NO} + 4 \text{ H}_2O$$

EXAMPLE 13.30

How many liters of N_2 at STP can be produced from 25.0 mL of 2.5 M HCN and 11.6 L of O_2 at STP? (Notice that amounts are given for both reactants. In such a case, one of the reactants will most surely limit the reaction. If this concept is not clear to you, go back and review Section 10.8.)

$$HCN(aq) + O_2(g) \longrightarrow N_2(g) + CO_2(g) + H_2O(l) \quad \text{(unbalanced)}$$

Solution Map:

$$\boxed{\text{L of } O_2} \longrightarrow \boxed{\text{moles of } O_2} \longrightarrow \boxed{\text{moles of } N_2} \longrightarrow \boxed{\text{L of } N_2}$$

$$\boxed{\text{mL of HCN solution}} \longrightarrow \boxed{\text{L of HCN solution}} \longrightarrow \boxed{\text{moles of HCN}} \longrightarrow$$

$$\boxed{\text{moles of } N_2} \longrightarrow \boxed{\text{L of } N_2}$$

Strategy and Solution:

Step 1.   Balance the chemical equation.

$$4 \text{ HCN(aq)} + 5 \text{ O}_2(g) \longrightarrow 2 \text{ N}_2(g) + 4 \text{ CO}_2(g) + 2 \text{ H}_2O(l)$$

Step 2.   Calculate how many liters of N_2 can be produced from each reactant. Remember that one mole of gas has a volume of 22.4 L at STP.

$$11.6 \text{ L O}_2 \times \frac{\text{mol O}_2}{22.4 \text{ L O}_2} \times \frac{2 \text{ mol N}_2}{5 \text{ mol O}_2} \times \frac{22.4 \text{ L N}_2}{\text{mol N}_2} = 4.64 \text{ L N}_2$$

$$25.0 \text{ mL solution} \times \frac{\text{L solution}}{1,000 \text{ mL solution}} \times \frac{2.5 \text{ mol HCN}}{\text{L solution}} \times$$

$$\frac{2 \text{ mol N}_2}{4 \text{ mol HCN}} \times \frac{22.4 \text{ L N}_2}{\text{mol N}_2} = 0.70 \text{ L N}_2$$

HCN is the limiting reactant; 0.70 liters of N_2 can be produced.

Exercise 13.32

How many grams of $Zn_3(PO_4)_2$ can be produced from 4.80 grams of Zn and 26.0 mL of 5.0 M H_3PO_4? (Remember to balance the equation.)

$$Zn(s) + H_3PO_4(aq) \longrightarrow Zn_3(PO_4)_2(s) + H_2(g)$$

Exercise 13.33

Calculate the volume of H_2 in liters at STP that can be produced from 2.50 grams of Mg and 25.0 mL of 12.0 M HCl. (Balance the equation.)

$$Mg(s) + HCl(aq) \longrightarrow MgCl_2(aq) + H_2(g)$$

Summary

We can characterize any solution by naming its solvent, its solute, and its concentration. Among the several ways of expressing concentration, the one most convenient to chemists is molarity. The molarity of a solution is the number of moles of solute in a liter of solution.

The solubility of a solute at a given temperature is the maximum amount of the solute that will dissolve in a given amount of solvent at that temperature. Solubilities vary widely and depend on the structures of both solute and solvent. Because water is the most common solvent, much chemistry is carried out in aqueous solution. Although ionic compounds usually dissociate when they dissolve, the solubilities of ionic compounds vary greatly.

In solutions, electrolytes form ions either by dissociation or ionization. Non-electrolytes do not form ions in solution.

Solutions provide a convenient means of running chemical reactions in a controlled and measurable way. In fact, some reactions can be run only in solution. There are three general ways in which ions react in solution: (1) formation of a precipitate, (2) formation of water, or (3) formation of a gas.

The physical properties of solutions usually differ from the physical properties of pure solvent. The colligative properties of a solution are those that depend on the number of solute particles present in the solution. The vapor pressure of a solution, a colligative property, is lower than the vapor pressure of the pure solvent. Consequently, the boiling point of a solution is higher and the freezing point lower than the corresponding values of the pure solvent. At any given concentration, an electrolyte will have a greater effect on the colligative properties of a solution than will a nonelectrolyte.

Stoichiometric calculations involving solutions are easy. They merely require converting volume and concentration values to numbers of moles before using the mole ratio method introduced in Chapter 10.

Key Terms

solvent	(p. 364)	precipitate	(p. 372)
solute	(p. 364)	net ionic equation	(p. 374)
aqueous solution	(p. 364)	concentration	(p. 376)
saturated	(p. 365)	concentrated	(p. 376)
solubility	(p. 365)	dilute	(p. 376)
supersaturated	(p. 365)	weight percent	(p. 376)
electrolyte	(p. 368)	molarity	(p. 379)
strong electrolyte	(p. 368)	molality	(p. 386)
ionization	(p. 368)	colligative properties	(p. 391)
weak electrolyte	(p. 368)	boiling point elevation	(p. 392)
solvated	(p. 369)	boiling point elevation constant	(p. 393)
hydrated	(p. 369)	freezing point depression	(p. 395)
hydrate	(p. 370)	freezing point depression constant	(p. 396)
anhydrous	(p. 370)		

Questions and Problems

Section 13.1 What Is a Solution?

Questions

1. Distinguish between the terms "solute" and "solvent."

2. Is it possible for a gas to dissolve another gas? A solid to dissolve another solid?

3. What is the solvent in air? What is the solvent in 14-karat gold (pure gold is 24-karat)?

4. List all the possible kinds of solutions made up of combinations of the three states of matter.

5. Name the solute and the solvent in the following systems. Explain the reasons for your choices.
   (a) seawater
   (b) radiator water and coolant
   (c) 6 grams of mercury and 3 grams of sodium (an amalgam)
   (d) 14-karat gold
   (e) bronze (roughly 80% Cu, 20% tin, with traces of other elements)

Section 13.2 Solubility; Saturated Solutions

Questions

6. Distinguish between the terms "saturated" and "supersaturated."

7. What effect does raising the temperature usually have on solubility? What might be the reasons for this effect?

8. What effect does pressure have on solubility?

9. Think of an example of a supersaturated solution that can be made in the kitchen and that results in something edible when seeded.

Section 13.3 Kinds of Substances That Dissolve in One Another

Questions

10. Explain what is meant by "like dissolves like."

11. Will table salt, NaCl, dissolve in vegetable oil? Explain.

12. Why will vinegar, a dilute solution of acetic acid, dissolve in water but not in oil?

13. Name a common polar solvent. Which of the following substances will dissolve in polar solvents: sugar, gasoline, vegetable oil, Na_2CO_3, H_2SO_4, H_2O, NH_3, CH_4?

Section 13.4 Electrolytes

Questions

14. Why are soluble ionic substances called electrolytes?

15. How can you experimentally determine whether a compound is a strong or weak electrolyte?

16. Give two examples of a strong electrolyte.

17. Are all electrolytes ionic compounds? Explain.

18. Which of the following compounds are strong electrolytes: sugar, acetic acid, alcohol, KI, H_2O, HCl, NH_3, $Ca(OH)_2$, AgCl, $BaSO_4$?

19. Arrange the following compounds in order of their increasing strength as electrolytes: sucrose, acetic acid, HCl, H_2O, CaS.

Section 13.5 Hydrates

Questions

20. Explain the process of hydration. Draw a picture of a hydrated Br^- ion.

21. How can you form solid hydrates?

Section 13.6 Solubility Rules

Problems

22. Which of the following ionic compounds will dissolve readily in water: NH_4NO_3, $Y(OH)_3$, $Th(CO_3)_2$, $PbSO_4$, KBr, BaS, NiS, Hg_2CO_3, $MnPO_4$?

23. Which substance will have the greater solubility in water: Na_2S or MgS? $FeSO_4$ or $CaSO_4$? $(NH_4)_3PO_4$ or Ag_3PO_4?

Section 13.7 Reactions of Ions in Solution

Question

24. What are the three general ways by which ions can be removed from solution?

Problems

25. State whether mixing solutions of the following pairs of compounds results in a reaction. Give the products, if any.
   (a) $KOH + FePO_4$
   (b) $CsCl + AgNO_3$
   (c) $BaSO_4 + CuBr_2$
   (d) $Ca(OH)_2 + KOH$

26. State whether mixing solutions of the following pairs of compounds results in a reaction. Give the products, if any.
   (a) $NaOH + HCl$
   (b) $K_2S + CuCl_2$
   (c) $Li_2CO_3 + HNO_3$
   (d) $(NH_4)_2SO_3 + NiBr_2$

27. Will mixing solutions of the following pairs of compounds result in a reaction? If reaction occurs, give the chemical equation for the reaction.
   (a) $NH_4NO_3 + K_2CO_3$
   (b) $Hg(NO_3)_2 + MgCO_3$
   (c) $RbI + Sr(OH)_2$
   (d) $CaS + (NH_4)_2CO_3$

28. Will mixing solutions of the following pairs of compounds result in a reaction? If reaction occurs, give the chemical equation for the reaction.
   (a) $AgNO_3 + NaBr$
   (b) $Na_2CO_3 + MnCl_2$
   (c) $NH_4NO_3 + CuCl_2$

Section 13.8 Net Ionic Equations

Problems

29. Write the net ionic equation for

$$2\,AgNO_3(aq) + Na_2SO_4(aq) \longrightarrow$$
$$Ag_2SO_4(s) + 2\,NaNO_3(aq)$$

30. Write the net ionic equation for

$$Na_2CO_3(aq) + 2\,HCl(aq) \longrightarrow$$
$$CO_2(g) + H_2O(l) + 2\,NaCl(aq)$$

31. Write the net ionic equations for any reactions that result in Problem 25.

32. Write the net ionic equations for any reactions that result in Problem 26.

Section 13.9 Concentration

Questions

33. What are the differences between molarity and weight percent? (There are more than one.)

34. Distinguish between the terms "dilute" and "concentrated."

35. What information is necessary to convert from weight percent to molarity and vice versa?

36. How does molality differ from molarity and from weight percent?

Problems

37. What is the weight percent of a solution prepared by mixing 65 grams of C_3H_6O into 110 grams of ethanol?

38. What is the weight percent of a solution prepared by dissolving 21.10 grams of NaOH in 55 mL of H_2O? (Remember that the density of water is 1.00 g/mL.)

39. The weight percent of a $Ca(OH)_2$ solution is 0.15%. How many grams of $Ca(OH)_2$ are necessary to mix with 1,050 grams of H_2O to make the solution?

40. A solution was made by dissolving 1.00 mole of K_2SO_4 in 1,950. grams of water. What was the weight percent of the finished solution?

41. How many grams of solvent would be necessary to prepare a 30.0% solution containing 23 grams of KNCS?

42. To make an aqueous solution that is 25.0% $LiNO_3$ and contains 0.71 moles of $LiNO_3$, how many grams of water must be used?

43. A solution that weighed 19.1 grams was evaporated to dryness; care was taken that no solute was lost. After the evaporation was complete, 3.25 grams of NaCl remained. What was the weight percent of NaCl in the solution?

44. An aqueous solution has a density of 1.08 g/ml and contains 14.0% by weight of HNO_3. How many mL of the solution must be taken to obtain 22 grams of HNO_3? To obtain 1.0 mole of HNO_3?

45. How many grams of $NiCl_2 \cdot 6\ H_2O$ must be used to prepare 150 grams of a 5.0% solution of $NiCl_2$?

46. What is the molarity of a solution prepared with 15 grams of C_5H_{12} and enough benzene to make 250.0 mL of solution?

47. How many grams of $CoCl_2$ are necessary to prepare 550 mL of a 1.5 M solution of Co^{2+}?

48. How many grams of Hg_2Cl_2 are necessary to prepare 850.0 mL of a 1.65 M solution of Cl^-?

49. Cupric sulfate is usually sold as the pentahydrate $CuSO_4 \cdot 5\ H_2O$. How many grams of this substance must be used to prepare 75 mL of 0.55 M Cu^{2+}?

50. How many moles of Ba^{2+} are contained in 175 mL of a solution that was prepared to be 0.33 M in $BaCl_2$? How many moles of Cl^- are in this solution?

51. Gaseous HBr is needed to prepare 300.0 mL of 4.00 M HBr. What volume of gas will we use at 25°C and 745 torr?

52. A solution contains 19.6 grams of $MnSO_4$ and 150.0 grams of water. The density of the solution is 1.12 g/mL. What is the weight percent of $MnSO_4$ in the solution? What is the molarity of $MnSO_4$ in the solution?

53. To prepare a photographic fixing solution, 45 grams of solid sodium thiosulfate ($Na_2S_2O_3$) are put into a 250.00-mL volumetric flask. Then 200.0 mL of hot water is added, and the solid is brought into solution. After the solution cools, enough water is added to fill the flask to 250.00 mL. What is the molarity of $Na_2S_2O_3$ in the final solution?

54. What is the molality of a solution that is prepared by mixing 3.6 moles of solute and 23.7 kg of solvent?

55. Calculate the molality of a solution that contains 45.2 grams of C_6H_6 and 455 grams of camphor.

56. What is the molality of a solution prepared by mixing 21.3 grams of KCl and 450 grams of H_2O? (Remember that KCl is a strong electrolyte.)

57. What is the molality of a solution that is 34.7% CCl_4?

58. Find the molality of a solution that is 19.3% C_6H_6.

59. A solution is 45% $C_6H_{12}O_6$ and has a density of 1.87 g/mL. Calculate the molality and the molarity of the solution.

60. A 7.3% solution of CO_2 has a density of 0.984 g/mL. Find the molarity and the molality of the solution. (An aqueous solution of CO_2 is a solution of carbonic acid. $H_2O(l) + CO_2(g) \longrightarrow H_2CO_3(aq)$)

Section 13.10 Dilution

Problems

61. A 2.5-mL sample of 1.50 M NaOH is diluted to 1.00 L. What is the molarity of the final solution?

62. A solution is 0.625 M in $MgBr_2$. To a 125-mL sample of this solution is added 500.0 mL of water. What is the concentration of Br^- in the final solution?

63. A 2.50 M solution of $CrBr_2$ was allowed to mix with enough solvent to make 3.60 L of solution. If we started with 750 mL of solution, what was the final molarity? What was the molarity of Cr^{2+} in that solution? Of Br^-?

64. Sulfuric acid, H_2SO_4, is sold in concentrated form at 18 M. What volume of the concentrated acid must be used to prepare 325 mL of a solution that is 1.10 M in H_2SO_4?

65. Concentrated HCl from suppliers is usually an aqueous solution that has a density of 1.198 g/mL and contains 40.0% HCl by weight. What volume of this HCl must be used to prepare 1.00 L of a 0.200 M solution?

66. A solution is prepared by mixing 23 mL of 0.35 M KNO_3 with 45 mL of 0.25 M KNO_3. What is the molarity of the final solution?

67. Calculate the final molarity of a solution that was prepared by mixing 350 mL of 2.5 M HCl with 50.0 mL of 8.0 M HCl.

Section 13.11 Colligative Properties of Solutions

Questions

68. List three colligative properties.

69. Explain why the vapor pressure of a solution is lower than the vapor pressure of the pure solvent.

70. Why will a 1.0 m solution of KI have a lower vapor pressure than a 1.0 m solution of ethylene glycol?

71. Why is the boiling point of a solution higher than that of the pure solvent?

72. Explain how you can determine the molar mass of a nonelectrolyte in the laboratory.

Problems

73. What is the boiling point elevation of a 4.7 m aqueous solution of CO_2?

74. Calculate the boiling point elevation of a 7.20 m solution with chloroform as the solvent.

75. What is the boiling point of a solution that contains 0.83 moles of solute dissolved in 145 grams of nitrobenzene? Assume the solute is a nonelectrolyte.

76. Find the boiling point of a solution that contains 3.75 moles of solute and 350.0 grams of acetic acid.

77. What is the boiling point of a solution containing 23.0 kg of NaCl and 50.0 kg of H_2O?

78. A solution contains 14 grams of unknown solute and 500.0 grams of nitrobenzene, and has a boiling point of 212 °C. What is the molar mass of the solute?

79. What is the freezing point depression of an 8.3 m aqueous solution of a nonelectrolyte?

80. Calculate the freezing point depression of a 2.0 m solution with benzene as the solvent.

81. What are the boiling point and freezing point of a solution that contains 83 grams of $C_2H_6O_2$ and 450 grams of water?

82. What is the freezing point of a solution containing 35 grams of NH_4NO_3 and 150 grams of water?

83. A solution contains 35 grams of an unknown solute and 150 grams of acetic acid. What is the molar mass of the solute if the solution freezes at 3.0°C?

Section 13.12 Stoichiometry Using Solutions

Problems

84. How many grams of Cr are necessary to react with 250 mL of 0.44 M HCl?

$$Cr + 2\ HCl \longrightarrow CrCl_2 + H_2$$

How many liters of H_2 measured at 755 torr and 25°C will be produced by the reaction?

85. An excess of Na_2CO_3 solution is added to 230 mL of 0.55 M $BaCl_2$ solution. What precipitates and how many grams of it result?

86. To a 50.0-mL solution of 0.00211 M $Ca(NO_3)_2$ are added 25 mL of 1.10 M H_2SO_4. How many grams of $CaSO_4$ are precipitated?

87. How many mL of 2.70 M $AgNO_3$ are necessary to form 16 grams of Ag_2CrO_4?

$$2\ AgNO_3(aq) + K_2CrO_4(aq) \longrightarrow$$
$$Ag_2CrO_4(s) + 2\ KNO_3(aq)$$

How many mL of 3.50 M K_2CrO_4 will be used in the process?

88. How many liters of hydrogen gas at 51°C and 1.05 atm can be produced from 15.6 grams of Mg and 150.0 mL of 5.40 M $HClO_4$, according to the following equation?

$$Mg(s) + 2\ HClO_4(aq) \longrightarrow Mg(ClO_4)_2(aq) + H_2(g)$$

Solutions to Exercises

13.1 K_2SO_4 is soluble.

13.2 (a) Reaction. $Sr_3(PO_4)_2$ is insoluble.
      (b) Reaction. H_2O is formed.
      (c) Reaction. CuS is insoluble.

13.3 $2\ OH^- + Mg^{2+} \longrightarrow Mg(OH)_2(s)$

13.4 $21.3\ g\ NaH_2PO_4 + 455\ g\ H_2O = 476\ g\ solution$

$$\frac{21.3\ g\ NaH_2PO_4}{476\ g\ solution} \times 100\% = 4.47\%\ NaH_2PO_4$$

13.5 $355\ g \times \dfrac{3.2\ g\ C_2H_5OH}{100\ g\ solution} = 11\ g\ C_2H_5OH$

13.6 $75 \text{ g solution} \times \dfrac{8.0 \text{ g SnCl}_2}{100 \text{ g solution}} \times \dfrac{\text{mol SnCl}_2}{189.615 \text{ g SnCl}_2} \times \dfrac{\text{mol SnCl}_2 \cdot 2 \text{ H}_2\text{O}}{\text{mol SnCl}_2} \times$

$\dfrac{225.646 \text{ g SnCl}_2 \cdot 2 \text{ H}_2\text{O}}{\text{mol SnCl}_2 \cdot 2 \text{ H}_2\text{O}} = 7.1 \text{ g SnCl}_2 \cdot 2 \text{ H}_2\text{O}$

13.7 $100 \text{ g solution} - 35 \text{ g NaCl} = 65 \text{ g solvent}$

$650 \text{ g solution} \times \dfrac{65 \text{ g solvent}}{100 \text{ g solution}} = 420 \text{ g solvent}$

13.8 $\dfrac{4.5 \text{ mol HCl}}{35 \text{ L solution}} = 0.13 \text{ M}$

13.9 $7.20 \text{ g NaC}_2\text{H}_3\text{O}_2 \times \dfrac{\text{mol NaC}_2\text{H}_3\text{O}_2}{82.034 \text{ g NaC}_2\text{H}_3\text{O}_2} = 0.0878 \text{ mol NaC}_2\text{H}_3\text{O}_2$

$500.0 \text{ mL solution} \times \dfrac{\text{L solution}}{1,000 \text{ mL solution}} = 0.5000 \text{ L solution}$

$\dfrac{0.0878 \text{ mol NaC}_2\text{H}_3\text{O}_2}{0.5000 \text{ L solution}} = 0.176 \ M \text{ NaC}_2\text{H}_3\text{O}_2$

13.10 $\dfrac{40.3 \text{ g Fe(ClO}_4)_2}{750 \text{ mL solution}} \times \dfrac{\text{mol Fe(ClO}_4)_2}{254.748 \text{ g Fe(ClO}_4)_2} \times \dfrac{1,000 \text{ mL solution}}{\text{L solution}} =$

$0.21 \text{ M Fe(ClO}_4)_2$

13.11 $250 \text{ mL solution} \times \dfrac{\text{L solution}}{1,000 \text{ mL solution}} \times \dfrac{0.45 \text{ mol NaOH}}{\text{L solution}} \times$

$\dfrac{39.9971 \text{ g NaOH}}{\text{mol NaOH}} = 4.5 \text{ g NaOH}$

13.12 $137 \text{ mL solution} \times \dfrac{\text{L solution}}{1,000 \text{ mL solution}} \times \dfrac{0.25 \text{ mol Cl}^-}{\text{L solution}} \times \dfrac{\text{mol CoCl}_3}{3 \text{ mol Cl}^-} \times$

$\dfrac{165.291 \text{ g CoCl}_3}{\text{mol CoCl}_3} = 1.9 \text{ g CoCl}_3$

13.13 $\dfrac{2.0 \text{ mol Na}_3\text{PO}_4}{\text{L solution}} \times \dfrac{3 \text{ mol Na}^+}{\text{mol Na}_3\text{PO}_4} = 6.0 \text{ M Na}^+$

13.14 $\dfrac{119.3 \text{ g Be}_3(\text{PO}_4)_2}{1.73 \text{ L solution}} \times \dfrac{\text{mol Be}_3(\text{PO}_4)_2}{216.979 \text{ g Be}_3(\text{PO}_4)_2} = 0.318 \text{ M Be}_3(\text{PO}_4)_2$

$\dfrac{0.318 \text{ mol Be}_3(\text{PO}_4)_2}{\text{L solution}} \times \dfrac{3 \text{ mol Be}^+}{\text{mol Be}_3(\text{PO}_4)_2} = 0.954 \text{ M Be}^+$

$\dfrac{0.318 \text{ mol Be}_3(\text{PO}_4)_2}{\text{L solution}} \times \dfrac{2 \text{ mol PO}_4^{3-}}{\text{mol Be}_3(\text{PO}_4)_2} = 0.636 \text{ M PO}_4^{3-}$

13.15 $\dfrac{35.2 \text{ g NaCl}}{100 \text{ g solution}} \times \dfrac{\text{mol NaCl}}{58.4425 \text{ g NaCl}} \times \dfrac{1.05 \text{ g solution}}{\text{mL solution}} \times \dfrac{1,000 \text{ mL solution}}{\text{L solution}} =$

6.32 M NaCl

13.16
$$K_2CO_3(aq) \longrightarrow 2\ K^+ + CO_3^{2-}$$

$$\frac{87.3\ g\ K_2CO_3}{100\ g\ solution} \times \frac{mol\ K_2CO_3}{138.2058\ g\ K_2CO_3} \times \frac{2\ mol\ K^+}{mol\ K_2CO_3} \times \frac{1.12\ g\ solution}{mL\ solution} \times$$

$$\frac{1{,}000\ mL\ solution}{L\ solution} = 14.1\ M\ K^+$$

13.17
$$\frac{12\ mol\ C_2H_5OH}{L\ solution} \times \frac{46.069\ g\ C_2H_5OH}{mol\ C_2H_5OH} \times \frac{L\ solution}{1{,}000\ mL\ solution} \times$$

$$\frac{mL\ solution}{0.894\ g\ solution} \times 100\% = 62\%\ C_2H_5OH$$

13.18
$$\frac{7.9\ g\ Cl_2}{20.0\ g\ CCl_4} \times \frac{mol\ Cl_2}{70.9054\ g\ Cl_2} \times \frac{1{,}000\ g\ CCl_4}{kg\ CCl_4} = 5.6\ m\ Cl_2$$

13.19
$$\frac{0.333\ g\ benzoic\ acid}{1.632\ g\ camphor} \times \frac{mol\ benzoic\ acid}{122.1\ g\ benzoic\ acid} \times \frac{1{,}000\ g\ camphor}{kg\ camphor} =$$

$$1.67\ m\ benzoic\ acid$$

13.20
$$100\ g\ solution - 13.8\ g\ C_6H_{14} = 86.2\ g\ solvent$$

$$\frac{13.8\ g\ C_6H_{14}}{86.2\ g\ solvent} \times \frac{mol\ C_6H_{14}}{86.177\ g\ C_6H_{14}} \times \frac{1{,}000\ g\ solvent}{kg\ solvent} = 1.86\ m\ C_6H_{14}$$

13.21
$$0.475\ mol\ C_{12}H_{22}O_{11} \times \frac{342.30\ g\ C_{12}H_{22}O_{11}}{mol\ C_{12}H_{22}O_{11}} = 163\ g\ C_{12}H_{22}O_{11}\ kg\ solvent$$

$$1{,}000\ g\ solvent + 163\ g\ C_{12}H_{22}O_{11} = 1163\ g\ solution$$

$$\frac{163\ g\ C_{12}H_{22}O_{11}}{1{,}163\ g\ solution} \times 100\% = 14.0\%$$

13.22
$$M_2 = 2.00\ M \times \frac{43\ mL}{155\ mL} = 0.55\ M$$

13.23
$$M_2 = 0.60\ M \times \frac{50.0\ mL}{500.0\ mL} = 0.060\ M$$

13.24
$$25.0\ mL + 155.0\ mL = 180.0\ mL$$

$$M_2 = 2.50\ M \times \frac{155.0\ mL}{180.0\ mL} = 2.15\ M$$

13.25
$$34.0\ mL\ solution \times \frac{L\ solution}{1{,}000\ mL\ solution} \times \frac{3.5\ mol\ HCl}{L\ solution} = 0.12\ mol\ HCl$$

$$78.5\ mL\ solution \times \frac{L\ solution}{1{,}000\ mL\ solution} \times \frac{1.4\ mol\ HCl}{L\ solution} = 0.11\ mol\ HCl$$

$$0.12\ mol\ HCl + 0.11\ mol\ HCl = 0.23\ mol\ HCl$$

$$34.0\ mL\ solution + 78.5\ mL\ solution = 112.5\ mL\ solution$$

$$\frac{0.23 \text{ mol HCl}}{112.5 \text{ mL solution}} \times \frac{1,000 \text{ mL solution}}{\text{L solution}} = 2.0 \text{ M HCl}$$

13.26 $4.0 \text{ mL solution} \times \dfrac{\text{L solution}}{1,000 \text{ mL solution}} \times \dfrac{10.0 \text{ mol NaCl}}{\text{L solution}} = 0.040 \text{ mol NaCl}$

$2.0 \text{ mL solution} \times \dfrac{\text{L solution}}{1,000 \text{ mL solution}} \times \dfrac{7.5 \text{ mol NaCl}}{\text{L solution}} = 0.015 \text{ mol NaCl}$

$0.040 \text{ mol NaCl} + 0.015 \text{ mol NaCl} = 0.055 \text{ mol NaCl}$

$4.0 \text{ mL solution} + 2.0 \text{ mL solution} = 6.0 \text{ mL solution}$

$\dfrac{0.055 \text{ mol NaCl}}{6.0 \text{ mL solution}} \times \dfrac{1,000 \text{ mL solution}}{\text{L solution}} = 9.2 \text{ M NaCl}$

13.27 $\dfrac{6.2 \text{ g KBr}}{42 \text{ g H}_2\text{O}} \times \dfrac{\text{mol KBr}}{119.002 \text{ g KBr}} \times \dfrac{1,000 \text{ g H}_2\text{O}}{\text{kg H}_2\text{O}} = 1.2 \text{ m KBr}$

$\Delta T_b = \dfrac{2 \text{ mol solute}}{\text{mol KBr}} \times \dfrac{1.2 \text{ mol KBr}}{\text{kg H}_2\text{O}} \times \dfrac{0.512 \text{ deg kg H}_2\text{O}}{\text{mol solute}} = 1.2 \text{ deg}$

$100.00°\text{C} + 1.2 \text{ deg} = 101.2°\text{C}$

13.28 $83.5°\text{C} - 80.1°\text{C} = 3.4 \text{ deg}$

$3.4 \text{ deg} = \text{m} \times \dfrac{2.53 \text{ deg kg benzene}}{\text{mol solute}}$

$\text{m} = \dfrac{1.3 \text{ mol solute}}{\text{kg benzene}}$

$32.0 \text{ g benzene} \times \dfrac{\text{kg benzene}}{1,000 \text{ g benzene}} \times \dfrac{1.3 \text{ mol solute}}{\text{kg benzene}} = 0.042 \text{ mol solute}$

$\dfrac{5.2 \text{ g solute}}{0.042 \text{ mol solute}} = 120 \text{ g/mol}$

13.29 $\dfrac{2.0 \text{ kg NaCl}}{15.0 \text{ kg H}_2\text{O}} \times \dfrac{1,000 \text{ g NaCl}}{\text{kg NaCl}} \times \dfrac{\text{mol NaCl}}{58.4425 \text{ g NaCl}} = 2.3 \text{ m}$

$\Delta T_f = \dfrac{2.3 \text{ mol NaCl}}{\text{kg H}_2\text{O}} \times \dfrac{2 \text{ mol solute}}{\text{mol NaCl}} \times \dfrac{1.86 \text{ deg kg H}_2\text{O}}{\text{mol solute}} = 8.6 \text{ deg}$

$0.00°\text{C} - 8.6 \text{ deg} = -8.6°\text{C}$

13.30 $6.7°\text{C} - 5.5°\text{C} = 1.2 \text{ deg}$

$1.2 \text{ deg} = \text{m} \times \dfrac{5.10 \text{ deg kg benzene}}{\text{mol solute}}$

$\text{m} = \dfrac{0.24 \text{ mol solute}}{\text{kg benzene}}$

$16.8 \text{ g benzene} \times \dfrac{\text{kg benzene}}{1,000 \text{ g benzene}} \times \dfrac{0.24 \text{ mol solute}}{\text{kg benzene}} = 0.0040 \text{ mol solute}$

$\dfrac{2.3 \text{ g solute}}{0.0040 \text{ mol solute}} = 580 \text{ g/mol}$

13.31 $\quad$ 2.83 g Cu $\times \dfrac{\text{mol Cu}}{63.546 \text{ g Cu}} \times \dfrac{8 \text{ mol HNO}_3}{3 \text{ mol Cu}} \times \dfrac{\text{L solution}}{3.50 \text{ mol HNO}_3} \times$

$$\dfrac{1{,}000 \text{ mL solution}}{\text{L solution}} = 33.9 \text{ mL solution}$$

13.32 $\quad$ $3 \text{ Zn(s)} + 2 \text{ H}_3\text{PO}_4\text{(aq)} \longrightarrow \text{Zn}_3(\text{PO}_4)_2\text{(s)} + 3 \text{ H}_2\text{(g)}$

4.80 g Zn $\times \dfrac{\text{mol Zn}}{65.39 \text{ g Zn}} \times \dfrac{1 \text{ mol Zn}_3(\text{PO}_4)_2}{3 \text{ mol Zn}} = 0.0245 \text{ mol Zn}_3(\text{PO}_4)_2$

26.0 mL solution $\times \dfrac{\text{L solution}}{1{,}000 \text{ mL solution}} \times \dfrac{5.0 \text{ mol H}_3\text{PO}_4}{\text{L solution}} \times \dfrac{1 \text{ mol Zn}_3(\text{PO}_4)_2}{2 \text{ mol H}_3\text{PO}_4} =$

$$0.065 \text{ mol Zn}_3(\text{PO}_4)_2$$

Zn is limiting.

$$0.0245 \text{ mol Zn}_3(\text{PO}_4)_2 \times \dfrac{386.1 \text{ g Zn}_3(\text{PO}_4)_2}{\text{mol Zn}_3(\text{PO}_4)_2} = 9.46 \text{ g Zn}_3(\text{PO}_4)_2$$

13.33 $\quad$ $\text{Mg(s)} + 2 \text{ HCl(aq)} \longrightarrow \text{MgCl}_2\text{(aq)} + \text{H}_2\text{(g)}$

2.50 g Mg $\times \dfrac{\text{mol Mg}}{24.3050 \text{ g Mg}} \times \dfrac{1 \text{ mol H}_2}{1 \text{ mol Mg}} = 0.103 \text{ mol H}_2$

25.0 mL solution $\times \dfrac{\text{L solution}}{1{,}000 \text{ mL solution}} \times \dfrac{12.0 \text{ mol HCl}}{\text{L solution}} \times$

$$\dfrac{1 \text{ mol MgCl}_2}{2 \text{ mol HCl}} = 0.150 \text{ mol H}_2$$

Mg is limiting.

$$0.103 \text{ mol H}_2 \times \dfrac{22.4 \text{ L H}_2}{\text{mol H}_2} = 2.31 \text{ L H}_2$$

Red cabbage juice is one of the many natural acid-base indicators. The pH values of the colored solutions from left to right are 1, 4, 7, 10, and 13.

Acids and Bases

I n chemistry, "acid" and "base" are two common and important words that apply to two categories of reactive substances. Although widely used, the two words are not so widely understood. How many people listening to a commercial saying that their stomach is "too acidic" or that a shampoo is "pH balanced" know what the words mean?

In this chapter we learn what acids and bases are, and we will make chemical calculations involving acidic and basic solutions. We find out how strong and weak acids and bases behave and how to distinguish them from one another. We will become familiar with the pH scale, the universal method of measuring the amount of acidity or basicity of solutions. We see how acids and bases react to create salts. We study titration, the method for making quantitative measurements of concentration by reacting two solutions together in the laboratory. At the end of the chapter, we discuss the preparation and uses of one acid and one base essential to modern civilization.

Once we have studied acids and bases and the reactions of acidic and basic solutions, we will have obtained some of the most crucial knowledge we need to practice the science of chemistry.

14.1 Definitions of Acid and Base

After studying this section, you should be able to:

- Identify a compound as an acid or a base.
- Explain the concept of conjugate acids and bases.
- Identify acid-base reactions.

Reactions in which water breaks up another species are called hydrolysis reactions (a term formed by the Greek suffix lysis, meaning loosening or decomposing with, and the prefix hydro, meaning water).

By its simplest definition, an **acid** is *a substance that produces H^+ ions in solution*. For example, HCl is a common acid that reacts with water according to the following equation.

$$HCl(g) + H_2O(l) \longrightarrow H_3O^+ + Cl^-$$

H^+ ions do not exist individually in the presence of water, because water molecules have unshared pairs of electrons that attract and capture H^+ to form H_3O^+ ions. H_3O^+ ions are called **hydronium ions.** H_3O^+ ions attract even more water molecules, making $H_5O_2^+$, $H_7O_3^+$, and so on. We will refer to aqueous H^+ as H_3O^+. For simplicity, some chemists use H^+ as a shorthand representation.

If we confine our use of the term "acid" to aqueous solutions, we define an acid as H_3O^+ *or any substance that produces H_3O^+ in solution.*

A **base** is commonly defined as OH^- *or any substance that will produce OH^- ions in aqueous solution.* As you may recall from Section 6.10, OH^- is hydroxide ion. When dissolved in water, metal hydroxides, such as NaOH, are bases.

$$NaOH \xrightarrow{\text{in } H_2O} Na^+ + OH^-$$

Aqueous solutions of acids and bases react vigorously to yield water. For instance, a solution of $HClO_4$ reacts with a solution of NaOH. We write the chemical equation for this acid-base reaction as

$$HClO_4(aq) + NaOH(aq) \longrightarrow NaClO_4(aq) + H_2O(l)$$

In aqueous solution, $HClO_4$ produces H_3O^+, NaOH produces OH^-, and these then react. The net ionic equation more accurately describes the reaction between the two solutions.

$$H_3O^+ + OH^- \longrightarrow 2\ H_2O$$

Recall that we studied reactions like this in Section 13.7. If you cannot predict the products of the reaction, go back and review that section.

The Swedish chemist Svante Arrhenius first proposed these simple definitions of acid and base in 1887. Although these definitions are useful, they are limited to reactions in aqueous solutions.

Because many acid-base reactions are run in nonaqueous solutions, more general acid-base definitions were suggested by J. N. Brønsted in Denmark and Thomas Lowry in England in 1923. They defined an acid as a *proton donor* and a base as a *proton acceptor*.

To understand this definition better, let's look at a chemical equation for a Brønsted-Lowry acid-base reaction.

$$\overparen{AH + B} \longrightarrow A^- + BH^+$$

In this equation, AH is an acid which transfers its proton to the base, B, resulting in the anion of the acid, A^-, and the new cation, BH^+. Acid-base reactions are, in fact, often called proton transfer reactions whether they take place in aqueous solution or not.

An example of an acid-base reaction without the presence of water occurs between HCl and NH_3 gases. When these gases mix, they react (Figure 14.1), and solid NH_4Cl is produced.

$$HCl(g) + NH_3(g) \longrightarrow NH_4Cl(s)$$

Figure 14.1 When HCl and NH_3 gases come in contact, they produce solid sodium chloride, NH_4Cl, which appears as white smoke.

In this reaction, there are no H_3O^+ and OH^- ions present in either the reactants or the products, but HCl can donate a H^+, and NH_3 can accept one, so the reaction occurs. In Brønsted-Lowry terms, this is an acid-base reaction. Protons from the HCl are transferred to NH_3 molecules. The resulting NH_4^+ and Cl^- ions combine to yield solid white NH_4Cl. HCl has donated protons, while NH_3 has accepted protons. HCl, therefore, is the acid, and NH_3 is the base.

The Brønsted-Lowry definitions can be generalized to include not only the reactants, but also the products. Consider the reaction between ammonia and water. Water, acting as a Brønsted-Lowry acid, donates protons to NH_3 molecules which act as a Brønsted-Lowry base.

$$NH_3 + H_2O \rightleftharpoons NH_4^+ + OH^-$$
$$\text{base} \quad \text{acid} \qquad \text{acid} \quad \text{base}$$

In a chemical equation, a double arrow is used to indicate a reversible reaction. We will learn more about reversible reactions in Chapter 15.

This reaction, however, is reversible. That is, after some of the products, NH_4^+ and OH^-, are created, they begin to react. In the reverse reaction, NH_4^+ donates protons, while OH^- accepts protons. NH_4^+ acts as an acid, and OH^- is a base.

NH_4^+ and NH_3 constitute what is known as a **conjugate acid-base pair.** NH_4^+ is the conjugate acid of NH_3, while NH_3 is the conjugate base of NH_4^+. H_2O and OH^- are also a conjugate acid-base pair. OH^- is the conjugate base of H_2O, while H_2O is the conjugate acid of OH^-.

Let's look at another example. As we have seen, HCl reacts with water to yield H_3O^+ and Cl^- ions.

$$HCl + H_2O \longrightarrow H_3O^+ + Cl^-$$
$$\text{acid} \quad \text{base} \qquad \text{acid} \quad \text{base}$$

HCl donates H^+ to H_2O. HCl is an acid. H_2O, acting as a base, accepts the protons. On the *product* side of the equation, H_3O^+ is an acid and Cl^- is a base. H_3O^+ and H_2O are a conjugate acid-base pair, while HCl and Cl^- are another acid-base pair.

Notice that in the reaction of NH_3 with water, H_2O acts as an acid, but in the reaction of HCl with water, H_2O acts as a base. Substances that can act either as an acid or a base are **amphoteric.** We will run into other substances that are amphoteric as we continue our study of acids and bases. According to the Brønsted-Lowry definition, a substance is not necessarily classified by what it *is* but rather by what it *does* in a particular situation.

neutral

EXAMPLE 14.1

Label the acids and bases in the following chemical equation for a reaction in aqueous solution.

$$HC_2H_3O_2 + H_2O \rightleftharpoons C_2H_3O_2^- + H_3O^+$$

Strategy and Solution:

Step 1. Identify which reactant loses protons and which reactant gains protons. In this example, $HC_2H_3O_2$ loses H^+, while H_2O gains H^+.

Step 2. Label the acid and the base on the left side of the equation.

Step 3. Determine which of the products will donate H^+ and which will accept H^+. In aqueous solution, H_3O^+ is always an acid (and OH^-, always a base). H_3O^+, then, is the acid, and $C_2H_3O_2^-$ must be the base. $C_2H_3O_2^-$ is the conjugate base of $HC_2H_3O_2$.

$$HC_2H_3O_2 + H_2O \rightleftharpoons C_2H_3O_2^- + H_3O^+$$
$$\text{acid} \qquad \text{base} \qquad \text{base} \qquad \text{acid}$$

Exercise 14.1

Label the acids and bases in the following chemical equations.

$$HNO_2 + OH^- \longrightarrow NO_2^- + H_2O$$
$$H_3O^+ + NH_3 \longrightarrow NH_4^+ + H_2O$$

In 1923, G. N. Lewis proposed the most general definitions of acids and bases. *An acid is any species (atom, ion, or molecule) that is strongly attracted to an unshared electron pair (a Lewis acid is an electron-pair acceptor). A base is any species that has an easily available electron pair (a Lewis base is an electron-pair donor).* According to this definition, individual hydrogen ions, H^+, or hydronium ions, H_3O^+, are acids because they are attracted to unshared electron pairs. OH^- is a base because it has attractive unshared electron pairs.

$$\left[:\overset{..}{\underset{..}{O}}-H \right]^-$$

Acid-base reactions that do not involve proton transfer are not discussed in greater detail in this chapter; however, reactions of Lewis acids and bases are important in modern chemical research and in industry.

14.2 Strong and Weak Acids and Bases

After studying this section, you should be able to:

- Explain the difference between strong and weak acids and bases.
- Explain what makes a given acid or base weak instead of strong.

As we have seen, the reaction of gaseous hydrogen chloride, HCl, with water is an example of an acid-base reaction that produces H_3O^+.

$$HCl(g) + H_2O(l) \longrightarrow H_3O^+ + Cl^-$$

The reaction is complete. Essentially no HCl molecules remain in the water, and the pressure of HCl gas is reduced essentially to zero. An acid, such as HCl, that is essentially 100% ionized in aqueous solution is called a **strong acid.**

There are acids, however, of which only a small fraction ionizes in water; these are called **weak acids.** Acetic acid, $HC_2H_3O_2$, is an example. If one mole of acetic acid is added to one liter of water, only a few thousandths of it will ionize; nearly all will remain in the solution as acetic acid molecules.

Strong acids react essentially completely to form H_3O^+. Only a fraction of a weak acid reacts to make H_3O^+. When we study Chapter 15, we will learn more about the strengths of acids and bases.

$$HC_2H_3O_2 + H_2O \rightleftharpoons C_2H_3O_2^- + H_3O^+$$
$$\text{acetic acid} \qquad\qquad \text{acetate ion}$$

If we were to mix one mole of NaOH with one liter of water, we would find that all of the NaOH would dissolve and dissociate. If we were to make a similar mixture of $Mg(OH)_2$ and water, however, only a small portion of the $Mg(OH)_2$ would dissolve, but the dissolved portion would nonetheless be totally dissociated into its ions. Because they are totally dissociated in solution, both NaOH and $Mg(OH)_2$ are said to be **strong bases,** regardless of solubility.

Another class of compounds is known as **weak bases.** Although these may be quite soluble in water, in solution they exist only partially in ionic form, leaving most of their molecules intact. An example is gaseous ammonia, NH_3.

$$NH_3(g) + H_2O(l) \rightleftharpoons NH_4^+ + OH^-$$

NH_3 gas is extremely soluble in water, but at any given time, only a small proportion of the NH_3 is in the form of ions; most remains in molecular form.

Weak bases do not completely ionize because they are only *weakly* attracted to protons. Moreover, in their reaction with water, they create a base, OH^-, which reacts readily with their conjugate acid. Let's analyze the NH_3 example.

$$\underset{\text{weak base}}{NH_3} + \underset{\text{weak acid}}{H_2O} \rightleftharpoons \underset{\text{weak acid}}{NH_4^+} + \underset{\text{strong base}}{OH^-}$$

Because of its unshared electron pair, NH_3 weakly attracts protons from water molecules, creating NH_4^+ and OH^-. In the reverse reaction, however, the base, OH^-, strongly attracts the proton from NH_4^+, producing NH_3 molecules.

Similarly, weak acids do not completely ionize because their reaction with water creates an acid, H_3O^+, which reacts readily with their conjugate bases. Acetic acid ($HC_2H_3O_2$) is a weak acid that reacts with water to form acetate ion ($C_2H_3O_2^-$) and H_3O^+.

$$\underset{\text{weak acid}}{HC_2H_3O_2} + \underset{\text{weak base}}{H_2O} \rightleftharpoons \underset{\text{weak base}}{C_2H_3O_2^-} + \underset{\text{strong acid}}{H_3O^+}$$

In the reverse reaction, H_3O^+ reacts vigorously with the weak base, $C_2H_3O_2^-$, to produce $HC_2H_3O_2$ again.

Let's look at the reaction of sulfuric acid with water.

$$H_2SO_4 + H_2O \longrightarrow H_3O^+ + HSO_4^-$$

The first step of the ionization of H_2SO_4 shown above goes essentially to completion. Because the conjugate base, HSO_4^-, has little attraction for H^+, the reverse reaction does not readily occur.

The second step, however, does not go to completion.

$$HSO_4^- + H_2O \rightleftharpoons H_3O^+ + SO_4^{2-}$$

The ion SO_4^{2-} has more attraction for H^+ than does HSO_4^- (after all, it has two negative charges, which are very attractive to H^+ ions). Because of this, SO_4^{2-} reacts vigorously with H_3O^+ in the reverse reaction. Not much SO_4^{2-} remains in solution; most of it has become HSO_4^-.

As a help in predicting how particular acid-base reactions might proceed, note and remember that generally the stronger an acid or base is, the more likely it is to dominate a reaction. Strong acids easily donate protons, and strong bases

The weaker an acid or base, the stronger its conjugate base or acid. The conjugate base, Cl^-, of the strong acid, HCl, is extremely weak.

Table 14.1   Common Acids and Bases in Aqueous Solution

	Acids		
Strong Acids		*Weak Acids*	
Perchloric	$HClO_4$	Carbonic	H_2CO_3
Hydrochloric	HCl	Acetic	$HC_2H_3O_2$
Sulfuric	H_2SO_4	Benzoic	$HC_7H_5O_2$
Nitric	HNO_3	Nitrous	HNO_2
Hydrobromic	HBr	Hydrofluoric	HF
Hydroiodic	HI		

	Bases		
Strong Bases		*Weak Bases*	
Sodium hydroxide	$NaOH$	Carbonate ion	CO_3^{2-}
Potassium hydroxide	KOH	Ammonia	NH_3
Barium hydroxide	$Ba(OH)_2$		
Calcium hydroxide	$Ca(OH)_2$		

Examples of weak acids and bases, with indications of their strength, appear in Table 15.1 in Section 15.12.

easily accept protons. The weaker the acid or base, the less easily it performs its function.

Table 14.1 lists some strong and weak acids and bases, all in aqueous solution.

14.3 Neutralization and the Formation of Salts

After studying this section, you should be able to:

- Write chemical equations for neutralization reactions.
- Explain what a neutral solution is.
- Identify a compound as a salt.

When aqueous solutions of acids and bases are mixed, the H_3O^+ and OH^- react rapidly to yield water, and the other ions usually remain unchanged. If the concentrations of H_3O^+ and OH^- are equal, the reaction is called **neutralization,** and the resulting solution will be **neutral.** An example of a neutralization is the reaction of equal numbers of moles of HCl and NaOH in aqueous solution.

$$HCl(aq) + NaOH(aq) \longrightarrow NaCl(aq) + H_2O(l)$$

We write this chemical equation in ionic form as

$$H_3O^+ + Cl^- + Na^+ + OH^- \longrightarrow 2\,H_2O + Cl^- + Na^+$$

The net ionic equation for neutralization should look familiar.

$$H_3O^+ + OH^- \longrightarrow 2\,H_2O$$

If the water is evaporated after neutralization, the remaining ions will be left as a crystalline solid. In the reaction of HCl and NaOH, for example, the Cl^- and Na^+ ions will be left behind in the form of NaCl. The ionic substances that result from the reaction of acids and bases are called **salts.** NaCl is a salt. Some other salts are KCl, Na_2SO_4, LiBr, BaI_2, and NH_4F.

Salts are ionic compounds.

Total neutralization occurs only if the number of H_3O^+ ions is the same as the number of OH^- ions. If a solution contains more H_3O^+ than OH^-, it is said to be **acidic.** If a solution contains more OH^- than H_3O^+, it is said to be **basic.**

EXAMPLE 14.2

Write the chemical equation for the neutralization of HNO_3 and KOH.

Solution: A neutralization will produce a salt and water.

$$HNO_3 + KOH \longrightarrow KNO_3 + H_2O$$

Exercise 14.2

Write the chemical equation for the neutralization reaction of H_2SO_4 and $Mg(OH)_2$.

14.4 Ionization of Water

After studying this section, you should be able to:

- Explain the mechanism by which water ionizes.

There is a limit to the strength of chemical bonds, even the strongest covalent ones. At any given moment, in any sample of molecules, at least a few molecules will have been knocked apart by the motion of the molecules around them. In any sample of water, some of the molecules are constantly being ionized.

$$H_2O \longrightarrow H^+ + OH^-$$

Because H^+ ions don't exist as individuals in the presence of water, our ionization equation is more correctly written as a proton transfer reaction between two water molecules.

$$H_2O + H_2O \rightleftharpoons H_3O^+ + OH^-$$

When water ionizes, one OH^- ion is produced for every H_3O^+ ion. Since the concentration of acid then equals the concentration of base, $[H_3O^+] = [OH^-]$, the water is neither acidic nor basic; it is neutral. (The same cannot usually be said, however, if anything is dissolved in the water.)

A hydroxide ion has a strong attraction for a positively charged proton, so the ionization reaction of water is reversible, and the water molecules are quickly restored. Nevertheless, at any particular instant, in any sample of liquid water, there are some ionized molecules. The amount is small, however; only 10^{-7} mol/L of water is ionized at any time in a sample of pure water at room temperature. That is, for pure water at 25°C,

$$[H_3O^+] = [OH^-] = 1 \times 10^{-7} \text{ M}$$

Remember from Section 13.9 that we can show the concentration of something in moles per liter by using brackets [ ]; thus $[H_3O^+]$ is the concentration of H_3O^+ in moles per liter.

Pure water is 10^{-7} M in OH^- and 10^{-7} M in H_3O^+.

One ten millionth of a mole per liter seems like a small amount to be concerned with, but the ionization of water plays an important part in many chemical and biological processes.

14.5 Aqueous Solutions of Acids and Bases

After studying this section, you should be able to:

* Calculate the molarity of OH^- in a solution, knowing the molarity of H_3O^+ and vice versa.

An acidic solution is more than 10^{-7} M in H_3O^+, while a basic solution is more than 10^{-7} M in OH^-.

Water ionizes so that $[H_3O^+] = [OH^-] = 10^{-7}$ M. When gaseous HCl is dissolved in water, the HCl molecules react with the water, and H^+ ions from the HCl become attached to H_2O molecules, forming H_3O^+. Although there is then an excess of H_3O^+ ions present, there is still a little OH^- in solution because of the ionization of water. Even though OH^- ions are present, the solution is still acidic. In an **acidic** solution, $[H_3O^+] > [OH^-]$. Similarly, a **basic** solution contains H_3O^+ ions, but these are few in number compared to the number of OH^- ions. In a basic solution, $[OH^-] > [H_3O^+]$. *Every aqueous solution contains both H_3O^+ and OH^- ions.*

In fact, for any aqueous solution, the molarity of H_3O^+ ions times the molarity of OH^- ions always equals 1.00×10^{-14}.

We will learn more about this relationship and its significance in Chapter 15.

$$[H_3O^+][OH^-] = 1.00 \times 10^{-14} = K_w$$

The constant 1.00×10^{-14} is given the symbol, K_w. This equation tells us that if the concentration of H_3O^+ goes up, the concentration of OH^- must go down and vice versa. The product of the two molar concentrations is always 1.00×10^{-14}.

EXAMPLE 14.3

What is the molarity of OH^- in a 0.00405 M solution of HCl? Is the solution acidic, basic, or neutral?

Solution Map: $\boxed{[\text{HCl}]} \longrightarrow \boxed{[\text{H}_3\text{O}]} \longrightarrow \boxed{[\text{OH}^-]}$

Strategy and Solution:

Step 1.   Calculate $[H_3O^+]$. Every mole of HCl dissolved contributes one mole of H_3O^+ to the solution. Thus,

$$[H_3O^+] = 0.00405 \text{ M}$$

Step 2.   Substitute the molarity of H_3O^+ into

$$[H_3O^+][OH^-] = 1.00 \times 10^{-14}$$

$$(0.00405)[OH^-] = 1.00 \times 10^{-14}$$

Step 3.   Rearrange the equation and solve for $[OH^-]$.

$$[OH^-] = \frac{1.00 \times 10^{-14}}{0.00405} = 2.47 \times 10^{-12} \text{ M}$$

Because HCl is a strong acid, the molarity of OH^- is very small compared to the molarity of H_3O^+, and the solution is acidic.

EXAMPLE 14.4

Determine the molarity of H_3O^+ in a solution in which $[OH^-] = 0.082$ M. Is the solution acidic, basic, or neutral?

Solution Map: $\boxed{[OH^-]} \longrightarrow \boxed{[H_3O^+]}$

Strategy and Solution:

Step 1. Substitute the molarity of OH^- into

$$[H_3O^+][OH^-] = 1.00 \times 10^{-14}$$

$$[H_3O^+][0.082] = 1.00 \times 10^{-14}$$

Step 2. Rearrange the equation and solve for $[H_3O^+]$.

$$[H_3O^+] = \frac{1.00 \times 10^{-14}}{0.082} = 1.2 \times 10^{-13} \text{ M}$$

Since $[OH^-] > [H_3O^+]$, the solution is basic.

> **Exercise 14.3**
>
> Calculate $[OH^-]$ in a 0.0500 M HNO_3 solution.

> **Exercise 14.4**
>
> What is the molarity of H_3O^+ in a 0.00003 M NaOH solution?

14.6 The pH Scale

After studying this section, you should be able to:

- Calculate the pH of a solution from the concentration of H_3O^+ or OH^-.
- Calculate the molarity of OH^- or H_3O^+ in a solution, knowing the pH.
- Explain why the pH value gets smaller as the acidity increases.

In laboratory reactions, industrial processes, living creatures, and countless other systems in which solution chemistry is important, the level of acidity is often crucial. Acidity in aqueous solutions, which is determined by the concentration of H_3O^+, is so often measured that it is reported as a special quantity called the **pH.**

The numerical value of the pH of a solution is defined as the negative logarithm (base 10) of the molarity of H_3O^+.

$$pH = -\log [H_3O^+]$$

Remember that the logarithm of a number is the power to which 10 must be raised to give that number. Thus the logarithm of 10^4 is 4.

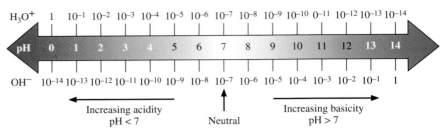

Increasing acidity
pH < 7

Neutral

Increasing basicity
pH > 7

Figure 14.2

If the H_3O^+ value is an integer power of 10, we can easily find the pH. For example, if $[H_3O^+]$ is 1×10^{-3} M, or simply 10^{-3} M, we take the log of 10^{-3}, which is -3, and change its sign. This gives us a pH of 3.

EXAMPLE 14.5

The molarity of H_3O^+ in a solution is 1×10^{-4} M. What is the pH of that solution?

Solution Map: $\boxed{[H_3O^+]}$ $\longrightarrow$ $\boxed{pH}$

Strategy and Solution: $pH = -\log [H_3O^+]$

Step 1.   Take the logarithm of the H_3O^+ concentration. The log of 10^{-4} is -4.

Step 2.   Change the sign of the log. This gives the pH. The pH is 4.

In a neutral solution,
$[H_3O^+] = [OH^-] = 10^{-7}$ M
and pH = 7.

The pH scale is most often used for $[H_3O^+]$ values between 1×10^{-1} M and 1×10^{-14} M, but it is not limited to these values. That is, pH usually lies between values of 1 and 14. Figure 14.2 shows the correlation among values of pH, $[H_3O^+]$, and $[OH^-]$. In pure water, where $[H_3O^+]$ and $[OH^-]$ are equal (each is 10^{-7} M), the pH is 7. Pure water is neutral. *Any* solution with a pH of 7 is neutral.

A pH value of less than 7 means that the $[H_3O^+]$ is greater than 10^{-7} M and therefore must be greater than the $[OH^-]$. The solution, then, is acidic. As acidity increases, the value of the pH decreases; a pH value of 1 or less indicates that the solution is extremely acidic. It is also possible to have negative pH values, although these are seldom used. If, for example, $[H_3O^+] = 10^1$ M, then the pH of the solution is the negative log of 10^1, or -1.

In an acidic solution,
$[H_3O^+] > [OH^-],$
$[H_3O^+] > 10^{-7}$ M,
and pH < 7.

A pH value greater than 7 means that the $[H_3O^+]$ is less than 10^{-7} M and less than the $[OH^-]$. Such a solution is basic. As a solution becomes more basic, the value of the pH increases. A pH value of 14 or greater indicates an extremely basic solution.

In a basic solution,
$[OH^-] > [H_3O^+],$
$[H_3O^+] < 10^{-7}$ M,
and pH > 7.

If the value of the H_3O^+ concentration of a solution is not an integer power of 10, the value of the pH will not be an integer number. For example, if $[H_3O^+] = 4.5 \times 10^{-3}$ M, the value of the pH will be the log of that number. Because $4.5 \times 10^{-3} = 10^{-2.35}$, the pH of the solution would be 2.35. Figure 14.3 shows the $[H_3O^+]$ and $[OH^-]$ values for pH values between 2 and 3.

To calculate the pH on a calculator, follow these guidelines:

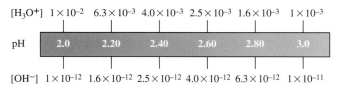

Figure 14.3

Guidelines for Using an Electronic Calculator to Find Logarithms

1. *Enter the $[H_3O^+]$.*
2. *Press the* **LOG** *key.*
3. *Press the* **+/−** *key to change the sign.*

Most chemists use electronic calculators to determine the logarithms of such numbers. The number of significant figures in the original concentration will equal the number of digits after the decimal in the pH value. For example, if the $[H_3O^+]$ is 2.5×10^{-5} M, then the pH value is 4.60. (The 4 in the pH value is not significant for the same reason that the −5 exponent in the concentration is not significant. Both are place holders.)

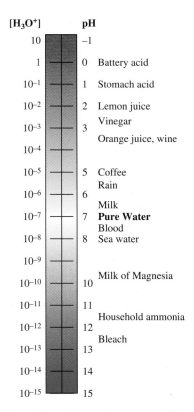

The acidity of several common substances. The red area indicates that the substance is acidic. The blue area indicates that the substance is basic.

EXAMPLE 14.6

What is the pH of a solution that is 5.3×10^{-3} M in H_3O^+? Is the solution acidic, basic, or neutral?

Solution Map: $\boxed{[H_3O^+]} \longrightarrow \boxed{pH}$

Strategy and Solution: $pH = -\log (5.3 \times 10^{-3})$

Step 1. Take the logarithm of the H_3O^+ concentration. Using an electronic calculator, we find that the log of 5.3×10^{-3} is −2.28.

Step 2. Change the sign of the log, which gives the pH. The pH is 2.28. Since the pH is less than 7, the solution is acidic.

Exercise 14.5

The sour taste of orange juice is caused by acidity. Find the pH of a glass of orange juice that has a $[H_3O^+]$ of 0.00032 M.

Exercise 14.6

Most wines are somewhat acidic. Calculate the pH of a wine that has $[H_3O^+]$ of 1.59×10^{-4} M.

If $[OH^-]$ is given for a solution, first find $[H_3O^+]$ and then determine the pH of the solution.

EXAMPLE 14.7

A solution of NaOH has $[OH^-]$ of 1.0×10^{-3} M. What is its pH? Is the solution acidic, basic, or neutral?

Solution Map: $\boxed{[OH^-]} \longrightarrow \boxed{[H_3O^+]} \longrightarrow \boxed{pH}$

Strategy and Solution:

Step 1.   Find $[H_3O^+]$.

$$[H_3O^+] = \frac{1.00 \times 10^{-14}}{[OH^-]} = \frac{1.00 \times 10^{-14}}{1.0 \times 10^{-3}} = 1.0 \times 10^{-11} \text{ M}$$

Step 2.   Find the pH.

$$pH = -\log[H_3O^+] = -\log(1.0 \times 10^{-11}) = 11.00$$

Since the pH is greater than 7, the solution is basic.

EXAMPLE 14.8

In a solution of $Ba(OH)_2$, $[OH^-] = 4.1 \times 10^{-4}$ M. What is the pH of the solution? Is the solution acidic, basic, or neutral?

Solution Map: $\boxed{[OH^-]} \longrightarrow \boxed{[H_3O^+]} \longrightarrow \boxed{pH}$

Strategy and Solution:

Step 1.   Find $[H_3O^+]$.

$$[H_3O^+] = \frac{1.00 \times 10^{-14}}{4.1 \times 10^{-4}} = 2.4 \times 10^{-11} \text{ M}$$

Step 2.   Find the pH.

$$pH = -\log(2.4 \times 10^{-11}) = 10.62$$

The pH is greater than 7; therefore the solution is basic.

Exercise 14.7

Find the pH of a 0.0030 M KOH solution.

Exercise 14.8

Milk of magnesia is a base that reacts with and neutralizes stomach acid. What is the pH of milk of magnesia, in which the $[OH^-]$ is 6.3×10^{-5} M?

It is frequently necessary to know the value of $[H_3O^+]$ or $[OH^-]$ when only the pH is known. The process of finding concentration from pH is the reverse of what we have just done.

To calculate $[H_3O^+]$ from pH using an electronic calculator, follow these guidelines.

Guidelines for Using an Electronic Calculator to Find [H₃O⁺] from the pH

1. *Enter the pH value.*
2. *Change the sign by using the* **+/−** *key.*
3. *Take the anitlog of the* −pH *by using the* **10ˣ** *key if your calculator has it or use the* **INV** *and* **LOG** *keys in that order.*

Let's study some examples.

EXAMPLE 14.9

The pH of a solution is 5.2. Is the solution acidic or basic? What is the molarity of H_3O^+ in the solution?

Solution Map: $\boxed{\text{pH}}$ $\longrightarrow$ $\boxed{[H_3O^+]}$

Strategy and Solution: Since the value of the pH is less than 7, the solution is acidic. The calculation of $[H_3O^+]$ can be done in 2 steps.

Step 1. Convert the pH to a concentration value given as a power of 10. Because pH is the negative log of $[H_3O^+]$ and because a log is a power to which we raise 10, we first change the sign of the pH value, then use it as an exponent of 10.

$$5.2 = -\log[H_3O^+]$$

Change the sign: -5.2

$$-5.2 = \log[H_3O^+]$$

Use this value as an exponent: $10^{-5.2}$.

$$10^{-5.2} = [H_3O^+]$$

Step 2. Determine the value of $10^{-5.2}$ with your calculator.

$$10^{-5.2} = 6 \times 10^{-6}$$
$$[H_3O^+] = 6 \times 10^{-6} \text{ M}$$

Exercise 14.9

A tomato has a pH of 4.15. What is the molarity of H_3O^+ in this tomato?

Exercise 14.10

A cleaning solution containing borax has a pH of 9.2. What is $[H_3O^+]$ in this solution?

EXAMPLE 14.10

Household ammonia has a pH of 11.71. What is the hydroxide ion concentration of a solution that has a pH of 11.71?

Solution Map:

Strategy and Solution:

Step 1.   Find $[H_3O^+]$.

$$11.71 = -\log[H_3O^+]$$

$$[H_3O^+] = 10^{-11.71} = 1.9 \times 10^{-12} \text{ M}$$

Step 2.   Calculate $[OH^-]$ using $[H_3O^+][OH^-] = 1.00 \times 10^{-14}$.

$$[OH^-] = \frac{1.00 \times 10^{-14}}{[H_3O^+]} = \frac{1.00 \times 10^{-14}}{1.9 \times 10^{-12}} = 5.3 \times 10^{-3} \text{ M}$$

Exercise 14.11

What is $[OH^-]$ in a solution that has a pH of 12.84?

Figure 14.4   An electronic pH meter.

Figure 14.5   Samples of pH paper.

There are several good ways to determine the pH of a solution directly. Electronic instruments rely on electrodes dipped into a solution to give immediate, accurate readings of pH levels (Figure 14.4). Paper strips impregnated with mixtures of dye will give instantaneous pH approximations when they are wet with a solution; these are improved versions of the old and honored litmus test, in which the litmus paper turns pink to indicate acidity and blue to indicate basicity (Figure 14.5).

The level of acidity is crucial in many kinds of chemical reactions, especially those of biological systems. The pH of human blood, for example, is 7.4, corresponding to a $[H_3O^+]$ of 4×10^{-8} M. Deviation from 7.4 by even one tenth of a pH unit can cause serious consequences, and deviations of more than two tenths can cause death. Plants can grow only in soils in which the pH of the moisture lies close to an optimum value. Farmers and horticulturists must therefore regulate soil pH. We will encounter pH and the regulation of acidity often as we continue our study of chemistry.

14.7 Titration

After studying this section, you should be able to:

- Perform chemical calculations based on titration data.

__Titration,__ one of the standard methods of analysis in chemistry laboratories, is used to determine the concentration of a given solution. During titration, one measures the volume of a solution that has a known concentration (the *known*) required to react exactly with a measured volume of a second solution having unknown concentration (the *unknown*). Using the measured volume, the chemical equation for the reaction that takes place during titration, and the concentration of the known solution, it is possible to calculate the unknown

(a)   (b)   (c)

Figure 14.6 Titration of an acid with a base. (a) The buret is filled with a basic solution of known concentration. (b) The basic solution is slowly added to a known volume of acidic solution. The concentration of acidic solution is unknown. (c) When the indicator dye changes color permanently, the number of added moles of OH^- equals the number of moles of H_3O^+ originally present in the flask.

concentration. In acid-base titration, one of the solutions is an acid and the other is a base. Figure 14.6 shows a titration in progress.

To titrate a solution, we first arrange a way to determine exactly when the unknown has been completely used up. A common method is to add an indicator dye to the unknown solution. The dye changes color as soon as any excess of the known solution is present (Figure 14.7). During the titration, a buret is used to add the known solution to the unknown, until the indicator changes color. The volumes of known and unknown solutions are recorded and any necessary calculations are made. An example will help you to understand titration.

Figure 14.7 Grape juice is a natural indicator that is red in acidic solution and green in basic solution.

EXAMPLE 14.11

To a sample of 125 mL of NaOH solution of unknown concentration were added 88.1 mL of 0.117 M HCl solution, which reacted exactly with all of the OH^- in the NaOH solution. What was the concentration of the NaOH solution?

Solution Map:

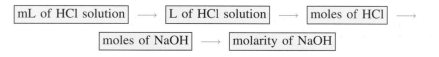

Strategy and Solution:

Step 1. Write and balance the chemical equation for the reaction.

$$HCl + NaOH \longrightarrow H_2O + NaCl$$

Step 2. Knowing the concentration and volume of the HCl solution, we can calculate the number of moles of HCl present in it.

$$88.1 \text{ mL HCl solution} \times \frac{\text{L solution}}{1,000 \text{ mL solution}} \times \frac{0.117 \text{ mol HCl}}{\text{L HCl solution}} =$$
$$1.03 \times 10^{-2} \text{ mol HCl}$$

Step 3. From the number of moles of HCl, we can find the number of moles of NaOH in the second solution. From the chemical equation, we know that one mole of HCl reacts with one mole of NaOH.

$$1.03 \times 10^{-2} \text{ mol HCl} \times \frac{1 \text{ mol NaOH}}{1 \text{ mol HCl}} = 1.03 \times 10^{-2} \text{ mol NaOH}$$

Step 4. Calculate the molarity of the NaOH solution. Divide the number of moles by the volume of the solution.

$$\frac{1.03 \times 10^{-2} \text{ mol NaOH}}{125 \text{ mL NaOH solution}} \times \frac{1,000 \text{ mL solution}}{\text{L solution}} = 0.0824 \text{ M NaOH}$$

Exercise 14.12

What is the concentration of a $Mg(OH)_2$ solution that requires 18.4 mL of a 0.45 M HNO_3 solution to titrate 10.0 mL?

Exercise 14.13

How many mL of 0.500 M $HClO_4$ are required to titrate 500.0 mL of a solution of $Ba(OH)_2$ that is 1.00 M?

Exercise 14.14

How many mL of 3.68 M $HClO_3$ are required to titrate 12.00 mL of 6.00 M KOH?

14.8 Normality

After studying this section, you should be able to:

- Find the normality of a solution, knowing the molarity.
- Perform calculations for titrations using normal solutions.

Suppose that we wish to determine the concentration of a solution of KOH. We will titrate it with an acid solution of known concentration. Which acid is best?

Any strong acid. We are interested in the reaction $H_3O^+ + OH^- \longrightarrow H_2O$, nothing more. Whether we use HCl, $HClO_4$, H_2SO_4, or some other strong acid makes no difference in the reaction; all we need to know is the concentration of H_3O^+ in the acidic solution. Whatever the acid, we will use the number of moles of H_3O^+ per liter of solution instead of using the number of moles of compound per liter.

The **normality** of a solution is defined as the number of equivalents per liter of solution, one **equivalent** being defined as *that amount of compound that will furnish, react with, or otherwise be equivalent to one mole of H_3O^+.* For example, one mole of HCl dissolves in one liter of solution to produce one mole of H_3O^+, that is, one equivalent. The normality of this HCl solution is one eq/L. The adjective that denotes normality is *normal*. This HCl solution is said to be one normal. The abbreviation for normality is N.

One mole of $Ba(OH)_2$ contains two moles of OH^-. The two moles of OH^- can react with two moles of H_3O^+; therefore there are two equivalents per mole of $Ba(OH)_2$. Let's study a few more examples.

Normality has the units eq/L soln.

EXAMPLE 14.12

How many equivalents are there in a mole of each of the following: HI, H_3PO_4, NaOH, Na_2SO_4?

Solution: For HI, the answer is easy. Each mole furnishes one mole of H_3O^+; therefore there is one equivalent per mole.

For H_3PO_4, there are three equivalents per mole because each mole of H_3PO_4 furnishes three moles of H_3O^+.

For NaOH, one equivalent per mole, since each mole of NaOH will react with one mole of H_3O^+.

For Na_2SO_4, two equivalents per mole. In an imaginary reaction, one mole of SO_4^{2-} would react with two moles of H_3O^+ to form H_2SO_4; therefore Na_2SO_4 has two equivalents per mole.

> **Exercise 14.15**
>
> How many equivalents are there in a mole of each of the following: $Al(OH)_3$, HBr, $Sr(OH)_2$, NH_3?

Since the equivalent/normality language is to be used in calculations involving reaction, we have to be careful to count only ions that actually react. For example, acetic acid has one equivalent of H_3O^+ per mole if the reaction with base is taken to completion, but, as we will soon learn, such reactions do not always go to completion. In making our calculations concerning normality, for now we will assume that reaction conditions are such that all of our substances react to completion.

EXAMPLE 14.13

How many equivalents are there in a mole of $NaHSO_4$?

Solution: This one is harder because to answer, we must first ask what the solution is to be used for. If we are to use the solution as a supply of H_3O^+, then clearly there is one equivalent per mole. But if we are to use the solution as a supply of SO_4^{2-}, then there are two equivalents per mole, since a mole of SO_4^{2-} contains two equivalents. Sometimes it is necessary to define the use of a solution before saying how many equivalents are in it.

EXAMPLE 14.14

A solution is made by dissolving 0.366 mole of H_2SO_4 in enough water to make 1.22 L of solution. What is the normality of the solution?

Solution Map: $\boxed{\text{moles of } H_2SO_4} \longrightarrow \boxed{\text{equivalents}} \longrightarrow \boxed{\text{normality}}$

Strategy and Solution:

Step 1. Find the number of equivalents of solute. Each mole of H_2SO_4 furnishes 2 equivalents.

$$0.366 \text{ mol } H_2SO_4 \times \frac{2 \text{ eq}}{\text{mol } H_2 \text{ } SO_4} = 0.732 \text{ eq}$$

Step 2. Divide the number of equivalents by the number of liters of solution.

$$\frac{0.732 \text{ eq}}{1.22 \text{ L solution}} = 0.600 \text{ eq/L or } 0.600 \text{ N}$$

Exercise 14.16

A solution is made by dissolving 0.442 mole of H_3PO_4 in enough water to make 550 mL of solution. What is the normality of the solution?

EXAMPLE 14.15

A solution is 0.750 *N* in H_3PO_4. How many moles of H_3PO_4 are in 41.2 mL of the solution?

Solution Map:

$\boxed{\text{mL of solution}} \longrightarrow \boxed{\text{L of solution}} \longrightarrow \boxed{\text{equivalents}} \longrightarrow$
$\boxed{\text{moles of } H_3PO_4}$

Solution:

$$41.2 \text{ mL solution} \times \frac{\text{L solution}}{1,000 \text{ mL solution}} \times \frac{0.750 \text{ eq}}{\text{L solution}} \times$$

$$\frac{\text{mol } H_3PO_4}{3 \text{ eq}} = 1.03 \times 10^{-2} \text{ mol } H_3PO_4$$

Exercise 14.17

A solution is 0.115 N in $Mg(OH)_2$. How many moles of $Mg(OH)_2$ are there in 125 mL of the solution?

EXAMPLE 14.16

A solution with a volume of 700.0 mL contains 35.5 grams of H_2SO_4. What is its normality?

Solution Map:

$$\boxed{\text{grams of } H_2SO_4} \longrightarrow \boxed{\text{moles of } H_2SO_4} \longrightarrow$$

$$\boxed{\text{equivalents}} \longrightarrow \boxed{\text{normality}}$$

Strategy and Solution:

Step 1. Find the number of equivalents of H_2SO_4 that were used. Use the molar mass and the number of equivalents per mole.

$$35.5 \text{ g } H_2SO_4 \times \frac{\text{mol } H_2SO_4}{98.079 \text{ g } H_2SO_4} \times \frac{2 \text{ eq}}{\text{mol } H_2SO_4} = 0.724 \text{ eq } H_2SO_4$$

Step 2. Divide the number of equivalents by the volume of solution.

$$\frac{0.724 \text{ eq}}{700.0 \text{ mL solution}} \times \frac{1{,}000 \text{ mL solution}}{\text{L solution}} = 1.03 \text{ eq/L or } 1.03 \text{ N}$$

> **Exercise 14.18**
>
> 2.5 grams of $Mg(OH)_2$ are dissolved in enough water to make 550 mL of solution. What is the normality of the solution?

With the equivalent so defined, one equivalent of a compound reacts exactly with one equivalent of another, regardless of what the mole ratios of the reaction may be. When titration is complete, *the number of equivalents of the substance being titrated is exactly equal to the number of equivalents of the titrating substance.* Let's look at an example.

EXAMPLE 14.17

A solution of H_2SO_4 having a volume of 38.1 mL is titrated with 22.4 mL of 2.55 N base solution. What is the normality of the H_2SO_4 solution?

Solution Map:

$$\boxed{\text{mL of base solution}} \longrightarrow \boxed{\text{L of base solution}} \longrightarrow$$

$$\boxed{\text{equivalents}} \longrightarrow \boxed{\text{normality of acid solution}}$$

Strategy and Solution:

Step 1. Find the number of equivalents.

$$22.4 \text{ mL solution} \times \frac{\text{L solution}}{1{,}000 \text{ mL solution}} \times \frac{2.55 \text{ eq}}{\text{L solution}} = 0.0571 \text{ eq}$$

Step 2.   Calculate the normality of the acid.

$$\frac{0.0571 \text{ eq}}{38.1 \text{ mL solution}} \times \frac{1,000 \text{ mL solution}}{\text{L solution}} = 1.50 \text{ N}$$

Exercise 14.19

25.0 mL of NaOH solution are titrated with 46.9 mL of 1.66 N H_2SO_4 solution. What was the normality of the NaOH solution?

Exercise 14.20

What is the normality of an acidic solution that required 12.8 mL of 3.7 N base solution to titrate 10.0 mL of it?

14.9 Reactions That Produce Acids and Bases

After studying this section, you should be able to:

• Predict whether an oxide will produce an acidic or basic solution.

Water reacts with active metals, such as those of the alkali family, to form hydroxides and liberate H_2. The reaction with K is an example (Figure 14.8).

$$2H_2O + 2K \longrightarrow 2KOH + H_2$$

The hydroxides of the alkali metals are highly soluble and totally dissociated in water. Other metals react to give only partially soluble hydroxides. Magnesium, for example, reacts with steam to produce magnesium hydroxide and hydrogen gas.

$$Mg + 2H_2O \longrightarrow Mg(OH)_2 + H_2$$

The oxides of metals and of nonmetals react with water in characteristically different ways. The oxides of metals usually produce basic solutions and are therefore known as basic oxides.

$$\underset{\text{sodium oxide}}{Na_2O} + H_2O \longrightarrow 2NaOH$$

$$\underset{\text{barium oxide}}{BaO} + H_2O \longrightarrow Ba(OH)_2$$

Figure 14.8   Potassium, K, reacts violently with water to produce an aqueous solution of KOH and hydrogen gas.

Both NaOH and $Ba(OH)_2$ are strong electrolytes. NaCH is quite soluble; $Ba(OH)_2$ is slightly soluble.

Although basicity is a typical property of metal oxides, there are certain circumstances in which some metals form acidic solutions. A few metals, such as aluminum, have oxides that can be either acidic or basic depending on the conditions in the solution.

Increasing acidic character ⟶

	I	II	III	IV	V	VI	VII
	Li_2O	BeO	B_2O_3	CO_2	N_2O_5	(O_2)	OF_2
	Na_2O	MgO	Al_2O_3	SiO_2	P_4O_{10}	SO_3	Cl_2O_7
	K_2O	CaO	Ga_2O_3	GeO_2	As_2O_5	SeO_3	Br_2O_7
	Rb_2O	SrO	In_2O_3	SnO_2	Sb_2O_5	TeO_3	I_2O_7
	Cs_2O	BaO	Tl_2O_3	PbO_2	Bi_2O_5	PoO_3	At_2O_7

Increasing basic character

Figure 14.9 The acidity of main-group oxides. In general, acidity increases from left to right and bottom to top in the periodic table.

The oxides of most nonmetals react with water to form acidic solutions (Figure 14.9). Sulfur trioxide, for example, reacts with water to produce the strong acid H_2SO_4.

$$SO_3 + H_2O \longrightarrow H_2SO_4$$

Not all nonmetal oxides form strong acids, however. The oxide of carbon, CO_2, forms H_2CO_3, a weak acid.

$$CO_2(g) + H_2O \longrightarrow H_2CO_3$$

14.10 Industrial Chemicals: H₂SO₄, NaOH, NaClO, and Na₂CO₃

Our study so far has focused on the principles of chemistry, although we have had some exposure to practical chemistry along the way. There is a third aspect of chemistry: industrial chemistry. The present human population (about 4.5 billion) can survive only if the necessary food and other life-supporting materials continue to be produced at a high rate. We no longer have the choice of going back to the farming society of the nineteenth century. Like it or not, most of the world is industrialized and must remain so unless the population is drastically reduced. Industrial societies depend on a continuous supply of enormous amounts of fertilizers, fuels, medicines, fabrics, and innumerable other materials. The production of these depends, in turn, on the supply of what are called *heavy chemicals* (heavy in terms of the total amount produced, not in density).

Sulfuric acid is a colorless, viscous, extremely corrosive liquid with a density of $1.8 \ g/cm^3$. The total mass of H_2SO_4 produced in the United States every year is about twice the mass of all the men, women, and children in the United States, but almost none of the acid reaches the consumer market because it is used in manufacturing processes.

Sulfuric acid is used primarily as a cheap, concentrated source of hydronium ions. Somewhat more than a third of the available H_2SO_4 is used to make fertilizers. An example is the preparation of the moderately soluble calcium dihydrogen phosphate, $Ca(H_2PO_4)_2$, from the natural mineral, $Ca_3(PO_4)_2$. Because it is nearly insoluble, $Ca_3(PO_4)_2$ is virtually useless to plants. $Ca(H_2PO_4)_2$, however, can be dissolved and thus absorbed by plants.

About 40 million tons of sulfuric acid are produced each year in the United States.

H_2SO_4 is used in the manufacture of other chemicals, as well as in the production of steel, petroleum, nonferrous metals, paint, plastics, and numberless other products.

Sulfur blocks stacked for shipment. The stack is so large that steps have been cut into it.

Approximately 22 billion pounds of NaOH are produced in the United States each year. That's about 88 pounds for each person in the United States.

Figure 14.10 In oxygen, sulfur burns brightly, producing sulfur dioxide, SO_2.

$$Ca_3(PO_4)_2 + 2\ H_2SO_4 \longrightarrow Ca(H_2PO_4)_2 + 2\ CaSO_4$$

Except for one stage, the manufacture of H_2SO_4 is straightforward. Sulfur is burned (or metal sulfides are roasted in air) to form sulfur dioxide—a reaction that occurs readily (Figure 14.10).

$$S(s) + O_2(g) \longrightarrow SO_2(g) \qquad \Delta H = -296\ kJ$$

The next step is more difficult because the reaction does not occur readily.

$$2\ SO_2(g) + O_2(g) \longrightarrow 2\ SO_3(g) \qquad \Delta H = -198\ kJ$$

A carefully selected set of conditions is necessary to make the process economically practical. These conditions are explained in Chapter 15. The resulting SO_3 is reacted with water to give H_2SO_4.

$$SO_3 + H_2O \longrightarrow H_2SO_4 \qquad \Delta H = -130\ kJ$$

Sodium hydroxide, NaOH, sometimes called lye, is the most common soluble strong base. It is used in large quantities to produce petroleum, soap, and paper. Because it is a strong, soluble base, NaOH finds uses where cheaper bases such as $Ca(OH)_2$ are unsatisfactory. In one application, NaOH is frequently used to react with weak acids. For example, in the natural gas and petroleum industries, NaOH is used to remove undesirable compounds of sulfur, such as H_2S, which would oxidize during burning to produce atmospheric acids.

$$2\ NaOH + H_2S \longrightarrow Na_2S + 2\ H_2O$$

NaOH will also remove sulfur dioxide.

$$2\ NaOH + SO_2 \longrightarrow Na_2SO_3 + H_2O$$

Because NaOH is made from NaCl (the process is explained in Chapter 16), the byproduct, Cl_2, is also produced in massive amounts. Much of this chlorine is used to make sodium hypochlorite, NaClO, which is the active material in household bleach, such as Clorox.

NaClO is used industrially to bleach paper and fibers, such as cotton, for making fabrics. NaClO can be made directly by electrically decomposing NaCl in aqueous solution, but most is prepared by reacting NaOH with Cl_2.

$$2\ NaOH + Cl_2 \longrightarrow NaCl + NaClO + H_2O$$

More recently, large tonnages of Cl_2 are being used to make starting materials for plastics. More than half of polyvinylchloride (PVC) is chlorine. If you are holding a plastic pen, the chances are that it is made of PVC.

Another heavy chemical is Na_2CO_3, sodium carbonate, often called soda ash. Sodium carbonate is used in the manufacture of "soda glass," the kind of glass that is used in jars, bottles, and many other kinds of inexpensive glass products (except window glass, which is usually made using K_2CO_3). Much sodium carbonate goes into the manufacture of sodium bicarbonate, $NaHCO_3$, household "baking soda." Baking soda is used in bakeries as well as in the kitchen for reacting with acids, such as lactic acid in milk, to make CO_2, the gas that makes bread rise.

Summary

Acids and bases play an essential role in chemistry, in industry, and in biological systems. As a rough and convenient definition, an aqueous acid solution is one that contains more H_3O^+ than OH^-. A basic solution contains more OH^- than H_3O^+. If the concentrations of H_3O^+ and OH^- are equal, as they are in pure water, the solution is neither acidic nor basic, but neutral. A more fundamental definition of acid-base behavior involves proton transfer directly, without the presence, necessarily, of H_3O^+ or OH^-. In proton transfer reactions, every acid has its conjugate base, the substance formed when it loses its proton. Similarly, every base becomes its conjugate acid when it receives a proton.

Not all acids and bases are strong electrolytes. Many dissociate only to a limited degree in aqueous solution, yielding solutions that contain significant amounts of the undissociated substance. Such acids and bases are said to be *weak*.

Acids and bases readily react with one another, creating water and salts. If the number of moles of H_3O^+ equals the number of moles of OH^-, the reaction is called a neutralization.

The pH scale is commonly used to assign values to the acidity or basicity of a solution. The pH is the negative logarithm of the H_3O^+ molarity.

The concentrations of solutions of acids, bases, and other compounds can be measured by titration. In this laboratory technique, a solution of known concentration is reacted with another of unknown concentration until the unknown is entirely consumed. A calculation involving the volumes of the known and unknown solutions and the concentration of the known yields the concentration of the unknown. Primarily for convenience in making titration calculations, concentrations are often measured in equivalents per liter of solution, where one equivalent of a substance is that amount that produces or reacts with one mole of H_3O^+. The number of equivalents in a liter of solution is defined as the normality of that solution.

Key Terms

acid	*(p. 410)*	neutralization	*(p. 414)*
hydronium ion	*(p. 410)*	neutral	*(p. 414)*
base	*(p. 410)*	salt	*(p. 415)*
conjugate acid-base pair	*(p. 411)*	acidic	*(p. 416)*
amphoteric	*(p. 411)*	basic	*(p. 416)*
strong acid	*(p. 412)*	pH	*(p. 417)*
weak acid	*(p. 412)*	titration	*(p. 422)*
strong base	*(p. 413)*	normality	*(p. 425)*
weak base	*(p. 413)*	equivalent	*(p. 425)*

Questions and Problems

Section 14.1 Definitions of Acid and Base

Questions

1. The hydrogen ion, H^+, does not exist in aqueous solution. Explain why.

2. Acid-base reactions are often called proton-transfer reactions. Why?

3. The polyatomic ion HPO_4^{2-} is amphoteric. Explain.

Problems

4. Label the acids and bases in the following chemical equations.
   (a) $HNO_3 + H_2O \longrightarrow H_3O^+ + NO_3^-$
   (b) $N_2H_4 + H_2O \longrightarrow N_2H_3^- + H_3O^+$
   (c) $N_2H_4 + HCl \longrightarrow N_2H_5^+ + Cl^-$
   (d) $HCN + H_3O^+ \longrightarrow H_2CN^+ + H_2O$

5. Label the acids and bases in the following chemical equations.
   (a) $H_2PO_4^- + OH^- \longrightarrow HPO_4^{2-} + H_2O$
   (b) $H_2PO_4^- + H_2O \longrightarrow H_3PO_4 + OH^-$
   (c) $NH_4^+ + OH^- \longrightarrow NH_3 + H_2O$
   (d) $HSO_4^- + H_2O \longrightarrow H_2SO_4 + OH^-$

Section 14.2 Strong and Weak Acids and Bases

Questions

6. Distinguish between the terms "strong acid" and "weak acid."

7. Explain how a concentrated solution can be weakly acidic.

8. Explain how a solution of a strong base can be only slightly basic.

9. Why are weak acids and bases only partially ionized in solution?

10. Label each of the following compounds as a weak or strong acid, weak or strong base, or none of these: acetic acid, $Mg(OH)_2$, KOH, HF, NaCl, H_2SO_4, HSO_4^-, HI, H_2O, HPO_4^{2-}.

Section 14.3 Neutralization and the Formation of Salts

Questions

11. Why must the number of moles of H_3O^+ and of OH^- be equal in a neutralization?

12. What are the ionic substances called that result from the reactions of acids and bases? Give two examples of such substances.

Problems

13. Complete and balance the neutralization equations for the following combinations of acids and bases.
    (a) $KOH + H_3PO_4 \longrightarrow$
    (b) $HC_2H_3O_2 + NaOH \longrightarrow$
    (c) $Ca(OH)_2 + HCl \longrightarrow$
    (d) $H_2SO_4 + KOH \longrightarrow$

14. Complete and balance the neutralization equations for the following combinations of acids and bases. In which cases, if any, are there no acid-base reactions?

(a) $HNO_3 + Mg(OH)_2 \longrightarrow$
(b) $Ba(OH)_2 + HC_2H_3O_2 \longrightarrow$
(c) $RbI + Sr(OH)_2 \longrightarrow$
(d) $CO_2 + H_2O + LiOH \longrightarrow$

Section 14.4 Ionization of Water

Questions

15. Explain the process of ionization of water. In pure water, what fraction of the molecules is ionized?

16. How can pure water be neutral if H_3O^+ and OH^- are present in every sample?

Section 14.5 Aqueous Solutions of Acids and Bases

Question

17. How can a solution of HCl be acidic if there are OH^- ions present?

Problems

18. What is $[OH^-]$ when $[H_3O^+]$ is 3.7×10^{-5} M? When it is 7.4×10^{-7} M?

19. What is $[H_3O^+]$ when $[OH^-]$ is 5.5×10^{-3} M? When it is 7.5×10^{-9} M?

20. What is $[OH^-]$ in a 0.0082 M HNO_3 solution?

21. What is the molarity of OH^- in a 0.10 M solution of $HClO_4$?

22. What is $[H_3O^+]$ in a 0.0073 M NaOH solution?

23. What is $[H_3O^+]$ in a 0.00060 M solution of $Ba(OH)_2$?

Section 14.6 The pH Scale

Questions

24. Explain the relationship between pH and the molarity of H_3O^+. Of OH^-.

25. What is the pH of pure water?

26. Is it possible to have a negative pH? Explain.

Problems

27. Find the pH of each of the following.
    (a) $[H_3O^+] = 8.4 \times 10^{-2}$ M
    (b) $[OH^-] = 1 \times 10^{-4}$ M
    (c) $[OH^-] = 9.5 \times 10^{-6}$ M

28. What is the pH of each of the following solutions?
    (a) $[H_3O^+] = 6.7 \times 10^{-2}$ M
    (b) $[OH^-] = 2.4 \times 10^{-11}$ M
    (c) $[OH^-] = 7.7 \times 10^{-5}$ M

29. Which solutions in Problem 27 are acidic?

30. Which solutions in Problem 28 are basic?

31. Calculate the pH of a 0.0045 M solution of HNO_3.

32. What is the pH of a 0.0035 M solution of $HClO_4$?

33. Find the pH of a 0.0029 M KOH solution.

34. What is the pH of a 0.00044 M solution of $Ba(OH)_2$?

35. Find $[H_3O^+]$ of solutions with the following pH values: 3.82, 12.9, 10.5, 4.7, 7.2.

36. Calculate $[H_3O^+]$ of solutions with the following pH values: 3.75, 9.28, 7.40, 4.81.

37. What is $[OH^-]$ of solutions with the following pH values: 12.7, 3.4, 5.2, 13.5, 9.8?

38. Find $[OH^-]$ of solutions with the following pH values: 1.48, 8.3, 2.8, 8.3.

39. A 15-gram sample of Ba metal was reacted with enough water to form 2.00 L of solution.

$$Ba + 2\,H_2O \longrightarrow Ba(OH)_2 + H_2$$

Calculate the final pH of the solution.

40. How many grams of Li_2O are necessary to make 250 mL of a solution that has a pH of 9.40?

$$Li_2O + H_2O \longrightarrow 2\,LiOH$$

41. What volume of concentrated (16 M) nitric acid, HNO_3, must be used to prepare 500.0 mL of a solution in which $[H_3O^+] = 2.25 \times 10^{-2}$ M?

Section 14.7 Titration

Questions

42. Explain the technique of titration.

43. Why is an indicator required in titration? What does it do?

Problems

44. What is the molarity of a solution of HNO_3 if 53.2 mL were required to titrate 100.0 mL of 2.05 M KOH?

45. A 10.0-mL sample of a NaOH solution that was 1.00 M in OH^- was used to prepare 500.0 mL of a solution. The solution was then titrated with 23.6 mL of an HCl solution of unknown concentration. What was the concentration of the HCl solution?

46. In a titration, 15.4 mL of a solution with a pH of 3.50 were used to titrate 10.0 mL of a KOH solution. Calculate the concentration of the KOH solution.

47. Because it is easily purified and handled and reacts readily, potassium acid phthalate, $KHC_8H_4O_4$, is used to find the concentrations of solutions of bases. In one experiment, a sample of $KHC_8H_4O_4$ weighing 3.176 grams reacted exactly with 43.25 mL of a solution of Na_2CO_3. What was the molarity of CO_3^{2-} in the solution? The molecular weight of $KHC_8H_4O_4$ is 204.22 g/mol.

$$2\,KHC_8H_4O_4 + Na_2CO_3 \longrightarrow 2\,NaKC_8H_4O_4 + H_2O + CO_2$$

48. A 25.00-mL sample of 0.442 M H_3PO_4 solution was titrated with 0.113 M $Ba(OH)_2$. What reaction occurred? What volume of the $Ba(OH)_2$ solution was needed?

Section 14.8 Normality

Questions

49. What is the advantage of using normality in titration calculations?

50. Define the term "equivalent." Is it possible to determine the number of equivalents in one mole of Na_2SO_4? Explain.

51. How many equivalents are there in one mole of each of these compounds: HNO_3, $HClO_4$, $Ca(OH)_2$, K_3PO_4?

52. How many equivalents are there in one mole of each of the following compounds: $H_2Cr_2O_7$, NH_4Cl, $Ba(OH)_2$, $NaClO_3$?

Problems

53. Calculate the normality of each of the following solutions.
   (a) 2.50 M HCl
   (b) 0.19 M KOH
   (c) 5.0 M H_3PO_4
   (d) 0.042 M $Ba(OH)_2$

54. What is the normality of each of the following solutions?
   (a) 0.50 M HNO_3
   (b) 3.5 M H_2SO_4
   (c) 0.024 M $Ca(OH)_2$
   (d) 0.0048 M $K_2Cr_2O_7$

55. What is the normality of a solution prepared by dissolving 5.8 moles of HCl in enough water to make 350.0 mL of solution?

56. Calculate the normality of a solution that contains 23.8 grams of KOH in 500.0 mL of solution.

57. How many moles of $Ba(OH)_2$ are there in 2.0 L of a 0.0028 N solution?

58. How many moles of H_2SO_4 are contained in 250.0 mL of 3.2 N H_2SO_4 solution?

59. Calculate the number of grams of NaOH in 2.0 L of 0.071 N NaOH.

60. What volume of 0.55 N $HC_2H_3O_2$ is required to titrate 125 mL of 0.025 N base solution?

61. Find the normality of an acidic solution that requires 14.72 mL of 4.5 N KOH to titrate 15.0 mL of the acid.

62. A base of unknown normality is used to titrate 15.0 mL of 2.5 N H_2SO_4. The volume of the base is 27.2 mL. What is the normality of the unknown?

63. 56.1 mL of an HNO_3 solution neutralizes 4.8 grams of $Ba(OH)_2$. What is the normality of the HNO_3 solution?

Section 14.9 Reactions That Produce Acids and Bases

Questions

64. What kind of solutions do most metal oxides form, acidic or basic? Write a chemical equation for the reaction of a metal oxide with water.

65. What kind of solutions do most nonmetal oxides form, acidic or basic? Write a chemical equation for the reaction of a nonmetal oxide with water.

Solutions to Exercises

14.1

$$HNO_2 + OH^- \longrightarrow NO_2^- + H_2O$$
$$\text{acid} \quad \text{base} \qquad \text{base} \quad \text{acid}$$
$$H_3O^+ + NH_3 \longrightarrow NH_4^+ + H_2O$$
$$\text{acid} \quad \text{base} \qquad \text{acid} \quad \text{base}$$

14.2

$$H_2SO_4 + Mg(OH)_2 \longrightarrow MgSO_4(aq) + 2\, H_2O$$

14.3

$$[OH^-] = \frac{1.00 \times 10^{-14}}{0.0500} = 2.00 \times 10^{-13}\ M$$

14.4

$$[H_3O^+] = \frac{1.00 \times 10^{-14}}{0.00003} = 3 \times 10^{-10}\ M$$

14.5

$$pH = -\log 0.00032 = 3.49$$

14.6

$$pH = -\log (1.59 \times 10^{-4}) = 3.799$$

14.7

$$[H_3O^+] = \frac{1.00 \times 10^{-14}}{3.0 \times 10^{-3}} = 3.3 \times 10^{-12}$$

$$pH = -\log 3.3 \times 10^{-12} = 11.48$$

14.8

$$[H_3O^+] = \frac{1.00 \times 10^{-14}}{6.3 \times 10^{-5}} = 1.6 \times 10^{-10}\ M$$

$$pH = -\log (1.6 \times 10^{-10}) = 9.80$$

14.9

$$[H_3O^+] = 10^{-4.15} = 7.1 \times 10^{-5}\ M$$

14.10

$$[H_3O^+] = 10^{-9.2} = 6 \times 10^{-10}\ M$$

14.11

$$[H_3O^+] = 10^{-12.84} = 1.4 \times 10^{-13}\ M$$

$$[OH^-] = \frac{1.00 \times 10^{-14}}{1.4 \times 10^{-13}} = 7.1 \times 10^{-2}\ M$$

14.12

$$2\, HNO_3 + Mg(OH)_2 \longrightarrow 2\, H_2O + MgCl_2$$

$$18.4\ \text{mL solution} \times \frac{\text{L solution}}{1{,}000\ \text{mL solution}} \times \frac{0.45\ \text{mol } HNO_3}{\text{L solution}} \times$$

$$\frac{1\ \text{mol } Mg(OH)_2}{2\ \text{mol } HNO_3} = 4.1 \times 10^{-3}\ \text{mol } Mg(OH)_2$$

$$\frac{4.1 \times 10^{-3} \text{ mol Mg(OH)}_2}{10.0 \text{ mL solution}} \times \frac{1,000 \text{ mL solution}}{\text{L solution}} = 0.41 \text{ M Mg(OH)}_2$$

14.13
$$2 \text{ HClO}_4 + \text{Ba(OH)}_2 \longrightarrow \text{Ba(ClO}_4)_2 + 2 \text{ H}_2\text{O}$$

$$500.0 \text{ mL solution} \times \frac{\text{L solution}}{1,000 \text{ mL solution}} \times \frac{1.00 \text{ mol Ba(OH)}_2}{\text{L solution}} \times$$

$$\frac{2 \text{ mol HClO}_4}{1 \text{ mol Ba(OH)}_2} = 1.00 \text{ mol HClO}_4$$

$$1.00 \text{ mol HClO}_4 \times \frac{\text{L solution}}{0.500 \text{ mol HClO}_4} \times \frac{1,000 \text{ mL solution}}{\text{L solution}} =$$

$$2.00 \times 10^3 \text{ mL solution}$$

14.14
$$\text{HClO}_3 + \text{KOH} \longrightarrow \text{KClO}_3 + \text{H}_2\text{O}$$

$$12.00 \text{ mL solution} \times \frac{\text{L solution}}{1,000 \text{ mL solution}} \times \frac{6.00 \text{ mol KOH}}{\text{L solution}} \times$$

$$\frac{1 \text{ mol HClO}_3}{1 \text{ mol KOH}} = 0.0720 \text{ mol HClO}_3$$

$$0.0720 \text{ mol HClO}_3 \times \frac{\text{L solution}}{3.68 \text{ mol HClO}_3} \times \frac{1,000 \text{ mL solution}}{\text{L solution}} = 19.6 \text{ mL solution}$$

14.15 Three, one, two, one

14.16
$$\frac{0.442 \text{ mol H}_3\text{PO}_4}{550 \text{ mL solution}} \times \frac{3 \text{ eq}}{\text{mol H}_3\text{PO}_4} \times \frac{1,000 \text{ mL solution}}{\text{L}} = 2.41 \text{ N}$$

14.17
$$125 \text{ mL solution} \times \frac{\text{L solution}}{1,000 \text{ mL solution}} \times$$

$$\frac{0.115 \text{ eq}}{\text{L solution}} \times \frac{\text{mol Mg(OH)}_2}{2 \text{ eq}} = 7.19 \times 10^{-3} \text{ mol Mg(OH)}_2$$

14.18
$$2.5 \text{ g Mg(OH)}_2 \times \frac{\text{mol Mg(OH)}_2}{58.3196 \text{ g Mg(OH)}_2} \times \frac{2 \text{ eq}}{\text{mol Mg(OH)}_2} = 0.086 \text{ eq}$$

$$\frac{0.086 \text{ eq}}{550 \text{ mL solution}} \times \frac{1,000 \text{ mL solution}}{\text{L solution}} = 0.16 \text{ N}$$

14.19
$$46.9 \text{ mL solution} \times \frac{\text{L solution}}{1,000 \text{ mL solution}} \times \frac{1.66 \text{ eq}}{\text{L solution}} = 0.0779 \text{ eq}$$

$$\frac{0.0779 \text{ eq}}{25.0 \text{ mL solution}} \times \frac{1,000 \text{ mL solution}}{\text{L solution}} = 3.12 \text{ N}$$

14.20
$$12.8 \text{ mL solution} \times \frac{\text{L solution}}{1,000 \text{ mL solution}} \times \frac{3.7 \text{ eq}}{\text{L solution}} = 0.047 \text{ eq}$$

$$\frac{0.047 \text{ eq}}{10.0 \text{ mL solution}} \times \frac{1,000 \text{ mL solution}}{\text{L solution}} = 4.7 \text{ N}$$

In large cities that enjoy abundant sunshine and have lots of traffic, nitrogen dioxide, NO₂, makes the sky reddish brown.

CHAPTER 15

Rates of Chemical Reactions and Chemical Equilibrium

I n previous chapters, we learned how to write chemical equations that describe the results of reactions. Given the atoms of two different elements, we can now predict what kind of chemical bond those atoms will form. With that information, we can make some reasonable guesses about the properties of a product. If a reaction goes to completion, we know how to calculate the amount of product formed.

There is an important *if* in the first paragraph: "*if* a reaction goes to completion." But many reactions do not go to completion. Some substances will not react at all. Other systems react so slowly that lifetimes pass before any product can be detected. We have learned about the reactions of individual atoms of the elements, but most atoms on earth are no longer individual; they are already combined. To understand a chemical reaction as it usually happens,

437

to be able to predict the yield, we must study the factors that determine the rate of reaction. We need to apply to chemical reaction the concept of dynamic equilibrium, which we first met in the context of vapor pressure. Our familiarity with the general concept of dynamic equilibrium will enable us to understand the characteristics of chemical equilibrium.

Chemical equilibrium is an important controlling factor in most reactions. Equilibrium conditions control not only the chemical and physical processes of all living organisms, but also the conditions that make life possible on earth. Indeed, they regulate much of the operation of the entire universe.

We begin the chapter by examining chemical equilibrium as the balance between the reaction rates of opposing chemical reactions. We explore the factors that control reaction rates and how those factors affect chemical equilibrium. We learn how to calculate the relative amounts of reactants and products in a chemical equilibrium, and we finally apply these techniques to real reactions.

15.1 Reversible Chemical Reactions

After studying this section, you should be able to:

- Write chemical equations for reversible reactions.

Let's consider the reaction of iodine and hydrogen to form hydrogen iodide. If we place some I_2 and some H_2 together in a container at 425°C, they will react to form HI.

$$H_2(g) + I_2(g) \longrightarrow 2\ HI(g)$$

On the other hand, if we put some HI in a container and heat it to 425°C, it will start to decompose and form H_2 and I_2—the reverse of the reaction of H_2 and I_2.

$$2\ HI(g) \longrightarrow H_2(g) + I_2(g)$$

Chemical equations in which both the forward and the reverse reactions are significant are often written with twin arrows.

$$\underset{\text{reactants}}{H_2(g) + I_2(g)} \underset{\text{reverse reaction}}{\overset{\text{forward reaction}}{\rightleftharpoons}} \underset{\text{products}}{2\ HI(g)}$$

Many chemical reactions are reversible in practice, and all reactions are in theory. The terms "reactant" and "product" are merely convenient names for the substances written on the left and right sides of the chemical equation. They are not always descriptive of how the reaction actually occurs.

15.2 Chemical Equilibrium

After studying this section, you should be able to:

- Explain the concept of chemical equilibrium.

Some chemical reactions run and finish in a fraction of a second; others may take centuries. The **reaction rate** is simply a measure of how fast a substance reacts. Reaction rates are often measured by the number of moles of substance that appear or disappear in some unit of time, say, a minute.

Starting again with H_2 and I_2 in a heated vessel, we watch the reaction run by following the appearance of HI.

$$H_2(g) + I_2(g) \rightleftharpoons 2 \, HI(g)$$

As the concentrations of H_2 and I_2 get lower and lower, the rate of reaction gets slower and slower. After a while, we notice that no more HI appears, although there is still quite a lot of unreacted H_2 and I_2. Why has the reaction stopped producing more HI? As the forward reaction of H_2 and I_2 proceeds, HI is formed. But HI reacts with itself in the reverse reaction to form H_2 and I_2. As the concentration of HI increases, the rate of this reverse reaction increases. At the same time, because H_2 and I_2 are being used up, the rate of the forward reaction slows. Eventually, the rate at which H_2 and I_2 react to form HI equals the rate at which the HI breaks down to form H_2 and I_2 (Figure 15.1). When the rate of the forward reaction equals the rate of the reverse reaction, we say that the system has reached **chemical equilibrium**. At equilibrium, the forward and reverse reactions continue to run, but the *amounts* of reactants and products remain constant.

We have seen systems at equilibrium before. The concept is not new to us. Vapor pressure above a liquid in a closed container reaches equilibrium when the rate of molecules leaving the liquid surface is the same as the rate of molecules rejoining the surface. Now we see exactly the same effect, only the competing processes are reversible chemical reactions. Let's look at some of the details of chemical equilibrium.

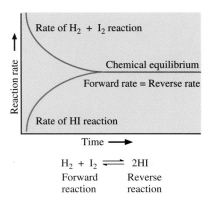

Figure 15.1   The rate of the reaction to form HI (blue line) becomes smaller and smaller as the reactants, H_2 and I_2, are used up. Meanwhile, the rate of the reverse reaction, forming H_2 and I_2 from HI (red line), increases as more HI is made. Eventually forward and reverse reaction rates become the same, and equilibrium is reached.

15.3 Collision Theory; Activation Energy

After studying this section, you should be able to:

- Predict relative rates of chemical reactions from relative activation energies.
- Name two major factors that determine whether a molecular collision will result in reaction.

To understand chemical equilibrium, we must first study the reaction rates that control equilibrium. To understand reaction rates, we need to know something about the way individual particles react.

Before molecules can react, they must come into contact, that is, collide. Consider a system of two gases mixed together in the same container. Chemists have determined that in a gas under ordinary conditions, each molecule collides with another about 10^{10} times every second. If reaction occurred in only one collision in every 1,000, reactions between gases would still be completed within millionths of a second. Some gas reactions, however, take years because collisions that produce reaction almost never take place. Reaction does not usually occur instantly because only a small proportion, often an extremely small proportion, of molecular collisions actually results in reaction. Even after decades of research, the details of the way in which molecules must come together to react is not fully understood.

Generally, several factors determine whether a molecular collision results in reaction. One essential factor is the energy with which the collision occurs. In

Only collisions having energy equal to or greater than the activation energy result in reaction.

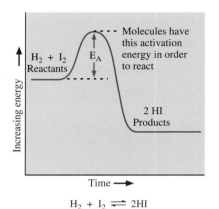

$$H_2 + I_2 \rightleftharpoons 2HI$$

Figure 15.2 Activation energy diagram of an exothermic reaction. The energy of the reactants is greater than the energy of the products. E_A is the activation energy of the forward reaction.

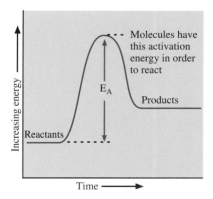

Figure 15.3 Activation energy diagram of an endothermic reaction. The energy of the products is greater than the energy of the reactants. The activation energy of the forward reaction, E_A, is greater than the activation energy of the reverse reaction.

many cases, as in our example of H_2 and I_2, collisions between molecules must be highly energetic if reaction is to occur. The existing bonds must be broken and new bonds made. The collision must squeeze the atoms in the molecules closely together so that the electrons mingle, so closely that it matters little which pairs of atoms were originally bonded. Because electrons in atoms repel each other, considerable energy is required to get the atoms close enough for reaction. The energy that colliding particles must have in order to react is called the **activation energy.** The chief reason that most collisions do not result in reaction is that the colliding particles do not possess the required activation energy.

Schematically, we can think of activation energy as a barrier over which reactants must pass in order to become products. Even if the overall reaction releases energy, the molecules must first gain the activation energy before reaction can take place. Figure 15.2 will help you visualize activation energy. All other things being equal, a higher activation energy means a slower reaction; a lower activation energy means a faster reaction.

We can learn something else of value from Figure 15.2. Notice that after H_2 and I_2 molecules have climbed the activation energy barrier and reacted, the resulting HI molecules have less energy than do those of H_2 and I_2; that is, energy is released, the reaction is *exothermic,* and ΔH is negative. Because the product has less energy than the reactants, the reverse reaction has a higher activation energy than does the forward reaction.

On the other hand, in the reaction shown in Figure 15.3, the energy of the *products* is higher than that of the reactants. The activation energy of the forward reaction is, therefore, greater than the activation energy of the reverse reaction. In this case, energy is absorbed, and ΔH is positive. The reaction is

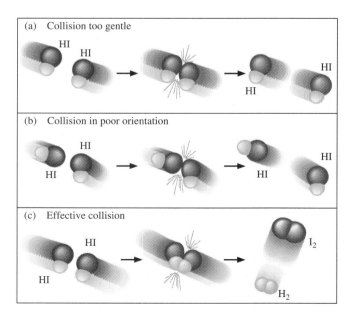

Figure 15.4 Three collisions of HI molecules, only one of which results in reaction to form H_2 and I_2. (a) Collision does not have enough energy, and molecules bounce apart. (b) Molecules are not oriented so that they can react. (c) Effective collision.

endothermic. Consider these relations, and study Figures 15.2 and 15.3 until you understand them both.

Another major factor affecting reaction rates is the arrangement of molecules during collision. So that the bonds in the product can form, the reactant molecules must collide in such a way that the atoms making up the product are close together during collision. Most collisions don't meet the exact requirements for reaction; therefore only a small proportion of collisions, even those with adequate energy, can result in reaction. Figure 15.4 illustrates three collisions, only one of which leads to reaction.

Two important factors that determine whether a collision will result in reaction are activation energy and the way the atoms are arranged during the collision.

15.4 **Effects of Concentration Changes**

After studying this section, you should be able to:

- Explain the effects of concentration changes on reaction rates.
- Predict the effects of concentration changes on chemical equilibrium.

The concentrations of reacting substances have an important effect on rates of reactions and therefore on chemical equilibrium. To understand the relation between concentration and rate, let's examine a simple imaginary reaction

$$A + B \longrightarrow C + D$$

For A and B to react, they must at some point collide. Suppose you had a one-liter vessel that contained 10 molecules of A and 500 molecules of B. The reaction forming AB would not be fast, because the B molecules would have difficulty finding A molecules with which to collide and react. If there were 100 A molecules, however, the likelihood of B molecules colliding with A molecules would increase, on the average, ten times. The reaction rate would similarly increase. Instead, if we were to increase the number of B molecules from 500 to 5,000, the rate of reaction would similarly increase ten times. The rate of the reaction is dependent on the number of both A molecules and B molecules: the more molecules, of either kind, the faster the rate of reaction.

An increase in the concentration of a reactant will increase the reaction rate.

As an illustration, we have taken a simple, one-collision system. Only a few actual reactions operate in a completely simple way. Many reactions take place in a series of steps, require a surface on which to occur, or have other complexities. For the sake of simplicity, and to allow us easily to approach the concepts of chemical equilibrium, we will confine our investigation of reaction rate to simple systems. H_2 and I_2 react in such a way.

In general, the rate of any simple, single-collision reaction is proportional to the product of the concentrations of the reactants.

Now let's consider what happens to a chemical equilibrium when concentrations of molecules change. Going back to our HI reaction, suppose we add some HI to the reaction system at equilibrium.

$$H_2(g) + I_2(g) \rightleftharpoons 2\ HI(g)$$

Because the concentration of HI suddenly becomes higher, the rate of the reverse reaction, the one that uses up HI, will temporarily be faster than that of the forward reaction. The system will no longer be at equilibrium.

As the reverse reaction continues running faster than the forward reaction, some HI will disappear and more H_2 and I_2 will be produced. The reaction of HI will begin to slow down and that of H_2 and I_2 to speed up. Eventually, the rates of the forward and reverse reactions will again become the same. A new

If you change a concentration, you will shift the equilibrium.

equilibrium is established. By adding HI, we have caused more H_2 and I_2 to be created and some (but not all) of the additional HI to be used up. The response of the reaction has been to counteract our change. We say that *the equilibrium has shifted to the left.*

More practical is the question of what happens to the HI equilibrium system if we *remove* some HI. The rate of the forward reaction does not at first change because the concentrations of H_2 and I_2 have not yet changed. The rate of the *reverse* reaction, however, is now less because the concentration of HI is lower. The system is no longer at equilibrium. As the forward reaction continues running, more H_2 and I_2 is used up to form HI, and the forward reaction slows down. When the rates of the forward and reverse reactions are once again equal, equilibrium is restored. The system has responded in such a way as to reduce the effect of the withdrawal of HI. *The equilibrium has shifted to the right.* If we were to continue to remove HI, the H_2 and I_2 would continue reacting to replace it until none of those two substances remained. Isolating and removing a product is one way to cause reversible reactions to proceed to completion in a desired direction. This process is quite often used in industry to obtain as much product as possible.

EXAMPLE 15.1

What happens to the equilibrium concentrations of SO_2, O_2, and SO_3 when O_2 is added to the following reaction?

$$2\ SO_2(g) + O_2(g) \rightleftharpoons 2\ SO_3(g)$$

Solution: Immediately the forward reaction rate increases, and the reaction is no longer at equilibrium. As O_2 is used up, the concentration of SO_2 decreases and the forward reaction slows down. Simultaneously, the concentration of SO_3 increases and the reverse reaction speeds up. When the rates of the forward and reverse reactions are equal, equilibrium is restored. At the new equilibrium, there is more O_2, more SO_3, and less SO_2 than at the original equilibrium; however, some of the O_2 we added has been used up. Remember always that the position of an equilibrium (that is, the relative concentrations of reactants and products) is determined by the competing rates of the forward and reverse reactions. Those rates, in turn, are affected by the concentrations of all the reactants and products.

EXAMPLE 15.2

Ammonia is at equilibrium with its elements in a vessel at 650°C.

$$N_2(g) + 3\ H_2(g) \rightleftharpoons 2\ NH_3(g)$$

More H_2 is added to this equilibrium system. Qualitatively, what will be the effect of this addition on the concentrations of N_2, H_2, and NH_3?

Solution: We know that adding H_2 to this system should cause the equilibrium to shift to the right (the direction that uses up some of the added H_2). Not enough reaction will occur, however, to use up all the added H_2. Therefore, the new concentration of H_2 will still be greater than it was originally. The reaction will create NH_3, so the new concentration of NH_3 will be greater than before.

The new concentration of N_2 will be less than before, because some of the N_2 originally present will be used up in reaction with the added H_2.

Exercise 15.1

The gaseous equilibrium system described by the following equation is present in a container at 1,000 K.

$$H_2O(g) + CO(g) \rightleftharpoons CO_2(g) + H_2(g)$$

To this system is added additional H_2O. What changes will have occurred in the concentrations of H_2O, CO, CO_2, and H_2 when equilibrium is restored?

EXAMPLE 15.3

For the chemical equilibrium shown in Exercise 15.1, answer the following questions.
(a) Will the equilibrium shift to the right or the left when some CO_2 is removed from the system?
(b) Will the equilibrium shift to the right or the left when some H_2 is added to the system?

Solution:
(a) When CO_2 is removed, the equilibrium shifts to the right.
(b) When H_2 is added, the equilibrium shifts to the left.

Exercise 15.2

Various changes are made to the chemical equilibrium shown below. For each change, state whether the equilibrium shifts to the left or the right.

$$NH_4^+ + C_2H_3O_2^- \rightleftharpoons NH_3 + HC_2H_3O_2$$

(a) NH_4^+ is added.
(b) $C_2H_3O_2^-$ is removed.
(c) $HC_2H_3O_2$ is removed.
(d) Equal amounts of NH_4^+ and $HC_2H_3O_2$ are added.

15.5 Effects of Temperature Changes

After studying this section, you should be able to:

• Explain the effects of temperature changes on reaction rates.
• Predict the effects of temperature changes on chemical equilibrium.

We learned in Chapter 2 that temperature is a measure of the *average kinetic energy* of molecules. Most molecules have energies close to the average, but at any given instant a few have especially high energies. We can see this graphically in Figure 15.5. In reactions with large activation energies, only the molecules with the highest energies have enough energy to react.

What happens to a reaction rate when the temperature of the reactants increases? A small increase in the temperature leads to a small increase in the *average* kinetic energy of the particles (Figure 15.6). Because of the shape of the energy distribution curve, however, a small increase in average kinetic energy results in a large increase in the number of molecules with high energy. Study Figure 15.6 to see why. An increase in the number of molecules with high energy means that more molecules will have sufficient energy to react.

An increase in kinetic energy means also that there is an increase in the average velocity of the particles. Correspondingly, there will be more collisions by each particle every second. More frequent collisions together with a higher proportion of very energetic collisions act to increase the rate of reaction substantially.

What all this means is that the rate of a reaction with high activation energy is highly sensitive to temperature. On the other hand, if the activation energy is low, and most molecules already have enough energy to get over the barrier, a temperature change will have less effect.

In a reversible reaction, therefore, temperature change often affects the forward and reverse reactions differently. An increase in temperature may increase

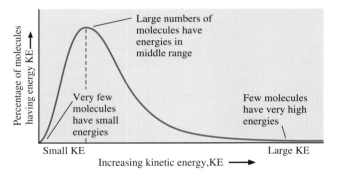

Figure 15.5 A graph of the energies of molecules versus the number having that energy. In this figure, a few molecules have little energy (left side of the curve), and a few have very high energies (shaded tail of the curve on the right side). Only those few molecules of high energy are able to react.

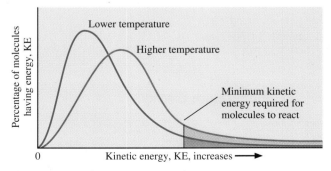

Figure 15.6 Energies of molecules in a reaction at low and higher temperatures. The blue curve is at the same temperature as in Figure 15.5, but the red curve is at a higher temperature. At the higher temperature, the number of molecules having enough energy to react is greater.

the rate of one reaction more than the rate of the other. Cooling the system may slow down one reaction more than the other. Because changes in the relative rates of forward and reverse reactions also change the equilibrium between reactants and products, temperature changes often produce equilibrium changes.

A quick way to decide how temperature will affect an equilibrium is simply to think of the energy of reaction as either a reactant or a product. If energy is absorbed by the reaction (ΔH positive), it can be considered a "reactant." Adding or taking away energy, therefore, will have the same effect as adding or taking away a reactant. If energy is released (ΔH negative), then adding or taking away energy has the same effect as adding or taking away a product. Raising the temperature is the same as adding energy. Lowering the temperature is the same as taking away energy. Just as it does with changes in concentration, the equilibrium will shift so as to counteract any change.

By changing temperature, chemists use the temperature sensitivity of chemical equilibria to get the maximum amounts of products from chemical reactions.

Consider an endothermic reaction.

$$A + B \rightleftharpoons C + D \qquad \Delta H = +100 \text{ kJ}$$

Written another way, the equation reads

$$A + B + 100 \text{ kJ} \rightleftharpoons C + D$$

Adding energy to this endothermic reaction (raising the temperature) will have the same effect as adding either A or B. It will shift the equilibrium to the right, and more C and D will be produced, using up some of the added energy and cooling the system. Running the reaction at lower temperatures is the same as removing A or B. The equilibrium will shift to the left; more A and B will be produced, more energy will be produced, and the temperature will rise again.

The reaction of phosphorus trichloride with chlorine yields phosphorus pentachloride and is exothermic.

$$PCl_3 + Cl_2 \rightleftharpoons PCl_5 \qquad \Delta H = -92.8 \text{ kJ}$$

that is,

$$PCl_3 + Cl_2 \rightleftharpoons PCl_5 + 92.8 \text{ kJ}$$

ΔH values are written for the amounts of reactants and products shown in the chemical equation, in this case, for one mole of each.

Adding energy (raising the temperature) will have the same effect as adding more PCl_5; more PCl_3 and Cl_2 will be produced. The equilibrium will shift to the left. Reducing the temperature will shift the equilibrium to the right, and more PCl_5 will be produced.

EXAMPLE 15.4

Nitrogen dioxide, NO_2, reacts with itself to produce N_2O_4, dinitrogen tetroxide; $\Delta H = -58$ kJ.

$$2 \ NO_2(g) \rightleftharpoons N_2O_4(g) + 58 \text{ kJ}$$

What will happen to the concentrations of NO_2 and N_2O_4 if we raise the temperature of the system at equilibrium? Will the equilibrium shift to the right or to the left?

Solution: Raising the temperature of this equilibrium system reduces the equilibrium concentration of N_2O_4 and increases that of NO_2. The equilibrium shifts

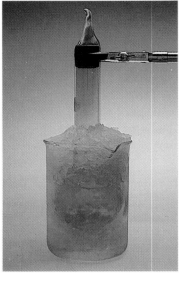

(a)   (b)

Temperature dependence of the NO_2–N_2O_4 reaction. (a) When a flask of the mixture is cooled in ice, more of the colorless N_2O_4 is formed, using up the reddish-brown NO_2. (b) The color of the mixture returns when the flask is placed in warm water.

to the left. An interesting lecture demonstration is based on this reaction. NO_2 is a brown gas; N_2O_4 is colorless. A flask containing the equilibrium mixture at $0°C$ is pale brown. If the flask is heated, the brown color intensifies because the equilibrium shift causes an increase in the concentration of NO_2.

Exercise 15.3

The decomposition of HBr into its elements is endothermic.

$$2\ HBr \rightleftharpoons H_2 + Br_2$$

What effect will raising the temperature have on the equilibrium concentration of HBr? Will the equilibrium shift to the right or the left?

15.6 Effects of Pressure and Volume Changes

After studying this section, you should be able to:

• Describe the effects of pressure and volume changes on the rates of reactions of gases.
• Describe the effects of pressure and volume changes on gaseous equilibria.

Reactions of liquids and solids are not much affected by changes in pressure. Reactions involving gases, however, are sometimes greatly affected by even small changes in pressure or volume.

Remember from Chapter 11 that the ideal gas law includes pressure, volume, number of moles, and temperature.

$$PV = nRT$$

We can rearrange the ideal gas law to look like this:

$$P = \frac{n}{V} RT$$

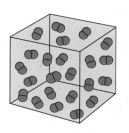

Figure 15.7 Each box has the same number of molecules, but because the box on the left has a greater volume, its concentration is less.

The quantity n/V has the units of moles per liter, that is, molar concentration units. When the volume decreases, n/V increases. A decrease in the volume of a sample of gas containing a constant number of moles increases its concentration. Similarly, an *increase* in the volume results in a *decrease* in the concentration. (See Figure 15.7.) As a result, the effect of volume changes on the reaction rates and chemical equilibria of gases can easily be predicted from what we already know about concentration changes.

Increasing the volume of a reaction vessel decreases the concentrations, while decreasing the volume increases the concentrations.

Let's look at an example. At constant temperature, we place mercury, mercuric oxide, and oxygen in a container whose volume we can change. We allow the reaction to reach equilibrium.

$$2 \text{ HgO(s)} \rightleftharpoons 2 \text{ Hg(l)} + \text{O}_2\text{(g)}$$

If we decrease the volume of the vessel, the concentration (n/V) of O_2 increases, and the reverse reaction speeds up. More HgO is formed, and some O_2 disappears. The equilibrium shifts to the left, using up O_2 and *decreasing* its concentration until equilibrium is again achieved. If we increase the volume, the O_2 concentration decreases. This will have the same effect as taking away some O_2. The reverse reaction will *slow,* and the equilibrium will shift to the *right*.

Here is the ideal gas equation again.

$$P = \frac{n}{V} RT$$

At constant temperature, the pressure of a gas is directly *proportional* to its molar concentration. Increasing the pressure is the same as increasing the concentration. Likewise, decreasing the pressure decreases the concentration.

Looking at the HgO reaction again, we see that increasing the pressure is equivalent to adding more O_2 to the reaction. The reverse reaction speeds up and the equilibrium shifts to the left. Decreasing the pressure shifts the equilibrium to the right.

Often a chemical equilibrium consists entirely of gases. What happens in this case when the pressure changes? When the pressure increases, all the molar concentrations increase. If the total number of moles of gas on the left side of the equation is the same as on the right, pressure change will have no effect. But if the total number of moles is different on the two sides, the equilibrium will shift toward the side with the smaller number of moles. Let's see why.

When NO_2 is at equilibrium with N_2O_4, the gases are in the same container and therefore have the same volume. Since the chemical equation has two moles of NO_2 and only one mole of N_2O_4, the concentration, (n/V), of NO_2 will be affected twice as much as that of N_2O_4.

$$2 \text{ NO}_2\text{(g)} \rightleftharpoons \text{N}_2\text{O}_4\text{(g)}$$

Chemists routinely optimize the yields of gas reactions by carefully adjusting both the pressure and the temperature conditions under which the reactions are run.

Because pressure and molar concentration are directly proportional, an increase in pressure is equivalent to increasing the concentration of NO_2 twice as much as the concentration of N_2O_4. An increase in pressure will therefore increase the rate of the forward reaction more than that of the reverse reaction, and the equilibrium will shift to the right. Because two NO_2 molecules will react and disappear for every one molecule of N_2O_4 produced, ultimately the total number of molecules (and therefore the pressure) decreases. Notice that an *increase* in total pressure will shift an equilibrium to the side with the *fewest* number of moles of gas, tending to *decrease* the pressure again. On the other hand, if we *decrease* the pressure, the equilibrium will shift to the side with the *greatest* number of moles of gas, ultimately *increasing* the pressure again. In either case the equilibrium shifts in such a way as to oppose the original change (Figure 15.8).

EXAMPLE 15.5

H_2 is needed to form ammonia using the Haber process. H_2 is commercially produced by the following reaction at high temperature.

$$CH_4(g) + H_2O(g) \rightleftharpoons CO(g) + 3\ H_2(g)$$

Which way will the equilibrium shift if the total pressure of the system is increased? Explain.

Solution: In this reaction two moles of gas appear on the left side of the equation, one each of CH_4 and H_2O. Four moles of gas appear on the right, one of CO and three of H_2. With an increase in total pressure, the equilibrium will shift to the side with the fewer number of moles, that is, to the left.

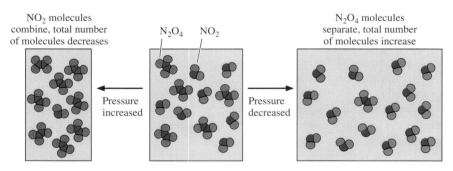

NO₂ molecules combine, total number of molecules decreases

N₂O₄   NO₂

N₂O₄ molecules separate, total number of molecules increase

Pressure increased

Pressure decreased

Figure 15.8 An equilibrium mixture of NO_2 and N_2O_4 is placed in a container (center box). If the pressure is increased (equivalent to decreasing the volume, the left box), NO_2 molecules combine to form N_2O_4. The total number of the molecules decreases, ultimately decreasing the pressure. On the other hand, if the pressure decreases (equivalent to increasing the volume, the right box), N_2O_4 reacts to produce NO_2. The total number of molecules increases, ultimately increasing the pressure.

Exercise 15.4

Will the equilibrium shift to the right or the left when the total pressure of the following equilibrium is doubled?

$$PCl_5(g) \rightleftharpoons PCl_3(g) + Cl_2(g)$$

Exercise 15.5

Given the chemical equilibrium

$$H_2O(g) + Cl_2O(g) \rightleftharpoons 2\ HOCl(g)$$

will the equilibrium shift to the right or the left when the total pressure of the system is lowered?

15.7 Le Chatelier's Principle

After studying this section, you should be able to:

- Use Le Chatelier's principle to explain the effects of a change in the conditions under which a chemical equilibrium exists.

We have seen that a chemical equilibrium responds to various externally caused changes by shifting in a direction that produces results opposite to those of the changes themselves. We have seen the reasons for these effects. Now we will see how all the effects can be considered together. The tendency for changes in conditions to produce changes in an equilibrium is formalized in a statement called **Le Chatelier's principle,** which was first proposed by Henri Le Chatelier:

If conditions are changed in a reaction system at dynamic equilibrium, reaction will take place in the direction that tends to counteract the change and bring the system again to equilibrium.

The response to changes in a system at equilibrium never proceeds far enough to counteract all of the change. When some reactant is added to a system, increasing its concentration, the equilibrium will shift to the right and use up some of the added reactant, but there will still be more of that reactant than in the original equilibrium. As we will soon learn, the actual amount of response to a given change can be determined precisely by calculation or experiment.

Both in the laboratory and in industry, Le Chatelier's principle is applied to drive reactions in a desired direction. Concentrations of reactants are kept as high as is practical, and products are removed continuously whenever possible. With exothermic reactions, energy is removed whenever possible. Endothermic reactions are run at high temperatures.

Profiles in Science

Henri Le Chatelier (1850–1936)

Because he was surrounded by science and engineering during his whole life, it is scarcely surprising that Henri Le Chatelier himself became an outstanding scientist. Henri's father was a distinguished engineer who created the French National Railway system, helped construct the first open hearth steel furnace, and was closely involved in establishing the aluminum industry in France. Henri's mother came from a family of artists and scientists. One of his brothers was an engineer, another was an inventor, and yet a third created high-temperature enamels for metalwork. Henri received his Bachelor of Science degree in 1867, completed his graduate study in 1872, and soon became a professor at the School of Mines. Not long afterward, he was appointed professor at the Sorbonne, the leading academic institution in France.

Le Chatelier devoted his middle years to the study of chemical equilibrium, establishing the laws of displacement of equilibrium by studying the energetics of reactions and establishing the principle that bears his name. But equilibrium was not his only interest. He accomplished the synthesis of ammonia from the elements, anticipating Haber, who is usually given credit for the process. He devised several of the methods still used in producing and analyzing steel, becoming the French Minister of Mines in the process. He directed more than a hundred graduate students, many of whom received doctorates in chemistry. His 500 scientific papers cover many fields in chemistry and metallurgy along with contributions in ceramics, high-temperature processes, hydraulics, explosives, and countless other subjects.

Le Chatelier was a great national French figure. His linking of science and industry was important to the nation, especially to national defense. His reputation was international; for example, U.S. President Woodrow Wilson consulted him when the National Research Council was established in 1916.

The list of honors bestowed on Le Chatelier would consume pages of print, but he is best known and remembered because of the simple but powerful Le Chatelier's principle. He died of a heart attack at age 86, shortly after giving a geometry lesson to some of his thirty-four grandchildren.

15.8 Catalysts

After studying this section, you should be able to:

• Explain how catalysts affect reaction rates.

Many systems are affected by the presence of **catalysts,** substances that change the rates of reactions and yet are not consumed by the reactions. Catalysts operate by providing pathways that have lower activation energies, as illustrated in Figure 15.9.

The sweet-smelling compound ethyl butyrate, $C_6H_{12}O_2$, can react with water to produce ethanol, C_2H_5OH, and butyric acid, $C_4H_7O_2$, a compound found in rancid butter.

$$C_6H_{12}O_2 + H_2O \longrightarrow C_4H_7O_2 + C_2H_5OH$$

With only water and ethyl butyrate present, the reaction runs so slowly that even after years there is hardly enough product to identify. But if a little H_3O^+ is

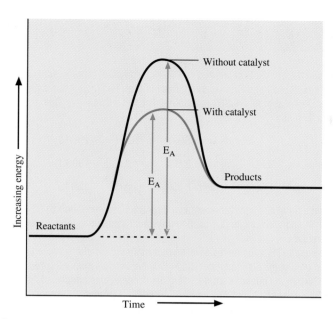

Figure 15.9 Activation energy diagram for an endothermic reaction. When a catalyst is added to this reaction, the activation energy is lowered.

present as a catalyst, the reaction will produce a significant amount of product in a few days. If some OH^- is added instead, the reaction will run to completion in a few hours. Finally, if the biological catalyst lipase is used, the reaction is over within a second or so.

The most remarkable occurrences of catalyzed reactions are in biological systems, in which complex reactions proceed at rapid rates. The entire chemistry of living beings operates with the assistance of catalysts called enzymes, of which lipase is one.

Although catalysts can change reaction rates, they affect the rates of both forward and reverse reactions equally. That is, the forward reaction will speed up at exactly the same rate as the reverse reaction. Because the rates of the forward and reverse reactions are equal, the chemical equilibrium will not shift with the addition of a catalyst. However, a catalyst can change the time it takes a reaction to reach equilibrium.

A chemical equilibrium will not shift with the addition of a catalyst.

15.9 Equilibrium Equation and Equilibrium Constant

After studying this section, you should be able to:

- Write an equilibrium equation, given a chemical equation.
- Calculate an equilibrium constant from experimental data.

We now know that chemical equilibrium is reached when the forward and reverse reaction rates are equal. Because reaction rates depend on concentration, equilibrium must also depend on concentration, regardless of the complexity of the individual reactions. For the equilibrium

$$H_2(g) + I_2(g) \rightleftharpoons 2\,HI(g)$$

Science in Action

The Manufacture of Sulfuric Acid

In Section 14.10, we first learned of the importance of sulfuric acid, H_2SO_4, which is manufactured in enormous quantities. In fact, H_2SO_4 is often called the most important chemical of all. H_2SO_4 is manufactured in a three-step operation beginning with sulfur and air.

$$S + O_2 \rightleftharpoons SO_2 \qquad \Delta H = -296 \text{ kJ}$$

$$2\,SO_2 + O_2 \rightleftharpoons 2\,SO_3 \qquad \Delta H = -198 \text{ kJ}$$

$$SO_3 + H_2O \rightleftharpoons H_2SO_4 \qquad \Delta H = -130 \text{ kJ}$$

This seems simple enough, but there is a difficulty. The oxidation of SO_2 to SO_3 is slow, and the equilibrium is not particularly favorable, although it does lie in the correct direction.

In the early 1800's, it was discovered that platinum would catalyze the oxidation of SO_2, but only some 50 years later did H_2SO_4 begin to be manufactured by a catalytic process. Over many subsequent decades, the process was continually improved, and now H_2SO_4 is manufactured this way in the "contact process" using a catalyst of vanadium oxide, V_2O_5.

The conditions under which the SO_2 oxidation is carried out are carefully chosen and maintained. Since the reaction is exothermic, Le Chatelier's principle tells us that its rate increases by running it at low temperatures. But lowering the temperature makes the slow reaction even slower, so a compromise must be made to find a temperature high enough to assure a good rate yet low enough to keep the equilibrium as favorable as possible. The situation is helped by two other factors, also stringently controlled. Since the reaction has three moles of gas on the left ($2\,SO_2 + 1\,O_2$) and only two ($2\,SO_3$) on the right, it can be enhanced by high pressure. The catalytic material is also carefully selected and fabricated. In the presently used process, the reaction does not benefit by removing SO_3 from the reaction vessel. Sulfuric acid manufacture is an excellent example of the application of chemical principles to processes that make civilization possible today.

---

at 425°C, for example, the following is found experimentally to be true.

$$\frac{[HI][HI]}{[H_2][I_2]} = 54$$

Study this remarkable statement for a few minutes. For the reaction that has reached equilibrium, it says: If you measure the actual concentration of HI in moles per liter, then square that value and divide it by the product of the molar concentrations of H_2 and I_2, you will get the number 54. It does not say how much H_2 or I_2 you started with or whether you might have added some HI or taken some out. It simply says that if the reaction has reached equilibrium, the relations of the molar concentrations of HI, H_2, and I_2 will be as stated—no matter what amounts were put into or removed from the system.

Suppose you begin with one mole each of H_2 and I_2, place them in a one-liter vessel, heat to 425°C, and wait until equilibrium is established. If you then measure the concentrations of all the species, you will find that $[I_2] = 0.214$ M, $[H_2] = 0.214$ M, and $[HI] = 1.572$ M. Let's put this set of values into the equilibrium equation for the reaction.

$$\frac{[HI]^2}{[H_2][I_2]} = \frac{(1.572)^2}{(0.214)(0.214)} = 54.0$$

Within the limits of the number of significant figures, our experiment behaved according to the equilibrium equation.

Let's try again. This time we will begin our reaction with 10.00 moles of H_2 and 0.500 mole of I_2 in a one-liter vessel. When we measure our concentrations after equilibrium is reached, we find that $[H_2] = 9.50$ M, $[I_2] = 0.00194$ M, and $[HI] = 0.996$ M. The system again meets the requirements of the equilibrium equation.

$$\frac{(0.996)^2}{(9.50)(0.00194)} = 53.8$$

If you could find some *pure* HI, put it in a vessel of any volume whatever, and heat it to 425°C, H_2 and I_2 would form until

$$\frac{[HI]^2}{[H_2][I_2]} = 54$$

Now, in any of these cases, suppose you were to remove some of the HI after equilibrium had been reached. The reaction of HI to form H_2 and I_2 would slow down because of the decrease in the concentration of HI. However, more HI would be formed and eventually equilibrium would be restored so that

$$\frac{[HI]^2}{[H_2][I_2]} = 54$$

For any chemical equilibrium at any fixed temperature, an **equilibrium constant** exists and can be written with its **equilibrium equation.** We have written the equation for the HI system in which the value of the constant is 54. No matter what amounts of the substances are initially present, added, or taken away, as long as all the substances of the chemical equilibrium are present, the reaction will run until the concentration relations given by the equilibrium equation are met. *For each temperature, the equilibrium constant for a reaction will have a particular value;* this can usually be determined by experiment.

An equilibrium constant will remain constant under any change in conditions except temperature.

To write the equilibrium equation for a reaction, place the product of the molar concentrations of the substances appearing on the right side of the chemical equation in the numerator (each molar concentration raised to the power corresponding to the coefficient of that substance in the chemical equation) and put the product of the molar concentrations of the substances on the left side of the chemical equation (again raised to the appropriate powers) in the denominator. As an example, for the reaction consisting of a moles of A and b moles of B reacting to give c moles of C and d moles of D,

$$a\ A + b\ B \rightleftharpoons c\ C + d\ D$$

the equilibrium constant is

$$K = \frac{[C]^c \times [D]^d}{[A]^a \times [B]^b}$$

Although we have written an explicit value, 54, for the equilibrium constant of the HI system, the equilibrium equation often appears with the letter K standing for the constant, especially if the numerical value is not known.

$$K = \frac{[HI]^2}{[H_2][I_2]} = 54$$

Notice that the equilibrium constant for this particular reaction has no units, although many equilibrium constants do have units. As a matter of convenience, we will do as most chemists do and usually not bother to write units for our equilibrium constants.

While we may not *write* the units on equilibrium constants, we must keep track of the *kinds* of units. When we discuss equilibria of gases, we will use units of mol/L, but we should keep in mind that gas concentrations (unlike those of liquids) can be and often are measured in terms of pressure. Remember that $PV = nRT$, and thus $n/V = P/RT$. If temperature and volume are known, gas pressures can be converted directly to concentrations in mol/L. Many times, pressures, rather than molar concentrations, are used with reactions of gases.

This is an ammonia manufacturing plant in The Netherlands. This plant can produce thousands of metric tons of ammonia per day.

EXAMPLE 15.6

Write the equilibrium equation for the Haber process.

$$3\,H_2(g) + N_2(g) \rightleftharpoons 2\,NH_3(g)$$

Strategy and Solution:

Step 1. Place the molar concentration of the product in the numerator. Write NH_3 in brackets ([ ]) and raise it to the power of 2.

Step 2. Place the molar concentrations of the reactants in the denominator. Write H_2 in brackets ([ ]) and raise it to the power of 3. Write N_2 also in brackets. In this case, the exponent for the molar concentration is 1 and is omitted. Set the fraction equal to K.

$$\frac{[NH_3]^2}{[H_2]^3\,[N_2]} = K$$

Although we have written the equilibrium constant for the NH_3 reaction, we do not necessarily know its value at any temperature. That information must be obtained experimentally. While we have not written the units on the constant, we should know what they are. For this particular constant they are $\dfrac{L^2}{mol^2}$.

EXAMPLE 15.7

The reaction to prepare methanol, CH_3OH, from hydrogen and carbon monoxide is

$$2\,H_2(g) + CO(g) \rightleftharpoons CH_3OH(g)$$

Write the equilibrium equation for this reaction.

Strategy and Solution:

Step 1.   The numerator is $[CH_3OH]$.

Step 2.   The denominator is $[H_2]^2$ multiplied by $[CO]$. The equilibrium equation is

$$\frac{[CH_3OH]}{[H_2]^2[CO]} = K$$

The units of this constant are $\dfrac{L^2}{mol^2}$.

Exercise 15.6

Write the equilibrium equation for the following reaction. State the units of the constant.

$$2\ NO(g) + Cl_2(g) \rightleftharpoons 2\ NOCl(g)$$

Exercise 15.7

Ozone, O_3, is an extremely important component of our atmosphere. Write the equilibrium equation for the formation of ozone from oxygen molecules.

$$3\ O_2(g) \rightleftharpoons 2\ O_3(g)$$

If the temperature of an equilibrium reaction changes, the equilibrium shifts and the value of the equilibrium constant changes. For example, the value of K for the HI reaction is 54 at 425°C, 45 at 490°C, and 59 at 394°C. When we give the value of an equilibrium constant, we must always specify the temperature for which it is given.

15.10 Interpreting the Values of Equilibrium Constants

After studying this section, you should be able to:

• Predict the extent of a reaction at equilibrium by looking at its equilibrium constant.

Equilibrium constants for different reactions have a wide range of values. What do such values mean? The equilibrium constant of the reaction of H_2 and Cl_2 to form HCl, for example, is 2×10^{33} at 25°C.

$$H_2(g) + Cl_2(g) \rightleftharpoons 2\ HCl(g)$$

$$K = \frac{[HCl]^2}{[H_2][Cl_2]} = 2 \times 10^{33} \text{ at } 25°C$$

An equilibrium constant greater than 1 indicates that the concentrations of the products are larger than the concentrations of the reactants.

Because the equilibrium constant is so large, we know that the numerator is much larger than the denominator and that the concentration of the product is much larger than the concentrations of the reactants. In fact, if we were to start with one mole each of H_2 and Cl_2 in a one-liter vessel and bring them to equilibrium at 25°C, there would be only 4×10^{-16} moles or 0.0000000000000004 moles each remaining of H_2 and Cl_2. For all practical purposes, the reaction goes to completion to the right as written.

If the equilibrium constant is small, we know that the numerator is smaller than the denominator and that little product is present at equilibrium; nearly all of the reactants remain. For example, the equilibrium constant for the reaction of O_2 and N_2 to give NO_2 is small (which is fortunate; otherwise there would be very little oxygen in the air).

When the equilibrium constant is very small, the concentrations of the products are very much smaller than the concentrations of the reactants.

$$N_2(g) + 2\ O_2(g) \rightleftharpoons 2\ NO_2(g)$$

$$K = \frac{[NO_2]^2}{[N_2][O_2]^2} = 8.3 \times 10^{-10} \text{ at } 25°C$$

When the constant is this small, the proportion of the product, in this case NO_2, must be small as well. Fairly significant concentrations of NO_2 are sometimes present in smoggy air, however, because NO_2 decomposes quite slowly, and equilibrium takes a long time to reach.

For the experiment in which we produced HI, we started with one mole each of H_2 and I_2, and approximately three quarters of each reactant was converted to HI. This result matches the rough guess we might have made just by looking at the value of the equilibrium constant, $K = 54$. Since that number is larger than 1, the numerator of the equilibrium constant must be larger than the denominator. The reaction, therefore, must run until [HI] is larger than either $[H_2]$ or $[I_2]$, and that is the result we calculated. Without making calculations, we can sometimes look at the values of equilibrium constants and guess how reactions will go, especially if the constants are especially large or small.

A very small equilibrium constant indicates that the concentrations of the reactants remain essentially unchanged.

15.11 Calculations Using Equilibrium Constants

After studying this section, you should be able to:

- Calculate the molar concentrations of products and reactants after equilibrium has been reached, given the initial molarities and the equilibrium constants.

Often it is necessary to calculate the actual amounts of reactants and products present at equilibrium. When equilibrium constants are small, we can easily and quite accurately estimate the amount of product formed when the system comes to equilibrium. Let's say we start with 2 moles of O_2 and 2 of N_2 in a one-liter flask and wait (a long time) until they reach equilibrium. Because the amount of NO_2 formed is tiny, we can approximate and say that almost none of the reactants was used up, that there are still 2 moles of each reactant in the vessel. We place these values in the equilibrium equation and solve for NO_2.

$$K = \frac{[NO_2]^2}{(2)(2)} = 8.3 \times 10^{-10}$$

Therefore,

$$[NO_2]^2 = (8.3 \times 10^{-10})(4) = 3 \times 10^{-9}$$

$$[NO_2] = \sqrt{3 \times 10^{-9}} = 6 \times 10^{-5} \text{ M}$$

A concentration of 6×10^{-5} M is small. Our assumption that only small amounts of O_2 and N_2 were used is valid. If we were to subtract 6×10^{-5} from 2, the answer, to the correct number of significant figures, would be 2.

If the temperature is not stated with an equilibrium constant, assume it to be 25°C.

Exercise 15.8

The equilibrium constant for the formation of ozone, $3 O_2(g) \rightleftharpoons 2 O_3(g)$, has the value of 1×10^{-48} at 25°C. If 1.00 mole of O_2 is placed in a 2.00-liter container at 25°C, what will the concentrations of O_2 and O_3 be at equilibrium?

15.12 Ionization Constant

After studying this section, you should be able to:

- Calculate ionization constants from experimental data.

We first encountered weak electrolytes in Section 13.4. Most weak electrolytes are either weak acids or weak bases. We now know enough to understand why weak electrolytes behave as they do. Consider ammonia, for example. Ammonia reacts with water to form NH_4^+ and OH^- ions.

$$NH_3 + H_2O \rightleftharpoons NH_4^+ + OH^-$$

We can write the following equilibrium equation for the reaction.

$$K = \frac{[NH_4^+][OH^-]}{[NH_3][H_2O]}$$

Water appears in the equilibrium equation. Weak acids and bases are usually used in dilute aqueous solutions, that is, in solutions that are mostly water. A liter of pure water contains 55.5 moles of water. The concentration of pure water, therefore, is 55.5 M. As a good approximation, we can say that in a dilute aqueous solution of a weak electrolyte (in which the concentration of water is always large), the amount of water used up in reaction with the weak electrolyte will not significantly change the concentration of the water. That is, the concentration of H_2O in such reactions can be assumed constant at 55.5 M. Let's rewrite the equilibrium equation for the ionization of ammonia.

The concentration of water in dilute solutions is essentially constant at 55.5 M.

$$K = \frac{[NH_4^-][OH^-]}{[NH_3](55.5 \text{ M})}$$

So that we will not have to rewrite 55.5 M every time we use an equilibrium equation for a weak electrolyte, we multiply both sides of the equation by 55.5

M, which gives us all the constant numbers on the left and all the variable molar concentrations on the right.

$$(55.5 \text{ M})K = \frac{[NH_4^+][OH^-]}{[NH_3]}$$

A constant number times a constant number gives a constant number.

$$(55.5)K = K_i$$

Equilibrium constants for weak electrolytes are written to include the molar concentration of water in the constant itself. In this form, they are called **ionization constants** and are generally given the symbol K_i. Ionization constants for weak acids are given the special symbol K_a, and those for weak bases, K_b. For ammonia, K_b is written as follows.

Because it is already included in the ionization constant, the equilibrium equations for K_i will never include the concentration of water as a reactant or product.

$$K_b = \frac{[NH_4^+][OH^-]}{[NH_3]} = 1.8 \times 10^{-5}$$

We can see that the ionization constant for ammonia is small. We already know that a small equilibrium constant means that little product is formed, so we can conclude that little NH_4^+ and OH^- result from the reaction of NH_3 and H_2O. Because a solution of ammonia yields relatively few OH^- ions in solution, ammonia is classified as a weak base.

For all of our calculations, assume that K_a and K_b are measured at 25°C.

Like the equilibrium constants we have already encountered, ionization constants are usually determined experimentally and calculated from the values found. Table 15.1 lists the ionization constants of several weak acids.

Table 15.1   Some Weak Acids and Their Ionization Constants at 25°C*

Acid	Formula	Ionization Reaction	K_a
Chlorous	$HClO_2$	$HClO_2 + H_2O \rightleftharpoons ClO_2^- + H_3O^+$	1.1×10^{-2}
Phosphoric	H_3PO_4	$H_3PO_4 + H_2O \rightleftharpoons H_2PO_4^- + H_3O^+$	7.1×10^{-3}
Nitrous	HNO_2	$HNO_2 + H_2O \rightleftharpoons NO_2^- + H_3O^+$	4.5×10^{-4}
Acetic	$HC_2H_3O_2$	$HC_2H_3O_2 + H_2O \rightleftharpoons C_2H_3O_2^- + H_3O^+$	1.8×10^{-5}
Hydrosulfuric	H_2S	$H_2S + H_2O \rightleftharpoons HS^- + H_3O^+$	9.1×10^{-8}
Hypochlorous	$HClO$	$HClO + H_2O \rightleftharpoons ClO^- + H_3O^+$	1.1×10^{-8}
Hydrocyanic	HCN	$HCN + H_2O \rightleftharpoons CN^- + H_3O^+$	2.0×10^{-9}

*Some of these acids can ionize a second time or more, as in

$$H_2PO_4^- + H_2O \rightleftharpoons HPO_4^{2-} + H_3O^+ \qquad K_{a_2} = 6.2 \times 10^{-8}$$

followed by

$$HPO_4^{2-} + H_2O \rightleftharpoons PO_4^{3-} + H_3O^+ \qquad K_{a_3} = 4.4 \times 10^{-13}$$

There is only one widely used weak base. It is ammonia.

$$NH_3 + H_2O \rightleftharpoons NH_4^+ + OH^- \qquad K_b = 1.8 \times 10^{-5}$$

For this reason, we will not require a table of ionization constants of weak bases.

EXAMPLE 15.8

Formic acid, HCO_2H, reacts with water to form H_3O^+ and formate ion, HCO_2^-.

$$HCO_2H + H_2O \rightleftharpoons HCO_2^- + H_3O^+$$

If a solution is made with 1.000 mole of formic acid and enough water to make 1.000 liter of solution, $[H_3O^+]$ is found to be 1.32×10^{-2} M and $[HCO_2H]$ to be 0.987 M at equilibrium. What is the ionization constant for formic acid?

Strategy and Solution:

Step 1. Write the equilibrium equation.

$$K_a = \frac{[HCO_2^-][H_3O^+]}{[HCO_2H]}$$

Step 2. Substitute the values for the concentrations. Notice that because one hydrogen ion is formed when each formate ion is formed, $[HCO_2^-] = [H_3O^+] = 1.32 \times 10^{-2}$ M. Now do the arithmetic.

$$K_a = \frac{(1.32 \times 10^{-2})(1.32 \times 10^{-2})}{0.987} = 1.77 \times 10^{-4}$$

Exercise 15.9

Calculate the ionization constant for the ionization of HPO_4^{2-} if a 0.0010 M solution in Na_2HPO_4 has a pH of 7.68. (Remember to calculate $[H_3O^+]$ from the pH.)

$$HPO_4^{2-} + H_2O \rightleftharpoons PO_4^{3-} + H_3O^+$$

Exercise 15.10

Find the ionization constant for the following reaction. Remember from Section 14.4 that $[H_3O^+] = [OH^-] = 1.00 \times 10^{-7}$ M. We use the symbol K_w to represent the ionization constant of water.

$$2\ H_2O \rightleftharpoons H_3O^+ + OH^-$$

15.13 Calculations Using Ionization Constants

After studying this section, you should be able to:

- Calculate $[H_3O^+]$ or $[OH^-]$ in solutions of weak acids and bases, given their molarities and ionization constants.

We can solve problems that involve weak acids and bases in the same way that we solve other equilibrium problems.

EXAMPLE 15.9

What is $[OH^-]$ in a solution of ammonia that is prepared by adding enough water to 1.0 mole of NH_3 to make 1.0 liter of solution? K_b for NH_3 is 1.7×10^{-5}.

Strategy and Solution:

Step 1.   Write the balanced chemical equation and the equilibrium equation.

$$NH_3 + H_2O \rightleftharpoons NH_4^+ + OH^-$$

$$K_b = \frac{[NH_4^+][OH^-]}{[NH_3]} = 1.7 \times 10^{-5}$$

Step 2.   If the problem is not already stated in terms of molarity, calculate the initial molarities. In this problem, we have been given the number of moles of NH_3 placed in 1.0 L; therefore $[NH_3]$ is 1.0 mol/1.0 L or 1.0 M.

Step 3.   Determine the equilibrium concentrations. We have not been given information about NH_4^+ or OH^-, but we know that for every mole of NH_3 that reacts, there is one mole of NH_4^+ and one mole of OH^-. We will use the letter x for the molarity of NH_3 that reacts and for the molarities of NH_4^+ and OH^- produced. Therefore, $[NH_4^+] = [OH^-] = x$ and $[NH_3] = 1.0 - x$.

Step 4.   Substitute terms into the equilibrium equation.

$$\frac{(x)(x)}{(1.0 - x)} = K_b = 1.7 \times 10^{-5}$$

Step 5.   Solve the equilibrium equation for the value of the unknown. To find the solution to this equation, we would normally have to use the quadratic equation. Fortunately, as we did in Section 15.11, we can assume that x is small because the value of K_b is small. The amount of ammonia used up in the reaction will therefore be small compared to the original amount. We can approximate and say that the amount of ammonia left is the same as the amount we started with and that $(1.0 - x) \approx 1.0$. When we substitute this into the equilibrium equation, the equation becomes easy to solve.

$$K_b = 1.7 \times 10^{-5} = \frac{(x)(x)}{(1.0 - x)} \approx \frac{(x)(x)}{1.0} = x^2$$

$$[OH^-] = x = \sqrt{1.7 \times 10^{-5}} = 4.1 \times 10^{-3}$$

We can now go back and see if our approximation was valid. We said that $(1.0 - x) \approx 1.0$. The value of x is 4.1×10^{-3}, or 0.0041. If we subtract 0.0041 from 1.0, we get 0.9959, which, to the correct number of significant figures, is 1.0. That tells us that our approximation was correct.

Although the approximation we just used is appropriate for most of the problems we will encounter, we need a way to decide whether to try it. If the values of the molarities in a problem are at least 10,000 times larger than the equilibrium constant to be used, it is ordinarily safe to assume that those molarities will not be significantly changed by any substances that are created or used by the reaction. In Example 15.9, the initial concentration of NH_3 was 1.0 M, and the value of the equilibrium constant was 1.7×10^{-5}, or 0.000017; that is, the molarity was more than 50,000 times larger than the constant.

EXAMPLE 15.10

What is $[H_3O^+]$ in a solution containing 0.030 moles of hypochlorous acid, HOCl, in 3.0 liters of solution? The ionization constant, K_a, of HOCl is 3.5×10^{-8}.

$$HOCl + H_2O \rightleftharpoons H_3O^+ + OCl^-$$

Strategy and Solution:

Step 1. Write the equilibrium equation.

$$K_a = \frac{[H_3O^+][OCl^-]}{[HOCl]}$$

Step 2. Convert given values to initial molarities.

$$\text{initial molarity of HOCl} = \frac{0.030 \text{ mol HOCl}}{3.0 \text{ L}} = 0.010 \text{ M}$$

Steps 3 and 4. Let x stand for $[H_3O^+]$ and for $[OCl^-]$. Notice that even though the initial concentration of HOCl is only 0.010 M, the ionization constant is still about 300,000 times smaller. We will make the approximation that the concentration of HOCl is unchanged by the reaction and remains 0.010 M. We substitute into the equilibrium equation.

$$K_a = 3.5 \times 10^{-8} = \frac{x^2}{0.010}$$

Step 5. Do the arithmetic.

$$x^2 = (3.5 \times 10^{-8})(0.010) = 3.5 \times 10^{-10}$$

$$x = [H_3O^+] = \sqrt{3.5 \times 10^{-10}} = 1.9 \times 10^{-5} \text{ M}$$

As a check, we compare x with the initial [HOCl]: $0.010 - 1.9 \times 10^{-5} = 0.010$ M. Our approximation is acceptable.

Exercise 15.11

Calculate $[H_3O^+]$ in an aqueous solution of 0.55 M $HC_2H_3O_2$.

15.14 Buffers

After studying this section, you should be able to:

- Explain how a buffer operates.
- Explain how to prepare an acid-base buffer.

In many industrial and biological systems, it is necessary to have precise control of ion concentrations, particularly of $[H_3O^+]$ and $[OH^-]$. For example, in human blood, $[H_3O^+]$ must be kept between 3.5×10^{-8} and 4.5×10^{-8}, that is, the pH must lie between 7.35 and 7.45. Blood and many other body fluids, as well as solutions used in the laboratory and in industry, are said to be buffered. A **buffer** is *a chemical equilibrium designed to control concentration by applying Le Chatelier's principle.*

A buffer consists of a weak acid and its conjugate base in solution.

Let's look at an acetic acid-acetate ion buffer, for example. To make the buffer, we start with 0.10 moles of acetic acid, $HC_2H_3O_2$, and 0.10 moles of acetate ion, $C_2H_3O_2^-$, and add enough water to make one liter of solution. The reaction rapidly reaches equilibrium, and the resulting pH is 4.74.

$$HC_2H_3O_2 \rightleftharpoons C_2H_3O_2^- + H_3O^+$$

Now if we were to add 0.01 moles of OH^- to one liter of the buffer solution, most of the OH^- would be consumed by reacting with $HC_2H_3O_2$. When the reaction again reached equilibrium, the new pH would be 4.83, only a slight change from the original. On the other hand, if we were to add 0.01 moles of H_3O^+ to one liter of the buffer, the H_3O^+ would react with $C_2H_3O_2^-$, and the new pH would be 4.66, again only a slight change from the original. In either case, the pH remains relatively constant because the changes in concentrations of $HC_2H_3O_2$ and $C_2H_3O_2^-$ are small. By contrast, if we add 0.01 moles of OH^- to a liter of pure water, the pH will change from 7.0 to 12.0.

In a buffer, the concentrations of all the other species in the reaction are large in comparison to the concentration of the controlled component. For pH values of 7 or less, an acid-base buffer contains a quantity of a weak acid and its conjugate base, that is large in comparison to the quantities of H_3O^+ and OH^- to be controlled. A buffer for pH values greater than 7 uses similar amounts of a weak base and its conjugate acid.

EXAMPLE 15.11

Explain how a buffer solution that is 1.5 M in H_3PO_4 and 1.5 M in $H_2PO_4^-$ responds to the addition of a small quantity, say 0.015 moles, of OH^-. $K_a = 7.1 \times 10^{-3}$. Would a buffer containing only 1 M H_3PO_4 and $H_2PO_4^-$ react much differently?

$$H_3PO_4 + H_2O \rightleftharpoons H_3O^+ + H_2PO_4^-$$

Solution: When a small amount of OH^- is added, it reacts with the H_3PO_4, creating a little more $H_2PO_4^-$, but the concentrations of $H_2PO_4^-$ and H_3PO_4 are changed only slightly. The pH shifts upward slightly. If the buffer had contained only 1.0 M H_3PO_4 and $H_2PO_4^-$, the result would have been almost the same. To have adequate control, it is necessary only that the quantities be quite large in comparison to the amount of acid or base likely to be added.

Exercise 15.12

A buffer solution containing 1 M NH_3 and 1 M NH_4^+ is prepared. (a) Write the chemical equation for the equilibrium reaction. (b) Describe what happens when 0.08 moles of OH^- are added to a liter of the buffer. (c) What happens when 0.05 moles of H_3O^+ are added.

15.15 **Solubility Product**

After studying this section, you should be able to:

- Calculate solubility product constants, given experimental data.
- Calculate the molar concentrations of ions present in saturated solutions of slightly soluble electrolytes, given the solubility product constants.

We have learned that some solid substances dissolve easily in water and that others are insoluble or only slightly soluble. When a compound has dissolved as much as possible, an equilibrium exists between the ions on the surface of the solid and those in solution. Under these conditions, the rate of the ions leaving the surface of the solid equals the rate of the ions leaving the solution to join the solid. Figure 15.10 illustrates this process.

We can take as an example the very slightly soluble compound silver chloride, AgCl, for which we can write a chemical equation to indicate the equilibrium between the solid and the ions in solution.

$$AgCl(s) \rightleftharpoons Ag^+ + Cl^-$$

As we can for any equilibrium, we write an equilibrium equation for the process.

$$K_{sp} = [Ag^+][Cl^-]$$

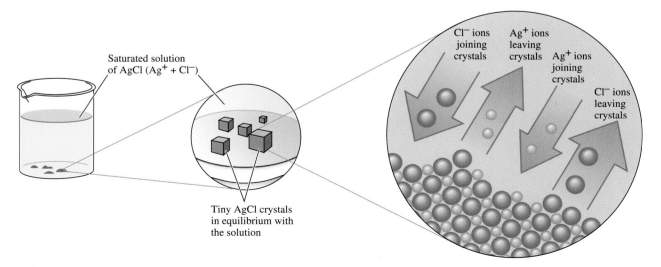

Figure 15.10   In this saturated solution of AgCl, equilibrium has been established. Every second, as many ions return to the solid as leave it, and the concentration of ions in solution remains constant.

Table 15.2 Some Slightly Soluble Compounds and Their Solubility Products at 25°C	
Solid	Solubility Product, K_{sp}
Lithium carbonate, Li_2CO_3	1.7×10^{-3}
Calcium sulfate, $CaSO_4$	2.45×10^{-5}
Silver bromate, $AgBrO_3$	5.8×10^{-5}
Magnesium carbonate, $MgCO_3$	2.6×10^{-5}
Calcium carbonate, $CaCO_3$	8.7×10^{-9}
Barium carbonate, $BaCO_3$	8.1×10^{-9}
Silver chloride, $AgCl$	1.6×10^{-10}
Barium sulfate, $BaSO_4$	1.1×10^{-10}
Copper(II) sulfide, CuS	8.7×10^{-36}

[AgCl] does not appear in the equilibrium equation because AgCl is a solid. When AgCl dissolves, it dissociates; therefore there is no undissociated AgCl in the solution. Because AgCl is not in the solution, [AgCl] has no meaning. The influence that solid AgCl has on the equilibrium is expressed by the constant itself. It is not uncommon for reactions to occur between solids and other substances, and such reactions all have equilibrium constants. In no case in which a solid is involved in a reaction is the solid ever included in the equilibrium equation.

K_{sp} is the **solubility product** (sometimes called the *ion product*) for a solid compound in equilibrium with a solution of its ions. Solubility products are used in exactly the same way that other equilibrium constants are used, except that the calculations are somewhat simplified because the only variable terms in the equation are the molar concentrations of the ions in solution.

Solubility product constants apply only to saturated solutions in contact with undissolved solid. Like most other equilibrium constants, solubility products are dependent on temperature; they usually increase as the temperature increases.

Table 15.2 lists some slightly soluble compounds and their solubility products.

EXAMPLE 15.12

Solid AgCl is shaken with water, in which some dissolves. When the molar concentrations of the dissolved ions are measured, it is found that

$$[Ag^+] = 1.3 \times 10^{-5} \text{ M}$$

$$[Cl^-] = 1.3 \times 10^{-5} \text{ M}$$

Calculate the value of the solubility product for AgCl.

Strategy and Solution:

Step 1.   Write the appropriate equations.

$$AgCl(s) \rightleftharpoons Ag^+ + Cl^-$$

$$K_{sp} = [Ag^+][Cl^-]$$

Step 2. Substitute the values of the molar concentrations and do the arithmetic.

$$K_{sp} = (1.3 \times 10^{-5})(1.3 \times 10^{-5}) = 1.7 \times 10^{-10}$$

Exercise 15.13

What is the value of the solubility product of $PbCrO_4$ if the CrO_4^{2-} concentration is 1.4×10^{-7} M in a saturated solution of the compound?

$$PbCrO_4(s) \rightleftharpoons Pb^{2+} + CrO_4^{2-}$$

(Hint: If $[CrO_4^{2-}]$ is 1.4×10^{-7} M, what must $[Pb^{2+}]$ be?)

EXAMPLE 15.13

The value of the solubility product of silver acetate, $AgC_2H_3O_2$, is 2×10^{-3}. At 25°C, what is the molar concentration of Ag^+ in a saturated solution of $AgC_2H_3O_2$? (This is the same as asking the solubility in moles per liter of $AgC_2H_3O_2$ at 25 °C.)

Strategy and Solution:

Step 1. Write the equations.

$$AgC_2H_3O_2 \rightleftharpoons Ag^+ + C_2H_3O_2^-$$

$$K_{sp} = [Ag^+][C_2H_3O_2^-] = 2 \times 10^{-3}$$

Step 2. Determine the molar concentrations of the ions. Let $x = [Ag^+] = [C_2H_3O_2^-]$. (Why are the molar concentrations of the two ions equal?)

Step 3. Substititute and do the arithmetic.

$$K_{sp} = (x)(x) = x^2 = 2 \times 10^{-3}$$

$$x = [Ag^+] = \sqrt{2 \times 10^{-3}} = 4 \times 10^{-2} \text{ M}$$

It is possible to calculate the solubility of a salt from its solubility product constant. It is also possible to calculate the solubility product constant from the solubility.

Exercise 15.14

Calculate the molar concentration of Cu^+ in a saturated solution of CuI if the value of the solubility product of CuI is 4×10^{-5}.

Summary

All chemical reactions are reversible. In some, the degree of reversibility is so small as to be unmeasurable; but in others, the reverse reactions occur to such an extent as to exert substantial influence. When the rates or speeds of forward and reverse reactions are equal, a chemical equilibrium is established. In a chemical equilibrium, the relative concentrations of products and reactants remain constant.

At equilibrium, the relative concentrations of products and reactants can be expressed by the equilibrium equation, the constant of which is called the equilibrium constant. The equilibrium constant is a property of a reaction at any particular temperature. We can use an equilibrium equation to calculate the molar concentrations of reactants and products under various conditions. Equilibrium constants for particular kinds of reactions are written in special ways and given individual names, such as *ionization constant* and *solubility product constant*.

Reactions at equilibrium respond to changes in conditions, for example, the addition or removal of reacting substances, or changes in temperature, volume, or pressure. Such responses are the subject of Le Chatelier's principle, which states that the response of an equilibrium to externally caused change is to reverse, to some extent, the effects of the change. Le Chatelier's principle is often applied to a reaction to shift an equilibrium in a desired direction.

An automatic and practical use of Le Chatelier's principle can be seen in buffers, which control the (small) concentration of one component in a chemical equilibrium by means of relatively large concentrations of other components. Buffers are essential to biological systems and industrial chemistry.

Key Terms

reaction rate	*(p. 438)*	equilibrium constant	*(p. 453)*
chemical equilibrium	*(p. 439)*	equilibrium equation	*(p. 453)*
activation energy	*(p. 440)*	ionization constant	*(p. 458)*
Le Chatelier's principle	*(p. 449)*	buffer	*(p. 462)*
catalyst	*(p. 450)*	solubility product	*(p. 464)*

Questions and Problems

Section 15.1 Reversible Chemical Reactions

Question

1. What distinguishes reactants from products in a chemical equation? Is this a fundamental distinction? Explain.

Section 15.2 Chemical Equilibrium

Questions

2. What is a reaction rate? In what units can a reaction rate be measured?

3. How can a reaction continue running and not change any concentrations?

4. Explain why many reactions do not go to completion. Does any reaction truly go to completion? Explain.

Section 15.3 Collision Theory; Activation Energy

Questions

5. Why, in the case of some reactions, do relatively few of the collisions between reactants result in reaction?

6. What is activation energy?

7. Other things being equal, why are reactions requiring high activation energy slower than those that require low activation energy?

8. Explain how the activation energies of the forward and reverse reactions of an equilibrium are related to ΔH.

Section 15.4 Effects of Concentration Changes

Questions

9. Explain how adding or removing one reactant or product can affect the rate of reaction.

10. Explain how adding or removing one reactant or product can affect the concentrations of all the species in a reaction at equilibrium.

Problems

11. In the equilibrium reaction,

$$C + D \rightleftharpoons E + F$$

what happens when some E is added to the reaction mixture? When some D is taken away?

Section 15.5 Effects of Temperature Changes

Questions

12. How does an increase in temperature affect the rate of a reaction? A decrease in temperature?

13. How do changes in temperature affect an endothermic reaction at equilibrium? An exothermic reaction?

14. Why do small changes in temperature often produce large changes in reaction rates?

15. Other things being equal, is a small temperature change likely to have a greater effect on a fast reaction or on a slow reaction? Explain.

Section 15.6 Effects of Pressure and Volume Changes

Questions

16. Explain how changing the volume of the container of a gas reaction can shift the equilibrium.

17. Name one circumstance under which changing the volume of a reaction vessel containing gases would *not* shift the equilibrium.

Section 15.7 Le Chatelier's Principle

Questions

18. Using the concentrations of A, B, C, and D in the equilibrium reaction shown, describe in detail why Le Chatelier's principle must in fact be valid if an equilibrium constant is to remain constant. As an example, suppose you were to remove some D at equilibrium. Why would the concentrations of A, B, and C change?

$$A + B \rightleftharpoons C + D$$

$$K = \frac{[C][D]}{[A][B]}$$

19. Various activation energy barriers and relative energies of the reactants and products in equilibrium systems are shown in the figures that follow. All the systems are at the same temperature.

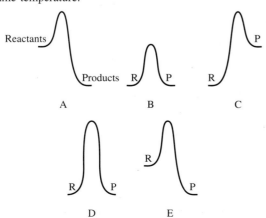

(a) In which system or systems will the equilibrium favor the reactants? Why?

(b) In which system or systems will the equilibrium favor the products? Why?

(c) Which reactions are likely to reach equilibrium fastest?

(d) Which reactions are exothermic? If there is more than one of these, which is more exothermic?

(e) Which reactions are endothermic? If there is more than one of these, which is more endothermic?

(f) In which reaction, if any, is there essentially no enthalpy change?

20. The following reactions are at equilibrium (reaction 1 is endothermic):

(1) $2\ HF(g) \rightleftharpoons H_2(g) + F_2(g)$

(2) $H_2O(g) + CO(g) \rightleftharpoons CO_2(g) + H_2(g)$

(3) $NH_4Cl(s) \rightleftharpoons NH_3(g) + HCl(g)$

(a) How will the equilibrium of reaction 1 be affected if H_2 is removed from the system? If HF is added to the system? If the temperature of the reaction mixture is lowered?

(b) In reaction 2, the concentration of CO rises as the temperature is raised. Is the reaction exothermic or endothermic? How would the equilibrium be affected if both CO_2 and H_2 were removed?

(c) Reaction 3 must be brought to a high temperature before enough NH_3 is present to be measured. What does this tell you about the reaction?

21. The substances in the reactions below are all gases.

(a) $A + B \rightleftharpoons C + D$ $\Delta H = -93$ kJ

(b) $D + E \rightleftharpoons F + G$ $\Delta H = +50$ kJ

(c) $H + I \rightleftharpoons 2\ J$ $\Delta H = 0$

Which reaction will have an equilibrium shift to the right (more product) if

(1) the pressure is increased at constant temperature and number of moles?

(2) the pressure is increased at constant volume and number of moles?

(3) the pressure is increased at constant temperature and volume?

(4) Which will have no shift of equilibrium under any pressure change whatever?

Explain each answer.

Section 15.8 Catalysts

Questions

22. Using activation energy diagrams, explain how a catalyst can increase the rate of a chemical reaction.

23. If a catalyst can change a reaction rate, why do catalysts not shift equilibria?

Section 15.9 Equilibrium Equation and Equilibrium Constant

Problems

24. Write equilibrium equations for the following reactions.
    (a) $2 NO(g) + Br_2(g) \rightleftharpoons 2 NOBr(g)$
    (b) $SO_2(g) + Cl_2(g) \rightleftharpoons SO_2Cl_2(g)$
    (c) $CH_4(g) + O_2(g) \rightleftharpoons CO_2(g) + 2 H_2O(g)$
    (d) $4 NH_3(g) + 7 O_2(g) \rightleftharpoons 4 NO_2(g) + 6 H_2O(g)$

25. Write equilibrium equations for the following reactions.
    (a) $2 CO(g) + O_2(g) \rightleftharpoons 2 CO_2(g)$
    (b) $2 NO(g) + O_2(g) \rightleftharpoons 2 NO_2(g)$
    (c) $4 HCl(g) + O_2(g) \rightleftharpoons 2 Cl_2(g) + 2 H_2O(g)$
    (d) $4 NH_3(g) + 5 O_2(g) \rightleftharpoons 4 NO(g) + 6 H_2O(g)$

26. In the preparation of sulfuric acid, O_2 and SO_2 react to give SO_3.

$$O_2(g) + 2 SO_2(g) \rightleftharpoons 2 SO_3(g)$$

In one experiment, O_2 and SO_2 were put into a vessel and allowed to come to equilibrium at 1,350 K. When the concentrations of the three substances were measured, $[SO_3] = 1.20$ M, $[O_2] = 0.45$ M, and $[SO_2] = 1.80$ M. Calculate the equilibrium constant for this reaction at this temperature.

27. Given $[CO] = 4.53$ M, $[O_2] = 5.17$ M, and $[CO_2] = 1.56 \times 10^{-3}$ M at a certain temperature, calculate the equilibrium constant for this reaction in which all the reacting species are gases.

$$2 CO(g) + O_2(g) \rightleftharpoons 2 CO_2(g)$$

28. The dissociation reaction of H_2S gas is in equilibrium at high temperature.

$$2 H_2S(g) \rightleftharpoons 2 H_2(g) + S_2(g)$$

The concentrations are $[H_2S] = 0.50$ M, $[H_2] = 0.10$ M, $[S_2] = 0.40$ M. What is the value of the equilibrium constant for this reaction?

29. The ammonia reaction is run in a 2.50-L vessel at a certain temperature.

$$N_2(g) + 3 H_2(g) \rightleftharpoons 2 NH_3(g)$$

The reaction results in the following numbers of moles of each substance: NH_3, 1.00 mole; N_2, 0.50 moles; H_2, 0.75 moles. What is the value of K for this reaction at this temperature?

30. A 1.00-L vessel contains 0.015 mole of Br_2, 0.037 moles of NO, and 0.027 moles of NOBr. What is the value of the equilibrium constant?

$$Br_2 + 2 NO \rightleftharpoons 2 NOBr$$

Section 15.10 Interpreting the Values of Equilibrium Constants

Question

31. If the value of an equilibrium constant is small, what can be said about how far the reaction will proceed? If it is large, what can be said?

Section 15.11 Calculations Using Equilibrium Constants

Questions

32. To avoid using quadratic or higher-order algebra when making calculations involving equilibrium constants, what kinds of approximations can be made?

33. Under what circumstances is it usually safe to make each of the approximations mentioned in your answer to question 33?

34. How can you tell if an approximation you made in a problem was indeed justified?

Problems

35. Calculate the equilibrium concentration of CO_2, given that $[CO] = 1.60$ M and $[O_2] = 4.50$ M. $K = 4.27 \times 10^{-2}$.

$$2 CO(g) + O_2(g) \rightleftharpoons 2 CO_2(g)$$

36. The equilibrium constant for the following reaction is 6.28×10^{-2} at 400 K.

$$4 HCl(g) + O_2(g) \rightleftharpoons 2 Cl_2(g) + 2 H_2O(g)$$

Given the following partial pressures in atm, calculate the equilibrium partial pressure of Cl_2: $HCl = 1.52$, $O_2 = 1.24 \times 10^{-2}$, $H_2O = 2.75 \times 10^{-4}$.

37. If the equilibrium constant for the following reaction is 9.81, what is the concentration of CH_4 when $[CO_2] = 4.66$ M, $[O_2] = 2.00$ M, and $[H_2O] = 2.75$ M?

$$CH_4(g) + O_2(g) \rightleftharpoons CO_2(g) + 2 H_2O(g)$$

Section 15.12 Ionization Constant

Questions

38. Why does the concentration of water not appear in equilibrium equations for ionization reactions?

39. Give an example of a chemical equilibrium in which the concentration of water must appear. Why must it appear here and not in the reactions mentioned in question 39?

40. What are K_a and K_b?

Problems

41. Write equilibrium equations for the following reactions.
    (a) $H_2CO_3(aq) + H_2O(l) \rightleftharpoons H_3O^+ + HCO_3^-$
    (b) $HNO_2(aq) + H_2O(l) \rightleftharpoons H_3O^+ + NO_2^-$

42. Calculate the ionization constant for the ionization of an imaginary acid with formula HA when a 0.0030 M acid solution has a pH of 4.2.

43. Calculate K_a for the sweetener saccharin if a 0.875 M solution of the compound has a pH of 5.87. (When you don't know the formula for a weak acid, write the ionization equation as $HA + H_2O \rightleftharpoons H_3O^+ + A^-$.)

44. Lysine is one of the amino acids. Find the K_a of lysine if a 0.225 M solution has a pH of 5.59.

45. Hydroxylamine, NH_2OH, is a weak base. Calculate its K_b if a 1.50 M solution has a pH of 10.1.

Section 15.13 Calculations Using Ionization Constants

Problems

46. Calculate $[H_3O^+]$ in an aqueous solution of each of the following weak acids prepared as a 1.00 *M* solution.
    (a) H_3BO_3 $K_{a1} = 4.4 \times 10^{-7}$ (Consider only the first ionization of H_3BO_3.)
    (b) HCN $K_a = 2 \times 10^{-9}$
    (c) HF $K_a = 6.7 \times 10^{-4}$
    (d) HIO $K_a = 3 \times 10^{-11}$

47. Calculate $[H_3O^+]$ in an aqueous solution of each of the following weak acids prepared as a 2.50 M solution. Consider only the first ionization in any case where there is more than one stage.
    (a) H_2S $K_{a1} = 9.1 \times 10^{-8}$
    (b) H_2O_2 $K_a = 2.4 \times 10^{-12}$
    (c) HN_3 $K_a = 1.2 \times 10^{-5}$
    (d) HCO_2H $K_a = 1.7 \times 10^{-4}$

48. Calculate $[NH_4^+]$ in a 1.25 M solution of NH_3. What is the pH of the solution? $K_b = 1.8 \times 10^{-5}$.

49. What is the pH of a 0.0250 M solution of HCN?

50. For glycine, an amino acid, $K_a = 1.67 \times 10^{-10}$. What is the pH of a 1.00 M solution?

51. Find the pH of the weak base, hydrazine, in a 1.00 M solution. $K_b = 1.7 \times 10^{-6}$.

Section 15.14 Buffers

Questions

52. Explain how a buffer operates.

More Difficult Problems

62. The table to the right gives the characteristics of five reactions, the rates of which are to be compared when the two reactants are first mixed. Assume that no products have yet appeared; thus there are not yet any reverse reactions. Which reaction is faster?
    (a) 1 or 2   (c) 1 or 4   (e) 4 or 5
    (b) 2 or 3   (d) 1 or 5   (f) 1 or 3

53. Why must a buffer include large concentrations of all the members of the equilibrium except the controlled species?

Section 15.15 Solubility Product

Questions

54. Explain the operation of the equilibrium in a saturated solution of a slightly soluble salt in contact with its solid.

55. When we write solubility product constants, why do we not show the term for the concentration of the solid?

Problems

Problems in this section pertain to saturated solutions at equilibrium.

56. Calculate the solubility product for $MgCO_3$ if $[Mg^{2+}] = 2.37 \times 10^{-3}$ M and for FeS if $[S^{2-}] = 2.24 \times 10^{-9}$ M.

57. A saturated solution of CaF_2 is found to have $[Ca^{2+}] = 2.05 \times 10^{-4}$. What is the value of the solubility product of CaF_2? (Remember that when one Ca^{2+} ion is formed in solution, two F^- ions also form.)

58. Calculate the solubility product for Ag_2SO_4 if $[SO_4^{2-}] = 2.52 \times 10^{-2}$ M.

59. Calculate the solubility product of $Mn(OH)_2$ if $[Mn^{2+}] = 1.09 \times 10^{-4}$ M.

60. Calculate the molar concentration of the metal ions for the following compounds.
    (a) $CaSO_4$ $K_{sp} = 2 \times 10^{-5}$
    (b) CuBr $K_{sp} = 4 \times 10^{-8}$
    (c) $MgCO_3$ $K_{sp} = 5.6 \times 10^{-6}$
    (d) $SrCrO_4$ $K_{sp} = 3 \times 10^{-5}$

61. Calculate the molar concentration of the metal ions for the following compounds:
    (a) $SrSO_4$ $K_{sp} = 2.8 \times 10^{-7}$
    (b) $PbCO_3$ $K_{sp} = 3 \times 10^{-14}$
    (c) $AgBrO_3$ $K_{sp} = 5.2 \times 10^{-5}$
    (d) CuCl $K_{sp} = 1 \times 10^{-6}$

Reaction	Activation Energy	Temperature	Concentrations of Reactants
1	Low	Low	1 mol/L each
2	High	Low	1 mol/L each
3	Low	High	1 mol/L each
4	Low	Low	1 mol/L of one reactant 3 mol/L of second reactant
5	Low	Low	3 mol/L of each

Solutions to Exercises

15.1 The concentrations of CO_2 and H_2 will increase while the concentration of CO will decrease. Not enough reaction will occur to use up all of the added H_2O; therefore the new concentration of $H_2O(g)$ will be greater than the original.

15.2 (a) right (b) left (c) right (d) neither

15.3 The concentration of HBr will decrease. The equilibrium will shift to the right.

15.4 The equilibrium will shift to the left.

15.5 The number of moles of gas is equal on both sides of the equation. The equilibrium will not shift.

15.6
$$K = \frac{[NOCl]^2}{[NO]^2[Cl_2]}$$

The units of K are L/mol.

15.7
$$K = \frac{[O_3]^2}{[O_2]^3}$$

15.8
$$[O_2] = \frac{1.00 \text{ mol}}{2.00 \text{ L}} = 0.500 \text{ M}$$

$$K = \frac{[O_3]^2}{[O_2]^3} = \frac{[O_3]^2}{(0.500)^3} = 1 \times 10^{-48}$$

$$[O_3]^2 = (1 \times 10^{-48})\,(0.500)^3 = 1 \times 10^{-49}$$

$$[O_3] = \sqrt{1 \times 10^{-49}} = 3 \times 10^{-25} \text{ M}$$

15.9
$$[H_3O^+] = 10^{-7.68} = 2.1 \times 10^{-8} \text{ M}$$

$$[HPO_4^{2-}] = 0.0010 - 2.1 \times 10^{-8} = 0.0010 \text{ M}$$

$$K_a = \frac{(2.1 \times 10^{-8})\,(2.1 \times 10^{-8})}{0.0010} = 4.4 \times 10^{-13}$$

15.10
$$K_w = [H_3O^+][OH^-] = (1.00 \times 10^{-7})(1.00 \times 10^{-7}) = 1.00 \times 10^{-14}$$

15.11
$$HC_2H_3O_2 + H_2O \rightleftharpoons H_3O^+ + C_2H_3O_2^-$$

$$K_a = \frac{[H_3O^+][C_2H_3O_2^-]}{[HC_2H_3O_2]}$$

$$K_a = 1.8 \times 10^{-5} = \frac{x^2}{0.55}$$

$$x^2 = (1.8 \times 10^{-5})(0.55) = 9.9 \times 10^{-6}$$

$$x = [H_3O^+] = \sqrt{9.9 \times 10^{-6}} = 3.1 \times 10^{-3} \text{ M}$$

15.12 (a) $NH_3 + H_2O \rightleftharpoons NH_4^+ + OH^-$
(b) 0.08 moles of base react with NH_4^+ to give H_2O and NH_3; the pH shifts slightly to a higher value.
(c) 0.05 moles of H_3O^+ react with NH_3 to give NH_4^+; the pH shifts to a slightly lower value.

15.13
$$[Pb^{2+}] = [CrO_4^{2-}] = 1.4 \times 10^{-7} \text{ M}$$
$$K_{sp} = (1.4 \times 10^{-7})(1.4 \times 10^{-7}) = 2.0 \times 10^{-14}$$

15.14
$$K_{sp} = 4 \times 10^{-5} = x^2$$
$$x = [Cu^+] = \sqrt{4 \times 10^{-5}} = 6 \times 10^{-3} \text{ M}$$

When a coiled copper wire is immersed in a solution containing silver ions, the copper reduces the silver ions, forming shiny needles of metallic silver. The resulting copper ions turn the solution blue.

Oxidation-Reduction Reactions

C hemical reactions involve the electrons of atoms. During oxidation-reduction (redox) reactions, which we first studied in Chapter 6, electrons are transferred from one chemical species to another. Now we review and strengthen what we have already learned about redox reactions. We also learn how to balance the sometimes difficult chemical equations that describe these reactions.

Electrons can travel through wires, creating an electric current. For this reason, oxidation-reduction reactions can generate an electric current and can be controlled by an electric current. In this chapter, we find out how this happens and also how batteries work.

16.1 Review of Oxidation-Reduction Reactions

In Sections 6.1 and 6.2, we learned that atoms can undergo oxidation or reduction. Reduction occurs when an atom gains electrons in an electron-transfer reaction; in this process, the atom is said to be reduced.

$$Cl_2 + 2\ e^- \longrightarrow 2\ Cl^- \qquad \text{reduction}$$

Oxidation occurs when an atom loses electrons; then the atom is said to be oxidized.

$$Na \longrightarrow Na^+ + e^- \qquad \text{oxidation}$$

Figure 16.1 In the green trees, a redox reaction, photosynthesis, is occurring. In the same picture, combustion, also a redox reaction, is taking place too.

The transfer of electrons requires both an electron donor and an electron acceptor. Therefore, there must be reduction every time there is oxidation. Oxidation-reduction reactions are often called **redox reactions** for short.

When sodium reacts with gaseous chlorine to produce sodium chloride, electrons are donated from sodium atoms to chlorine atoms. (For simplicity, we'll look at the reaction of a single pair of atoms.)

$$Na \cdot + : \overset{\cdot\cdot}{Cl} : \longrightarrow Na^+ + : \overset{\cdot\cdot}{\underset{\cdot\cdot}{Cl}} : {}^-$$

Anything that takes electrons from a substance *oxidizes* that substance and is called an **oxidizing agent.** Anything that donates electrons to a substance *reduces* that substance and is said to be a **reducing agent.** As it oxidizes something, an oxidizing agent gets reduced. As it reduces something, a reducing agent gets oxidized. In the chemical equation above, the reducing agent is sodium, which reduces chlorine. At the same time, chlorine, which is the oxidizing agent, oxidizes sodium.

In some redox reactions, electron transfer occurs in such a way that it is easy to see where the electrons come from and where they go, that is, to find out which substance is the oxidizing agent and which is the reducing agent. In other reactions, however, it is not as obvious where electrons come from or where they go—or even how many electrons are involved in the reaction. In still other redox reactions, electrons may not be entirely transferred; there may be merely a change in the amount of electron sharing.

In Section 7.1 we learned how to assign oxidation numbers. Knowing how to assign oxidation numbers allows us to understand reactions in which electric charges play a part and to balance difficult chemical equations systematically.

16.2 Balancing Redox Equations

After studying this section, you should be able to:

- Write oxidation and reduction half-reactions, given a redox equation.
- Balance redox equations both for the number of atoms and for the charges.

We have learned in previous chapters how to balance simple chemical equations. Some equations are so easy that you can balance them immediately; others require a little thought and perhaps some trial and error. However, many equations, especially those involving redox reactions, are so difficult to balance that the trial and error method can take hours. To save ourselves time, we need an organized approach.

Redox reactions are common. Combustion, photosynthesis, taking power from a battery, and corrosion are examples of common redox reactions.

Complex redox equations can be balanced by any of several methods, all of which depend on one fact: In electron-transfer reactions, no electrons are created or destroyed. *The number of electrons gained must equal the number of electrons lost.*

Balancing Method Using Oxidation Numbers

One stepwise method of balancing equations involves assigning oxidation numbers to atoms. We will balance the equation for the oxidation of aluminum by bromine as an example.

Guidelines for Balancing Redox Equations Using Oxidation Numbers

1. *Write the reactants and products in equation form.* Below the symbol for each kind of atom on each side of the equation, write the oxidation number for that atom. (How to assign oxidation numbers is explained in Section 7.1.)

$$\underset{0}{Al} + \underset{0}{Br_2} \longrightarrow \underset{+3,\,-1}{AlBr_3}$$

Notice that the oxidation number of Al changes from 0 to +3. The oxidation number of Br changes from 0 to −1.

2. *Separate the chemical equation into two half-reactions, one for the substance that is being oxidized (the one that has atoms in which the oxidation number increases) and one for the substance that is being reduced.* Include in the half-reactions only the atoms that are being oxidized and reduced. Write the oxidation number of each atom as you would write the charge on an ion. (The oxidation number of the ion is the same as its charge.) If necessary, balance the half-reactions. Here we must balance the second reaction by placing a 2 in front of the Br^-.

$$Al \longrightarrow Al^{3+} \qquad \text{oxidation}$$

$$Br_2 \longrightarrow 2\ Br^- \qquad \text{reduction}$$

Aluminum foil, Al, and liquid bromine, Br_2, react vigorously to produce $AlBr_3$. The aluminum is oxidized and the bromine is reduced during reaction.

3. *Remembering that oxidation is the loss of electrons and reduction the gain of electrons, add to each half-reaction the number of electrons required to bring about the oxidation number change that occurs in that half-reaction.* In the reduction half-reaction, the electrons are added to the left side. In the oxidation half-reaction, they are added to the right. In this example, Al is oxidized to Al^{3+}, and 3 e^- must be added to the right side of the equation. Br_2 is reduced; 2 e^- must be added to the left side.

$$Al \longrightarrow Al^{3+} + 3\ e^- \qquad \text{oxidation}$$

$$Br_2 + 2\ e^- \longrightarrow 2\ Br^- \qquad \text{reduction}$$

When the electrons are added correctly, the charges balance, that is, the total number of charges is the same on each side of each equation. In this case, each side of the oxidation half-reaction has a total charge of 0, while each side of the reduction half-reaction has a total charge of −2.

4. *Multiply all the terms in each half-reaction by a number that will make the number of electrons in each the same.* (In many cases, only one half-reaction will be multiplied.)

$$2\ Al \longrightarrow 2\ Al^{3+} + 6\ e^- \qquad \text{oxidation}$$

$$3\ Br_2 + 6\ e^- \longrightarrow 6\ Br^- \qquad \text{reduction}$$

For both half-reactions to have the same number of electrons, the oxidation half-reaction must be multiplied by 2, while the reduction half-reaction must be multiplied by 3.

5. *Add together the two new half-reactions and cancel out the terms that appear on both sides.* If the process has been done correctly, the numbers of electrons will cancel each other and will not appear in the final equation.

$$2 \text{ Al} + 3 \text{ Br}_2 + \cancel{6 \text{ e}^-} \longrightarrow 2 \text{ Al}^{+3} + 6 \text{ Br}^- + \cancel{6 \text{ e}^-}$$

Replace and balance any elements or substances left out when the half-reactions were first written. Recombine any substances that were separated.

$$2 \text{ Al} + 3 \text{ Br}_2 \longrightarrow 2 \text{ AlBr}_3$$

6. *Check the numbers of atoms of each kind on both sides of the equation to be sure no mistake has been made.* In any chemical equation, it is necessary to balance not only the numbers of atoms of each kind, but also the total number of charges. This means that if there are ions involved in the reaction, the totals of all the charges on each side of the equation must be equal. When checking an equation, be sure to count charges, too.

$$2 \text{ Al} + 3 \text{ Br}_2 \longrightarrow 2 \text{ AlBr}_3$$

If you follow the procedure, a redox equation will practically balance itself.

The atoms balance and the charge adds to zero on each side of the equation.

EXAMPLE 16.1

Balance the following redox equation.

$$\text{KClO}_4 + \text{H}_3\text{AsO}_3 \longrightarrow \text{KCl} + \text{H}_3\text{AsO}_4$$

Strategy and Solution:

Step 1.   Assign oxidation numbers to all the elements.

$$\begin{array}{cccc} \text{KClO}_4 & + & \text{H}_3\text{AsO}_3 & \longrightarrow & \text{KCl} & + & \text{H}_3\text{AsO}_4 \\ +1, +7, -2 & & +1, +3, -2 & & +1, -1 & & +1, +5, -2 \end{array}$$

The oxidation number for Cl changes from $+7$ to -1. The oxidation number for As changes from $+3$ to $+5$.

Step 2.   Write the half-reactions, including only the species that undergo a change in oxidation number.

$$\text{Cl}^{7+} \longrightarrow \text{Cl}^{1-} \qquad \text{reduction}$$

$$\text{As}^{3+} \longrightarrow \text{As}^{5+} \qquad \text{oxidation}$$

Step 3.   Add the appropriate numbers of electrons to balance the oxidation numbers with electron charges.

$$\text{Cl}^{7+} + 8 \text{ e}^- \longrightarrow \text{Cl}^{1-} \qquad \text{reduction}$$

$$\text{As}^{3+} \longrightarrow \text{As}^{5+} + 2 \text{ e}^- \qquad \text{oxidation}$$

Step 4.   Multiply each of the two half-reactions by a number that will make the number of electrons in each half-reaction equal. In this case, we need only multiply the oxidation half-reaction by 4.

$$\text{Cl}^{7+} + 8 \text{ e}^- \longrightarrow \text{Cl}^{1-} \qquad \text{reduction}$$

$$4 \, (\text{As}^{3+} \longrightarrow \text{As}^{5+} + 2 \text{ e}^-) = 4 \text{ As}^{3+} \longrightarrow 4 \text{ As}^{5+} + 8 \text{ e}^- \qquad \text{oxidation}$$

Step 5. Add the two half-reactions together and cancel out the terms that appear on both sides. Replace the elements that were taken out in step 2.

$$KClO_4 + 4\ H_3AsO_3 + \cancel{8\ e^-} \longrightarrow KCl + 4\ H_3AsO_4 + \cancel{8\ e^-}$$

Step 6. Check: The number of atoms of each kind is the same on both sides of the equation. There is no charge on either side of this equation.

$$KClO_4 + 4\ H_3AsO_3 \longrightarrow KCl + 4\ H_3AsO_4$$

EXAMPLE 16.2

Balance the following equation. (Notice that equations need not be in compound form to be useful or to be balanced. The equation that follows is written in ionic form. Unless, for some reason, you need to write your equations in complete form, it is much easier to write them ionically.)

$$Cr^{2+} + Cl_2 \longrightarrow Cr^{3+} + Cl^-$$

Strategy and Solution: Aha, you say! This one is easy: Simply place a 2 in front of the Cl^-. True, the equation is then balanced for numbers of atoms, but the total charge on the left of the equation does not equal that on the right. As it stands, the equation balances for charge but not for numbers of atoms. Put a 2 in front of Cl^- and the equation is balanced for numbers of atoms but not for charge. We must always balance both charges and atoms.

Step 1. Assign oxidation numbers.

$$\begin{array}{ccccc} Cr^{2+} & + & Cl_2 & \longrightarrow & Cr^{3+} & + & Cl^- \\ +2 & & 0 & & +3 & & -1 \end{array}$$

Steps 2 and 3. Write the half-reactions and add electrons where needed.

$$Cr^{2+} \longrightarrow Cr^{3+} + 1\ e^- \quad \text{oxidation}$$

$$Cl_2 + 2\ e^- \longrightarrow 2\ Cl^- \quad \text{reduction}$$

Step 4. Multiply the equations by numbers that make the numbers of electrons the same. We need only multiply the oxidation half-reaction by 2.

$$2\ Cr^{2+} \longrightarrow 2\ Cr^{3+} + 2\ e^- \quad \text{oxidation}$$

$$Cl_2 + 2\ e^- \longrightarrow 2\ Cl^- \quad \text{reduction}$$

Step 5. Add the two half-reactions together. Cancel terms.

$$2\ Cr^{2+} + Cl_2 + \cancel{2\ e^-} \longrightarrow 2\ Cr^{3+} + 2\ Cl^- + \cancel{2\ e^-}$$

$$2\ Cr^{2+} + Cl_2 \longrightarrow 2\ Cr^{3+} + 2\ Cl^-$$

Step 6. Check. The atoms of each kind balance, and so do the charges. A total of four positive charges is shown on the right side as well as on the left side.

When liquid bromine is added to solid red phosphorus, a vigorous reaction takes place, producing PBr$_3$.

Figure 16.2 Colorless oxalic acid, H$_2$C$_2$O$_4$, is being titrated with the intensely purple oxidizing agent, potassium permanganate, KMnO$_4$, forming products that have little or no color. When all the oxalic acid is used up, the purple color of excess MnO$_4^-$ in the flask signals the end of the titration.

> **Exercise 16.1**
>
> Balance the following equations.
> (a) $Cu^{2+} + CN^- \longrightarrow Cu^+ + (CN)_2$
> (b) $P + Br_2 \longrightarrow PBr_3$

Balancing Redox Reactions with Acid or Base

Another method of balancing equations takes advantage of the fact that many reactions occur in acidic or basic solutions. This method does not require oxidation numbers to be assigned to all the atoms in the equation. We will examine the steps for this method as we balance the reaction of permanganate ion with oxalic acid (Figure 16.2).

$$MnO_4^- + H_2C_2O_4 \longrightarrow Mn^{2+} + CO_2$$

Guidelines for Balancing Redox Equations Containing Acid or Base

1. *Divide the equation into two half-reactions, recognizing that elements other than oxygen and hydrogen must appear on both sides of each equation. (Often, oxygen and hydrogen will appear only on one side of the equation at this stage.)* This equation has the following half-reactions.

$$MnO_4^- \longrightarrow Mn^{2+}$$

Note that Mn appears on both sides of the equation; however, O does not.

$$H_2C_2O_4 \longrightarrow CO_2$$

Although C and O appear on both sides of the equation, H does not.

2. *Balance all the atoms except oxygen and hydrogen.*

$$MnO_4^- \longrightarrow Mn^{2+}$$

(Mn appears once on each side of the equation.)

$$H_2C_2O_4 \longrightarrow CO_2$$

(In this half-reaction, two C atoms appear on the left and one on the right. A 2 is placed in front of CO$_2$ to balance the C.)

$$H_2C_2O_4 \longrightarrow 2\ CO_2$$

3. *In acidic solution, because H$_2$O and H$^+$ are available, add H$_2$O to the side that needs oxygen then add H$^+$ to the side that needs hydrogen. In basic solution, because OH$^-$ and H$_2$O are available, add OH$^-$ to the side that needs oxygen and add H$_2$O to the side that needs hydrogen.*

We are told that this reaction takes place in acid. To the first half-reaction we add H$_2$O to the right side and H$^+$ to the left side.

$$MnO_4^- + H^+ \longrightarrow Mn^{2+} + H_2O$$

To the second half-reaction we add H$^+$ to the right side. The O balances.

$$H_2C_2O_4 \longrightarrow 2\ CO_2 + H^+$$

4. *Balance the hydrogen and oxygen in each half-reaction.* Use the trial-and-error method if necessary.

$$MnO_4^- + 8\,H^+ \longrightarrow Mn^{2+} + 4\,H_2O$$

$$H_2C_2O_4 \longrightarrow 2\,CO_2 + 2\,H^+$$

Although H^+ ions exist as hydronium ions, H_3O^+ in aqueous solution, we will write H^+ to simplify balancing redox equations.

5. *Balance charges on each side of each half-reaction by adding electrons, e^-.* Be sure that the total charge is the same on both sides of the equation.

$$MnO_4^- + 8\,H^+ + 5\,e^- \longrightarrow Mn^{2+} + 4\,H_2O \qquad \text{reduction}$$

The total charge is +2 on each side of the reduction half-reaction.

$$H_2C_2O_4 \longrightarrow 2\,CO_2 + 2\,H^+ + 2\,e^- \qquad \text{oxidation}$$

The total charge is 0 on each side of the oxidation half-reaction.

6. *Multiply all the terms in each half-reaction by a number that will make the number of electrons used in each half-reaction the same.* In this example, multiply the reduction half-reaction by 2 and multiply the oxidation half-reaction by 5.

$$2\,MnO_4^- + 16\,H^+ + 10\,e^- \longrightarrow 2\,Mn^{2+} + 8\,H_2O$$

$$5\,H_2C_2O_4 \longrightarrow 10\,CO_2 + 10\,H^+ + 10\,e^-$$

7. *Add the two half-reactions together and cancel out the terms that appear on both sides.* If the process has been done correctly, the numbers of electrons will cancel each other and should not appear in the final equation.

$$2\,MnO_4^- + 16\,H^+ + 5\,H_2C_2O_4 + \cancel{10\,e^-} \longrightarrow 2\,Mn^{2+} + 8\,H_2O + 10\,CO_2 + 10\,H^+ + \cancel{10\,e^-}$$

$$2\,MnO_4^- + 16\,H^+ + 5\,H_2C_2O_4 \longrightarrow 2\,Mn^{2+} + 8\,H_2O + 10\,CO_2 + 10\,H^+$$

Now cancel out the H^+ that appear on both sides. There are 10 H^+ on the right that will cancel with 10 H^+ of the 16 H^+ on the left.

$$2\,MnO_4^- + 6\,H^+ + 5\,H_2C_2O_4 \longrightarrow 2\,Mn^{2+} + 8\,H_2O + 10\,CO_2$$

You may wish to change H^+ to H_3O^+. If you do, add an equal number of H_2O to the opposite side of the equation.

$$2\,MnO_4^- + 6\,H_3O^+ + 5\,H_2C_2O_4 \longrightarrow 2\,Mn^{2+} + 14\,H_2O + 10\,CO_2$$

8. *Check to be sure that all the atoms and charges balance.* The total charge is +4 on each side of the equation.

EXAMPLE 16.3

Dichromate ion, $Cr_2O_7^{2-}$, is a powerful oxidizing agent that is frequently used in laboratory reactions. Balance the following equation.

$$Cr_2O_7^{2-} + I^- + H_2SO_4 \longrightarrow Cr_2(SO_4)_3 + I_2 + SO_4^{2-}$$

Strategy and Solution:

Step 1. Separate the equation into two half-reactions. Select a reactant, say, $Cr_2O_7^{2-}$, and its corresponding products, in this case $Cr_2(SO_4)_3$ and SO_4^{2-}.

Begin your first half-reaction with these. Recognizing that in a half-reaction each element except oxygen and hydrogen must appear on both sides of the equation, add H_2SO_4 to the left side. Now, except for H, all the elements represented on the left also appear on the right.

$$Cr_2O_7^{2-} + H_2SO_4 \longrightarrow Cr_2(SO_4)_3 + SO_4^{2-}$$

That's the first half-reaction. What remains are the I^- and I_2. These make the other half-reaction.

$$I^- \longrightarrow I_2$$

Step 2. Balance all the atoms except oxygen and hydrogen. In this first half-reaction, SO_4^{2-} must appear four times on each side. In the second half-reaction, the iodine must appear on the both sides in multiples of two.

$$Cr_2O_7^{2-} + 4\ H_2SO_4 \longrightarrow Cr_2(SO_4)_3 + SO_4^{2-}$$

$$2\ I^- \longrightarrow I_2$$

Step 3. The reaction takes place in acid, H_2SO_4. In acidic solution, add H_2O to the side that needs oxygen and H^+ to the side that needs hydrogen.

$$Cr_2O_7^{2-} + 4\ H_2SO_4 + H^+ \longrightarrow Cr_2(SO_4)_3 + SO_4^{2-} + H_2O$$

The second half-reaction needs neither oxygen nor hydrogen.

Step 4. Balance the hydrogen and oxygen in each half-reaction. Use the trial-and-error method if necessary.

$$Cr_2O_7^{2-} + 4\ H_2SO_4 + 6\ H^+ \longrightarrow Cr_2(SO_4)_3 + SO_4^{2-} + 7\ H_2O$$

Step 5. Balance the charge on each side of the half-reactions by adding electrons, e^-. Make sure that the total charge is the same on both sides of the equation.

$$Cr_2O_7^{2-} + 4\ H_2SO_4 + 6\ H^+ + 6\ e^- \longrightarrow Cr_2(SO_4)_3 + SO_4^{2-} + 7\ H_2O \quad \text{reduction}$$

$$2\ I^- \longrightarrow I_2 + 2\ e^- \quad \text{oxidation}$$

Step 6. Multiply all the terms in each half-reaction by a number that will make the number of electrons used in each half-reaction the same. In this example, multiply the oxidation half-reaction by 3.

$$Cr_2O_7^{2-} + 4\ H_2SO_4 + 6\ H^+ + 6\ e^- \longrightarrow Cr_2(SO_4)_3 + SO_4^{2-} + 7\ H_2O$$

$$6\ I^- \longrightarrow 3\ I_2 + 6\ e^-$$

Step 7. Add the two new half-reactions together and cancel out the terms that appear on both sides. If the process has been done correctly, the numbers of electrons will cancel each other and should not appear in the final equation.

$$Cr_2O_7^{2-} + 4\ H_2SO_4 + 6\ H^+ + 6\ I^- \longrightarrow Cr_2(SO_4)_3 + SO_4^{2-} + 7\ H_2O + 3\ I_2$$

Step 8. Check to be sure that all the atoms and charges balance.

Exercise 16.2

Balance the following equations. The reactions take place in acidic solution.
(a) $Sn^{2+} + ClO_4^- \longrightarrow Cl^- + Sn^{4+}$
(b) $Cu + H_2AsO_4^- \longrightarrow AsH_3 + Cu^{2+}$

EXAMPLE 16.4

The following reaction takes place in basic solution. Balance the equation using half-reactions.

$$Mg + H_2O \longrightarrow Mg(OH)_2 + H_2$$

Strategy and Solution:

Step 1. Divide the equation into two half-reactions, recognizing that the same elements other than oxygen and hydrogen must appear on both sides of the equation.

$$Mg \longrightarrow Mg(OH)_2$$

Notice that only H_2O and H_2 remain. The second half-reaction must consist of these.

$$H_2O \longrightarrow H_2$$

Step 2. Balance all the atoms except oxygen and hydrogen. In this example, they are already balanced, so we can skip this step.

Steps 3 and 4. This reaction takes place in base. In basic solution, add H_2O to the side that needs hydrogen and add OH^- to the side that needs oxygen. Balance the oxygen and hydrogen.

$$Mg + 2\ OH^- \longrightarrow Mg(OH)_2$$

$$2\ H_2O \longrightarrow H_2 + 2\ OH^-$$

Step 5. Balance the charge on each side of the half-reactions by adding electrons, e^-. Be sure that the total charge is the same on both sides of the equation.

$$Mg + 2\ OH^- \longrightarrow Mg(OH)_2 + 2\ e^- \qquad \text{oxidation}$$

$$2\ H_2O + 2\ e^- \longrightarrow H_2 + 2\ OH^- \qquad \text{reduction}$$

Step 6. Multiply all the terms in each half-reaction by a number that will make the number of electrons used in each half-reaction the same. In this example, the number of electrons in each is the same.

Step 7. Add the two new half-reactions together and cancel out the terms that appear on both sides. If the process has been done correctly, the numbers of electrons will cancel each other and should not appear in the final equation.

The reaction of aluminum metal with NaOH solution produces bubbles of hydrogen gas.

$$Mg + 2\,H_2O \longrightarrow Mg(OH)_2 + H_2$$

Step 8. Check to be sure that all the atoms and charges balance.

Exercise 16.3

Balance the following equation without assigning oxidation numbers. (Recall that NH_3 is a weak base.)

$$CuO + NH_3 \longrightarrow Cu + N_2 + H_2O$$

Exercise 16.4

Unlike most metals, aluminum reacts with bases. Balance the following equation without assigning oxidation numbers.

$$Al + NaOH + H_2O \longrightarrow NaAl(OH)_4 + H_2$$

16.3 Some Important Redox Reactions

After studying this section, you should be able to:

● Write balanced chemical equations for some common but important redox reactions.

In this section, we review several redox reactions important to life and civilization. Although the indiscriminate memorization of chemical reactions is not of much use, do study this section carefully. For practice, write the oxidation numbers of the reactants and products. If you understand and remember what follows, you will gain a feeling for everyday, working chemistry.

Reactions Used Frequently in Chemistry Laboratories

Reduction by Active Metals

Active metals, usually Na, Li, Al, or Zn, are used as reducing agents in the laboratory. A common reaction is that of Zn with acid (Figure 16.3).

$$Zn + 2\,H_3O^+ \longrightarrow Zn^{2+} + H_2 + 2\,H_2O$$

In this reaction, Zn is oxidized to Zn^{2+}, and hydrogen is reduced from $+1$ to 0.

Oxidation by Dichromate or Permanganate Ions

When chemists want to oxidize a substance, the oxidizing agent of choice is often either dichromate ion, $Cr_2O_7^{2-}$, or permanganate ion, MnO_4^-. Both are strong, well-behaved oxidizing agents, the strength of which depends on the concentration of the acid in the solution. We saw a reaction involving each of these reagents when we were balancing equations. Here is another example.

Figure 16.3 The reaction of zinc metal with acid readily produces H_2 gas. When chemists need a little H_2, they almost automatically think of adding some zinc to an acid.

Dichromate ion will oxidize bromide ion to the elemental halogen, even though bromine is itself a strong oxidizing agent.

$$Cr_2O_7^{2-} + 6\ Br^- + 14\ H_3O^+ \longrightarrow 2\ Cr^{3+} + 3\ Br_2 + 7\ H_2O$$

Oxidation by Nitrate Ion

Nitrate ion, usually in the form of nitric acid, is frequently used as an oxidizing agent. For example, nitric acid will oxidize any metal except Au, Pt, Rh, and Ir. The reaction is so sensitive to the concentration of H_3O^+ that variations in the concentration of the acid itself will cause nitric acid to react differently with the same substances. Here are two reactions of nitric acid with copper. In concentrated solution (Figure 16.4), the nitrogen in nitrate is reduced from the +5 oxidation state to +4.

$$Cu + 4\ H_3O^+ + 2\ NO_3^- \longrightarrow Cu^{2+} + 2\ NO_2 + 2\ H_2O$$

In dilute solution, however, the nitrogen is reduced much further, all the way to the +2 state, and the proportion of acid to copper changes.

$$3\ Cu + 8\ H_3O^+ + 2\ NO_3^- \longrightarrow 3\ Cu^{2+} + 2\ NO + 4\ H_2O$$

How far nitrate ion is reduced also depends on the strength of the reducing agent. Because Zn is a stronger reducing agent than Cu, the reaction with Zn takes the nitrogen all the way to the −3 oxidation state.

$$4\ Zn +\ 10\ H_3O^+ + NO_3^- \longrightarrow 4\ Zn^{2+} + NH_4^+ + 3\ H_2O$$

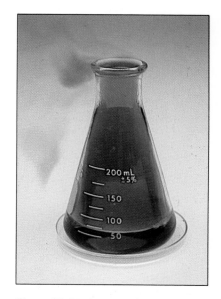

Figure 16.4   Copper metal and nitric acid, HNO_3, react to form brown gaseous nitrogen dioxide, NO_2. The resulting solution turns blue-green due to the presence of Cu^{2+} ions.

Industrial Redox Reactions

Here are three examples of important industrial redox reactions.

Combustion

The most common redox reaction is probably that of the combustion of any burnable material with atmospheric O_2. Compounds containing carbon and hydrogen are most often oxidized to produce CO_2, H_2O, and energy. The combustion of octane, one of the compounds in gasoline, is an example.

$$2\ C_8H_{18} + 25\ O_2 \longrightarrow 16\ CO_2 + 18\ H_2O$$

The Haber Process for Making Ammonia

Among the most important compounds essential to nature and to our industrial world are those of nitrogen (Figure 16.5). But, despite the enormous amount of N_2 in the air, compounds of nitrogen are in short supply. The problem is that N_2 is difficult either to reduce or oxidize. Commercially, most nitrogen compounds have their beginning in the Haber process for converting N_2 to NH_3 by using a catalyst at high pressure.

$$3\ H_2(g) + N_2(g) \xrightarrow[\text{500°C, 300 atm}]{\text{catalyst}} 2\ NH_3(g)$$

Figure 16.5   Nitrogen can be added to soil by spraying with ammonia gas.

The Ostwald Oxidation of Ammonia with O_2 and a Pt Catalyst

Much of the use of nitrogen is in compounds in which the N has positive oxidation numbers, as in NO_3^-. An essential step in preparing such compounds is the oxidation of NH_3 in the Ostwald process.

$$4\ NH_3 + 5\ O_2 \xrightarrow{\text{Pt catalyst}} 4\ NO + 6\ H_2O$$

The Water–Gas Reactions for Making H_2

There are several commercial processes for making H_2. Which one is used in a particular situation depends on several factors, including raw material supply. In one process, the starting materials are water and coke for carbon. The process consists of two reactions, both run at high temperatures. In the first reaction, steam is passed over red-hot coke, yielding carbon monoxide.

$$H_2O + C \longrightarrow H_2 + CO$$

The second reaction separates the CO from the H_2. More steam is added, and the mixture is passed over a metallic catalyst. The CO then reacts with the steam to form CO_2 and more H_2. The H_2 in the original mixture remains unreacted.

$$CO + H_2O \xrightarrow{\text{catalyst}} H_2 + CO_2$$

Biological Redox Reactions

Fermentation

Fermentation is used to make many kinds of products, including ethanol, which is used in beverages. The reaction of glucose is accelerated by a biological

Profiles in Science

Wilhelm Ostwald (1853–1932)

Born in 1853, the son of a Latvian barrel maker, Ostwald was interested in experimentation. While still a boy, he designed and made hundreds of fireworks, experimented with collages, and made simple optical devices.

As a college undergraduate, he began the serious study of chemistry. While in graduate school, he was the first person to measure the rates of chemical reactions. During the following years, he made substantial contributions to the chemistry of redox reactions, particularly those involving electric current.

Ostwald's fame, however, arose from his study of catalysis (see Section 15.8). In 1901, he proposed the definitive description of catalysis as we know it today. While studying the catalytic oxidation of ammonia, Ostwald developed the process that bears his name. For his catalysis work, including the ammonia oxidation, he received the Nobel Prize in 1909.

Before the First World War, Germany was badly deficient in nitrogen compounds, and much effort was expended to develop a local supply. Beginning with ammonia from the Haber process, Germany used the Ostwald method to produce more than 50 million pounds of nitric acid in 1917 alone. Most of this went into munitions used in World War I.

Now the Ostwald process is used worldwide to produce nitric acid for fertilizer and for the chemical industry, as well as for munitions.

---

catalyst, called an enzyme. In this reaction, both the reducing agent and the oxidizing agent are carbon; some carbon atoms are oxidized and some are reduced. A redox reaction in which the same species is both reduced and oxidized is called a **disproportionation reaction.**

$$C_6H_{12}O_6 \xrightarrow{\text{catalyst}} 2\ C_2H_5OH + 2\ CO_2$$
$$\text{glucose} \qquad\qquad \text{ethylalcohol}$$

Respiration

In all aerobic living organisms, glucose is oxidized by oxygen (just as in combustion) to produce energy.

$$C_6H_{12}O_6 + 6\ O_2 \longrightarrow 6\ CO_2 + 6\ H_2O + \text{energy}$$

In this reaction, carbon is oxidized from an oxidation number of zero to $+4$, while oxygen from O_2 molecules is reduced from zero to -2.

Photosynthesis

Plants use the energy of sunlight to create glucose by a process opposite to that of respiration.

$$6\ CO_2 + 6\ H_2O + \text{energy} \longrightarrow C_6H_{12}O_6 + O_2$$

Photosynthesis, a process performed by plant life, can be considered the ultimate energy source for life on earth.

Science in Action

Starting with a purified oxide ore, pure metals can be produced using a reducing agent. An example is the "thermite reaction" of Fe_2O_3 and metallic aluminum (Figure 16.6).

$$Fe_2O_3 + Al \longrightarrow 2 Fe + Al_2O_3$$

The process is quite exothermic, so much so that it is sometimes used to produce melted iron for welding.

In many cases, carbon or carbon monoxide is used as the reducing agent to purify metals. Most of the iron used for steel making is made in a blast furnace, where a form of coal called coke furnishes the carbon for the reaction with oxygen to form CO. The carbon monoxide then reacts with the iron ore.

$$Fe_2O_3 + 3 CO \longrightarrow 2 Fe + 3 CO_2$$

Solid materials—ore, coke, and limestone—are constantly added to the top of the furnace. Oxygen is blown through from the bottom, burning the coke and producing a temperature of about 1,600°C, and the molten iron is continually drawn off from a tap in the bottom of the furnace. A blast furnace, usually as high as a several-story building, will produce about two million pounds of iron daily. Once started, blast furnaces are operated around-the-clock for months, stopping only for necessary maintenance.

Reduction of Metal Ores with Reducing Agents

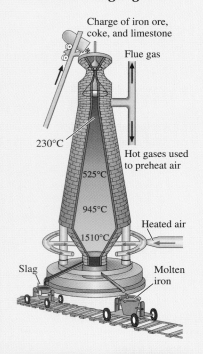

A blast furnace.

Figure 16.6 The thermite reaction releases so much energy that white-hot molten iron is the product.

Nitrogen Fixation

Bacteria in the roots of certain plants such as beans, peas, clover, and alfalfa are able to "fix" nitrogen from the air. The process, which is not fully understood, involves the action of a biological compound called nitrogenase. Nitrogen compounds, which are estimated to amount to as much as 10^8 tons per year, are added to the soil when the plants decay. Such compounds are essential to plant and animal growth. In this overall process, nitrogen begins with an oxidation number of zero and finishes (as a nitrate) with oxidation number +5.

The amount of nitrogen "fixed" by industrial processes such as the Haber process is only 2×10^7 tons per year or about a fifth as much as that produced by natural means.

Metallurgical Redox Reactions

Smelting

The smelting of metal ores to obtain elemental metals is a reduction process. In the case of ores that are oxides, such as Fe_2O_3, a reducing agent is added to the ore, and the mixture is heated. With other ores, such as the sulfides, simple roasting in air is sufficient. One of the sources of copper is the ore chalcocite, CuS, which, on roasting, yields an impure copper metal called "blister copper."

$$Cu_2S + O_2 \longrightarrow 2\,Cu + SO_2$$

Corrosion

The redox reaction of metals with components of the atmosphere, destroying the metal and producing oxides and other compounds, is called corrosion. Billions of dollars are spent annually to replace metal parts that have corroded. The corrosion called rust is a major problem. Rust is iron(III) oxide, and the rusting of iron is oxidation by oxygen in the air.

$$4\,Fe(s) + 3\,O_2(g) \longrightarrow 2\,Fe_2O_3(s)$$

Most metals form thin coatings of oxides, which often protect them from further oxidation. Aluminum, an extremely active metal, rapidly oxidizes upon contact with air. However, a thin coat of Al_2O_3 forms and binds tightly to the aluminum surface, excluding air and preventing further oxidation. The oxide coat of iron, on the other hand, is slightly permeable to O_2, which allows rusting to continue. Serious rusting usually takes place by an accelerated process in which the electron transfer from Fe to O involves an electric current flowing from the iron metal to spots of oxygen-containing rust. Moisture is necessary for the process, and the rate of the reaction is greatly increased by the presence of ions such as Cl^-, which form reactive chloro complexes with the iron. Because of this, the presence of salt in the air near the ocean or salt on the road to melt snow and ice accelerates the oxidation process.

Destruction of property through corrosion causes losses of up to a hundred billion dollars yearly in the United States.

16.4 Redox Reactions in Batteries

After studying this section, you should be able to:

- Describe how a battery functions.
- Identify the fundamental parts of a battery.

Science in Action

Explosives

M ost of us think of dynamite and demolitions or of warfare when we hear the word "explosive," but explosives have many peacetime uses, too. They are used in mining of all kinds, in construction work (for preparing building sites and for driving studs into concrete), in space exploration (explosive bolts separate the shuttle from its boosters and do many other jobs), and in industry (where, among other jobs, explosions force metal into molds, making intricately shaped parts). The list goes on and on.

A dictionary defines an explosion as a sudden, violent change of a solid or liquid into a quickly expanding gas. Although the largest explosions are nuclear, most are chemical reactions between a fuel and an oxidizing agent. Simple explosions, such as a house blowing up because of escaping gas or the new fuel-air bomb used by the military, are reactions of oxygen and some compound of hydrogen and carbon, producing carbon dioxide and water. But explosions can also be produced by single compounds, the molecules of which contain both a fuel and an oxidizing agent.

Although it was evidently known in China more than 20 centuries ago, gunpowder, the earliest known explosive, was introduced to the Western world in Germany about 1,250 AD. In gunpowder, the oxidizing agent is "saltpetre" (potassium nitrate, KNO_3), which reacts rapidly with sulfur and charcoal, producing mostly CO_2, K_2S, and N_2. In 1847, nitroglycerine, $C_3H_5O_3(NO_2)_3$, was discovered. This is a tricky, sensitive, dangerous, oily liquid, which explodes violently when jarred. Alfred Nobel of Sweden first produced nitroglycerine commercially and, by mixing it with sawdust and other inert materials, invented dynamite. The wealth he gained from this discovery enabled him to establish the Nobel Prizes. In nitroglycerine, the oxidizing agent is the nitro group, NO_2. Other nitro compounds, such as TNT and nitrocellulose (the major component of modern gunpowder), react in a similar manner.

Figure 16.7 When a strip of zinc metal is placed in a $CuSO_4$ solution, copper metal forms on the zinc strip. The blue color of the solution fades. Simultaneously, zinc metal enters the solution as Zn^{2+} ions.

Let's examine the oxidation-reduction reaction that occurs when a strip of zinc metal is placed in a solution of copper sulfate, $CuSO_4$ (Figure 16.7). The solution is initially blue, but as the reaction proceeds, the blue color lightens, the amount of zinc metal becomes less, and orange-red metallic copper appears. The reaction occurs spontaneously, releasing energy.

$$Zn + Cu^{2+} \longrightarrow Zn^{2+} + Cu + energy$$

The zinc metal reacts to release Zn^{2+}, while copper metal forms from Cu^{2+} ions.

The half-reactions are

$$Zn \longrightarrow Zn^{2+} + 2\ e^- \qquad oxidation$$

$$Cu^{2+} + 2\ e^- \longrightarrow Cu \qquad reduction$$

Two electrons are transferred from each Zn atom to a Cu^{2+} ion, forming a Cu atom.

Recall that electricity, or electric current, results from the flow of electric charge or electrons through a wire or other conductor. By connecting the two half-reactions of our experiment in such a way that transferred electrons must flow through an exterior wire, it is possible to create an electric current. Figure 16.4 illustrates how this can be accomplished. In this arrangement, electrons are carried through the wire, rather than being handed over directly as in an ordinary electron-transfer reaction.

The two half-reactions proceed in what are known as **half-cells.** Physically, the half-cells have metal strips or plates called **electrodes** that are connected by an external wire that conducts electrons between them. Figure 16.4 shows the zinc electrode of one half-cell and the copper electrode of the other.

The electrodes are immersed in aqueous solutions of electrolytes, to which we were first introduced in Chapter 13. Recall that electrolytes dissociate into ions, which conduct electricity in solution. A solution of Na_2SO_4, which is placed in the container with the zinc electrode, doesn't take part in the redox reaction but is used only to provide conducting ions. $CuSO_4$ solution is placed in the container with the copper electrode. This blue solution conducts and also furnishes copper ions, which are reduced during reaction. Conduction between the half-cells is accomplished by various ions, which must be able to move between the cells. One of the ways in which this is made possible is by a salt bridge, as pictured in Figure 16.4. The salt bridge prevents the flow of solutions but allows the passage of ions. A complete electric circuit results; the wire and the electrodes conduct electrons, and the electrolytic solutions conduct ions.

At the Zn electrode, Zn atoms undergo oxidation, each atom giving up two electrons to become Zn^{2+} ions in solution.

$$Zn \text{ (metal)} \longrightarrow Zn^{2+} + 2 \text{ e}^- \quad \text{oxidation}$$

An electrode at which oxidation takes place is called the **anode** (where the anions go).

The electrons flow through the wire into the Cu electrode, which becomes negatively charged. Cu^{2+} ions migrate from the $CuSO_4$ solution to the Cu electrode and accept electrons to become Cu metal, which forms or "plates out" on the Cu electrode.

$$Cu^{2+} + 2 \text{ e}^- \longrightarrow Cu(\text{metal}) \quad \text{reduction}$$

It is popularly thought that water conducts electricity readily, but this is not so. Pure H_2O conducts hardly at all, but even dilute solutions of ions conduct easily.

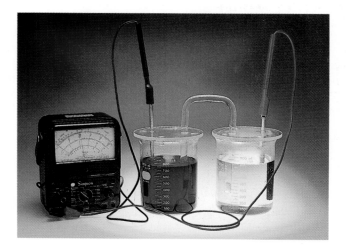

Figure 16.8   We can make a battery by using the Cu–Zn pair. On the left, a bar of copper dips into a solution containing Cu^{2+} ions; on the right, a Zn bar dips into a Na_2SO_4 solution. Ions pass through a salt bridge, the tube connecting the beakers. Electrons pass through an external circuit, in this case, wires and a voltmeter.

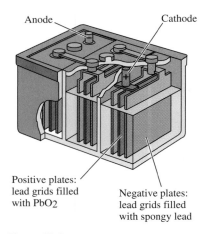

Anode   Cathode

Positive plates:
lead grids filled
with PbO2

Negative plates:
lead grids filled
with spongy lead

Figure 16.9   An automotive battery.

A negative electrode (to which cations are attracted) is called the **cathode.** The Cu^{2+} cations undergo reduction at the cathode.

The overall reaction occurs spontaneously with the release of energy. The chemical energy of this reaction causes electrons to flow through the wire, and the resulting electricity can be used as an energy source. The Cu^{2+}/Zn system is a **battery,** a device that converts the chemical energy of an electron-transfer reaction to electric energy. Many spontaneous redox reactions can be used to make batteries; this particular one with its Cu^{2+}/Zn system is called a *Daniell cell.* Such batteries were used in the United States to provide electricity for early telegraph systems. We now have far better batteries, and the Daniell cell has become a laboratory curiosity.

Automotive batteries are composed of a number of connected, individual batteries or cells. An automotive battery is shown in Figure 16.9. The electrodes are stacks of waffle-like plates made of a lead alloy. The recesses of half the plates contain pure lead; PbO_2 fills the recesses of the others. The electrolytic solution is aqueous H_2SO_4. When the battery is being discharged (when electricity is being produced), PbO_2, Pb, and H_3O^+ all react. Solid $PbSO_4$, the product of both half-reactions, forms a coating on both sets of plates. The oxidation half-reaction at the anode is

$$\underset{0}{Pb} + SO_4^{2-} \longrightarrow \underset{+2}{PbSO_4} + 2\ e^-$$

while the reduction half-reaction at the cathode is

$$\underset{+4}{PbO_2} + SO_4^{2-} + 4\ H_3O^+ + 2\ e^- \longrightarrow \underset{+2}{PbSO_4} + 2\ H_2O$$

(So that you can follow the changes, the oxidation numbers for Pb in its different states are written under the symbols.)

A battery or a mechanical device called a generator can supply electricity, but neither supplies the actual electric charges. Batteries and generators are merely pumps; they cause existing electric charges to flow through conductors, in the same way that a water pump can cause existing water to flow through a garden hose. Just as a sample of water can be pumped over and over through a closed loop of hose, so can electric charges flow through a continuous loop, the electric circuit. A switch is just a way to open and close the circuit, to allow or interrupt the flow of electricity.

16.5 Electrolytic Cells

After studying this section, you should be able to:

- Describe how electrolytic cells function.
- Explain the process of electroplating.

If, by some external means, say a generator, electricity is forced *backward* through a battery, the electron-transfer reaction will occur in the opposite direction. That is, forcing electric charge through the half-cells in a direction opposite to that of the normal flow reverses the reaction. This is what we call recharging a battery. To recharge an automotive battery, electrons are forced into the anode by an external source, reversing the two half-reactions. This causes $PbSO_4$ to dissolve and creates new Pb and PbO_2 on the plates.

The reverse of the Daniell cell reaction would be

$$Cu + Zn^{2+} + energy \longrightarrow Cu^{2+} + Zn$$

Energy is necessary to reverse the reaction. If the generator is switched off, the reaction will proceed in the normal direction and become a battery once again, producing energy.

Cells that use electric energy to cause redox reactions to take place are called **electrolytic cells,** and such a reaction is called **electrolysis.**

Electrolysis is important in laboratory experiments and especially in industrial processes. For example, nearly all the metallic copper that is used to make electric wire, water pipe, pennies, and countless other products is electrolytically refined. A block of raw copper serves as an anode, which slowly disappears as the copper atoms lose their electrons and become cations. The cations then migrate to the cathode, gain electrons, and plate out as pure copper metal. If the raw copper contains atoms of less-active metals such as Ag, these atoms refuse to give up their electrons and never go into solution. Atoms of more active metals such as Zn easily give up their electrons but do not accept electrons at the cathode; they stay in the solution as cations. Thousands of tons of copper are purified in this way every year.

In Chapter 14, we discussed the importance and some of the chemistry of sodium hydroxide, NaOH. Now we have the information needed to discuss its manufacture. NaOH is often made from metallic sodium formed in an electrolytic cell like that diagrammed in Figure 16.10. Electric current from an

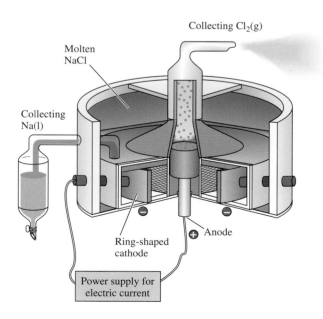

Collecting $Cl_2(g)$

Molten NaCl

Collecting Na(l)

Ring-shaped cathode

Anode

Power supply for electric current

Figure 16.10   Molten NaCl at high temperature is commercially electrolyzed to yield chlorine gas, Cl_2, and sodium metal, Na, which is liquid at these temperatures.

Science in Action

Electroplating and Electroforming

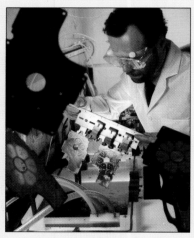

In this picture, the brown objects held by clips are being inserted into the blue solution of $CuSO_4$, where copper will be electrolytically deposited on them to form electronic circuits.

For more than 150 years, electrolytic reactions have been used to electroplate metal coatings onto other metals. More recently, with significant advances even in the last decade, the method has also been used to deposit metal coatings on nonmetallic materials such as plastics.

In the electroplating process, the object to be electroplated, the cathode, is immersed in a container of the plating solution, which contains positive ions of the metal to be deposited. The anode is a bar of the plating metal. Electric current is passed through the cell. Metal ions move to the cathode, receive electrons, and become metal atoms that stick to the surface. In this way, a layer of metal is built up. Metal atoms in the anode give up electrons, becoming positive ions that replace those reduced at the cathode. Thus the concentration of metal ions in solution remains constant.

Automobile bumpers, made of ordinary steel, are electroplated, usually in three steps. After careful and specialized cleaning, the bumper is first plated with a thin coat of copper to assure uniformity of further layers of plating. The bumper is removed from the copper solution, cleaned, and given a preliminary polish. Then a thick coat of nickel, perhaps 0.1 mm thick, is plated over the copper. Finally, a thin cosmetic coat of chromium is added, and the bumper is given its final polish. Properly done, such a plating will last decades under severe weather conditions.

Many different kinds of metal are used for electroplating. Among those in common use are cadmium (which is inexpensive and is used for preventing bolts, screws, and other small metal parts from rusting), chromium (used to give a hard, shiny, unreactive surface when cost is less of a factor), silver and gold (used in tableware, jewelry, and other expensive products), and copper.

Many plastic objects are now electroplated, but only fairly recently has it been possible to make plastic surfaces conduct electricity satisfactorily so that they can act as cathodes.

Also fairly recently, a process called **electroforming** has come into wide use. A conducting mold cavity, or sometimes a molded wax figure, is heavily plated with a metal. Then the mold is taken off, or the wax is melted out, leaving only the metal plate in the desired shape. The empty shell of plated metal can then be filled with plaster or melted metal to make the final object.

Because the reaction products of NaCl electrolysis react with electrodes made of metal, carbon cathodes and anodes are used. These conduct electricity and are chemically inactive.

exterior source is passed through melted NaCl. This causes Na^+ ions to move to the negative electrode (cathode) and become reduced by accepting electrons.

$$Na^+ + e^- \longrightarrow Na \text{ (in metallic form)}$$

The molten metallic Na rises to the top of the molten NaCl, is removed, and is later reacted with water to give NaOH and hydrogen gas.

$$2\ Na + 2\ H_2O \longrightarrow 2\ NaOH(aq) + H_2$$

As the Na^+ ions are being reduced to Na, Cl^- anions are being attracted to the anode and oxidized by giving up electrons (the oxidation step),

$$Cl^- \longrightarrow Cl \text{ (atom)} + e^-$$

and Cl_2 gas is formed.

$$2\ Cl \text{ (atoms)} \longrightarrow Cl_2(g)$$

The Cl_2 gas bubbles off from around the anode and is captured for use in the processes discussed in Chapter 14.

Sometimes this electrolysis reaction is carried out in aqueous solution, in which case the Na reacts instantly to form NaOH. Special precautions are then taken to keep the Cl_2 from reaching the resulting basic solution, with which it would readily react.

Billions of metal products, such as the exposed metal on cars, much silverware, fountain pen clips, electrical contacts in satellites, paper clips, and "tin" cans for food, are **electroplated,** that is, they are electrically coated with another metal. Electroplating (Figure 16.11) is an electrolytic reaction.

"Tin" cans are not tin at all. Most are steel with a very thin plating of tin.

As we have learned long since, chemical reactions are often accompanied by a release of energy. A battery separates the oxidation and reduction half-reactions, sending the transferred electrons through a wire. The difference between a battery and, for example, an electroplating reaction, is this: In a battery, the electron transfer occurs spontaneously with the release of energy, while the electroplating reaction *absorbs* energy. Charging a dead battery is the electrolytic process of running the battery reaction in reverse.

Summary

Oxidation-reduction reactions, reactions in which electron transfer occurs, constitute a large proportion of all chemical reactions. When a chemical species *loses* electrons, it is oxidized, and the process is called oxidation. When it *gains* electrons, it is reduced. Gaining electrons reduces the charge of a particle, which is why the term "reduction" refers to electron gain.

Because electrons can flow through conductors, the electrons that are transferred in oxidation-reduction reactions can be directed through wires. Batteries and electrolytic cells (such as electroplating units) are examples of useful applications of oxidation-reduction reactions.

Balancing the equation for a complex oxidation-reduction reaction can be greatly simplified by dividing the reaction into half-reactions.

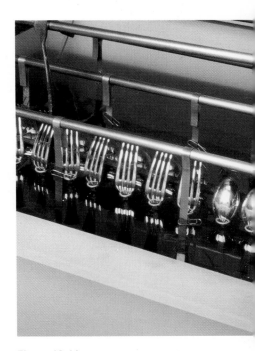

Figure 16.11 Electroplating tableware.

Key Terms

redox reaction	*(p. 474)*	electrode	*(p. 489)*	electrolytic cell	*(p. 491)*
oxidizing agent	*(p. 474)*	anode	*(p. 489)*	electrolysis	*(p. 491)*
reducing agent	*(p. 474)*	cathode	*(p. 490)*	electroplated	*(p. 493)*
half-cell	*(p. 489)*	battery	*(p. 490)*		

Questions and Problems

Section 16.2 Balancing Redox Equations

Questions

1. Why must the charges as well as the atoms balance in redox equations?

2. Which steps are the same for both methods of balancing redox equations?

Problems

3. Balance the following equations using the oxidation number method.
   (a) $Cr^{2+} + H_3O^+ + NO_3^- \longrightarrow Cr^{3+} + NO_2^- + H_2O$
   (b) $SO_4^{2-} + I^- + H_3O^+ \longrightarrow SO_2 + I_2 + H_2O$
   (c) $IO_3^- + N_2H_4 \longrightarrow N_2 + I^- + H_2O$
   (d) $H_2O + PH_4^+ + Sn^{4+} \longrightarrow Sn^{2+} + P_4 + H_3O^+$
   (e) $IO_3^- + NH_3 \longrightarrow N_2 + I_2 + OH^- + H_2O$

4. Balance the following equations using the oxidation number method.
   (a) $H_2O_2 + CH_3OH \longrightarrow CO_2 + H_2O$
   (b) $H_2O + CrO_4^{2-} + Fe^{2+} \longrightarrow Cr^{3+} + Fe^{3+} + OH^-$
   (c) $H_2O + H_2CO \longrightarrow CO_2 + H_2$
   (d) $MnO_4^- + FeO + H_2O \longrightarrow Fe_2O_3 + MnO_2 + OH^-$
   (e) $H_3O^+ + MnO_4^- + SCN^- \longrightarrow Mn^{2+} + HCN + H_2O + SO_4^{2-}$

5. Name the oxidizing agent and the reducing agent in each of the reactions in Problems 3 and 4.

6. Balance the following redox equations for reactions taking place in acidic solution.
   (a) $Cu + HNO_3 \longrightarrow Cu^{2+} + NO_2$
   (b) $MnO_4^- + H_2S \longrightarrow MnO_2 + S$
   (c) $Pb + PbO_2 + H_2SO_4 \longrightarrow PbSO_4$
   (d) $H_2Cr_2O_7 + Zn \longrightarrow Cr^{3+} + Zn^{2+}$

7. Balance the following redox equations for reactions taking place in acidic solution.
   (a) $HNO_3 + Zn \longrightarrow Zn^{2+} + NO$
   (b) $H_2O_2 + CH_3CH_2OH \longrightarrow CO_2 + H_2O$
   (c) $C_2H_6 + O_2 \longrightarrow CO_2 + H_2O$
   (d) $As_2O_3 + CrO_4^{2-} \longrightarrow H_3AsO_4 + Cr^{3+}$
   (e) $NO_2 \longrightarrow HNO_3 + NO$

8. Balance the following redox equations for reactions taking place in basic solution.
   (a) $K_2S + I_2 + KOH \longrightarrow K_2SO_4 + KI$
   (b) $Bi(OH)_3 + Cu \longrightarrow Bi + Cu^{2+}$
   (c) $NO_2^- + Br_2 \longrightarrow NO_3^- + Br^-$
   (d) $P_4 + NaOH \longrightarrow PH_3 + NaH_2PO_2$
   (e) $Ag + CN^- + O_2 \longrightarrow Ag(CN)_2^- + OH^-$

9. Balance the following redox equations for reactions taking place in basic solution.
   (a) $NH_3 + O_2 \longrightarrow NO$
   (b) $Br_2 \longrightarrow Br^- + BrO^-$

(c) $SnO_2^{2-} + Bi^{3+} \longrightarrow SnO_3^{2-} + Bi$
(d) $Cl^- + MnO_4^- \longrightarrow ClO_4^- + MnO_4^{2-}$

Section 16.4 Batteries

Questions

10. When constructing a battery, why would you separate the solutions with a salt bridge?

11. What characteristics of a reaction are necessary for it to be used as a battery?

12. Why must batteries be operated in two separate half-cells?

13. At what electrode of a battery does oxidation take place? Reduction?

Section 16.5 Electrolytic Cells

Questions

14. What is electrolysis?

15. In an electrolytic cell, at which electrode does an oxidation reaction occur? A reduction reaction? Explain.

16. Define and distinguish clearly between *cathode, anode, cation,* and *anion.*

17. Which are attracted to the cathode, anions or cations? Which to the anode?

18. Draw a diagram of an electrolytic cell having positive and negative platinum electrodes. Show the directions of migration and the destinations of positive and negative ions in the solution. Write the half-reactions that occur at the electrodes. The overall reaction is
$$Cd + 2\ Fe(NO_3)_3 \longrightarrow Cd(NO_3)_2 + 2\ Fe(NO_3)_2$$

19. Describe the difference between an electrolytic reaction and that of a battery.

Solutions to Exercises

16.1 (a)

$$2(Cu^{2+} + e^- \longrightarrow Cu^+) \text{ reduction}$$
$$\underline{2\ CN^- \longrightarrow 2\ e^- + (CN)_2 \text{ oxidation}}$$
$$2\ Cu^{2+} + 2\ CN^- \longrightarrow 2\ Cu^+ + (CN)_2$$

(b)

$$2\ (P \longrightarrow P^{3+} + 3\ e^-) \text{ oxidation}$$
$$\underline{3\ (Br_2 + 2\ e^- \longrightarrow 2\ Br^-) \text{ reduction}}$$
$$2\ P + 3\ Br_2 \longrightarrow 2\ PBr_3$$

16.2 (a)

$$4 \ (Sn^{2+} \longrightarrow Sn^{4+} + 2 \ e^-) \ \text{oxidation}$$

$$\underline{ClO_4^- + 8 \ H^+ + 8 \ e^- \longrightarrow Cl^- + 4 \ H_2O \ \text{reduction}}$$

$$4 \ Sn^{2+} + 8 \ H^+ + ClO_4^- \longrightarrow 4 \ Sn^{4+} + Cl^- + 4 \ H_2O$$

(b)

$$4(Cu \longrightarrow Cu^{2+} + 2 \ e^-) \ \text{oxidation}$$

$$\underline{H_2AsO_4^- + 9 \ H^+ + 8 \ e^- \longrightarrow AsH_3 + 4 \ H_2O \ \text{reduction}}$$

$$4 \ Cu + H_2AsO_4^- + 9 \ H^+ \longrightarrow 4 \ Cu^{2+} + AsH_3 + 4 \ H_2O$$

16.3

$$3 \ (CuO + H_2O + 2 \ e^- \longrightarrow Cu + 2 \ OH^-) \ \text{reduction}$$

$$\underline{2 \ NH_3 + 6 \ OH^- \longrightarrow N_2 + 6 \ H_2O + 6 \ e^- \ \text{oxidation}}$$

$$3 \ CuO + 2 \ NH_3 \longrightarrow 3 \ Cu + N_2 + 3 \ H_2O$$

16.4

$$2(3 \ OH^- + NaOH + Al \longrightarrow NaAl(OH)_4 + 3 \ e^-) \ \text{oxidation}$$

$$\underline{6 \ e^- + 6 \ H_2O \longrightarrow 3 \ H_2 + 6 \ OH^- \ \text{reduction}}$$

$$2 \ NaOH + 2 \ Al + 6 \ H_2O \longrightarrow 2 \ NaAl(OH)_4 + 3 \ H_2$$

A Review of Mathematics

A beginning chemistry student must have strong arithmetic skills and should be comfortable with simple algebraic operations. A brief review of algebra, however, can be helpful for those students who need a ready reference. In addition to algebra, this appendix has discussions of percents, exponents, logarithms, and graphing. Its purpose is to supplement the discussions in the text and to give you more practice in the form of examples, exercises, and problems.

A.1 Percent

In chemistry as in many other fields of study, you will be asked to report a portion of something as a percent or you will be given data in terms of percent. The percent of something is defined as 100% multiplied by the ratio of the part to the total.

The percent of something is a fraction whose denominator is not stated but is understood always to be 100.

Let's look at apples as an example. If there are 42 apples and 16 are green, what is the percent of green apples in the sample? The ratio of green apples to all the apples is

$$\frac{16 \text{ green apples}}{42 \text{ apples}}$$

Now, multiply the ratio by 100%.

$$\frac{16 \text{ green apples}}{42 \text{ apples}} \times 100\% = 38\%$$

The percent of green apples is, therefore, 38%.

EXAMPLE A.1

You have 100 marbles and 24 of them are red. What is the percent of red marbles?

A.1

Solution:

$$\frac{24 \text{ red marbles}}{100 \text{ marbles}} \times 100\% = 24\%$$

Exercise A.1

A sample of 1,247 fruit flies contains 358 red-eyed flies. What is the percentage of red-eyed flies?

EXAMPLE A.2

Change $\frac{7}{10}$ to percent.

Solution:

$$\frac{7}{10} \times 100\% = 70\%$$

Exercise A.2

Change each of the following to percent: $\frac{3.0}{25}$, 0.22, $\frac{2}{9}$.

A.2 Algebra

In algebra, letters or other symbols take the place of numbers being calculated. The symbols are either variables that stand for quantities that can change or represent constants that stand for fixed quantities. The last few letters of the alphabet, x, y, and z, are often used for variables, while the earlier letters, along with the letter k or K, are often used for constants. There are also many exceptions to these practices.

The relations between algebraic symbols are expressed in equations. An equation says that one mathematical expression is equal to another in number and kind. Both sides of an equation must have the same units. For example, although the numbers may be equal, it is incorrect to say that three cats equal three dogs.

A simple equation is $y = 1 + 3$, which says that the quantity on the left of the equation, y, is the same as the quantity on the right, $1 + 3$. In this case, the value of y is 4, and whatever y refers to, there must be four of them.

The operations of algebra consist primarily of setting up equations with symbols representing both known and unknown quantities, then solving the equations to find the values of the unknowns. To reach a solution, we must rearrange the equation to isolate the desired variable on one side of the equation without making the equality statement untrue. There are several ways to make such changes:

Guidelines for Rearranging Algebraic Equations

1. *Both sides of the equation can be multiplied or divided by the same entity (we take entity to mean a number, constant, or variable).*

2. *The same entity can be added to, or subtracted from, both sides of the equation.*

3. *Both sides of the equation can be raised to the same power.*

Powers are discussed in the next section.

EXAMPLE A.3

Solve for x: $2x - 5 = 1$.

Strategy and Solution:

Step 1. Place all the terms containing x on one side of the equation and all those not containing x on the other. Do this by adding 5 to both sides.

$$2x - 5 + 5 = 1 + 5$$

$$2x = 6$$

Step 2. Divide both sides by 2.

$$\frac{2x}{2} = \frac{6}{2}$$

$$x = 3$$

EXAMPLE A.4

Solve for z: $2z - 15 = 10 - 3z$.

Strategy and Solution:

Step 1. Place all the terms containing z on one side of the equation and all those not containing z on the other. Do this by adding 15 to both sides.

$$2z - 15 + 15 = 10 - 3z + 15$$

$$2z = 25 - 3z$$

Step 2. Add $3z$ to both sides.

$$2z + 3z = 25 - 3z + 3z$$

$$5z = 25$$

Step 3. Divide both sides by 5.

$$\frac{5z}{5} = \frac{25}{5}$$

$$z = 5$$

Exercise A.3

Solve for y: $\frac{y}{3} + 4 = 6$.

Exercise A.4

Solve for y:

$$\frac{2(3y^2 - 15y - 36)}{2y} = 3y - 1$$

Equations often contain symbols for constants and variables other than the one being solved for. The solution to such an equation will contain many or all of those symbols.

EXAMPLE A.5

Solve for x: $\frac{x}{2} + \frac{y}{3} = 5a$.

Strategy and Solution:

Step 1. Find a common denominator. The lowest common denominator of 3 and 2 is 6, so we use 6. Multiply the first fraction by $\frac{3}{3}$ and the second fraction by $\frac{2}{2}$.

$$\frac{3}{3} \times \frac{x}{2} + \frac{y}{3} \times \frac{2}{2} = 5a$$

$$\frac{3x}{6} + \frac{2y}{6} = 5a$$

Step 2. To remove the fraction, multiply both sides of the equation by 6.

$$6\left(\frac{3x}{6} + \frac{2y}{6}\right) = 5a \times 6$$

$$3x + 2y = 30a$$

Step 3. Isolate x by subtracting $2y$ from both sides.

$$3x + 2y - 2y = 30a - 2y$$

$$3x = 30a - 2y$$

Step 4. Divide both sides by 3.

$$\frac{3x}{3} = \frac{30a - 2y}{3}$$

$$x = 10a - \frac{2}{3}y$$

Exercise A.5

Solve for z: $4z + 16x = 12a + 20d$.

A.3 **Exponents**

An exponent is a simple way to show that a number has been repeatedly multiplied by itself. The number of times a number has been multiplied by itself, that is, *the power to which it has been raised,* is indicated at the top right of the number by using a superscript. Here are several examples.

$$2 \times 2 \times 2 = 2^3$$
$$2 \times 2 \times 2 \times 2 = 2^4$$
$$2 \times 2 \times 2 \times 2 \times 2 = 2^5$$

In the first example, three 2's are multiplied together; thus the superscript, or exponent, is 3. The exponents for the other two examples are 4 and 5.

Any number raised to a power of 1 is simply the number itself, and any number raised to a power of 0 is equal to 1.

$$8^1 = 8$$
$$8^0 = 1$$

The examples we have looked at so far all have positive exponents, but an exponent can also be negative. A number raised to a negative power is equal to a fraction that has 1 as a numerator and the number raised to the power as the denominator.

$$9^{-1} = \frac{1}{9^1} = \frac{1}{9}$$

$$2^{-3} = \frac{1}{2^3} = \frac{1}{8}$$

The reverse of raising a number or fraction to a power is called taking the root. Asking, "What is the fourth root of 16?" is equivalent to asking, "What is the number that, multiplied by itself four times, equals 16?" The fourth root of 16 is 2 because $2^4 = 16$. The symbol for a root is a radical sign, $\sqrt[4]{16}$. A fractional exponent is an easier way to express the root. Our example looks like this:

$$\sqrt[4]{16} = 16^{1/4} = 2$$

An exponent can indicate both a power and a root at the same time. The exponent 3/2 directs us to take the second root (often called the square root) and then raise that answer to the third power. Here is an example.

$$9^{3/2} = (9^{1/2})^3 = (3)^3 = 27$$

Calculators often have keys for taking square roots, but getting higher roots of numbers is more difficult and is ordinarily done with an x^y key.

Exercise A.6

Calculate $(4.2)^4$ and $64^{1/2}$.

A.4 Logarithms

To serve as a brief, handy way to express certain kinds of numerical values, a system has been developed in which numbers can be given entirely in terms of exponents of a chosen number. Most often, the chosen number is 10. In this system, the exponent of 10 is called the *common logarithm* or *log*. For example, the logarithm of 10^4 is the exponent, 4. This is written as $\log 10^4 = 4$. In chemistry several important kinds of information, such as pH, are expressed exclusively with logarithms.

EXAMPLE A.6

What is the log of 1,000?

Solution:

Step 1.   Put the number in scientific notation.

$$1{,}000 = 10^3$$

Step 2.   The exponent of 10^3 is 3. Thus

$$\log 1{,}000 = 3$$

If the number for which the log is to be taken is less than 1, the logarithm will be negative.

EXAMPLE A.7

What is the log of 0.001?

Solution:

Step 1.   Put the number in scientific notation.

$$0.001 = 10^{-3}$$

Step 2.   The exponent of 10^{-3} is -3. Thus

$$\log 0.001 = -3$$

Most often you will be asked for the logarithm of a number that is not ten multiplied by itself some number of times, for instance, 2,000, which is $10 \times 10 \times 10 \times 2$. In such a case, the logarithm will not be an integer and will not be easy to figure out in your head. Before the days of electronic calculators, it was usual to look up difficult logarithms in mathematical tables. Now it is easier to calculate them. Section 14.6 gives a set of guidelines for using a calculator to obtain logarithms for pH values. A general version of those guidelines is repeated here.

Guidelines for Using an Electronic Calculator To Find Logs

1. *Enter the number for which the log is to be taken.*
2. *Press the "LOG" key.* Some calculators require that a function key be pressed before the log key will function.

EXAMPLE A.8

What is the logarithm of 325? (What is log 325?)

Solution:

Step 1.   Enter the number for which the log is to be taken.
Step 2.   Press the "LOG" key.

$$\log 325 = 2.512$$

Exercise A.7

What is the log of 90,200? Find log 511.

Exercise A.8

What is the log of 0.0327? Find $\log 6.0 \times 10^{-9}$.

If you have the log of a number, you can find the number itself by taking what is called the antilog; use the log as an exponent of 10.

EXAMPLE A.9

What is the antilog of 3?

Solution: Raise 10 to the power of 3.

$$10^3 = 1,000$$

This is expressed as

$$\text{antilog } 3 = 1,000$$

EXAMPLE A.10

What is the antilog of −6?

Solution: Raise 10 to the power of −6.

$$\text{antilog } -6 = 10^{-6} = 0.000001$$

Guidelines for Using an Electronic Calculator To Find Antilogs

1. *Enter the number for which the antilog is to be taken.*
2. *Press the "10ˣ" key.* Some calculators require that a function key be pressed before the "10^x" key will function. Other calculators may require that a key labeled "inv" and the log key be pressed to obtain an antilog.

EXAMPLE A.11

What is the antilog of 2.512?

Solution:

Step 1.   Enter the number for which the antilog is to be taken.
Step 2.   Press the "10^x" key.

$$\text{antilog } 2.512 = 325$$

Exercise A.9

What is the antilog of 1.773? Of 5.3304? Of -2.666?

Appendix A Problems

Section A.1 Percent

1. Change each of the following fractions to percent. Assume the fractions to be exact. Carry your answer to three digits.
$$\frac{1}{3}, \frac{1}{6}, \frac{1}{9}, \frac{2}{5}, \frac{5}{9}.$$

2. Change each of the following fractions to percent. Assume the fractions to be exact. Carry your answer to three digits.
$$\frac{1}{4}, \frac{1}{7}, \frac{1}{5}, \frac{3}{7}, \frac{9}{11}.$$

3. Change each of the following to percent: 0.067, 0.121, 0.000335, 9.3/11.5.

4. Change each of the following to percent: 0.197, 0.000022, 0.055, 2.2/5.3, 113.3/442.11.

Section A.2 Algebra

5. Solve the following for the unknown variable, x, y, or z:
(a) $2(4x - 5) = 5x - 1$
(b) $4y = y + \dfrac{(10y - 7)}{3}$
(c) $z(z - 1/2) = 2z$
(d) $x + a = \dfrac{4x - 51}{7}$

6. Solve the following for the unknown variable, x, y, or z:
(a) $3(y - 1) = 4y - 6$
(b) $7x = 2(x + 1) + 1$
(c) $5z + b = 3\left(2z + \dfrac{b}{5}\right)$

7. Solve the following for the unknown variable:
(a) $3 = \dfrac{(10y + 6)}{y}$
(b) $5x^2 = x(4x - 4)$
(c) $9 - z = \dfrac{4z^3 - 6z^2}{z^2}$

8. Solve the following for the unknown variable:
(a) $y^2 + 3y = 6y$
(b) $2^z + \dfrac{7}{4} = 2$
(c) $x^2 + 3x = 5x$

9. Solve the following for the unknown variable:
(a) $3y^2 = \dfrac{(2 - 7y^3)}{3y}$
(b) $3^x - 4y = 2\left(\dfrac{9}{2} - 2y\right)$
(c) $x(9x + 1) = 31x^2$

10. Solve the following for the indicated variable:
(a) Solve for x in terms of a, b, and y.
$$\frac{ax}{b} = 3a^2b + \frac{y}{2b}$$
(b) Solve for p in terms of n, K, and z.
$$ap = K\left(\frac{2na^2}{31} + za\right)$$
(c) Solve for v in terms of n, M, j, and x.
$$(v + 2)8M = \frac{(4M + j)}{16} + nx + M^-1$$

Section A.3 Exponents

11. Calculate the following: 2.25^3, $8.8^{3/2}$, 0.023^4, 438^{-2}.

12. Calculate the following: 6.6^4, $21^{5/2}$, 0.0044^{-3}.

Section A.4 Logarithms

13. Find the log of the following: 10, 10^3, 10^{3a}, 100.221, 1000, 0.0001, 3.1623, 316.23.

14. Find the log of the following: 100, 1/10, 10^4, 10^{-5q}, 0.00001, 5, 0.005.

15. Find the log of the following: 3.16×10^4, 7.7×10^{-6}, 0.00012.

16. Find the log of the following: 5.0×10^6, 1.99×10^{-7}, 310,000.

17. Find the antilogs of the following: 3.5, -6.7, 2.02, -1.2.

18. Find the antilogs of the following: 2.4, -5.5, -3.09, 0.6.

Solutions to Appendix A Exercises

A.1 $\dfrac{358 \text{ red-eyed flies}}{1{,}247 \text{ flies}} \times 100\% = 28.7\%$

A.2 $\dfrac{3.0}{25} \times 100\% = 12\,\%$

$0.22 \times 100\% = 22\%$

$\dfrac{2}{9} \times 100\% = 20\%$

A.3 $\dfrac{y}{3} + 4 - 4 = 6 - 4$

$3\left(\dfrac{y}{3}\right) = 2 \times 3$

$y = 6$

A.4 $\dfrac{2(3y^2 - 15y - 36)}{2y} = 3y - 3$

$3y^2 - 15y - 36 = y(3y - 3)$

$3y^2 - 15y - 36 = 3y^2 - 3y$

$15y - 15y - 36 = -3y + 15y$

$\dfrac{-36}{12} = \dfrac{12y}{12}$

$-3 = y$

A.5 $4z + 16x - 16x = 12a + 20d - 16x$

$\dfrac{4z}{4} = \dfrac{12a + 20d - 16x}{4}$

$z = 3a + 5d - 4x$

A.6 311; 8

A.7 4.955; 2.708

A.8 -1.485; -8.22

A.9 59.3; 2.140×10^{-5}; 2.16×10^{-3}

Graphing

Ore of the principle requirements of a good scientist is to be able to interpret data once they are obtained. Methods of trying to understand the
result of measurements vary from just looking at the numbers and hoping for
inspiration to trying mathematical combinations to see whether the numbers will
fit. Because pictures are often easier to understand than numbers, one of the
best ways to interpret data is to make a graph. Although there are many ways,
some quite complex, of drawing and interpreting graphs, we'll look at a simple,
yet powerful, technique of making two-dimensional graphs.

Because you will probably make several errors when making a graph, we
recommend always using a pencil so you can easily erase.

Guidelines for Graphing

1. *For axes, draw two intersecting lines at right angles on a piece of graph
   paper.* Some types of graph paper have lines for axes already drawn on the
   paper.
2. *Select which variable will fit nicely on one of the axes.* The best scale for an
   axis is in units of 1, 2, 5, 10, or some multiple of these numbers.
3. *Label the axes and indicate the scale for each.*
4. *Plot the data by locating a value of one of the variables on its axes.*
5. *Move up or over to the location of the corresponding value of the other
   variable; put a small dot (data point) there.* Sometimes data points are circled,
   so that they are clearly shown.
6. *Repeat these steps for all values of the variables.*
7. *If the data points lie in a line, draw a straight line that best "fits" all the
   data points; if the data points lie on a curve, draw a curve through all the
   data points.*

To understand these guidelines better, let's do a graph together. Table B.1
lists the results of a series of measurements taken to verify Charles' law. Turn
to page 313 for a description of the experiment and a drawing of the apparatus
(Figure 11.16). A student carefully heated the apparatus, taking the temperature
and measuring the volume every few minutes.

Table B.1 Temperature-Volume Data Taken by a Student

Temperature (°C)	Volume (mL)
5	82
21	86
32	91
48	56
65	100
81	106
93	107
110	113
123	117
140	122

Study these numbers for a moment. Do they make any sense to you? Probably not much. You can see that the volume increases as the temperature is raised, but that's about all. Now let's graph the data and see if we can learn more.

(Figure B.1) In this experiment, the temperature ranges from 5°C to 140°C. Looking at our graph paper, we find that 140° will fit nicely on the page if we let each square equal 5°. We label the horizontal axis "Temperature (°C)" and indicate the scale by numbering each 50° line from 0 to 150. Similarly, we find that the values of the volume range from 82 to 122 mL. We let each box on the vertical axis equal 10 mL. We label the vertical axis "Volume (mL)" and indicate the scale with appropriate numbers.

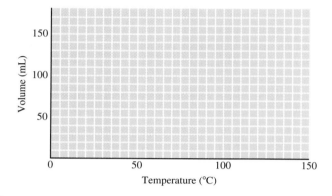

Figure B.1

Now we plot the data (Figure B.2). We find the location of 5 on the horizontal axis and move up to 82 on the vertical axis. We put a small dot there. Now we plot the next set of values, 21 degrees and 86 mL. We do this for all the measurements.

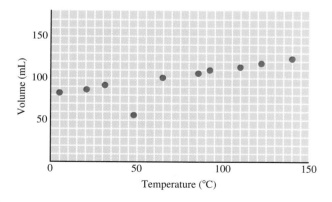

Figure B.2

Notice that nearly all the data points lie on a line. We draw a straight line through as many data points as possible (Figure B.3), being aware that not all the data points will fall on the line. Now our graph is finished.

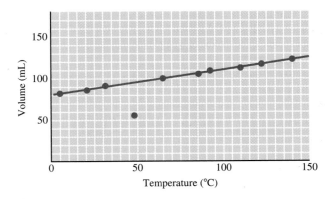

Figure B.3

A straight line graph is called a *linear* graph; it obeys the algebraic equation

$$y = mx + b$$

where x is the variable on the horizontal axis, y is the variable on the vertical axis, m is the slope of the line, and b is the place where the line intercepts the vertical axis. The linear equation is a *proportionality*, indicating that the two plotted variables are *proportional* to one another.

By making a graph of our data, we have converted them from a set of almost meaningless numbers to a picture that tells us much more than the numbers themselves will. Scientists almost instinctively begin to draw graphs when they see experimental data.

Appendix B Problems

1. Construct a graph with the vertical axis numbered from 0 to 100 in steps of 10 and the horizontal axis numbered from 0 to 20 in steps of 5. Plot the following pairs of numbers and draw the curve that best fits them.

Vertical	Horizontal
5	0.9
16	2.7
27	5.0
31	5.9
40	7.2
55	9.9
60	10.1
72	13.0
81	14.6
97	17.5

2. Construct a graph from the following pairs of numbers. Choose numbers for axes so that the curve is properly displayed.

Vertical	Horizontal
75	0.043
650	0.38
1420	0.80
2850	1.63
4090	2.34
4900	2.80

Vapor Pressure of Water at Various Temperatures

Temperature (°C)	Vapor Pressure (torr)	Temperature (°C)	Vapor Pressure (torr)
−10	2.1	26	25.2
0	4.6	27	26.7
1	4.9	28	28.3
2	5.3	29	30.0
3	5.7	30	31.8
4	6.1	31	33.7
5	6.5	32	35.7
6	7.0	33	37.7
7	7.5	34	39.9
8	8.0	35	42.2
9	8.6	36	44.6
10	9.2	37	47.1
11	9.8	38	49.7
12	10.5	39	52.4
13	11.2	40	55.3
14	12.0	41	58.3
15	12.8	42	61.5
16	13.6	43	64.8
17	14.5	44	68.3
18	15.5	45	71.9
19	16.5	46	75.7
20	17.5	47	79.6
21	18.7	48	83.7
22	19.8	49	88.0
23	21.1	50	92.5
24	22.4	51	97.2
25	23.8	52	102.1

Answers to Problems

Chapter 1

5. 4; 1; 4; 3; 7; 6; 6

7. (a) 182; (b) 0.3; (c) 1.6; (d) 2.8; (e) 1.0

11. 2.25×10^{-1}; 2.5; 4.4163×10^4; 2.000019×10^6; 3.21×10^{-6}

13. 3.6×10^{-2}; 1.9×10^2; 1.6×10^{-3}; 1.4×10^6

15. (a) 2.8×10^2 (b) 8.9×10^{-10} (c) 2.17×10^{-3} (d) 2.09×10^{-12} (e) 1.5×10^{-2}

19. 7.9 gal

23. 18 spoonfuls

25. Yes, she has enough money for 7 days.

27. 9 trips

29. 3.1×10^3 lines

31. 28 qt

33. 7.7×10^6 gal

39. (a) 1.4 mi (b) 128 cm (c) 2.23×10^{-5} mm (d) 1.433×10^8 Å (e) 162 in. (f) 1.83 ft (g) 1.7 in. (h) 2.9 m (i) 3.0 m (j) 51.7 km (k) 1.0×10^2 ft (l) 8.7 in.

41. 1,902 km

43. Yes, because there are 615 km between cities.

45. 3×10^{-5} m; 3×10^{-2} mm

47. 1.8×10^{-6} cm; 5×10^{-6} cm; 1.8×10^{-5} mm; 5×10^{-5} mm

51. (a) 23.7 qt (b) 200 mL (c) 0.913 pt (d) 7 fl oz (e) 0.474 qt (f) 3.7×10^{-2} m^3

53. 5.89×10^5 cm^3

55. 0.25 L

57. Yes, the hollow cube holds 125 L.

59. $0.89

65. (a) 70.3 kg (b) 97 lb (c) 1.41×10^3 g (d) 2.9×10^4 mg (e) 0.49 oz (f) 3.4×10^2 g

67. 5.75×10^4 mg

69. 2.5×10^4 g

71. 1.4×10^2 lb

73. No, 10.0 grams are needed.

75. 60 francs

79. 55°C

83.

g/cm^3	kg/m^3	lbs/gal	g/L
1.000	1,000.	8.34	1,000.
3.31×10^{-4}	0.331	2.76×10^{-3}	0.331
0.719	719	6.002	719
2.55×10^{-3}	2.55	2.13×10^{-2}	2.55

85. 696 cm^3

87. 60 cm^3; 10 g/cm^3

89. 265 g

91. 2.2×10^4 in.

93. 3×10^2 ft, yes

95. 372 mL

97. 2.4×10^5 mg nitrogen

99. 4.5 g/cm^3

Chapter 2

11. 3.9×10^4 cal; 1.6×10^2 kJ

13. 78 kJ

15. 0.23 kJ

17. 6.0 dL

21. 3.9×10^{-2} kJ

23. 40.°C

35. phospohorus, barium, palladium, tin

37. Au, Ag, Fe, Pb, Co

39. solid: Na, Ti, B, C, P, Rh, Cs, I, Si, As; liquid: Br; gas: Cl

47. 2 nitrogen atoms, 8 hydrogen atoms, 1 sulfur atom, 4 oxygen atoms; 1 carbon atom, 2 sulfur atoms; 2 cesium atoms, 1 sulfur atom; 9 nitrogen atoms, 12 hydrogen atoms, 1 iron atom, 6 carbon atoms; 2 carbon atoms, 7 hydrogen atoms, 1 sulfur atom, 1 nitrogen atom

Chapter 3

21. sodium 11; cobalt 27; gold 79; lithium 3; polonium 84; potassium 19; magnesium 12; mercury 80; silicon 14; silver 47; titanium 22; tin 50; fluorine 9; iron 26
27. 58; 60; 62; 63; 64; 65; 66; 68
29. (a) $^{16}_{8}E$, $^{18}_{8}E$, $^{15}_{8}E$, oxygen; $^{14}_{7}E$, $^{13}_{7}E$, $^{15}_{7}E$, nitrogen; $^{14}_{6}E$, carbon   **(b)** $^{14}_{7}E$ and $^{14}_{6}E$; $^{15}_{7}E$ and $^{15}_{8}E$;   **(c)** 8 neutrons, $^{16}_{8}E$, $^{15}_{7}E$, $^{14}_{6}E$; 7 neutrons, $^{14}_{7}E$, and $^{15}_{8}E$;   **(d)** same as (a);   **(e)** same as (a)

Chapter 4

13. 72: 162
15. 18; s, p, and d
17. $n = 2$, s, 2; $n = 4$, p, 5; $n = 7$, s, 1; $n = 4$, d, 7
19. (a) $4p^5$ **(b)** $3d^2$ **(c)** $2s^1$ **(d)** $6f^{10}$
37.

F   1s 2s 2p

Mg   1s 2s 2p 3s

Ga   1s 2s 2p 3s 3p 4s 3d 4p

V   1s 2s 2p 3s 3p 4s 3d

39. S
41. Zn $1s^2 2s^2 2p^6 3s^2 3p^6 4s^2 3d^{10}$
K $1s^2 2s^2 2p^6 3s^2 3p^6 4s^1$
Br $1s^2 2s^2 2p^6 3s^2 3p^6 4s^2 3d^{10} 4p^5$
Te $1s^2 2s^2 2p^6 3s^2 3p^6 4s^2 3d^{10} 4p^6 5s^2 4d^{10} 5p^4$
43. Se, 2 electrons are paired.
45. S; none; K, Cr, Cu; C; Rb, Nb, Mo, Ru, Rh, Ag
47. (a) Br, I   **(b)** Co   **(c)** N, P   **(d)** Ne, Ar
(e) Ti, Zr
57. ·N̈·   :Ö·   ·C̈·   ·Ca·
59. C, Si, Ge, Sn, Pb

Chapter 5

9. Ar: 3, VIIIA; As: 4, VA; Sn: 5, IVA; Ir: 6, VIIIB; Sr: 5, IIA
21. Metal: Cs; Nonmetals: C, S, Cl, and Ne; Metalloids: As and Si
23. Rb·   :Se·   :S·   ·Ï·   ·As·   ·B·   H·   He: :Ne:

25. M − 2: group VIA; M + 2: group IIA
29. (a) F, As, Ga, K   **(b)** Cl, Br, Rb, Cs   **(c)** Ar, Al, Na, Cs   **(d)** O, P, Sn   **(e)** Cu, V, Ta, U

Chapter 6

9. 3, 1, 1, 2, 2, 3
11. N $1s^2 2s^2 2p^3$
N^{3-} $1s^2 2s^2 2p^6$
Cl $1s^2 2s^2 2p^6 3s^2 3p^5$
Cl$^-$ $1s^2 2s^2 2p^6 3s^2 3p^6$
Br $1s^2 2s^2 2p^6 3s^2 3p^6 4s^2 3d^{10} 4p^5$
Br$^-$ $1s^2 2s^2 2p^6 3s^2 3p^6 4s^2 3d^{10} 4p^6$
O $1s^2 2s^2 2p^4$
O^{2-} $1s^2 2s^2 2p^6$
S $1s^2 2s^2 2p^6 3s^2 3p^4$
S^{2-} $1s^2 2s^2 2p^6 3s^2 3p^6$
P $1s^2 2s^2 2p^6 3s^2 3p^3$
P^{3-} $1s^2 2s^2 2p^6 3s^2 3p^6$
13.   ·N̈· $\left[:\ddot{N}:\right]^{3-}$:C̈l· $\left[:\ddot{C}l:\right]^{1-}$

·B̈r· $\left[:\ddot{B}r:\right]^{1-}$ ·Ö· $\left[:\ddot{O}:\right]^{2-}$

:S̈· $\left[:\ddot{S}:\right]^{2-}$ ·P̈· $\left[:\ddot{P}:\right]^{3-}$

15. K$^+$; Al^{3+}; Ca^{2+}; Na$^+$; Mg^{2+}; Sr^{2+}; Be^{2+}
17. Ba^{2+}; no ions; Br$^-$; Be^{2+}; Bi^{3+}
19. (a) Na$^+$, Cl, P, Na, S^{2-} **(b)** C^{4+}, Be^{2+}, Li, O^{2-}, C^{4-}
(c) Al^{3+}, Mg^{2+}, Ar, S, P^{3-} **(d)** N^{3+}, Ne, F$^-$, O^{2-}, N^{3-}
29. (a) MgO   **(b)** CsCl   **(c)** AlH$_3$ **(d)** K$_2$S
31. (a) Y$_2$X$_3$ **(b)** YX$_4$ **(c)** Y$_3$X   **(d)** YX
39. (a) S   **(b)** N   **(c)** O   **(d)** Br   **(e)** Sb   **(f)** H
(g) C
41. (a) H—C, C; C—N, N   **(b)** C—H, C; C—N, N; N—H, N   **(c)** C—Cl, Cl; C—H, C; C—C, the C bonded to the most electronegative elements, C—F F + B, Br; C—Br, Br; C—F, F   **(d)** C—H, C; C—O, O   **(e)** C—H, C; C—O, O; O—H, O   **(f)** S—O, O; O—H, O; S—F, F
47.

:F̈—C̈—F̈:   P:P   :F̈—S̈—S̈—F̈:

49. H—C≡N: H—C≡C—H :N̈=Ö—F̈: :Ö—S̈=Ö:

51. H—S̈—H :F̈—N̈—F̈: :Ö—N—N—Ö:
                    :F̈:   :Ö: :Ö:

53. H—Ö—N̈=Ö:

(structural formulas for acetaldehyde and aminoborane)

55. (structural formulas)

57.

$$\left[\begin{array}{c} :\ddot{O}: \\ :\ddot{O}-P-\ddot{O}: \\ :\ddot{O}: \end{array}\right]^{3-} \quad \left[\begin{array}{c} :\ddot{O}-B=\ddot{O}: \\ :\ddot{O}: \end{array}\right]^{3-} \quad \left[\begin{array}{c} :\ddot{F}: \\ :\ddot{F}-B-\ddot{F}: \\ :\ddot{F}: \end{array}\right]^{-}$$

$$\left[\begin{array}{c} :\ddot{O}-\ddot{S}-\ddot{S}-\ddot{O}: \\ :\ddot{O}:\ :\ddot{O}: \end{array}\right]^{2-}$$

59.

	AsO_4^{3-}	SO_4^{2-}	Br^-	CO_3^{2-}	$C_2O_4^{2-}$	HPO_4^{2-}
K^+	K_3AsO_4	K_2SO_4	KBr	K_2CO_3	$K_2C_2O_4$	K_2HPO_4
Sr^{2+}	$Sr_3(AsO_4)_2$	$SrSO_4$	$SrBr_2$	$SrCO_3$	SrC_2O_4	$SrHPO_4$
Fe^{3+}	$FeAsO_4$	$Fe_2(SO_4)_3$	$FeBr_3$	$Fe_2(CO_3)_3$	$Fe_2(C_2O_4)_3$	$Fe_2(HPO_4)_3$
H^+	H_3AsO_4	H_2SO_4	HBr	H_2CO_3	$H_2C_2O_4$	H_3PO_4
Cu^+	Cu_3AsO_4	Cu_2SO_4	$CuBr$	Cu_2CO_3	$Cu_2C_2O_4$	Cu_2HPO_4
Zn^{2+}	$Zn_3(AsO_4)_2$	$ZnSO_4$	$ZnBr_2$	$ZnCO_3$	ZnC_2O_4	$ZnHPO_4$

65. nonpolar; polar; polar; polar; polar; polar

Chapter 7

3. +2; +4; 0; +3; +4
5. Ba^{2+}, C^{3+}, O^{2-}; H^+, B^{3+}, O^{2-}; I^{5+}, O^{2-}; Br^{5+}, O^{2-}; Mn^{7+}, O^{2-}; Al^{3+}, C^{2-}, H^+, O^{2-}
7. K^+, Br^-; Li^+, O^{2-}; Ca^{2+}, S^{2-}; Be^{2+}, I^-; Al^{3+}, Se^{2-}; Ba^{2+}, O^{2-}
15. cesium bromide, strontium iodide, sodium phosphide, calcium carbide, beryllium nitride, silver oxide

17. KCl, BeS, K_2O, BaF_2, CaI_2, MgSe
19. tin(IV) oxide, lead(IV) oxide, iron(III) oxide, arsenic(III) iodide, mercury(I) iodide, copper(II) boride
21. iron(III) fluoride, cobalt(III) oxide, cobalt(III) chloride, copper(II) oxide, mercury(I) oxide
23. $PbCl_4$, CuO, FeI_3, SnO_2, AsF_3
25. stannic oxide, plumbic oxide, ferric oxide, arsenous iodide, mercurous iodide, cupric boride
27. ferric fluoride, cobaltic oxide, cobaltic chloride, cuprous oxide, mercurous oxide
29. $FeCl_3$, PbO_2, $HgBr_2$, CuO, CoI_2
31. bromine pentafluoride, boron tribromide, dinitrogen pentasulfide, dihydrogen trisulfide, iodine heptafluoride, carbon disulfide
33. SO_2, S_4N_2, AsI_2, P_4Se_3, BN
37. H_2S, HCl, HI
43. lead(IV) sulfate, copper(II) chlorite, tin(II) bicarbonate, iron(III) phosphate, mercury(II) nitrate, lead(II) chromate
45. calcium nitrite, potassium chlorate, cesium sulfite, magnesium hydroxide, barium cyanide, potassium bromate
47. $Mn(NO_2)_2$, $CoPO_4$, $CuIO_3$, $Pb(SO_4)_2$, $Fe(HCO_3)_3$
49. $Fe_2(CrO_4)_3$, $Cu(NO_2)_2$, $Pb(C_2H_3O_2)_2$, $Sn(CO_3)_2$, Cu_2SO_4, $FeSO_3$
51. hypoiodous acid, iodous acid, iodic acid, periodic acid
53. HNO_2, $H_2C_2O_4$, $HClO_2$, H_3PO_3, $HClO_4$
55. $LiNa_2PO_4$, $MgKPO_4$, $CuHSO_3$, $NH_4HCr_2O_7$

Chapter 8

9. (a) 28.0 (b) 107.86 (c) 69.7
11. 152.0
13. 64.92
15. 190.9
19. 1.8×10^{24} atoms
21. 2.6×10^{24} molecules
23. 0.500 mol
25. 3.0×10^{20} pages; 5.9×10^{12} houses

27. 1.2×10^{22} m^3

31. 6.634×10^{-23} g; 2.280×10^{-22} g; 6.655×10^{-23} g; 1.994×10^{-23} g; 8.634×10^{-23} g; 9.786×10^{-23} g; 3.271×10^{-22} g

33. 10.811 g/mol; 1.0079 g/mol; 85.4678 g/mol; 14.0067 g/mol; 44.9559 g/mol; 77.904 g/mol; 118.710 g/mol; 18.9984 g/mol

35. 32.066 g

37. 1.35×10^{24} atoms

39. 9.6×10^{-4} g Pb

41. (a) 1.65 mol Mg   (b) 1.64 mol Pb   (c) 1.45 mol Mn   (d) 0.011 mol He   (e) 0.524 mol S

43. (a) 0.889 g He   (b) 0.0146 g S   (c) 580. g Hg   (d) 137 g Ag

45. 394 g Li

47. 2.20 mol Pb; 2.27 mol Hg; 114 mol He

49. 4.6 mol Ne; 0.86 mol Ag; 1.2 mol As

53. 44.010; 34.0146; 30.0061; 28.010

55. 98.079; 46.069; 261.3337; 452.8

61. 191.207; 450.19; 78.541

63. 28.010 g

65. (a) 1.27 mol H_2SO_4 (b) 6.30 mol NaOH   (c) 4.1×10^{-4} mol K_2CrO_4 (d) 5.55×10^{-3} mol CCl_4 (e) 0.0603 mol MnO_2

67. (a) 2.73 g $K_2Cr_2O_7$ (b) 152 g HCl   (c) 3.1×10^5 g CO   (d) 14.4 g $BaBr_2$ (e) 3.55×10^{-3} g Al_2O_3

69. 4.892 Zanthu amu/atom Ni

71. 7.20×10^{24} atom ^{12}C/Zanthu mol

Chapter 9

7. 14.6 mol N

9. 0.750 mol H_2CO_3

11. 1 mol $C_6H_{12}F_2$

13. (a) 1.67 mol O   (b) 0.321 mol O   (c) 3.81 mol O   (d) 6.33×10^5 mol O   (e) 7.11×10^{-2} mol O

15. 34.1 g S

17. 41.77 g Ba

19. (a) 5.01 g O   (b) 91.1 g O   (c) 97.0 g O   (d) 2.52×10^3 g O

23. 42.6% Na, 11% C, 46.2% O

25. 24.7% K, 34.8% Mn, 40.5% O

27. 54.6% C, 41.5% O, 4% H

29. 75.9% C, 17.7% N, 6.4% H

31. 1.60% H, 22.2% N, 76.2% O

35. C_8H_{18}

37. Cs_2S_2, Co_3S_4, K_2S_2, Al_2S_3, SiS_2

39. 17.323% Cl; 0.522; 3.398×10^{-22} g; 1.5 mol $C_{12}H_9OCl$

41. 24.425% Ca, 17.0722% N, 58.503% O; 82.2453%

N, 17.7547% H; 69.943% Fe, 30.057% O; 21.643% Mg, 21.638% C, 56.976% O

47. C_2H_6O

49. $C_{18}H_{21}NO_3$

51. CrO

55. $C_2H_4O_2$

57. $C_{15}H_{10}O_5$

59. 20.19 g C, 2.19 g H, 2.35 g N, 2.67 g O; $C_{10}H_{13}NO$; $C_{20}H_{26}N_2O_2$

61. $C_9H_8O_4$; $C_9H_8O_4$

63. 24.1 g X/mol X, Mg; 10. g Z/mol Z, B

Chapter 10

3. (a) $2 H_2 + O_2 \longrightarrow 2 H_2O$ (b) $N_2 + 3 I_2 \longrightarrow 2 NI_3$ (c) $HCl + NaOH \longrightarrow NaCl + H_2O$ (d) $2 KClO_3 \longrightarrow 2 KCl + 3 O_2$ (e) $CuCl_2 + H_2S \longrightarrow CuS + 2 HCl$

5. (a) $2 Na + Cl_2 \longrightarrow 2 NaCl$ (b) $SiO + 2 C \longrightarrow SiC + CO$ (c) $CaCO_3 \longrightarrow CaO + CO_2$ (d) $4 NH_3 + 5 O_2 \longrightarrow 4 NO + 6 H_2O$ (e) $MnO_2 + 4 HCl \longrightarrow MnCl_2 + Cl_2 + 2 H_2O$

11. (a) 1.40 mol Al_2O_3 (b) 1.40 mol Cu_2S (c) 1.40 mol $(NH_4)_2SO_4$ (d) 5.60 mol $HClO_4$

13. 2.06 mol Fe

15. 1.93 mol O_2

17. (a) 6.74 mol C   (b) 20.2 mol HCl   (c) 27.0 mol NaOH   (d) 11.2 mol O_2

19. 0.314 mol N_2

21. 0.215 mol H_2

23. 0.0940 mol Fe_2O_3; 0.282 mol CO

25. 476.0 g NO_2

27. 34.5 g PbO

29. 18.1 g $KClO_3$

31. 5.96 g $CaCl_2$

33. 68.33%

35. 72.46%

37. 72.0%

41. 2.87 mol PbO

43. 1.7 mol Fe

45. 1.46 mol $MnCl_2$

47. (a) C   (b) FeS_2 (c) $C_3H_6O_2$ (d) H_2SO_4

49. 13.7 g CH_4; 88.9 g $Al(OH)_3$

51. 6.87 g P_4

53. 380 g C_2H_5OH; 370 g CO_2; 490 mL C_2H_5OH

55. 3.6 g H_2O; 4.2 L

57. 1.70 L O_2

59. 0.81 g O_2; 1.5 mL Hg

Chapter 11

5. (a) 700 mm Hg   (b) 1.26 atm   (c) 27 atm

15. (a) 730. mL   **(b)** 600. mL   **(c)** 136 mL
17. 0.960 atm
19. 10.5 torr
21. 7.10 L
23. The original number of moles is not given.
29. 1.70 atm
31. $-156\,°C$
33. 2.05×10^3 torr; 374 torr
39. $-115\,°C$
41. 36 K; 456 K
43. 4.71 L
45. 986 torr
47. 106 K $= -167\,°C$
49. 930. mL
51. 513 torr
53. 0.0821 L atm/mol K
55. $6.93 \times 10^3\,°C$
57. 96.5 K
59. 6.43 L; 8.16 L
61. 23.9 atm
63. 0.122 mol
67. 0.0223 mol
69. 4.97 L
71. 81 L
73. 52.1 g/mol
75. 44.0 g/mol
77. 0.482 g/L
79. 1.36 g/L
81. 1.25 g/L
83. 0.698 g/L
85. 30.0 g/mol
91. 928 torr
93. 6.57 atm
95. 1.58 atm
97. 1.65 atm
99. 0.717 atm
101. 5.75 L CO_2; 11.5 L H_2O
103. 4.50 L
105. 7.00 L
107. 16.4 g/mol
109. 0.337 g P_4H_{10}
111. 16.4 atm
115. 0.13 mol
117. 0.636 atm
119. 1.5 atm

Chapter 12

21. 30.3 kJ
23. 8.76 kJ; 0.489 kJ; temporary dipoles are weaker than ionic bonds.

25. 43 kJ; 3.1 kJ; hydrogen bonds are stronger than temporary dipoles.
27. 1.37 kJ/g
47. 1.04 kJ
49. 4.50 kJ; 5.08 kJ

Chapter 13

23. Na_2S; $FeSO_4$; $(NH_4)_3PO_4$
25. (a) $Fe(OH)_3$ **(b)** $AgCl$ **(c)** no reaction
(d) no reaction
27. (a) no reaction
(b) $Hg(NO_3)_2(aq) + MgCO_3(aq) \longrightarrow Mg(NO_3)_2(aq) + HgCO_3(s)$
(c) no reaction
(d) $CaS(aq) + (NH_4)_2CO_3(aq) \longrightarrow CaCO_3(s) + (NH_4)_2S(aq)$
29. $2\,Ag^+ + SO_4^{2-} \longrightarrow Ag_2SO_4(s)$
31. (a) $Fe^{3+} + 3\,OH^- \longrightarrow Fe(OH)_3(s)$
(b) $Ag^+ + Cl^- \longrightarrow AgCl(s)$
37. 37%
39. 1.6 g $Ca(OH)_2$
41. 54 g
43. 17.0%
45. 14 g $NiCl_2 \cdot 6\,H_2O$
47. 110 g $CoCl_2$
49. 10. g $CuSO_4 \cdot 5\,H_2O$
51. 30.0 L
53. 1.1 M
55. 1.27 m
57. 3.45 m
59. 4.5 m; 4.7 M
61. 0.0038 M
63. 0.52 M; 0.52 M; 1.04 M
65. 0.0153 L
67. 3.2 M
73. 2.4 deg
75. $241\,°C$
77. $108.06\,°C$
79. 15 deg
81. $101.5\,°C$; $-5.6°C$
83. 69 g/mol
85. 25 g $BaCO_3$
87. 36 mL; 14 mL

Chapter 14

5. (a) $HNO_3\ +\ H_2O\ \longrightarrow\ H_3O^+\ +\ NO_3^-$
     acid   base   acid   base
(b) $N_2H_4\ +\ H_2O\ \longrightarrow\ N_2H_3^-\ +\ H_3O^+$
     acid   base   base   acid
(c) $N_2H_4\ +\ HCl\ \longrightarrow\ N_2H_5^+\ +\ Cl^-$
     base   acid   acid   base

(d) $HCN + H_3O^+ \longrightarrow H_2CN^+ + H_2O$
  base   acid   acid   base

13. (a) $3\ KOH + H_3PO_4 \longrightarrow K_3PO_4 + 3\ H_2O$
(b) $HC_2H_3O_2 + NaOH \longrightarrow NaC_2H_3O_2 + H_2O$
(c) $Ca(OH)_2 + 2\ HCl \longrightarrow CaCl_2 + 2\ H_2O$
(d) $H_2SO_4 + 2\ KOH \longrightarrow K_2SO_4 + 2\ H_2O$

19. 1.8×10^{-12} M; 1.3×10^{-6} M
21. 1.0×10^{-13} M
23. 8.3×10^{-12} M
27. (a) 1.08   (b) 10.0   (c) 8.96
29. (a)
31. 2.35
33. 11.47
35. 1.5×10^{-4} M; 1×10^{-13} M; 3×10^{-11} M; 2×10^{-5} M; 6×10^{-8} M
37. 5×10^{-2} M; 2×10^{-11} M; 2×10^{-9} M; 0.3 M; 6×10^{-5} M
39. 13.04
41. 0.70 mL
45. 0.424 M
47. 0.1798 M
53. 2.50 N; 0.19 N; 15.0 N; 0.084 N
55. 16.6 N
57. 0.0028 mol $Ba(OH)_2$
59. 5.7 g NaOH
61. 4.4 N
63. 1.0 N

Chapter 15

11. The equilibrium shifts to the left, increasing the concentrations of C and D while decreasing the concentration of F. Some of the added E will be used.

25. (a) $K = \dfrac{[CO_2]^2}{[CO]^2\,[O_2]}$

(b) $K = \dfrac{[NO_2]^2}{[NO]^2\,[O_2]}$

(c) $K = \dfrac{[Cl_2]^2\,[H_2O]^2}{[HCl]^4\,[O_2]}$

(d) $K = \dfrac{[NO]^4\,[H_2O]^6}{[NH_3]^4\,[O_2]^5}$

27. 2.29×10^{-8}
29. 30.
35. 0.701
37. 1.80
41. (a) $K_a = \dfrac{[H_3O^+]\,[HCO_3^-]}{[H_2CO_3]}$ (b) $K_a = \dfrac{[H_3O^+]\,[NO_2^-]}{[HNO_2]}$
43. 1.9×10^{-12}
45. 7×10^{-9}
47. (a) 4.8×10^{-4} M   (b) 2.4×10^{-6} M

(c) 5.5×10^{-3} M   (d) 2.1×10^{-2} M
49. 5.15
51. 11.11
57. 3.45×10^{-11} M
59. 5.18×10^{-12}
61. (a) 5.3×10^{-4} M   (b) 2×10^{-7} M   (c) 7.2×10^{-3} M   (d) 1×10^{-3} M

Chapter 16

3. (a) $2\ Cr^{2+} + NO_3^- + 2\ H_3O^+ \longrightarrow 2\ Cr^{3+} + NO_2^- + 3\ H_2O$ (b) $SO_4^{2-} + 2\ I^- + 4\ H_3O^+ \longrightarrow SO_2 + I_2 + 6\ H_2O$
(c) $2\ IO_3^- + 3\ N_2H_4 \longrightarrow 2\ I^- + 3\ N_2 + 6\ H_2O$
(d) $16\ H_2O + 4\ PH_4^+ + 6\ Sn^{4+} \longrightarrow 6\ Sn^{2+} + P_4 + 16\ H_3O^+$
(e) $6\ IO_3^- + 10\ NH_3 \longrightarrow 5\ N_2 + 3\ I_2 + 6\ OH^- + 12\ H_2O$
5. oxidizing agents: NO_3^-; SO_4^{2-}; IO_3^-; Sn^{4+}; IO_3^-; H_2O_2; CrO_4^{2-}; H_2O; MnO_4^-; MnO_4^-; reducing agents: Cr^{2+}, I^-, N_2H_4, PH_4^+, NH_3, CH_3OH; Fe^{2+}, H_2CO, FeO, SCN^-
7. (a) $6\ H_3O^+ + 2\ HNO_3 + 3\ Zn \longrightarrow 3\ Zn^{2+} + 2\ NO + 10\ H_2O$ (b) $6\ H_2O_2 + CH_3CH_2OH \longrightarrow 2\ CO_2 + 9\ H_2O$ (c) $2\ C_2H_6 + 7\ O_2 \longrightarrow 4\ CO_2 + 6\ H_2O$
(d) $20\ H_3O^+ + 3\ As_2O_3 + 4\ CrO_4^{2-} \longrightarrow 6\ H_3AsO_4 + 4\ Cr^{3+} + 21\ H_2O$ (e) (a) $H_2O + 3\ NO_2 \longrightarrow 2\ HNO_3 + NO$
9. (a) $4\ NH_3 + 5\ O_2 \longrightarrow 4\ NO + 6\ H_2O$ (b) $2\ OH^- + Br_2 \longrightarrow Br^- + BrO^- + H_2O$ (c) $6\ OH^- + 3\ SnO_2 + 2\ Bi^{3+} \longrightarrow 3\ SnO_3^{2-} + 2\ Bi + 3\ H_2O$ (d) $8\ OH^- + Cl^- + 8\ MnO_4^- \longrightarrow ClO_4^- + 8\ MnO_4^{2-} + 4\ H_2O$

Appendix A

1. 33.3%, 16.6%, 11.1%, 40.0%, 55.5%
3. 6.7%, 1.21%, 0.0335%, 80.9%
5. (a) $x = 3$ (b) $y = 7$ (c) $z = 2\,1/2$
(d) $x = \dfrac{(7a - 51)}{3}$
7. (a) $y = -\dfrac{6}{7}$ (b) $x = 4$ (c) $z = 3$
9. (a) $y = 1/2$ (b) $x = 2$ (c) $x = 1/2$
11. 11.4, 26, 2.8×10^{-7}, 20.9
13. 1, 3, $3a$, 2.000959, 3, -4, 0.50000, 2.50000
15. 4.50, -5.11, 3.92
17. 3.2×10^3, 1.9×10^{-7}, 1×10^2, 6×10^{-2}

Photo Credits

unnum. fig. p. 162 right, Julius Weber

unnum. fig. p. 166 bottom, Burndy Library

unnum. fig. p. 166 top, Bettmann Archives

Chapter 7

Chapter opening photo, unnum. figs. pp. 199, 201 bottom, 205, Charles D. Winters

unnum. figs. pp. 196, 201 top, Marna G. Clarke

unnum. fig. p. 197, Leon Lewandowski

Chapter 8

Chapter opening photo, Fig. 8–1, unnum. fig. p. 215, Charles D. Winters

unnum. fig. p. 212, Dave Pierce, Finnigan MAT-San Jose

unnum. fig. p. 213, Leon Lewandowski

unnum. fig. p. 216, National Foundation for History of Chemistry Pictorial Collection

Chapter 9

Chapter opening photo, unnum. figs. pp. 238, 245, 247, 250, Charles D. Winters

unnum. fig. p. 235, Ann Duncan/Tom Stack and Associates

unnum. fig. p. 249, Union Carbide Corporation

Chapter 10

Chapter opening photo, unnum. figs. pp. 238, 245, 247, 250, Charles D. Winters

unnum. fig. p. 235, National Center for Atmospheric Research/National Science Foundation

unnum. fig. p. 249, Bettmann Archives

Chapter 11

Chapter opening photo, PhotoResearchers © 1986 Wetmore

Fig. 11–2, RH photos

Figs. 11–4, 11–20, unnum. fig. p. 305, Charles D. Winters

unnum. fig. p. 296, Douglas Faulkner/Photoresearchers

unnum. fig. p. 301, Oak Ridge National Laboratories

unnum. fig. p. 304, Royal Society of London

unnum. fig. p. 313, Leon Lewandowski

unnum. fig. p. 314, The Granger Collection, New York

Chapter 12

Chapter opening photo, courtesy of NASA

unnum. figs. pp. 343, 345, Charles Steele

Figs. 12–12abc, unnum. figs. pp. 347, 351, Charles D. Winters

unnum. fig. p. 353, Gemological Institute of America

unnum. fig. p. 354, Leon Lewandowski

Chapter 13

Chapter opening photo, unnum. figs. pp. 373 top, 377, 398, Charles Steele

Figs. 13–1, 13–2, 13–4, 13–6, 13–7, 13–8, unnum. figs. pp. 370 left, 374, 389, Charles D. Winters

unnum. figs. pp. 370 right, 373 bottom, Marna G. Clarke

unnum. fig. p. 393, Union Carbide Corporation

unnum. fig. p. 371, Ann Duncan/Tom Stack and Associates

Chapter 14

Chapter opening photo, Fig. 14–1, Charles Steele

Fig. 14–4, courtesy of Beckman Instruments

Fig. 14–6, 14–10, Charles D. Winters

Fig. 14–7, Leon Lewandowski

Fig. 14–8, The World of Chemistry, Program 7

unnum. fig. p. 430, C. B. Jones/Taurus Photos

Chapter 15

Chapter opening photo, National Center for Atmospheric Research/National Science Foundation

unnum. fig. p. 446, Charles D. Winters

unnum. fig. p. 454, courtesy of Dresser Industries/Kellogg Company

unnum. fig. p. 450, Science Source PhotoResearchers

Chapter 16

Chapter opening photo, James Morgenthaler

Fig. 16–1, Richard P. Smith/Tom Stack and Associates

Fig. 16–2, 16–3, 16–4, unnum. figs. pp. 475, 478, 482 bottom, 483, Charles D. Winters

Fig. 16–7, Reed and Barton Silversmiths

unnum. fig. p. 482 top, Charles Steele

unnum. fig. p. 484, Grant Heilman from Grant Heilman

unnum. fig. p. 487 top, Brian Parker/Tom Stack and Associates

unnum. fig. p. 487 bottom, M. D. Ippolito

unnum. fig. p. 492, General Electric Company

unnum. fig. p. 485, Science Source PhotoResearchers

Glossary

A

Absolute temperature scale Temperature scale starting at absolute zero and divided into units of kelvins instead of degrees. The freezing and boiling temperatures of water on this scale are 273 and 373.

Absolute zero Theoretically the lowest temperature possible. Equal to $-273.2°C$.

Acceleration The rate of change of velocity.

Accuracy Closeness with which a measured result approaches the true value.

Acid Substance that yields hydrogen ions in aqueous solution; substance that donates protons in a chemical reaction.

Actinide Any of the elements with atomic numbers 89 through 102.

Activation energy In a chemical reaction, the energy barrier that reactants must overcome in order to become products.

Actual yield The actual amount of product formed by a particular chemical reaction, as distinguished from theoretical yield.

Alchemy A medieval blend of mysticism and experimentation with chemicals in an attempt to develop an "elixir" that would prolong life or produce gold from other substances.

Alkali metals Elements Li, Na, K, Rb, Cs, and Fr, located in the leftmost column of the periodic table, Group IA. Each atom of an alkali metal has one valence electron.

Alkaline earth metals Elements Be, Mg, Ca, Sr, Ba, Ra, located in the column second from the left in the periodic table, Group IIA. Each atom of an alkaline earth metal has two valence electrons.

Alpha particle High-speed helium nucleus, He^{2+}, usually emitted by a radioactive element.

Alpha rays Stream of alpha particles.

Amphoteric compound Substance that can act either as an acid or a base, depending on conditions.

Angstrom A unit of length equal to 10^{-10} m.

Anhydrous Without water.

Anion A negative ion.

Anode In an electrolytic cell, the electrode where oxidation takes place.

Aqueous solution Solution in which the solvent is water.

Atmosphere Unit of pressure based on the pressure of the atmosphere at sea level. One atmosphere equals 760 torr.

Atom Smallest particle of matter that has the characteristics of an element.

Atomic mass On the ^{12}C scale, the mass of an individual atom of any isotope of any element.

Atomic mass scale Relative scale of atomic masses based on a ^{12}C mass of exactly 12.

Atomic mass unit Unit of mass that is exactly 1/12 of the mass of an atom of ^{12}C.

Atomic number Number of protons in the nucleus of an atom of an element.

Atomic theory Theory that all matter is composed of individual particles called atoms.

Atomic weight Actually a term that refers to mass. On the ^{12}C scale, a weighted average of the atomic masses of the naturally occurring isotopes of an element.

Aufbau principle In the ground state of an atom, energy levels and sublevels fill in order of increasing energy.

Avogadro's hypothesis Theory stating that equal volumes of gases at the same temperature and pressure contain equal numbers of molecules, no matter what kinds of molecules make up the gases.

Avogadro's law States that at constant temperature and pressure, the volume of a gas sample is directly proportional to the number of moles of gas in the sample.

Avogadro's number Number of atoms in exactly 12 grams of ^{12}C. The number of individual items in one mole of those items. The value of Avogadro's number is 6.0221367×10^{23}.

B

Balanced chemical equation Chemical equation in which there are the same number of electric charges and the same number of atoms of each kind on both sides of the equation.

Barometer Device for measuring atmospheric pressure.

Base   Substance able to accept protons in a chemical reaction. Substance that yields hydroxide ions. OH^-, in aqueous solutions.

Battery   Device that converts chemical energy to electric energy by means of an oxidation-reduction reaction.

Binary compound   Compound consisting of only two elements.

Biochemistry   Study of the chemical composition and the chemical reactions of living organisms.

Bohr atom   Theoretical model of an atom consisting of a nucleus with electrons orbiting around it.

Boiling   Rapid vaporization that occurs when the vapor pressure of a liquid equals or exceeds the pressure on its surface.

Boiling point (boiling temperature)   Temperature at which the vapor pressure of a liquid equals the pressure exerted on it by the atmosphere or another gas.

Boiling point elevation   Difference in boiling temperatures of a solution and its pure solvent.

Boiling point elevation constant   For any particular solvent, the constant relating the amount of boiling point elevation to the molality of the solution.

Boiling temperature (boiling point)   Temperature at which the vapor pressure of a liquid equals the pressure exerted on it by the atmosphere or another gas.

Bond angle   Angle between the covalent bonds uniting any three atoms.

Boyle's Law   States that for a sample of an ideal gas at constant temperature, the product of the pressure and volume is a constant, $PV = k$.

Buffer   A chemical equilibrium designed to control concentration by applying Le Chatelier's principle. Most often a buffer resists change in pH, using a weak acid and its conjugate base or a weak base and its conjugate acid.

C

Calorie   Amount of energy required to raise the temperature of 1 gram of pure water from 14.5°C to 15.5°C.

Catalyst   Substance that increases the rate of a reaction by lowering the activation energy. A catalyst remains unchanged by reaction.

Cathode   In an electrolytic cell, the electrode where reduction takes place.

Cation   Positive ion.

Celsius scale   Temperature scale that defines the freezing and boiling temperatures of water at 0°C and 100°C, respectively (both at 1 atm).

Centimeter   Metric unit of length equal to 0.01 meter.

Charles' law   States that the volume of a sample of gas at constant pressure is directly proportional to its absolute temperature, $V = kT$.

Chemical analysis   Measurement to determine what elements are in a compound or what the components of a mixture are (qualitative analysis) or to determine what the relative amounts of the components of a compound or mixture are (quantitative analysis).

Chemical bond   Force that holds the atoms together in a chemical compound.

Chemical equation   Statement that describes a chemical reaction by giving the formulas and relative amounts of reactants and products.

Chemical equilibrium   Dynamic state in which the rate of the forward reaction equals the rate of the reverse reaction. In a reaction at equilibrium, the concentrations of the reactants and products do not change.

Chemical family (chemical group)   Group of elements that have similar physical and chemical properties.

Chemical formula   Combination of element symbols and subscripts indicating which elements are present and the relative numbers of atoms of each element in a compound.

Chemical group (chemical family)   Group of elements that have similar physical and chemical properties.

Chemical property   Property of a substance that can be determined only by chemical reaction.

Chemical reaction   Process in which one or more substances (reactants) are transformed into different substances (products).

Chemistry   The science that deals with the composition, structure, and properties of substances, the transformations that substances undergo, and the energies of those transformations.

Chromatography   Physical process that separates components of a mixture. Chromatography depends on the strength by which surfaces adsorb different substances.

Coefficient   In a chemical equation, the number preceding the symbol of a chemical species.

Colligative properties   Solution properties that depend on the number of solute particles in solution and the nature of the solvent.

Combined gas law $\dfrac{PV}{T} = k.$

Compound   Pure substance that is composed of elements and is characterized by a chemical formula.

Concentrated solution   Solution containing a large proportion of solute; opposite of dilute.

Concentration   Relative amounts of solute and solvent in a solution.

Condensation   Process in which a gas or vapor becomes a liquid or solid.

Condensed states   Liquids and solids.

Conjugate acid   Acid formed by adding H^+ ions to a base.

Conjugate base   Base formed by removing H^+ ions from an acid.

Conversion factor   Quantitative mathematical relation that changes a value given in one set of units to an equivalent value in a different set of units.

Coordinate covalent bond   Covalent bond in which both electrons in the shared pair are furnished by only one of the bonded atoms.

Covalent bond   Chemical bond in which a pair of electrons is shared by two atoms in a molecule.

Covalent crystal   Crystal in which all the atoms are connected by covalent bonds; also called covalent solid.

Covalent solid   Crystal in which all the atoms are connected by covalent bonds; also called covalent crystal.

Crystal   Solid in which the component particles (atoms, ions, or molecules) are organized in a pattern and held in place by chemical bonds or other forces.

Cubic centimeter   Metric unit of volume equivalent to a cube with sides of 1 cm. The same as a milliliter.

D

***d* block**   Elements with their highest energy electrons in *d* sublevels.

Dalton's law of partial pressures   States that, in a mixture of gases, the partial pressure exerted by each gas equals the pressure that the gas would exert if alone in the container at the same temperature. The total pressure of a mixture of gases equals the sum of the partial pressures. $P_t = P_1 + P_2 + P_3. \ldots$

Degree   Unit of temperature. Not all degrees are the same size; for example, a Fahrenheit degree does not equal a Celsius degree.

Density   Ratio of the mass of a substance to its volume.

Diatomic molecule   Molecule that contains two atoms.

Diluent   Additional solvent added to a solution.

Dilute solution   Solution that is not concentrated. Solutions containing less than a few grams per liter of solute are ordinarily considered dilute.

Dipole (electric dipole)   Separation of electric charge in a particle such that one portion has a positive charge and another portion a negative charge.

Dipole forces   Intermolecular forces caused by the interaction of the dipoles of two or more molecules.

Directly proportional (proportional)   Two variables are proportional when one variable is equal to the product of a constant number and the other variable.

Disproportionation   Reaction in which the same substance is both oxidized and reduced.

Dissociation   Separation of ionic compounds into individual ions upon dissolving in solution.

Distillation   Purification process that involves a change of state from liquid to vapor and back to liquid.

Double bond   Two covalent bonds (two shared electron pairs) between one pair of atoms.

E

Electric charge   Fundamental property of matter such that an electrically charged particle will exert a force on another electrically charged particle.

Electric circuit   Continuous loop of matter that conducts electricity.

Electric current   Flow of electric charge through a conductor.

Electric dipole (dipole)   Separation of electric charge in a particle such that one portion has a positive charge and another portion a negative charge.

Electrode   In a battery, electrolytic cell, or electronic device, a conductor from which or into which electrons flow.

Electrolysis   Chemical reaction caused by electric energy.

Electrolyte   Substance that, when dissolved in water, conducts electricity.

Electrolytic cell   An arrangement in which an oxidation-reduction reaction is caused by means of electric energy.

Electron   Fundamental particle with about 1/1838 the mass of a proton and a single negative electric charge of 1.6×10^{-19} coulombs.

Electron affinity   Measure of the attraction of an atom for an extra electron brought from far away and added to the atom.

Electron configuration   List of occupied energy levels and sublevels of an atom and the number of electrons in each.

Electron transfer reaction (oxidation-reduction reaction)   Chemical reaction in which an electron is transferred from one species (atom, molecule, or ion) to another.

Electronegativity   Relative attractiveness of an atom for the pair of electrons by which it is covalently bonded to another atom.

Electroform   To create an object by electrolytic deposition of a metal on a prepared shape.

Electroplate   To coat an object with metal by means of electrolytic deposition.

Element Substance that cannot be separated into simpler substances by chemical processes. A collection of atoms all with the same atomic number.

Empirical formula (simplest formula) Represents the smallest integer ratio of atoms in a compound.

Empirical weight Term that refers to mass. The sum of the atomic weights of the elements in a compound as indicated by the empirical formula, allowing for the fact that some elements appear more than once in the empirical formula.

Endothermic process Reaction or other process that absorbs energy.

Energy Ability to do work.

Energy level One of the quantized, allowable amounts of energy that an electron in an atom can possess.

Enthalpy Energy content of a sample of matter.

Enzyme Biological catalyst.

Equilibrium constant Numerical value that indicates the extent to which a reversible reaction proceeds. A temperature-dependent function containing the concentrations of the reactants and products of the reaction.

Equilibrium equation The relation equating reactant and product concentrations at equilibrium to a constant.

Equivalent That amount of substance that will supply, replace, react with, or otherwise be equivalent to one mole of H_3O^+.

Error Difference between a measured value and the true value of what is measured.

Evaporation Process by which a liquid slowly vaporizes and the resulting vapor molecules escape into the surrounding gas.

Exothermic process Reaction or other process that releases energy.

Extensive property Property, the value of which depends on the amount of the sample taken.

F

***f* block** Elements that have their highest energy electrons in f sublevels.

Fahrenheit scale Temperature scale that defines the freezing and boiling temperatures of water at 32°F and 212°F, respectively (both at 1 atm).

Family (chemical family) Group of elements that have similar physical and chemical properties.

Faraday Amount of electric charge possessed by one mole of electrons.

Formula weight Term that refers to mass. The sum of the atomic weights of the elements in a compound as indicated by the formula, allowing for the fact that some elements appear more than once in the formula.

Frasch process Process by which sulfur is mined. Steam is forced into wells, bringing melted sulfur to the surface.

Freezing Process by which a liquid turns into a solid.

Freezing point (freezing temperature) Temperature at which a solid is in equilibrium with its liquid. This is the same as the melting point or temperature.

Freezing point depression Difference in freezing temperature between a solution and its solvent.

Freezing point depression constant For any particular solvent, the constant relating the amount of freezing point depression to the molality of the solution.

Freezing temperature (freezing point) Temperature at which a solid is in equilibrium with its liquid. This is the same as the melting temperature or point.

G

Gas State of matter characterized by the fact that a sample will take both the shape and the volume of its container.

Gay-Lussac's law States that the pressure of a fixed mass of gas at constant volume is directly proportional to its absolute temperature, $P = kT$.

Glass Solid in which the particles are not organized in a pattern.

Gram One thousandth of a kilogram. The kilogram is the SI unit of mass.

Graphite Solid form of the element carbon.

Ground state State of an atom when all the electrons occupy the lowest energy levels possible.

H

Half-cell Half of an electrochemical cell in which either the oxidation or the reduction reaction takes place.

Halogens Elements F, Cl, Br, I, and At, located in the column second from the right of the periodic table, Group VIIA. Each atom of the halogens has seven valence electrons.

Heat Transfer of energy between a sample of matter and another sample in contact with it.

Heat of condensation Energy released when one gram of a gas condenses, usually to a liquid.

Heat of fusion Amount of energy required to melt one gram of any given substance.

Heat of sublimation Amount of energy required to sublime one gram of any given solid.

Heat of vaporization   Amount of energy required to vaporize one gram of any given liquid.

Heisenberg uncertainty principle   Theory that it is not possible to know simultaneously the exact position and exact momentum of any particle of matter.

Heterogeneous mixture   Mixture composed of substances divided into distinct regions with definite boundaries.

Homogeneous mixture (solution)   Mixture that is the same throughout. A sample taken from this mixture will have exactly the same composition as any other sample taken from the mixture.

Hund's rule   States that each orbital in a sublevel receives a single electron before any orbital in the same sublevel receives a second electron.

Hydrate   Stoichiometric combination of water molecules and molecules or ions of a compound.

Hydrated   In aqueous solution, a molecule or ion associated with one or more water molecules.

Hydride ion   Hydrogen atom with a negative charge, H^-.

Hydrogen bond   Unusually strong dipole–dipole interaction between a covalently bonded hydrogen atom of a molecule and any of the atoms N, O, or F in an adjacent molecule.

Hydrogen ion   Hydrogen atom with a positive charge, H^+.

Hydrolysis   Reaction of something with water.

Hydronium ion   Hydrogen ion associated with a water molecule, H_3O^+.

Hydroxide ion   Negative ion containing a hydrogen and an oxygen atom, OH^-.

I

Ideal gas (perfect gas)   Imaginary gas that perfectly obeys the ideal gas law. Under ordinary conditions, many gases approach ideal behavior.

Ideal gas constant R, the constant that appears in the ideal gas equation. One of the values of R is 0.08206 L atm/mol K.

Idea gas law   States that for an ideal gas, $PV = nRT$.

Intensive property   Property that is independent of the amount of sample taken. Temperature, for example, is an intensive property, as is density.

Inversely proportional   If the product of two variables is equal to a constant number, the variables are said to be inversely proportional.

Ion   Atom or group of atoms possessing an electric charge.

Ionic bond   Electric force uniting the ions in an ionic solid.

Ionic compound   Compound composed of ions.

Ionic solid   Crystal of ionic compound.

Ionization   Process of becoming an ion. Example: $Na \rightarrow Na^+ + e^-$.

Ionization constant   Equilibrium constant describing the extent to which a weak electrolyte is ionized in aqueous solution.

Ionization energy   Amount of energy required to remove an electron from a gaseous ion or molecule.

Ionizing solvent   Solvent, such as water, that reacts with a molecular compound to produce ions.

Isotopes   Two or more atoms that have the same atomic number but different mass numbers.

Isotopic abundance   Relative amounts of the different isotopes in a sample of an element.

J

Joule   The SI unit of energy; a calorie is 4.18 joules.

K

Kelvin   A unit of measuring temperature in the absolute scale, which begins at absolute zero. A Kelvin is the same size as a Celsius degree.

Kelvin scale (absolute scale)   Temperature scale starting at absolute zero and divided into units of kelvins instead of degrees. The freezing and boiling temperatures of water on this scale are 273 and 373, respectively.

Kilogram   SI unit of mass. One kilogram is 1,000 grams.

Kilojoule   One thousand joules.

Kilometer   A metric unit of length equal to 1,000 meters.

Kinetic energy   Energy associated with motion, $K.E. = \frac{1}{2}mv^2$.

Kinetic molecular theory   States that the molecules of a gas are in rapid, random movement and that their collisions with the walls of their vessel produce pressure.

L

Lanthanide   Any of the elements with atomic numbers 57 through 70.

Law of conservation of matter   States that no matter can be created or destroyed.

Law of conservation of energy   States that no energy can be created or destroyed.

Law of definite proportions   States that a pure sample of compound always contains its elements in a fixed proportion by mass.

Le Chatelier's principle   States that if an external change is made in the intensive properties of a system at equilibrium, the system will respond so as to counteract the effects of the change, thus reaching a new position of equilibrium.

Lewis diagram Diagram of an atom or molecule, showing the valence electrons.

Limiting reactant (limiting reagent) Reactant that is completely consumed during a chemical reaction.

Limiting reagent (limiting reactant) Reactant that is completely consumed during a chemical reaction.

Linear molecule Molecule in which a straight line can be drawn through the centers on all the atoms; molecule that has chemical bonds only at 180° angles.

Liquid Substance that will take the shape of its container but will not necessarily occupy the entire volume of the container.

Liter Metric unit of volume equal to 0.001 cubic meter or 1,000 cubic centimeters.

M

Macroscopic Large enough to be visible.

Manometer Device similar to a barometer for measuring the pressures of gases.

Mass Amount of matter in a sample.

Mass number Sum of the number of protons and neutrons in a nucleus.

Matter Anything that has mass and occupies space.

Melting Process by which a solid becomes a liquid.

Melting point (melting temperature) Temperature at which a solid is in equilibrium with its liquid phase. This is the same as the freezing point or freezing temperature.

Melting temperature (melting point) Temperature at which a solid is in equilibrium with its liquid phase. This is the same as the freezing temperature or freezing point.

Metal Any of the elements listed on the left of the periodic table that are shiny, ductile, malleable, and tend to lose electrons in chemical reactions.

Metallic bond "Sea" of free electrons holding metal ions together in a metallic crystal.

Metalloid Elements B, Si, GE, AS, Te, and Po that are neither entirely metallic nor nonmetallic in character.

Meter SI unit of length.

Metric system System of measurement that originated in France. SI is a subset of the metric system. Metric subunits are all related by powers of ten.

Milligram Metric unit of mass equal to 0.001 gram.

Milliliter Metric unit of volume equal to 0.001 liter.

Millimeter Metric unit of length equal to 0.001 meter.

Molality Concentration expressed as the number of moles per kilogram of solvent.

Molar mass Mass in grams of one mole of an element or compound.

Molarity Concentration unit of moles of solute per liter of solution.

Mole Sample containing Avogadro's number of items. For example, a mole of an element contains 6.0221367×10^{23} atoms of the element.

Mole ratio Any ratio of the numbers of moles of substances (reactants or products) in a chemical equation.

Molecular compound Compound consisting of individual, discrete molecules.

Molecular crystal Crystal composed of individual molecules held together by van der Waals forces.

Molecular formula Chemical formula that specifies the actual numbers of the atoms of the different elements in a molecule.

Molecular weight Term that refers to mass. The sum of the atomic weights of the elements in a molecule allowing for the fact that some elements appear more than once in the molecular formula.

Molecule Discrete group of atoms held together by covalent bonds. The smallest individually stable unit of an element or compound.

N

Natural law Concise statement of a physical relation that is always the same under the same conditions.

Net ionic equation Chemical equation written for the reactions of ionic compounds in aqueous solution, showing only the ions, precipitates, and molecules that actually participate in the reaction.

Neutral solution Solution that has equal concentrations of H_3O^+ and OH^- ions.

Neutralization Process of making a neutral solution (one that is neither acidic nor basic) from one that is acidic or basic.

Neutron One of the fundamental atomic particles located in the nucleus of an atom. It has approximately the same mass as a proton but no electric charge.

Noble gas One of the group of essentially unreactive gases at the far right side of the periodic table (Group VIII). Noble gas atoms each have eight valence electrons.

Nonmetal Any of the elements listed on the far right side of the periodic table that do not conduct electricity, shatter rather than bend, and are dull in appearance.

Normal boiling temperature Boiling temperature of a liquid at one atmosphere pressure.

Normal melting temperature (normal freezing temperature) Temperature at which a solid is in equilibrium with its liquid state at one atmosphere pressure.

Normality Concentration term expressed as the number of equivalents per liter of solution.

Nucleus Atomic center containing all the positive charge and nearly all the mass of an atom.

O

Octet rule States that in losing or gaining electrons, atoms tend to form ions that have eight valence electrons.

Orbital Sub-sublevel that can be occupied by an electron in an atom. An electron in any particular orbital spends most of its time in a particular region of an atom.

Ostwald oxidation Catalytically encouraged reaction of NH_3 with O_2 to form NO and eventually NO_2.

Oxidation Chemical reaction resulting in the loss of one or more electrons by an atom, ion, or molecule.

Oxidation number (oxidation state) Number assigned to each atom in the formula of a compound, used to keep track of electrons lost and gained in oxidation-reduction reactions.

Oxidation state (oxidation number) Number assigned to each atom in the formula of a compound, used to keep track of electrons lost and gained in oxidation-reduction reactions.

Oxidation-reduction reaction (electron transfer reaction) Chemical reaction in which one or more electrons are transferred from an atom, molecule, or ion to another.

Oxidizing agent Substance that oxidizes another substance in an oxidation-reduction reaction.

P

p block Area of the periodic table characterized by elements with their highest energy electrons in p sublevels.

Partial pressure Pressure exerted by an individual gas in a mixture, behaving as if it were in the container alone.

Pascal SI unit of pressure. One atm equals 101,325 Pa.

Pauli exclusion principle Theory that no two electrons can have the same set of quantum numbers.

Percent composition List of the percentages of the elements in a compound.

Percentage yield Ratio of the actual yield of a reaction to the theoretical yield, expressed as a percentage.

Period Horizontal row of elements in the periodic table.

Periodic table of the elements Table of the known elements listed in order of atomic number and arranged according to physical and chemical properties.

pH Method of designating the acidity or basicity of a solution. The negative logarithm of the molar concentration of the hydronium ion, $-\log [H_3O^+]$.

Phase In a sample of matter, a region with distinct boundaries and a unique set of properties. Also used to denote one of the states of matter, as in "liquid phase."

Physical change Change in the shape, appearance, or state of a substance. In a physical change no substance is destroyed nor is any new substance created.

Physical property Property that can be measured without chemical reaction, that is, without changing the chemical nature of the substance.

Physics Science devoted to the study of matter and energy.

Polar covalent bond Covalent bond in which the pair of electrons is unequally shared between the two bonded atoms.

Polyatomic ion Ion that contains more than one atom.

Potential energy Energy of an object due to its position in relation to forces exerted on it.

Precipitate Solid product produced by chemical reaction in solution.

Precision Closeness of agreement between successive measurements of the same value.

Pressure Force per unit area.

Principal quantum number Quantum number, designated by n, assigned to the main, or principal energy level of an electron.

Product New chemical substance created from reactants in a chemical reaction.

Proportional (directly proportional) Two variables are proportional when one variable is equal to the product of a constant number and the other variable.

Proton One of the fundamental particles located in the nucleus of an atom. The proton has a mass of about 1 amu and a single positive charge.

Proton transfer reaction Another term for acid-base reaction. A process in which a proton is transferred from an acid to a base.

Pure substance Sample that cannot be separated into simpler substances by physical changes. A pure substance has a definite composition and a definite set of properties.

Q

Quantized Restricted to particular values or amounts. Not continuous.

Quantum number Representation of the main energy level or one of the three kinds of sublevels in an atom.

Quantum mechanics (wave mechanics)   Theory of atomic structure stating that electrons in atoms behave as waves rather than as particles.

R

Reactant   In a chemical reaction, a starting substance from which products are created.

Reaction rate   Rate at which a reaction proceeds, usually measured in terms of the rate of disappearance of reactants or the rate of appearance of products.

Redox reaction   Oxidation-reduction or electron transfer reaction.

Reducing agent   Substance that loses electrons in a redox reaction.

Reduction   Process of gaining electrons in a redox reaction.

Representative element   Element in any of the Groups IA, IIA, IIIA, IVA, VA, VIA, VIIA, and the noble gases.

Rounding   Process of removing digits from a value in order to achieve the proper number of significant figures.

S

s block   Two leftmost columns of the periodic table characterized by elements having valence electrons located in only s sublevels.

Salt   Product of the reaction between an acid and a base. An ionic compound containing the cation of the base and the anion of the acid.

Saturated solution   Solution that will dissolve no more of the solute with which it is said to be saturated at any particular temperature.

Science   Process by which humans seek in an organized way to understand and explain the natural world.

Scientific method   Process to gain knowledge through observation, experimentation, testing, and theorizing.

Scientific notation   Method of representing a numerical quantity by using a number equal to or greater than 1 and less than 10 multiplied by 10 raised to an appropriate power. Example: 5.1×10^{13}.

SI (Système Internationale)   System of preferred base units to be used for measuring quantities. SI is a subset of the metric system.

Significant figures   Those digits in a reported value that are known to be accurate, plus one additional digit that is approximate.

Simplest formula (empirical formula)   Represents the smallest integer ratio of atoms in a compound.

Single bond   Shared pair of electrons between two atoms.

Solid   State of matter in which a sample has both its own volume and its own shape.

Solubility   Amount of a solute dissolved in a given amount of solvent at saturation at a specified temperature.

Solubility product   Equilibrium constant that describes the concentrations of the ions of a compound in its saturated solution.

Solute   Substance that is being dissolved in a solution. The component of a solution that is present in the smaller amount.

Solution   Homogeneous mixture.

Solvated   State in which a particle interacts closely with one or more molecules of solvent in a solution.

Solvent   In a solution, the substance that has the same phase as the resulting solution or is present in the greatest amount. The substance in which another is being dissolved.

Space-filling model   As distinguished from a Tinkertoy or stick-and-ball model, a model of a molecule that shows the true volumes of the atoms or ions in their relative positions.

Specific heat   Amount of energy required to raise the temperature of one gram of a substance one Celsius degree.

Spectroscopy   Study of the energies of photons emitted or absorbed by atoms and molecules.

Standard molar volume   Volume of one mole of an ideal gas at STP (one atmosphere pressure and 0°C); its value is 22.414 L.

Standard temperature and pressure (STP)   Pressure of 760 torr and temperature of 0°C.

Stock system   Convention by which chemical compounds are named, giving oxidation states as Roman numerals. As an example, iron chloride in which the iron atoms are in the highest oxidation state, +3, is named as iron(III) chloride.

Stoichiometric calculation   Calculation involving the quantities of elements or compounds in a chemical reaction.

STP (standard temperature and pressure)   Pressure of 760 torr and temperature of 0°C.

Strong acid   Acid that is entirely, or almost entirely, ionized and dissociated in aqueous solution.

Strong base   Base that is entirely, or almost entirely, ionized and dissociated in aqueous solution.

Strong electrolyte   Substance that is completely ionized or dissociated in solution.

Sublimation   Transformation of a solid directly into a vapor without first becoming a liquid. Sublimation is sometimes used as a purification process.

Supersaturation   State of a solution that has dissolved more solute than it can contain when saturated. Supersaturated solutions are not stable.

Système Internationale (SI)   System of prescribed base units to be used for measuring quantities. SI is a subset of the metric system.

T

Technology   Applied science.

Temperature   Measure of the average kinetic energy of the individual particles in a sample.

Temporary dipole force   Intermolecular force arising from temporary electron imbalances in molecules and the adjacent dipoles induced by those imbalances. One of the kinds of van der Waals forces.

Theory   Reasonable explanation of a natural law.

Theoretical yield   Amount of product of a chemical reaction calculated on the basis that all of the limiting reactant is converted to product.

Tinkertoy model   Ball-and-stick model of a molecule, designed to show the bond structure of the molecule.

Titration   Process in which the concentration of a known amount of a solution is measured by reacting it with a measured amount of a second solution of known concentration.

Torr   Metric unit of pressure, equivalent to the height of 1 mm of mercury.

Transition element   Any of the elements with atomic numbers 21 through 30, 39 through 48, 71 through 80, and 103 through 109.

Triple bond   Three shared pairs of electrons between two atoms.

U

Unit   Standard quantity, such as a kilometer or a second, used to express the results of measurements.

Unit analysis   Procedure for setting up calculations by means of arranging conversion factors so that the result is expressed in the proper units.

V

Valence electrons   Electrons with the highest n in an atom.

Value   Result of a quantitative measurement, usually expressed as a numerical quantity and a set of units.

van der Waals forces   Dipole–dipole forces or temporary dipole forces.

Vapor   Gaseous phase of a substance in contact with its condensed state. A vapor is simply a gas in particular circumstances.

Vapor pressure   Pressure exerted by a vapor in equilibrium with its condensed state at a particular temperature.

Vaporization   Transformation of a condensed state (usually a liquid) into a vapor.

Velocity   Rate of change of position with respect to time in a given direction.

Volatile   Having a tendency to evaporate or vaporize easily.

W

Wave mechanics (quantum mechanics)   Theory of atomic structure stating that electrons in atoms behave as waves rather than as particles.

Weak acid   Acid that is only partially ionized in aqueous solution.

Weak base   Base that is only partially ionized in solution.

Weak electrolyte   Electrolyte that only partially ionizes in solution, leaving the other part in molecular form.

Weight   Force by which a given mass is attracted to the earth or to some other body.

Weight percent   Ratio between the mass of one element in a sample of a compound to the total mass as expressed in percentage terms. For solutions, the mass of solute divided by the mass of solution times 100%.

Index

In the list below, an italicized letter preceding a page number indicates that reference is made to a definition, *d*; a Profile in Science, *h*; a Science in Action essay, *s*; a table, *t*; or a figure, *f*.